Lecture Notes in Computer Science   5050

Commenced Publication in 1973
Founding and Former Series Editors:
Gerhard Goos, Juris Hartmanis, and Jan van Leeuwen

Jacek M. Zurada   Gary G. Yen   Jun Wang (Eds.)

Computational Intelligence: Research Frontiers

IEEE World Congress on Computational Intelligence, WCCI 2008
Hong Kong, China, June 1-6, 2008
Plenary/Invited Lectures

 Springer

Volume Editors

Jacek M. Zurada
University of Louisville, Computational Intelligence Laboratory
Louisville, KY 40292, USA
E-mail: jacek.zurada@louisville.edu

Gary G. Yen
Oklahoma State University
Stillwater, OK 74078, USA
E-mail: gyen@okstate.edu

Jun Wang
The Chinese University of Hong Kong Shatin
New Territories Hong Kong, China
E-mail: jwang@mae.cuhk.edu.hk

Library of Congress Control Number: 2008927523

CR Subject Classification (1998): H.3, H.2.8, H.2, I.2.3, I.2.6, H.4, I.5.3

LNCS Sublibrary: SL 1 – Theoretical Computer Science and General Issues

ISSN   0302-9743
ISBN-10   3-540-68858-7 Springer Berlin Heidelberg New York
ISBN-13   978-3-540-68858-7 Springer Berlin Heidelberg New York

Springer is a part of Springer Science+Business Media

springer.com

© Springer-Verlag Berlin Heidelberg 2008
Printed in Germany

Typesetting: Camera-ready by author, data conversion by Scientific Publishing Services, Chennai, India
Printed on acid-free paper   SPIN: 12278381   06/3180   5 4 3 2 1 0

Preface

The 2008 IEEE World Congress on Computational Intelligence (WCCI 2008), held during June 1–6, 2008 in Hong Kong, China, marked an important milestone in advancing the paradigms of the new fields of computational intelligence. As the fifth event in the series that has spanned the globe (Orlando-1994, Anchorage-1998, Honolulu-2002, Vancouver-2006), the congress offered renewed and refreshing focus on the progress in nature-inspired and linguistically motivated computation. Most of the congress's program featured regular and special technical sessions that provided participants with new insights into the most recent developments in the field.

As a tradition, in addition to the parallel technical sessions, WCCI holds a series of plenary and invited lectures which are not included in the congress proceedings. As its predecessors, at WCCI 2008, 20 expert speakers shared their expertise on broader, if not panoramic, topics spanning a diverse spectrum of computational intelligence in the areas of neurocomputing, fuzzy systems, evolutionary computation, and adjacent areas. Thanks to their time and expertise, we endeavored to offer this volume to attendees directly at the congress and the general public afterwards.

In this preface, we provide an overview of the volume with 18 chapters divided into four topical sections. Section 1, "Machine Learning and Brain Computer Interface," explores mathematical learning theory and places them side by side with efforts to translate the theoretical advances to real-world tools. Section 2, "Fuzzy Modeling and Control," features papers in novel approaches to fuzzy modeling and their use in such diversified scenarios as human and social environments, but also for knowledge acquisition, fuzzy inference and intelligent industrial environments. Section 3 contains contributions to "Computational Evolution" and provides new observations on multi-objective optimization and cooperation/competition in biological and robotic environments as well as standards in modeling adaptive systems. Finally, Section 4 entitled "Applications" is dedicated to successful applications of computational intelligence technologies to a number of practical problems such as management and mining of large sets of data, the cocktail party problem, and fault-tolerant evolvable hardware.

The papers presented here are highly focused on relevant research achievements, yet for most part they are also highly tutorial in nature. The result is a balanced contribution to the field of computational intelligence that should serve the community as a survey and a reference, but also as an inspiration for future advancement of the state of the art of our field.

Chapter 1, authored by the plenary speaker Christopher M. Bishop, "A New Framework for Machine Learning," discusses the emergence of a powerful new framework for building sophisticated real-world applications using machine learning within the last five years. The cornerstones of this approach are (a) the adoption of a Bayesian viewpoint, (b) the use of graphical models to represent complex probability distributions, and (c) the development of fast, deterministic inference algorithms, such as variational Bayes and expectation propagation, which provide efficient solutions to inference and learning problems in terms of local message passing algorithms. The paper discusses the key ideas behind this new framework, and highlights the benefits compared with traditional methods such as neural networks and support vector machines. The paper is illustrated with examples of large-scale applications.

Chapter 2, authored by Kristin P. Bennett, Gautam Kunapuli, Jing Hu and Jong-Shi Pang, "Optimization and Machine Learning," exploits the interrelationships between optimization and machine learning. Appreciable progress has been made over the years in machine learning by ingeniously formulating machine learning problems into convex optimization problems with one or more hyper-parameters. In this chapter, Bennett and her collaborators take it to the next level, by treating machine learning problems as Stackelberg games. The resulted bilevel optimization problem allows for efficient, systematic search of large numbers of hyper-parameters. Discussions are also made about the recent progress in solving these bilevel problems and many interesting optimization challenges that remain.

Chapter 3, authored by Lei Xu, "Bayesian Ying Yang System, Best Harmony Learning, and Gaussian Manifold-Based Models" provides an elaborated introduction on the newly developed Bayesian Ying-Yang (BYY) harmony learning. Proposed in 1995 by Xu and systematically developed in the past decade, this model consists of a two-pathways featured BYY system as a general framework for unifying a number of learning models, and a best Ying-Yang harmony principle as a general theory for parameter learning and model selection. The BYY harmony learning leads not only to a criterion that outperforms typical model selection criteria in a two-phase implementation, but also to a model selection made during parameter learning, with significantly reduced computing cost.

Chapter 4, authored by Benjamin Blankertz, Michael Tangermann, Florin Popescu, Matthias Krauledat, Siamac Fazli, Márton Dónaczy, Gabriel Curio, and Klaus-Robert Müller, "Toward Brain Computer Interfacing," aims at making use of brain signals for the control of objects, spelling, gaming, to name a few aspects of the Brain Computer Interfacing (BCI). The authors provide a brief overview of the BCI project from a machine learning and signal processing perspective. Based on EEG signals, the authors take the audiences all the way from the measured signal, the preprocessing and filtering, and the classification to the respective applications. BCI as a new channel for man–machine communication is discussed in detail in a clinical setting and for gaming applications.

Chapter 5, authored by Shiro Usui and Yoshihiro Okumura, "Basic Scheme of Neuroinformatics Platform: XooNIps," features the international cooperation project in the emerging field of neuroinformatics. The Neuroinformatics Japan Center at RIKEN Brain Science Institute in Japan was established in 2005 and created a neuroinformatics base-platform "XooNIps." XooNIps features better scalability, extensibility, and customizability to operate under various site policies in the general community and can be easily customized to support different databases and portals. It provides a framework for successfully accumulating, sharing, and making public resources which were once difficult to accumulate, share and make public. The author elaborates at length on various details as to how the platform was created.

Chapter 6, authored by Witold Pedrycz, "Collaborative Architectures of Fuzzy Modeling," proposes a concept of Collaborative Computational Intelligence, and specifically a collaborative fuzzy model, in facing the rapidly emerging needs to deal with distributed sources of data. In the context of collaborative fuzzy modeling,

Pedrycz brings forward the concept of *experience–consistent* fuzzy system identification showing how fuzzy models built on a basis of limited data can benefit from taking advantage of the past experience.

Chapter 7, authored by Ronald Yager, "Human Behavioral and Social Network Modeling Using Soft Computing," identifies a common vocabulary understandable by both parties that is essential to man–machine cooperation. Yager has drawn upon structures from granular computing, particularly fuzzy sets, to provide this capability. The author then focuses on some tools useful for the fusion of information and question answering in the context of man–machine cooperation. Additionally, the fusion of probabilistic and possibilistic information is also investigated.

Chapter 8, authored by the plenary speaker Takeshi Yamakawa and by Takanori Koga, "Bio-inspired Self-Organizing Relationship Network as Knowledge Acquisition Tool and Fuzzy Inference Engine," focuses on the human brain that facilitates the topological mapping of the external complex information (multiple-dimensional) to the cortex (two-dimensional) by sensing and watching. The system is named self-organizing relationship (SOR) network. A set of units on the competitive layer of the SOR network after learning exhibits a set of typical input-output characteristics of the system of interest and thus the network achieves the knowledge acquisition (IF-THEN rules) from the raw data. The evaluation for each data necessary for the learning of the SOR network is possibly intuitive and deterministic. The paper also discusses applications of SOR network.

Chapter 9, authored by Hani Hagras, "Type-2 Fuzzy Logic Controllers: A Way Forward for Fuzzy Systems in Real-World Environment," reviews type-1 Fuzzy Logic Controllers (FLCs) that have been applied to date with great success to many different applications. However, for many real-world applications, there is a need to cope with large amounts of uncertainties. A type-2 FLC using type-2 fuzzy sets can handle such uncertainties to produce a better performance. In this chapter, the author introduces the interval type-2 FLCs and how they present a way forward for fuzzy systems in real-world environments and applications that face high levels of uncertainties.

Chapter 10, authored by the plenary speaker David B. Fogel, "The Burden of Proof – Part II," discusses standards of evidence in scientific work. The very term "standards" suggests they should be consistent, but they are not. Often, well-known "facts" or claims turn out to be wrong, disagreements over the interpretation of data and methods yield to political motivations. The paper discusses the standards of evidence in scientific work, with particular emphasis on evolutionary computation and modeling complex adaptive systems. Evidence shows that some seemingly simple systems are really quite complicated. In other cases, adjusting assumptions about a model leads to results that are at significant variance from what is commonly accepted. The implications of accepting well-known models of these systems are explored.

Chapter 11, authored by Dario Floreano, Sara Mitri, Andres Perez-Uribe, and Laurent Keller, "Evolution of Cooperation in Biological and Robotic Societies," examines the evolutionary methods that lead to the emergence of altruistic cooperation within robot society. The authors present four evolutionary algorithms that derive from biological theories on the evolution of altruism and compare them systematically in two experimental scenarios where altruistic cooperation can lead to a significant performance enhancement.

Chapter 12, authored by Günter Rudolph and Hans-Paul Schwefel, "Simulated Evolution Under Multiple Criteria Conditions Revisited," discusses how many well-crafted multi-objective evolutionary algorithms have seen great promise; yet most sophisticated algorithms today have somehow lost their character of mimicking nature found in organic evolution. The author reviews a couple of more bio-inspired designs without forgetting to foster the interdisciplinary dialogue with natural scientists. The presentation is elegant and far-reaching.

Chapter 13, authored by Kay Chen Tan and Chi Keong Goh, "Handling Uncertainties in Evolutionary Multi-objective Optimization," discusses how many of the studies on Evolutionary Multiobjective Optimization (EMO) assume that the problem is deterministic. In this chapter, the challenges faced in handling three different forms of uncertainties in EMO are discussed by the authors, including (1) noisy objective functions, (2) dynamic MO fitness landscape, and (3) robust MO optimization.

Chapter 14, authored by the plenary speaker James C. Bezdek and his coauthor Jacalyn M. Huband, "Visual Cluster Validity," introduces preliminary information about cluster validity, and then offers a short history of visual cluster validity. Emphasis is placed on methods that reorder and image a relational version of the data. VAT, sVAT and co-VAT, a family of algorithms that build visual representations of square and rectangular relation data are then reviewed. Finally, a new method is presented of visual cluster validity (for the square case) based on (human) comparison of a pair of reordered relation matrices, D* and D (U*). D* is built by the VAT algorithm; and D(U*) is built by applying a transformation to any crisp or fuzzy partition of the data. Examples of the technique are discussed.

Chapter 15, authored by the plenary speaker Teuvo Kohonen, "Data Management by Self-Organizing Maps," discusses the self-organizing map (SOM) as an automatic data-analysis method. It is widely applied to clustering problems and data exploration in industry, finance, natural sciences, and linguistics. The most extensive applications, exemplified in this paper, can be found in the management of massive textual data bases. The SOM organization, a kind of similarity diagram of the models, makes it possible to obtain an insightful view of the global metric relationships of data, especially of high-dimensional data items. A new aspect introduced in this paper is that an input item can even more accurately be represented by a linear mixture of a few best-matching models. This becomes possible by a least-squares fitting procedure where the coefficients in the linear mixture of models are constrained to nonnegative values.

Chapter 16, authored by DeLiang Wang and Guoning Hu, "Cocktail Party Processing," discusses the speech segregation, or so-called the cocktail party problem coined by Cherry in 1953, that has proven to be extremely challenging. The authors propose a computational auditory scene analysis (CASA) approach to this daunting challenge. This monaural approach performs auditory segmentation and grouping in a two-dimensional time-frequency representation that encodes proximity in frequency and time, periodicity, amplitude modulation, and onset/offset. This CASA approach, according to the authors, has led to major advances towards solving the cocktail party problem and many other applications.

Chapter 17, authored by Bernadette Bouchon-Meunier, Maria Rifqi, and Marie-Jeanne Lesot, "Similarities in Fuzzy Data Mining: From a Cognitive View to Real-World

Applications," covers similarity as a key concept for all attempts to construct human-like automated systems since it is very natural in the human process of categorization, underlying many natural capabilities such as language understanding, pattern recognition or decision-making. It is essential to any data mining tasks. In this chapter, the authors study the use of similarities in data mining, basing their discourse on cognitive approaches of similarity. Bouchon-Meunier and her co-authors then focus on fuzzy logic that provides interesting tools for data mining mainly because of its ability to represent imperfect information, which is crucial when databases are complex, large and contain heterogeneous, imprecise, vague, uncertain or incomplete data.

Chapter 18, authored by Garrison W. Greenwood, "Attaining Fault Tolerance Through Self-Adaptation: The Strengths and Weaknesses of Evolvable Hardware Approaches," discusses how self-adaptive systems autonomously change their behavior to compensate for faults or to improve their performance. Evolvable hardware, which combines evolutionary algorithms with reconfigurable hardware, is often proposed as the cornerstone for systems that use self-adaptation for fault recovery. This author describes how fault-tolerant systems are built. The advantages and disadvantages of intrinsic evolvable hardware fault recovery methods are highlighted and design guidelines are presented.

We would first like to thank the contributors of the volume for making the chapters available in time. We also acknowledge the help of Qingshan Liu of The Chinese University of Hong Kong for his assistance in formatting the manuscripts. In addition we thank Anna Kramer and Ursula Barth from Springer for their assistance in producing the final version of the volume. Last but not least, the editors are indebted to Alfred Hofmann, Springer's Editorial Director of Computer Science, for his continued encouragement and support of this project. Without their contributions and help, this book would not have been possible.

Jacek M. Zurada
Gary G. Yen
Jun Wang

WCCI 2008 Organizing Committee

General Chair

Jun Wang, The Chinese University of Hong Kong, Hong Kong

Program Chairs

Derong Liu (IJCNN 2008), University of Illinois- Chicago, USA
Gary G. Feng (FUZZ-IEEE 2008), City University of Hong Kong, Hong Kong
Zbigniew Michalewicz (CEC 2008), University of Adelaide and SolveIT Software,
  Australia

Program Co-chairs

Robert Kozma (IJCNN 2008), University of Memphis, USA
Jerry M. Mendel (FUZZ-IEEE 2008), University of Southern California, USA
Robert G. Reynolds (CEC 2008), Wayne State University, USA

Special Sessions Chairs

Jagath C. Rajapakse (IJCNN 2008), Nanyang Technological University, Singapore
Xiao-Jun Zeng (FUZZ-IEEE 2008), University of Manchester, UK
Yuhui Shi (CEC 2008), Xi'an Jiaotong - Liverpool University, China
Byoung-Tak Zhang (Emerging Areas), Seoul National University, Korea

Plenary Sessions Chair

Jacek M. Zurada, University of Louisville, USA

Invited Sessions Chair

Gary G. Yen, Oklahoma State University, USA

Panel Sessions Chair

Jennie Si, Arizona State University, USA

Poster Sessions Chair

Wen Yu, National Polytechnic Institute (CINVESTAV-IPN), Mexico

Tutorials Chairs

Wlodzislaw Duch, Nicholaus Copernicus University, Poland
Russell Eberhart, Indiana University Purdue University Indianapolis, USA
Qiang Shen, University of Wales, UK

Workshops Chairs

Irwin K.C. King, The Chinese University of Hong Kong, Hong Kong
Yangmin Li, University of Macau, Macao

Competitions Chairs

Leszek Rutkowski, Technical University of Czestochowa, Poland
Isabelle Guyon, ClopiNet, USA
Philip Hingston, Edith Cowan University, Australia

Finance Chair

Lizhi Liao, Hong Kong Baptist University, Hong Kong

Registration Chairs

Daniel W.C. Ho, City University of Hong Kong, Hong Kong
Yat Wing Liu, IEEE Hong Kong Section, Hong Kong

Publicity Chairs

Amit Bhaya (South America), Federal University of Rio de Janeiro, Brazil
Salim Bouzerdoum (Oceania), University of Wollongong, Australia
Soo-Young Lee (Asia), Korea Advanced Institute of Science and Technology, South
   Korea
Ling Guan (North America), Ryerson University, Canada
Jianwei Zhang (Europe), University of Hamburg, Germany

Publications Chairs

Zeng-Guang Hou, Chinese Academy of Sciences, China
Nian Zhang, South Dakota School of Mines and Technology, USA

Exhibits Chair

Edgar N. Sanchez, CINVESTAV, Mexico

Social Events Chair

Zhi-Qiang Liu, City University of Hong Kong, Hong Kong

International Liaison Chairs

Yeung Yam, The Chinese University of Hong Kong, Hong Kong
David Dapeng Zhang, Hong Kong Polytechnic University, Hong Kong

Local Arrangements Chairs

James T.-Y. Kwok, Hong Kong University of Science and Technology, Hong Kong
Peter Tam, IEEE Hong Kong Section, Hong Kong
Eric W.-M. Yu, Hong Kong Applied Science and Technology Research Institute,
    Hong Kong

Student Activities Chair

Rami Abielmona, Larus Technologies, Canada

Table of Contents

Machine Learning and Brain Computer Interface

Fuzzy Modeling and Control

Computational Evolution

Applications

A New Framework for Machine Learning

Christopher M. Bishop

Microsoft Research, Cambridge, U.K
Christopher.Bishop@microsoft.com
http://research.microsoft.com/~cmbishop

Abstract. The last five years have seen the emergence of a powerful new framework for building sophisticated real-world applications based on machine learning. The cornerstones of this approach are (i) the adoption of a Bayesian viewpoint, (ii) the use of graphical models to represent complex probability distributions, and (iii) the development of fast, deterministic inference algorithms, such as variational Bayes and expectation propagation, which provide efficient solutions to inference and learning problems in terms of local message passing algorithms. This paper reviews the key ideas behind this new framework, and highlights some of its major benefits. The framework is illustrated using an example large-scale application.

1 Introduction

In recent years the field of machine learning has witnessed an important convergence of ideas, leading to a powerful new framework for building real-world applications. The goal of this paper is to highlight the emergence of this new viewpoint, and to emphasize its practical advantages over previous approaches. This paper is not, however, intended to be comprehensive, and no attempt is made to give accurate historical attribution of all the many important contributions. A much more detailed and comprehensive treatment of the topics discussed here, including additional references, can be found in [5].

The new framework for machine learning is built upon three key ideas: (i) the adoption of a Bayesian viewpoint, (ii) the use of probabilistic graphical models, and (iii) the application of fast, deterministic inference algorithms. In Section 2 we give a brief overview of the key concepts of Bayesian statistics, illustrated using a simple curve-fitting problem. We then discuss the use of probabilistic graphical models in Section 3. Inference and learning problems in graphical models can be solved efficiently using local message-passing algorithms, as described in Section 4. The new framework for machine learning is then illustrated using a large-scale application in Section 5, and finally in Section 6 we give some brief conclusions.

2 Bayesian Methods

The Bayesian interpretation of probabilities provides a consistent, indeed optimal, framework for the quantification of uncertainty [2,3,13]. In pattern recognition

J.M. Zurada et al. (Eds.): WCCI 2008 Plenary/Invited Lectures, LNCS 5050, pp. 1–24, 2008.

and machine learning applications, uncertainty arises both through noise processes on the observed variables, as well as through the unknown values of latent variables and model parameters. The adoption of a Bayesian viewpoint therefore provides a principled formalism through which all sources of uncertainty can be addressed consistently. In principle, it involves no more than the systematic application of the sum rule and the product rule of probability.

Machine learning models can be divided into *parametric* and *non-parametric*, according to whether or not they are based on models having a prescribed number of adjustable parameters. Most applications to date have been built using parametric models, and indeed this will be the focus of this paper. However, many of the same points emphasized here apply equally to non-parametric techniques.

Consider a model governed by a set of parameters which we group into a vector $\mathbf{w}$. If we denote the training data set by $\mathcal{D}$, then a central quantity is the conditional probability distribution $p(\mathcal{D}|\mathbf{w})$. When viewed as a function of $\mathbf{w}$ this is known as the *likelihood function*, and it plays a central role both in conventional (frequentist) and Bayesian approaches to machine learning. In a frequentist setting the goal is to find an estimator $\mathbf{w}^{\star}$ for the parameter vector by optimizing some criterion, for example by maximizing the likelihood. A significant problem with such approaches is *over-fitting* whereby the parameters are tuned to the noise on the data, thereby degrading the generalization performance.

In a Bayesian setting we express the uncertainty in the value of $\mathbf{w}$ through a probability distribution $p(\mathbf{w})$. This captures everything that is known about the value of $\mathbf{w}$, aside from the information provided by the training data, and is usually known as the prior distribution. The contribution from the training data is expressed through the likelihood function, and this can be combined with the prior using Bayes' theorem to give the posterior distribution

$$p(\mathbf{w}|\mathcal{D}) = \frac{p(\mathcal{D}|\mathbf{w})p(\mathbf{w})}{p(\mathcal{D})}. \tag{1}$$

Here the denominator is given by

$$p(\mathcal{D}) = \int p(\mathcal{D}|\mathbf{w})p(\mathbf{w})\,d\mathbf{w} \tag{2}$$

and can be viewed as the normalization factor which ensures that the posterior distribution $p(\mathbf{w}|\mathcal{D})$ in (1) integrates to one. It also plays a central role in model selection, as we shall discuss shortly.

In order to illustrate the use of Bayesian methods in machine learning, we consider the problem of fitting a set of noisy data points using a polynomial function. Although this example involves only a single input variable, a single output variable, and a simple parametric model, it captures most of the important concepts underpinning real-world applications of more sophisticated multivariate non-linear models.

The polynomial function itself can be written in the form

$$y(x, \mathbf{w}) = w_0 + w_1 x + w_2 x^2 + \ldots + w_M x^M = \sum_{j=0}^{M} w_j x^j \qquad (3)$$

where M is the order of the polynomial.

First we consider briefly a conventional, non-Bayesian, approach to this problem. Figure 1 shows the training data and the function from which the data is generated, along with the result of fitting several polynomials of different order by minimizing the sum-of-squares error between the polynomial predictions and the data point, defined by

$$E(\mathbf{w}) = \frac{1}{2} \sum_{n=1}^{N} \{y(x_n, \mathbf{w}) - t_n\}^2 \qquad (4)$$

where t_n denotes the training set target value corresponding to an input value of x_n

It can be seen that if the order of the polynomial is too low ($M = 0$, 1) then the result is a poor representation of the underlying sinusoidal curve. Equally if the order of the polynomial is too high ($M = 9$) then the result is again poor due to over-fitting. The best approximation arises from a model of intermediate complexity ($M = 3$). This is confirmed by looking at the root-mean-square error, defined by

$$E_{\mathrm{RMS}} = \sqrt{2E(\mathbf{w}^\star)/N} \qquad (5)$$

on both the training set and an independent test set, as shown in Figure 2. The best generalization performance (i.e. the smallest test set error) occurs for models of intermediate complexity.

Now consider a Bayesian approach to this problem. If we assume that the data has Gaussian noise, then the likelihood function takes the form

$$p(\mathcal{D}|\mathbf{w}) = \prod_{n=1}^{N} \mathcal{N}\left(t_n|y(x_n, \mathbf{w}), \beta^{-1}\right) \qquad (6)$$

where β is the precision (inverse variance) of the noise process. Here $\mathcal{N}\left(t|\mu, \sigma^2\right)$ denotes a Gaussian distribution over the variable t, with mean μ and variance σ^2.

For simplicity we consider a Gaussian prior distribution of the form

$$p(\mathbf{w}|\alpha) = \mathcal{N}(\mathbf{w}|\mathbf{0}, \alpha^{-1}\mathbf{I}) = \left(\frac{\alpha}{2\pi}\right)^{(M+1)/2} \exp\left\{-\frac{\alpha}{2}\mathbf{w}^{\mathrm{T}}\mathbf{w}\right\} \qquad (7)$$

where α is the precision of the distribution. Using Bayes' theorem (1) it is then straightforward to evaluate the posterior distribution over $\mathbf{w}$, which also takes the form of a Gaussian.

The posterior distribution is not itself of interest, but it plays a crucial role in making predictions for new input values. These predictions are governed by

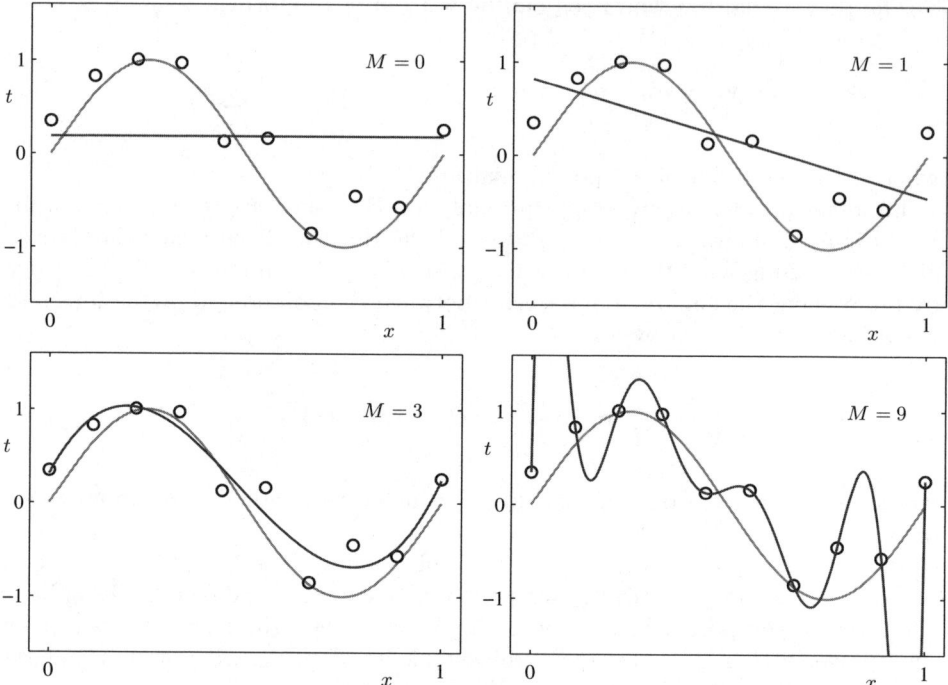

Fig. 1. Plot of a training data set of $N = 10$ points, shown as blue circles, each comprising an observation of the input variable x along with the corresponding target variable t. The green curve shows the function $\sin(2\pi x)$ used to generate the data. Our goal is to predict the value of t for some new value of x, without knowledge of the green curve. The red curves show the result of fitting polynomials of various orders M using least squares.

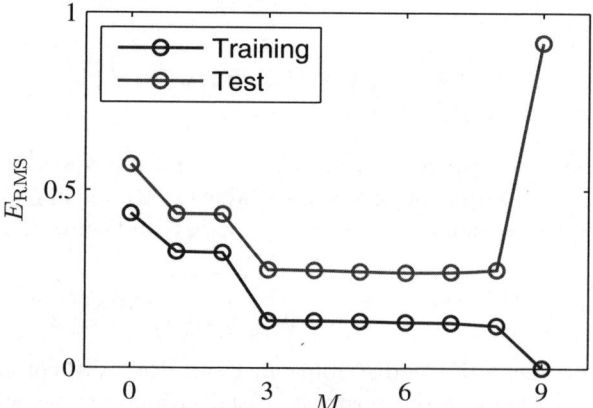

Fig. 2. Graphs of the root-mean-square error, defined by (5), evaluated on the training set and on an independent test set for various values of M

the predictive distribution, which is obtained from the sum and product rules of probability in the form

$$p(t|x, \mathcal{D}) = \int p(t|x, \mathbf{w}) p(\mathbf{w}|\mathcal{D}) \, d\mathbf{w}$$
$$= \mathcal{N}\left(t|m(x), s^2(x)\right) \tag{8}$$

where the mean and variance are given by

$$m(x) = \beta \phi(x)^{\mathrm{T}} \mathbf{S} \sum_{n=1}^{N} \phi(x_n) t_n \tag{9}$$

$$s^2(x) = \beta^{-1} + \phi(x)^{\mathrm{T}} \mathbf{S} \phi(x). \tag{10}$$

Here the matrix $\mathbf{S}$ is given by

$$\mathbf{S}^{-1} = \alpha \mathbf{I} + \beta \sum_{n=1}^{N} \phi(x_n) \phi(x_n)^{\mathrm{T}} \tag{11}$$

where $\mathbf{I}$ is the unit matrix, and we have defined the vector $\phi(x)$ with elements $\phi_i(x) = x^i$ for $i = 0, \ldots, M$.

Figure 3 shows a plot of the predictive distribution when the training set comprises $N = 4$ data points. Note that the variance of the predictive distribution is itself a function of the input variable x. In particular, the uncertainty in the predictions is smallest in the neighbourhood of the training data points. This intuitively pleasing result follows directly from the adoption of a Bayesian treatment.

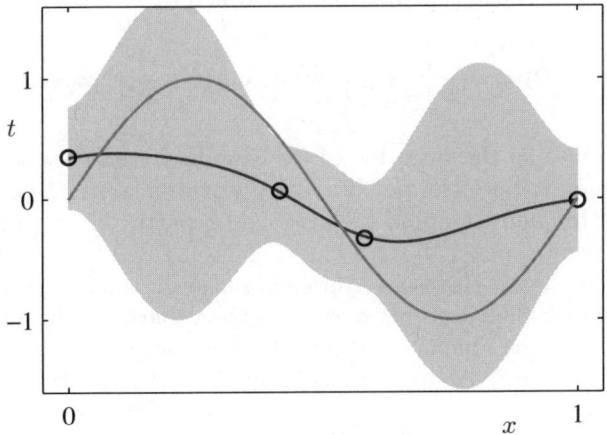

Fig. 3. Plot of the predictive distribution (8) with $N = 4$ training data points. The red curve shows the mean of the predictive distribution, while the red shaded region spans one standard deviation either side of the mean.

As can be seen from (10), the variance of the predictive distribution comprises two terms. The first corresponds to the noise on the training data and represents an irreducible level of uncertainty in predicting the value of t for a new input value x. The second term represents the uncertainty in predictions arising from the uncertainty in the model parameters $\mathbf{w}$. If we observe more data points, this latter uncertainty will decrease, as can be seen in Figure 4.

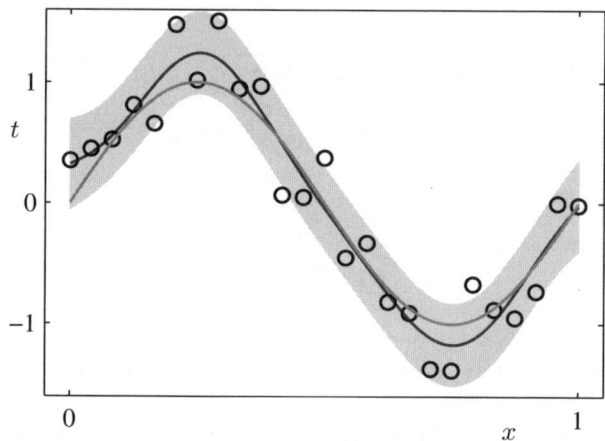

Fig. 4. As in Figure 3 but with $N = 25$ data points. The residual uncertainty in the predictive distribution is due mainly to the noise on the training data.

As an aside, suppose we make a frequentist point estimate of the model parameters by maximizing the posterior distribution. Equivalently we can maximize the logarithm of the posterior distribution, which takes the form

$$\ln p(\mathbf{w}|\mathcal{D}) = -\frac{\beta}{2}\sum_{n=1}^{N}\{t_n - \mathbf{w}^{\mathrm{T}}\phi(\mathbf{x}_n)\}^2 - \frac{\alpha}{2}\mathbf{w}^{\mathrm{T}}\mathbf{w} + \text{const.} \qquad (12)$$

which we recognise as the negative of the standard sum-of-squares error function with a quadratic weight penalty (regularization term). Thus we see how a conventional frequentist approach arises as a particular approximation to a Bayesian treatment.

We have seen that a Bayesian approach naturally makes predictions in the form of probability distributions over possible values, conditioned on the observed input variables. This is substantially more powerful than simply making point predictions as in conventional (non-Bayesian) machine learning approaches [5, pages 44–46].

Another major advantage of the Bayesian approach is that it automatically addresses the question of model complexity and model comparison. In conventional approaches based on a point estimate of the model parameters, it is common to optimize the model complexity to achieve a balance between too simple a model (which performs poorly on both training data and test data) and one which is too complex (which over-fits the training data and makes poor predictions on test data). This is usually addressed using data hold-out techniques such as cross-validation, in which part of the training data is kept aside in order to compare models of different complexity and to select the one which has the best generalization performance. Such cross-validations methods are wasteful of valuable training data, and are often computationally expensive due to the need for multiple training runs.

The Bayesian view of model comparison involves the use of probabilities to represent uncertainty in the choice of model, along with a consistent application of the sum and product rules of probability. Suppose we wish to compare a set of L models $\{\mathcal{M}_i\}$ where $i = 1, \ldots, L$. Here a model refers to a parametric representation for the probability distribution over the observed data $\mathcal{D}$, along with a prior distribution for the parameters. We shall suppose that the data is generated from one of these models but we are uncertain which one. Our uncertainty in the choice of model is expressed through a prior probability distribution $p(\mathcal{M}_i)$. Given a training set $\mathcal{D}$, we then wish to evaluate the posterior distribution

$$p(\mathcal{M}_i|\mathcal{D}) \propto p(\mathcal{M}_i)p(\mathcal{D}|\mathcal{M}_i). \qquad (13)$$

The prior allows us to express a preference for different models. Let us simply assume that all models are given equal prior probability. The interesting term is the *model evidence* $p(\mathcal{D}|\mathcal{M}_i)$ which expresses the preference shown by the data for different models. It is sometimes also called the *marginal likelihood* because it can be viewed as a likelihood function over the space of models, in which the parameters have been marginalized out. For a model governed by a set of parameters $\mathbf{w}$, the model evidence is given, from the sum and product rules of probability, by

$$p(\mathcal{D}|\mathcal{M}_i) = \int p(\mathcal{D}|\mathbf{w}, \mathcal{M}_i)p(\mathbf{w}|\mathcal{M}_i)\, \mathrm{d}\mathbf{w}. \qquad (14)$$

Recall that this term arises as the normalization factor in Bayes' theorem (1) for the parameters.

We can gain insight into Bayesian model comparison, and understand how the marginal likelihood can favour models of intermediate complexity, by considering Figure 5. Here the horizontal axis is a one-dimensional representation of the space of possible data sets, so that each point on this axis corresponds to a specific data set. We now consider three models $\mathcal{M}_1$, $\mathcal{M}_2$ and $\mathcal{M}_3$ of successively increasing complexity. Imagine running these models generatively to produce example data sets, and then looking at the distribution of data sets that result. Any given model can generate a variety of different data sets since the parameters are governed by a prior probability distribution, and for any choice of the parameters there may be random noise on the target variables. To generate a particular data set from a specific model, we first choose the values of the parameters from their prior distribution $p(\mathbf{w})$, and then for these parameter values we sample the data from $p(\mathcal{D}|\mathbf{w})$. A simple model (for example, based on a first order polynomial) has little variability and so will generate data sets that are fairly

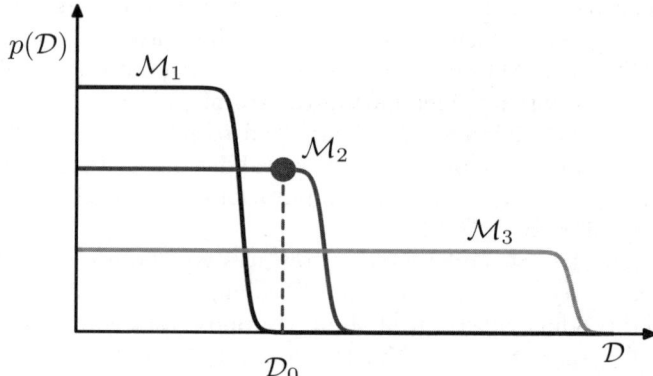

Fig. 5. Schematic illustration of the distribution of data sets for three models of different complexity, in which $\mathcal{M}_1$ is the simplest and $\mathcal{M}_3$ is the most complex. Note that the distributions are normalized. In this example, for the particular observed data set $\mathcal{D}_0$, the model $\mathcal{M}_2$ with intermediate complexity has the largest evidence.

similar to each other. Its distribution $p(\mathcal{D})$ is therefore confined to a relatively small region of the horizontal axis. By contrast, a complex model (such as a ninth order polynomial) can generate a great variety of different data sets, and so its distribution $p(\mathcal{D})$ is spread over a large region of the space of data sets. Because the distributions $p(\mathcal{D}|\mathcal{M}_i)$ are normalized, we see that the particular data set $\mathcal{D}_0$ can have the highest value of the evidence for the model of intermediate complexity. Essentially, the simpler model cannot fit the data well, whereas the more complex model spreads its predictive probability over too broad a range of data sets and so assigns relatively small probability to any one of them.

Returning to the polynomial regression problem, we can plot the model evidence against the order of the polynomial, as shown in Figure 6. Here we have assumed a prior of the form (7) with the parameter α fixed at $\alpha = 5 \times 10^{-3}$. The form of this plot is very instructive. Referring back to Figure 1, we see that the $M = 0$ polynomial has very poor fit to the data and consequently gives a relatively low value for the evidence. Going to the $M = 1$ polynomial greatly improves the data fit, and hence the evidence in Figure 6 is significantly higher. However, in going to $M = 2$, the data fit is improved only very marginally, due to the fact that the underlying sinusoidal function from which the data is generated is an odd function and so has no even terms in a polynomial expansion. Indeed, Figure 2 shows that the residual data error is reduced only slightly in going from $M = 1$ to $M = 2$. Because this richer model suffers a greater complexity penalty, the evidence actually falls in going from $M = 1$ to $M = 2$. When we go to $M = 3$ we obtain a significant further improvement in data fit, as seen in Figure 1, and so the evidence is increased again, giving the highest overall evidence for any of the polynomials. Further increases in the value of M produce only small improvements in the fit to the data but suffer increasing complexity penalty, leading overall to a decrease in the evidence values. Looking again at

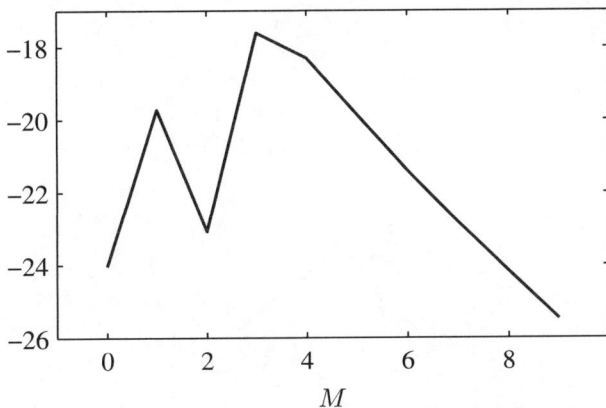

Fig. 6. Plot of the model evidence versus the order of the polynomial, for the simple curve fitting problem

Figure 2, we see that the generalization error is roughly constant between $M = 3$ and $M = 8$, and it would be difficult to choose between these models on the basis of this plot alone. The evidence values, however, show a clear preference for $M = 3$, since this is the simplest model which gives a good explanation for the observed data.

3   Graphical Models

The use of Bayesian methods in machine learning amounts to a consistent application of the sum and product rules of probability. We could therefore proceed to formulate and solve complicated probabilistic models purely by algebraic manipulation. However, it is highly advantageous to augment the analysis using diagrammatic representations of probability distributions, called *probabilistic graphical models*. These offer several useful properties:

1. They provide a simple way to visualize the structure of a probabilistic model and can be used to design and motivate new models.
2. Insights into the properties of the model can be obtained by inspection of the graph.
3. Complex computations, required to perform inference and learning in sophisticated models, can be expressed in terms of graphical manipulations, in which underlying mathematical expressions are carried along implicitly.

A graph comprises *nodes* (also called *vertices*) connected by *links* (also known as *edges* or *arcs*). In a probabilistic graphical model, each node represents a random variable (or group of random variables), and the links express probabilistic relationships between these variables.

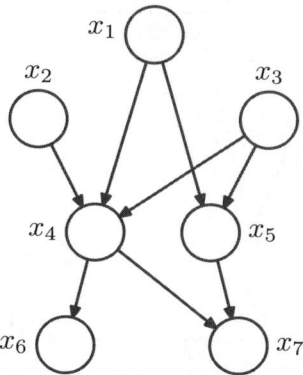

Fig. 7. Example of a directed graph describing the joint distribution over variables $x_1, \ldots, x_7$. The corresponding decomposition of the joint distribution is given by (16).

There are two main types of graphical model in widespread use, corresponding to directed graphs (in which the links have a directionality indicated by arrows) and undirected graphs (in which the links are symmetrical). In both cases the graph expresses the way in which the joint distribution over all of the random variables can be decomposed into a product of factors each depending only on a subset of the variables, but the relationship between the graph and the factorization is different for the two types of graph.

Consider first the case of directed graphs, also known as *Bayesian networks* or *belief networks*. An example is shown in Figure 7. If there is a link going from a node a to a node b, then we say that node a is the *parent* of node b. The graph specifies that the joint distribution factorizes into a product over all nodes of a conditional distribution for the variables at that node conditioned on the states of its parents

$$p(\mathbf{x}) = \prod_{k=1}^{K} p(x_k|\mathrm{pa}_k) \tag{15}$$

where pa_k denotes the set of parents of x_k, and $\mathbf{x} = \{x_1, \ldots, x_K\}$. For the specific case of the graph shown in Figure 7, the factorization takes the form

$$p(x_1)p(x_2)p(x_3)p(x_4|x_1, x_2, x_3)p(x_5|x_1, x_3)p(x_6|x_4)p(x_7|x_4, x_5). \tag{16}$$

A specific, and very familiar, example of a directed graph is the hidden Markov model, which is widely used in speech recognition, handwriting recognition, DNA analysis, and other sequential data applications, and is shown in Figure 8. The joint distribution for this model is given by

$$p(\mathbf{x}_1, \ldots, \mathbf{x}_N, \mathbf{z}_1, \ldots, \mathbf{z}_N) = p(\mathbf{z}_1)\left[\prod_{n=2}^{N} p(\mathbf{z}_n|\mathbf{z}_{n-1})\right]\prod_{n=1}^{N} p(\mathbf{x}_n|\mathbf{z}_n). \tag{17}$$

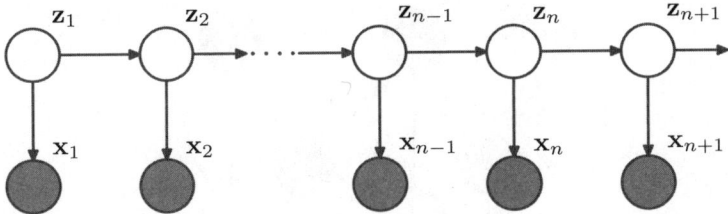

Fig. 8. The directed graph corresponding to a hidden Markov model. It represents the joint distribution over a set of observed variables $\mathbf{x}_1, \ldots, \mathbf{x}_N$ in terms of a Markov chain of hidden variables $\mathbf{z}_1, \ldots, \mathbf{z}_N$. Exactly the same graph also describes the Kalman filter.

Here $\mathbf{x}_1, \ldots, \mathbf{x}_N$ represent the observed variables (i.e. the data). In a graphical model the observed variables are denoted by shading the corresponding notes. The variables $\mathbf{z}_1, \ldots, \mathbf{z}_N$ represent latent (or hidden) variables which are not directly observed but which play a key role in the formulation of the model. In the case of the hidden Markov model the latent variables are discrete, while the observed variables may be discrete or continuous according to the particular application.

We can also consider the graph in Figure 8 for the case in which both the hidden and observed variables are Gaussian, in which case it describes the *Kalman filter*, a model which is widely used for tracking applications [21]. This highlights an important property of graphical models, namely that a particular graph describes a whole family of probability distributions which share the same factorization properties.

One of the powerful aspects of graphical models is the ease with which new models can be constructed, incorporating appropriate domain knowledge in the process. For example, Figure 9 shows an extension of the hidden Markov model which expresses the notion that there are two independent latent processes, and that at each time step the observed variables have distributions which are

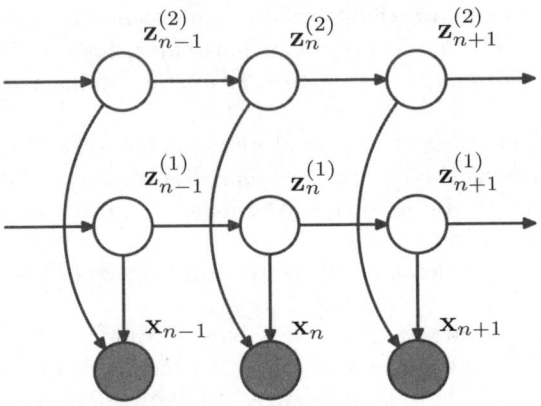

Fig. 9. A factorial hidden Markov model

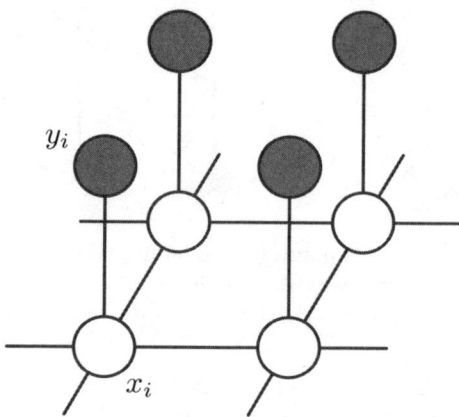

Fig. 10. An undirected graphical model representing a Markov random field for image de-noising, in which x_i is a binary latent (hidden) variable denoting the state of pixel i in the unknown noise-free image, and y_i denotes the corresponding value of pixel i in the observed noisy image

conditioned on the states of both of the corresponding latent variables. This is known as a *factorial hidden Markov model.*

Similarly, the Kalman filter can be extended to give a *switching state space model.* This has multiple Markov chains of continuous linear-Gaussian latent variables, each of which is analogous to the latent chain of the standard Kalman filter, together with a Markov chain of discrete variables of the form used in a hidden Markov model. The output at each time step is determined by stochastically choosing one of the continuous latent chains, using the state of the discrete latent variable as a switch, with the distribution of the observation at each step conditioned on the state of the corresponding continuous hidden variable.

Many other models can easily be constructed in this way. The key point is that new models can be formulated simply by drawing the corresponding graphical model, and prior knowledge from the application domain can be expressed through the structure of the graph. In particular, missing links in the graph determine the *conditional independence properties* of the joint distribution [5, Section 8.2].

The second major class of graphical model is based on *undirected graphs.* A well-known example is the *Markov random field,* illustrated in Figure 10. This graphical structure can be used to solve image processing problems such as de-noising and segmentation.

As with directed graphs, an undirected graph specifies the way in which the joint distribution of all variables in the model factorizes into a product of factors each involving only a subset of the variables. To understand this factorization we need to introduce the concept of a *clique,* which is defined as a subset of the nodes in a graph such that there exists a link between all pairs of nodes in the subset. In other words, the set of nodes in a clique is fully connected.

If we denote a clique by C and the set of variables in that clique by $\mathbf{x}_C$, then the joint distribution is written as a product of *potential functions* $\psi_C(\mathbf{x}_C)$ over the cliques of the graph

$$p(\mathbf{x}) = \frac{1}{Z} \prod_C \psi_C(\mathbf{x}_C). \tag{18}$$

Here the quantity Z, sometimes called the *partition function*, is a normalization constant and is given by

$$Z = \sum_{\mathbf{x}} \prod_C \psi_C(\mathbf{x}_C) \tag{19}$$

which ensures that the distribution $p(\mathbf{x})$ given by (18) is correctly normalized. By considering only potential functions which satisfy $\psi_C(\mathbf{x}_C) \geqslant 0$ we ensure that $p(\mathbf{x}) \geqslant 0$. In (19) we have assumed that $\mathbf{x}$ comprises discrete variables, but the framework is equally applicable to continuous variables, or a combination of the two, in which the summation is replaced by the appropriate combination of summation and integration.

Because we are restricted to potential functions which are strictly positive it is convenient to express them as exponentials, so that

$$\psi_C(\mathbf{x}_C) = \exp\{-E(\mathbf{x}_C)\} \tag{20}$$

where $E(\mathbf{x}_C)$ is called an *energy function*, and the exponential representation is called the *Boltzmann distribution*. The joint distribution is defined as the product of potentials, and so the total energy is obtained by adding the energies of each of the cliques.

In the case of Figure 10 there are three kinds of cliques, those that involve a single hidden variable, those that involve two adjacent hidden variables connected by a link, and those that involve one hidden and one observed variable, again connected by a link. An example of an energy function for such a model takes the form

$$E(\mathbf{x}, \mathbf{y}) = h \sum_i x_i - \beta \sum_{\{i,j\}} x_i x_j - \eta \sum_i x_i y_i \tag{21}$$

which defines a joint distribution over $\mathbf{x}$ and $\mathbf{y}$ given by

$$p(\mathbf{x}, \mathbf{y}) = \frac{1}{Z} \exp\{-E(\mathbf{x}, \mathbf{y})\}. \tag{22}$$

Directed and undirected graphs together allow most models of practical interest to be constructed. Which type of graph is more appropriate will depend on the application. Generally speaking, directed graphs are good at expressing causal relationships between variables. For example, if we have set of diseases and a set of symptoms then we can use a directed graph to capture the notion that the symptoms are caused by the diseases, and so there will be arrows (directed edges) going from disease variable nodes to symptom variable nodes. Undirected graphs, however, are better at expressing correlations between variables.

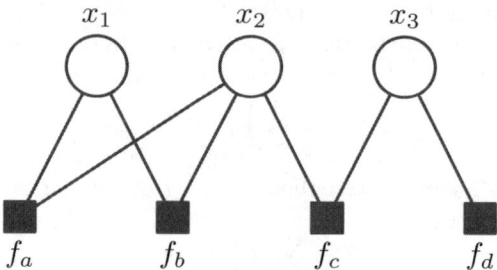

Fig. 11. Example of a factor graph, which corresponds to the factorization (23)

For example, in a model for segmenting an image into foreground and background we know that neighbouring pixels are very likely to share the same label (they are usually either both foreground or both background) and this can be captured using an undirected graph of the form shown in Figure 10.

It is often convenient to work with a third form of graphical representation known as a *factor graph*. We have seen that both directed and undirected graphs allow a global function of several variables to be expressed as a product of factors over subsets of those variables. Factor graphs make this decomposition explicit by introducing additional nodes for the factors themselves in addition to the nodes representing the variables.

Consider, for example, a distribution that is expressed in terms of the factorization

$$p(\mathbf{x}) = f_a(x_1, x_2) f_b(x_1, x_2) f_c(x_2, x_3) f_d(x_3). \tag{23}$$

This can be expressed by the factor graph shown in Figure 11. Factor graphs provide a useful representation for inference algorithms, discussed in the next section, as they allow both directed and undirected graphs to be treated in a unified way.

4   Approximate Inference

Having formulated a model in terms of a probabilistic graph we now need to learn the parameters of the model, and to use the trained model to make predictions. Some of the nodes in the graph correspond to observed variables representing the training data, and we are interested in finding the posterior distribution of other nodes, representing variables whose value we wish to predict, conditioned on the training data. We refer to this as an inference problem. The remaining nodes in the graph represent other latent, or hidden, variables whose values we are not directly interested in.

In a Bayesian setting, the model parameters are also random variables and are therefore represented as nodes in the graph. Computing the posterior distribution of those parameters is therefore just another inference problem!

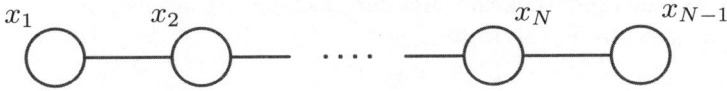

Fig. 12. A simple undirected graph comprising a chain of nodes, used to illustrate the solution of inference problems

Consider first the problem of performing inference on a simple chain of nodes shown in Figure 12. The cliques of this graph comprise pairs of adjacent nodes connected by links. Thus the joint distribution of all of the variables is given by

$$p(\mathbf{x}) = \frac{1}{Z}\psi_{1,2}(x_1, x_2)\psi_{2,3}(x_2, x_3) \cdots \psi_{N-1,N}(x_{N-1}, x_N). \tag{24}$$

Let us consider the inference problem of finding the marginal distribution $p(x_n)$ for a specific node x_n that is part way along the chain. Note that, for the moment, there are no observed nodes. By definition, the required marginal is obtained by summing the joint distribution over all variables except x_n, so that

$$p(x_n) = \sum_{x_1} \cdots \sum_{x_{n-1}} \sum_{x_{n+1}} \cdots \sum_{x_N} p(\mathbf{x}). \tag{25}$$

In a naive implementation, we would first evaluate the joint distribution and then perform the summations explicitly. The joint distribution can be represented as a set of numbers, one for each possible value for $\mathbf{x}$. Because there are N variables each with K states, there are K^N values for $\mathbf{x}$ and so evaluation and storage of the joint distribution, as well as marginalization to obtain $p(x_n)$, all involve storage and computation that scale exponentially with the length N of the chain.

We can, however, obtain a much more efficient algorithm by exploiting the conditional independence properties of the graphical model. If we substitute the factorized expression (24) for the joint distribution into (25), then we can rearrange the order of the summations and the multiplications to allow the required marginal to be evaluated much more efficiently. Consider for instance the summation over x_N. The potential $\psi_{N-1,N}(x_{N-1}, x_N)$ is the only one that depends on x_N, and so we can perform the summation

$$\sum_{x_N} \psi_{N-1,N}(x_{N-1}, x_N) \tag{26}$$

first to give a function of x_{N-1}. We can then use this to perform the summation over x_{N-1}, which will involve only this new function together with the potential $\psi_{N-2,N-1}(x_{N-2}, x_{N-1})$, because this is the only other place that x_{N-1} appears. Similarly, the summation over x_1 involves only the potential $\psi_{1,2}(x_1, x_2)$ and so can be performed separately to give a function of x_2, and so on.

If we group the potentials and summations together in this way, we can express the desired marginal in the form

$$
p(x_n) = \frac{1}{Z}
$$

$$
\underbrace{\left[\sum_{x_{n-1}} \psi_{n-1,n}(x_{n-1}, x_n) \cdots \left[\sum_{x_2} \psi_{2,3}(x_2, x_3) \left[\sum_{x_1} \psi_{1,2}(x_1, x_2) \right] \right] \cdots \right]}_{\mu_\alpha(x_n)}
$$

$$
\underbrace{\left[\sum_{x_{n+1}} \psi_{n,n+1}(x_n, x_{n+1}) \cdots \left[\sum_{x_N} \psi_{N-1,N}(x_{N-1}, x_N) \right] \cdots \right]}_{\mu_\beta(x_n)}. \tag{27}
$$

The key concept that we are exploiting is that multiplication is distributive over addition, so that

$$
ab + ac = a(b + c) \tag{28}
$$

in which the left-hand side involves three arithmetic operations whereas the right-hand side reduces this to two operations.

Let us work out the computational cost of evaluating the required marginal using this re-ordered expression. We have to perform $N - 1$ summations each of which is over K states and each of which involves a function of two variables. For instance, the summation over x_1 involves only the function $\psi_{1,2}(x_1, x_2)$, which is a table of $K \times K$ numbers. We have to sum this table over x_1 for each value of x_2 and so this has $O(K^2)$ cost. The resulting vector of K numbers is multiplied by the matrix of numbers $\psi_{2,3}(x_2, x_3)$ and so is again $O(K^2)$. Because there are $N - 1$ summations and multiplications of this kind, the total cost of evaluating the marginal $p(x_n)$ is $O(NK^2)$. This is linear in the length of the chain, in contrast to the exponential cost of a naive approach.

We can now give a powerful interpretation of this calculation in terms of the passing of local *messages* around on the graph. From (27) we see that the expression for the marginal $p(x_n)$ decomposes into the product of two factors times the normalization constant

$$
p(x_n) = \frac{1}{Z}\mu_\alpha(x_n)\mu_\beta(x_n). \tag{29}
$$

We shall interpret $\mu_\alpha(x_n)$ as a message passed forwards along the chain from node x_{n-1} to node x_n. Similarly, $\mu_\beta(x_n)$ can be viewed as a message passed backwards along the chain to node x_n from node x_{n+1}. Note that each of the messages comprises a set of K values, one for each choice of x_n, and so the product of two messages should be interpreted as the point-wise multiplication of the elements of the two messages to give another set of K values.

The message $\mu_\alpha(x_n)$ can be evaluated recursively because

$$\mu_\alpha(x_n) = \sum_{x_{n-1}} \psi_{n-1,n}(x_{n-1}, x_n) \left[\sum_{x_{n-2}} \cdots \right]$$
$$= \sum_{x_{n-1}} \psi_{n-1,n}(x_{n-1}, x_n)\mu_\alpha(x_{n-1}) \tag{30}$$

with an analogous result for the backward messages.

This example shows how the local factorization implied by the graphical structure allows the cost of exact inference to be reduced from being exponential in the length of the chain to being linear. A similar result holds for more complex graphs provided they have a tree structure (i.e. they do not have any loops). This exact inference technique is known as *belief propagation*.

A well-known special case of this message-passing algorithm is the forward-backward algorithm for inferring the posterior distribution of the hidden variables in a hidden Markov model [1,19]. Similarly, the forward recursions of the Kalman filter [14] and the backward recursions of the Kalman smoother [20] are also special cases of this result.

For most practical applications, however, this efficient exact solution of inference problems is no longer tractable. This lack of tractability arises either because the graph is no longer a tree, or because the individual local marginalizations no longer have an exact closed-form solution.

We therefore need to find approximate inference algorithms which can yield good results with reasonable computational cost. For a long time the only generally applicable method was to use Markov chain Monte Carlo sampling techniques. Unfortunately, this approach tends to be computationally costly and does not scale well to real-world applications involving large data sets. One of the most important advances in machine learning in recent years has therefore been the development of fast, approximate inference algorithms. Like the exact inference algorithms for trees discussed above, these can all be expressed in terms of local message passing on the corresponding graphical model, and this leads naturally to efficient software implementations. They are often called 'deterministic' algorithms because they provide analytical expressions for the posterior distribution, in contrast to Monte Carlo methods which yield their results in the form of a set of samples drawn from the posterior distribution. Here we introduce briefly some of the most prominent deterministic inference techniques. It should be emphasized, however, that there are many other such algorithms and new ones continue to be developed.

One of the simplest such algorithms is called *loopy belief propagation* [9] and simply involves applying the standard belief propagation equations, derived for tree-structured graphs, to more general graphical models. Although this is an ad-hoc procedure, and has no guarantee of convergence, it is often found to yield good results, and indeed gives state-of-the-art results for decoding certain kinds of error-correcting codes [4,8,10,15,16].

A more principled approach to approximate inference is to define a family of approximating distributions (whose members are simpler in some sense than the true posterior distribution) and then to seek the optimal member of that family by minimizing a suitable criterion which measures the dissimilarity between the approximate distribution and the exact posterior distribution. Different algorithms arise according to the simplifying assumptions in the approximate posterior, and according to the choice of criterion.

The *variational Bayes* method defines the dissimilarity between the true posterior distribution $p(\mathbf{Z}|\mathbf{X})$ and the approximating distribution $q(\mathbf{Z})$ to be the Kullback-Leibler divergence given by

$$\mathrm{KL}(q\|p) = -\int q(\mathbf{Z}) \ln \left\{ \frac{p(\mathbf{Z}|\mathbf{X})}{q(\mathbf{Z})} \right\} \, d\mathbf{Z} \tag{31}$$

where $\mathbf{Z}$ represents the set of all non-observed variables in the problem and $\mathbf{X}$ represents the observed variables.

The approximating distribution can be chosen to have a simple analytical form. For example, it might be a Gaussian, whose mean and covariance are then optimized so as to minimize the KL divergence with respect to the (non-Gaussian) true posterior distribution. A more flexible framework arises, however, if we assume a specific factorization for the approximating distribution $q(\mathbf{Z})$, without any restriction on the functional form of the factors. Suppose we partition the elements of $\mathbf{Z}$ into disjoint groups that we denote by $\mathbf{Z}_i$ where $i = 1, \ldots, M$. We then assume that the q distribution factorizes with respect to these groups, so that

$$q(\mathbf{Z}) = \prod_{i=1}^{M} q_i(\mathbf{Z}_i). \tag{32}$$

If we substitute (32) into (31) we can then minimize the KL divergence with respect to one of the factors $q_j(\mathbf{Z}_j)$, keeping the remaining factors fixed. This involves a free-form functional optimization performed using the calculus of variations, and gives the result

$$\ln q_j^\star(\mathbf{Z}_j) = \mathbb{E}_{i \neq j}[\ln p(\mathbf{X}, \mathbf{Z})] + \mathrm{const} \tag{33}$$

where $p(\mathbf{X}, \mathbf{Z})$ is the joint distribution of hidden and observed variables, and the expectation is taken over all groups of variables $\mathbf{Z}_i$ for $i \neq j$. The additive constant corresponds to the normalization coefficient for the distribution. In order to apply this approach in practice, the factors $q_i(\mathbf{Z}_i)$ are first suitably initialized, and then they are updated in turn using (33) until a suitable convergence criterion is satisfied.

As an illustration of the variational Bayes method, consider the toy problem shown in Figure 13. Here the green contours show the true posterior distribution $p(\mu, \tau)$ over the mean μ and precision (inverse variance) τ for a simple inference problem involving a Gaussian distribution [5, Section 10.1.3]. For the sake of illustration, suppose that this distribution is intractable to compute and so we wish to find an approximation using variational inference. The approximating

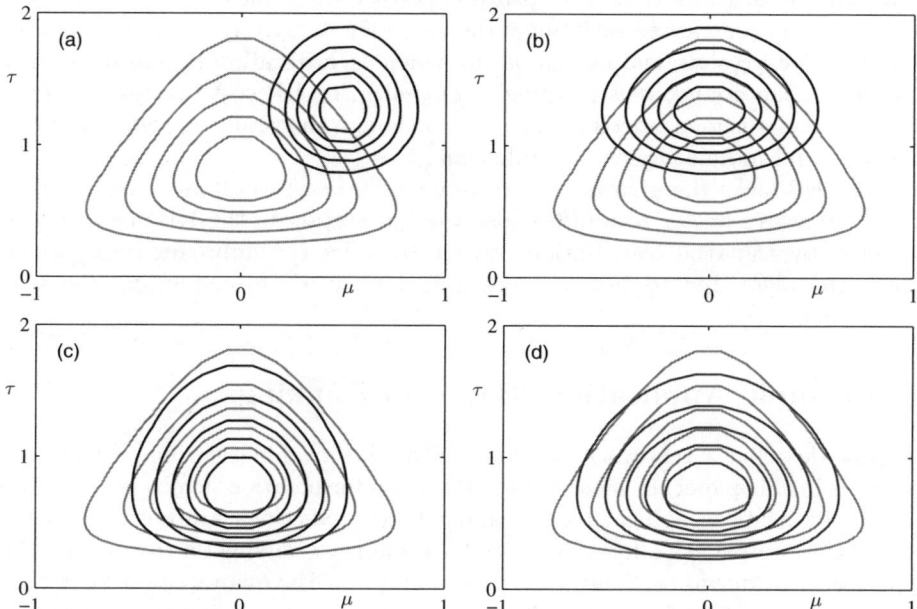

Fig. 13. An illustration of the variational Bayes inference algorithm. See the text for details.

distribution is assumed to factorize so that $q(\mu, \tau) = q_\mu(\mu)q_\tau(\tau)$. The blue contours in Figure 13(a) show the initialization of this factorized distribution. In Figure 13(b) the distribution $q_\mu(\mu)$ has been updated using (33) keeping $q_\tau(\tau)$ fixed. Similarly, in Figure 13(c) the distribution $q_\tau(\tau)$ has been updated keeping $q_\mu(\mu)$ fixed. Finally, the optimal factorized solution, to which the iterative scheme converges, is shown by the red contours in Figure 13(d).

If we consider distributions which are expressed in terms of probabilistic graphical models, then the variational update equations (33) can be cast in the form of a local message-passing algorithm. This makes possible the construction of general purpose software for variational inference in which the form of the model does not need to be specified in advance [6].

Another widely used approximate inference algorithm is called *expectation propagation* or *EP* [17,18]. As with the variational Bayes method discussed so far, this too is based on the minimization of a Kullback-Leibler divergence but now of the reverse form $\mathrm{KL}(p\|q)$, which gives the approximation rather different properties. EP again makes use of the factorization implied by a graphical model, and again the update equations for determining the approximate posterior distribution can be cast as a local message-passing algorithm. Although each step is minimizing a specific KL divergence, the overall algorithm does not optimize a unique quantity globally. However, for approximations which lie within the *exponential family* of distributions, if the iterations do converge, the resulting solution will be a stationary point of a particular energy function [17], although

each iteration of EP does not necessarily decrease the value of this energy function. This is in contrast to variational Bayes, which iteratively maximizes a lower
bound on the log marginal likelihood, in which each iteration is guaranteed not
to decrease the bound. It is possible to optimize the EP cost function directly,
in which case it is guaranteed to converge, although the resulting algorithms can
be slower and more complex to implement.

While EP lacks the guaranteed convergence properties of variational Bayes,
it can often give better results because the integration in the KL divergence is
weighted by the true distribution, rather than by the approximation, which
causes the algorithm to take a more global view of approximating the true
distribution.

5   Example Application: Bayesian Ranking

We now describe a real-world application of the machine learning framework
discusses in this paper. It is based on a Bayesian formulation, which is expressed
as a probabilistic graphical model, and where predictions are obtained using
expectation propagation formulated as local message-passing on the graph. The
application is known as *TrueSkill*[TM] [12], and is a Bayesian system for rating
player skill in competitive games. It can be viewed as a generalization of the well
known Elo system, which is used for example in Chess, and which was adopted
by the World Chess Federation in 1970 as an international standard. With the
advent of online gaming, the importance of skill rating systems has increased
significantly because the quality of the online experience of millions of players
each day is at stake.

Elo assigns each player i a skill rating s_i, and the probability of the possible
game outcomes is modelled as a function of the two players skills s_1 and s_2. In
a particular game each player exhibits a performance

$$p_i \sim \mathcal{N}(p_i|s_i, \beta^2) \tag{34}$$

which is normally distributed around their skill value with fixed variance β^2. The
probability that player 1 wins is given by the probability that their performance
exceeds that of player 2, so that

$$P(p_1 > p_2|s_1, s_2) = \Phi\left(\frac{s_1 - s_2}{\sqrt{2}\beta}\right) \tag{35}$$

where Φ is the cumulative density of a zero-mean, unit-variance Gaussian. The
Elo system then provides an update equation for the skill ratings which causes
the observed game outcome to become more likely, while preserving the constraint $s_1 + s_2 = \text{const}$. There is a variant of the Elo system which replaces the
cumulative Gaussian with a logistic sigmoid. In Elo, a player's rating is regarded
as provisional as long as it is based on less than a fixed number of, say, 20 games.
This problem was addressed previously by adopting a Bayesian approach, known
as *Glicko*, which models the belief about a player's rating using a Gaussian with
mean μ and variance σ^2 [11].

An important new application of skill rating systems are multiplayer online games, which present the following challenges:

1. Game outcomes often refer to teams of players, and yet a skill rating for individual players is needed for future matchmaking.
2. More than two players or teams compete, such that the game outcome is a permutation of teams or players rather than just a winner and a loser.

TrueSkill addresses both of these challenges in the context of a principled Bayesian approach.

Each player has a skill distribution which is Gaussian $s_i \sim \mathcal{N}(s_i|\mu_i, \sigma_i^2)$, and the performance of a player is again a noisy version of their skill given by (34). The performance t_j of team j is modelled as the sum of the performances of the players comprising that team, and the ordering of the team performances gives the ordering of the match results. This model can be expressed as the factor graph shown in Figure 14. Draws can be incorporated into the model by requiring a non-zero margin, governed by a parameter ϵ, between the performance of two teams in order to achieve a victory.

Skill estimates need to be reported after each game, and so an online learning scheme is used known as *Gaussian density filtering*, which can be viewed as a special case of expectation propagation. The posterior distribution is approximated by a Gaussian, and forms the prior distribution for the next game.

Extensive testing of *TrueSkill* demonstrates significant improvements over Elo [12]. In particular, the number of games which need to be played in order to determine accurate values for player skills can be substantially less (up to an order of magnitude) compared to Elo. This is illustrated in Figure 15 which shows the evolution of the skill ratings for two players over several hundred games, based on data collected during beta testing of the Xbox title 'Halo 2'. We see that *TrueSkill* exhibits good convergence within the first 10–20 games, whereas the Elo estimates are continuing to change significantly even after 100 games.

Intuitively, the reason for the faster convergence is that knowledge of the uncertainty in the skill estimates modulates the magnitude of the updates in an optimal way. For instance, informally, if a player with a skill rating of 120 ± 20 beats a player of rating 130 ± 2 then the system can make a substantial increase in the rating of the winning player. Elo by contrast does not have access to uncertainty estimates and so makes numerous small corrections over many games in order to increase the skill estimate for a particular player.

Xbox 360 Live is Microsoft's online console gaming service, allowing players to play together across the world on hundreds of different game titles. *TrueSkill* has been globally deployed as the skill rating system for *Xbox 360 Live*, analyzing millions of game outcomes resulting from billions of hours of online play. It processes hundreds of thousands of new game outcomes per day, making it one of the largest applications of Bayesian inference to date. *TrueSkill* provides ratings and league table information to the players, and is used to perform real-time matchmaking.

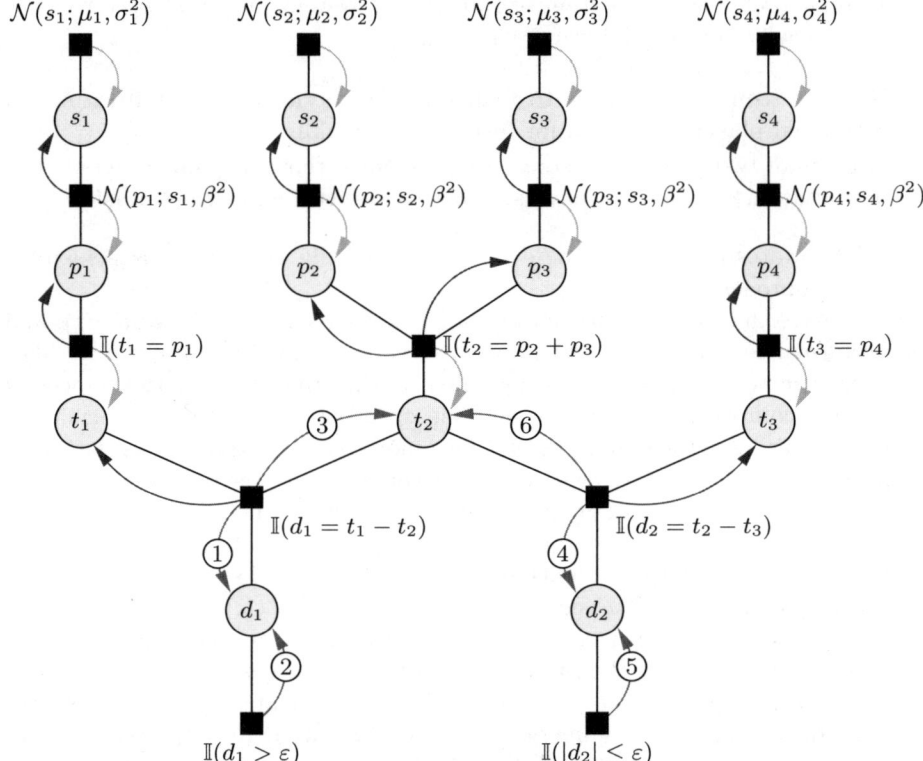

Fig. 14. An example TrueSkill factor graph, for two matches involving three teams and four players. The arrows indicate the optimal message passing schedule.

The graphical model formulation of *TrueSkill* makes it particularly straightforward to extend the model in interesting ways. For example, we can allow the skills to evolve in time by adding additional links to the graph corresponding to Gaussian temporal dynamics [12]. Similarly, the use of full expectation-propagation updates allows information to be propagated backwards in time (smoothing) as well as forwards in time (filtering). This permits a full analysis of historical data, for example a comparison of the strength of different chess players over a period of 150 years [7].

6   Discussion

In this paper we have highlighted the emergence of a new framework for the formulation and solution of problems in machine learning. The three main ingredients, namely a Bayesian approach, the use of graphical models, and use of approximate deterministic inference algorithms, fit very naturally together. In a fully Bayesian setting every unknown variable is given a probability distribution and hence corresponds to a node in a graphical model, and deterministic

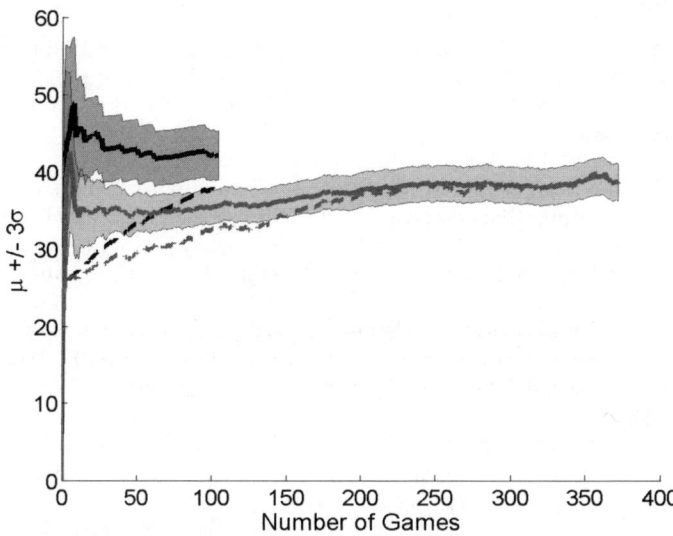

Fig. 15. Convergence trajectories for two players comparing Elo (dashed lines) with *TrueSkill* (solid lines). Note that the latter includes uncertainty estimates, shown by the shaded regions.

approximation algorithms, which provide efficient solutions to inference problems, can be cast in terms of messages passed locally between nodes of the graph.

In many conventional machine learning applications, the formulation of the model, and the algorithm used to perform learning and make predictions, are intertwined. One feature of the new framework is that there is a beneficial separation between the formulation of the model in terms of its graphical structure, and the solution of inference problems using local message passing. Research into new inference algorithms can proceed largely independently of the particular application, while domain experts can focus their efforts on the formulation of new application-specific models. Indeed, general purpose software can be developed[1] which implements a range of alternative inference algorithms for broad classes of graphical structures. Another major benefit of the new framework is that it allows fully Bayesian methods to be applied to large scale applications, something which was previously not feasible.

We live in an increasingly data-rich world, with ever greater requirements to extract useful information from that data. The framework for machine learning reviewed in this paper, which scales well to large data sets, offers the opportunity to develop many new and exciting applications for machine learning in the years ahead.

---

[1] One example is *Infer.Net* which is described at:
   http://research.microsoft.com/mlp/ml/Infer/Infer.htm

References

1. Baum, L.E.: An inequality and associated maximization technique in statistical estimation of probabilistic functions of Markov processes. Inequalities 3, 1–8 (1972)
2. Berger, J.O.: Statistical Decision Theory and Bayesian Analysis, 2nd edn. Springer, Heidelberg (1985)
3. Bernardo, J.M., Smith, A.F.M.: Bayesian Theory. Wiley, Chichester (1994)
4. Berrou, C., Glavieux, A., Thitimajshima, P.: Near Shannon limit error-correcting coding and decoding: Turbo-codes (1). In: Proceedings ICC 1993, pp. 1064–1070 (1993)
5. Bishop, C.M.: Pattern Recognition and Machine Learning. Springer, Heidelberg (2006)
6. Bishop, C.M., Spiegelhalter, D., Winn, J.: VIBES: A variational inference engine for Bayesian networks. In: Becker, S., Thrun, S., Obermeyer, K. (eds.) Advances in Neural Information Processing Systems, vol. 15, pp. 793–800. MIT Press, Cambridge (2003)
7. Dangauthier, P., Herbrich, R., Minka, T., Graepel, T.: Trueskill through time: Revisiting the history of chess. In: Advances in Neural Information Processing Systems, vol. 20 (2007), `http://books.nips.cc/nips20.html`
8. Frey, B.J.: Graphical Models for Machine Learning and Digital Communication. MIT Press, Cambridge (1998)
9. Frey, B.J., MacKay, D.J.C.: A revolution: Belief propagation in graphs with cycles. In: Jordan, M.I., Kearns, M.J., Solla, S.A. (eds.) Advances in Neural Information Processing Systems, vol. 10. MIT Press, Cambridge (1998)
10. Gallager, R.G.: Low-Density Parity-Check Codes. MIT Press, Cambridge (1963)
11. Glickman, M.E.: Parameter estimation in large dynmaical paird comparison experiments. Applied Statistics 48, 377–394 (1999)
12. Herbrich, R., Minka, T., Graepel, T.: Trueskill^{tm}: A Bayesian skill rating system. In: Advances in Neural Information Processing Systems, vol. 19, pp. 569–576. MIT Press, Cambridge (2007)
13. Jaynes, E.T.: Probability Theory: The Logic of Science. Cambridge University Press, Cambridge (2003)
14. Kalman, R.E.: A new approach to linear filtering and prediction problems. Transactions of the American Society for Mechanical Engineering, Series D, Journal of Basic Engineering 82, 35–45 (1960)
15. MacKay, D.J.C., Neal, R.M.: Good error-correcting codes based on very sparse matrices. IEEE Transactions on Information Theory 45, 399–431 (1999)
16. McEliece, R.J., MacKay, D.J.C., Cheng, J.F.: Turbo decoding as an instance of Pearl's 'Belief Ppropagation' algorithm. IEEE Journal on Selected Areas in Communications 16, 140–152 (1998)
17. Minka, T.: Expectation propagation for approximate Bayesian inference. In: Breese, J., Koller, D. (eds.) Proceedings of the Seventeenth Conference on Uncertainty in Artificial Intelligence, pp. 362–369. Morgan Kaufmann, San Francisco (2001)
18. Minka, T.: A family of approximate algorithms for Bayesian inference. Ph.D. thesis, MIT (2001)
19. Rabiner, L.R.: A tutorial on hidden Markov models and selected applications in speech recognition. Proceedings of the IEEE 77(2), 257–285 (1989)
20. Rauch, H.E., Tung, F., Striebel, C.T.: Maximum likelihood estimates of linear dynamical systems. AIAA Journal 3, 1445–1450 (1965)
21. Zarchan, P., Musoff, H.: Fundamentals of Kalman Filtering: A Practical Approach, 2nd edn. AIAA (2005)

Bilevel Optimization and Machine Learning

Kristin P. Bennett[1], Gautam Kunapuli[1], Jing Hu[1], and Jong-Shi Pang[2]

[1] Dept. of Mathematical Sciences, Rensselaer Polytechnic Institute[*], Troy, NY, USA
{bennek, kunapg, huj}@rpi.edu
[2] Dept. of Industrial and Enterprise Systems Engineering, University of Illinois at
Urbana Champaign, Urbana Champaign, IL, USA
jspang@uiuc.edu

Abstract. We examine the interplay of optimization and machine learning. Great progress has been made in machine learning by cleverly reducing machine learning problems to convex optimization problems with one or more hyper-parameters. The availability of powerful convex-programming theory and algorithms has enabled a flood of new research in machine learning models and methods. But many of the steps necessary for successful machine learning models fall outside of the convex machine learning paradigm. Thus we now propose framing machine learning problems as Stackelberg games. The resulting bilevel optimization problem allows for efficient systematic search of large numbers of hyper-parameters. We discuss recent progress in solving these bilevel problems and the many interesting optimization challenges that remain. Finally, we investigate the intriguing possibility of novel machine learning models enabled by bilevel programming.

1   Introduction

Convex optimization now forms a core tool in state-of-the-art machine learning. Convex optimization methods such as Support Vector Machines (SVM) and kernel methods have been applied with great success. For a learning task, such as regression, classification, ranking, and novelty detection, the modeler selects a convex loss and regularization functions suitable for the given task and optimizes for a given data set using powerful robust convex programming methods such as linear, quadratic, or semi-definite programming. But the many papers reporting the success of such methods frequently gloss over the critical choices that go into making a successful model. For example, as part of model selection, the modeler must select which variables to include, which data points to use, and how to set the possibly many model parameters. The machine learning problem is reduced to a convex optimization problem only because the boundary of what is considered to be part of the method is drawn very narrowly. Our goal here is to expand the mathematical programming models to more fully incorporate the

[*] This work was supported in part by the Office of Naval Research under grant no. N00014-06-1-0014. The authors are grateful to Professor Olvi Mangasarian for his suggestions on the penalty approach.

J.M. Zurada et al. (Eds.): WCCI 2008 Plenary/Invited Lectures, LNCS 5050, pp. 25–47, 2008.

entire machine learning process. Here we will examine the bilevel approach first developed in [1].

Consider support vector regression (SVR). In SVR, we wish to compute a linear regression function that maps the input, $\mathbf{x} \in R^n$, to a response $y \in R$, given a training set of (input, output) pairs, $\{(\mathbf{x}_i, y_i), i = 1 \ldots \ell\}$. To accomplish this, we must select hyper-parameters including the two parameters in the SVR objective and the features or variables that should be used in the model. Once these are selected, the learned function corresponds to the optimal solution of a quadratic program. The most commonly used and widely accepted method for selecting these hyper-parameters is still cross validation (CV).

In CV, the hyper-parameters are selected to minimize some estimate of the out-of-sample generalization error. A typical method would define a grid over the hyper-parameters of interest, and then do 10-fold cross validation for each of the grid values. The inefficiencies and expense of such a grid-search cross-validation approach effectively limit the desirable number of hyper-parameters in a model, due to the combinatorial explosion of grid points in high dimensions.

Here, we examine how model selection using out-of-sample testing can be treated as a Stackelberg game in which the leader sets the parameters to minimize the out-of-sample error, and the followers optimize the in-sample errors for each fold given the parameters. Model selection using out-of-sample testing can then be posed as an optimization problem, albeit with an "inner" and an "outer" objective. The main idea of the approach is as follows: the data is partitioned or bootstrapped into training and test sets. We seek a set of hyper-parameters such that when the optimal training problem is solved for each training set, the loss over the test sets is minimized. The resulting optimization problem is a bilevel program. Each learning function is optimized in its corresponding training problem with fixed hyper-parameters—this is the inner (or lower-level) optimization problem. The overall testing objective is minimized—this is the outer (or upper-level) optimization problem.

We develop two alternative methods for solving the bilinear programs. In both methods, the convex lower level problems are replaced by their Karush-Kuhn-Tucker (KKT) optimality conditions, so that the problem becomes a mathematical programming problem with equilibrium constraints (MPEC). The equivalent optimization problem has a linear objective and linear constraints except for the set of equilibrium constraints formed by the complementarity conditions. In our first approach, the equilibrium constraints are relaxed from equalities to inequalities to form a nonlinear program (NLP) that is then solved by a state-of-the-art general-purpose nonlinear programming solver, FILTER. In the second approach, the equilibrium constraints are treated as penalty terms and moved to the objective. The resulting penalty problem is then solved using the successive linearization algorithm for model selection (SLAMS). Further performance enhancements are obtained by stopping SLAMS at the first MPEC-feasible solution found, a version we term EZ-SLAMS.

Our successful bilevel programming approaches offer several fundamental advantages over prior approaches. First, recent advances in bilevel programming in the optimization community permit the systematic treatment of models based on popular loss functions used for SVM and kernel methods with many hyper-parameters. In addition to the ability to simultaneously optimize many hyper-parameters, the bilevel programming approach offers a broad framework in which a wide class of novel machine learning algorithms can be developed. Amenable problems with many parameters are pervasive in data analysis. The bilevel pro-gramming approach can be used to address feature selection, kernel construction, semi-supervised learning, and models with missing data.

This paper illustrates the bilevel approach applied to support vector regres-sion. Additional information can be found in [1]. Discussion of the extensions of the bilevel model selection method to other problems can be found in [2].

2   Bilevel Optimization

First, we briefly review bilevel optimization. Bilevel optimization problems are a class of constrained optimization problems whose constraints contain a *lower-level* optimization problem that is parameterized by a multi-dimensional de-sign variable. In operations research literature, the class of bilevel optimiza-tion problems was introduced in the early 1970s by Bracken and McGill [3]. These problems are closely related to the economic problem of the Stackelberg game, whose origin predates the work of Bracken and McGill. In the late 1980s, bilevel programming was given a renewed study in the extended framework of a *mathematical program with equilibrium constraints* (MPEC) in [4], which is an extension of a bilevel program with the optimization constraint replaced by a finite-dimensional variational inequality [5].

The systematic study of the bilevel optimization problem and its MPEC ex-tension attracted the intensive attention of mathematical programmers about a decade ago with the publication of a focused monograph by Luo, Pang and Ralph [4], which was followed by two related monographs [6,7]. During the past decade, there has been an explosion of research on these optimization problems. See the annotated bibliography [8], which contains many references. In general, bilevel programs/MPECs provide a powerful computational framework for deal-ing with parameter identification problems in an optimization setting. As such, they offer a novel paradigm for dealing with the model selection problem de-scribed in the last section. Instead of describing a bilevel optimization problem in its full generality, we focus our discussion on its application to CV for model selection.

3   A Bilevel Support-Vector Regression Model

We focus on a bilevel support-vector regression (SVR) problem and use it to illustrate the kind of problems that the bilevel approach can treat. Specifi-cally, suppose that the regression data are described by the ℓ points $\Omega :=$ $\{(\mathbf{x}_1, y_1), \ldots, (\mathbf{x}_\ell, y_\ell)\}$ in the Euclidean space $\Re^{n+1}$ for some positive integers

ℓ and n. Consider the regression problem of finding a function $f^* : \Re^n \to \Re$ among a given class that minimizes the regularized risk functional

$$R[f] \equiv P[f] + \frac{C}{\ell} \sum_{i=1}^{\ell} L(y_i, f(\mathbf{x}_i)),$$

where L is a loss function of the observed data and model outputs, P is a regularization operator, and C is the regularization parameter. Usually, the ε-insensitive loss $L_\varepsilon(y, f(\mathbf{x})) = \max\{|y - f(\mathbf{x})| - \varepsilon, 0\}$ is used in SVR, where $\varepsilon > 0$ is the *tube parameter*, which could be difficult to select as one does not know beforehand how accurately the function will fit the data. For linear functions: $f(\mathbf{x}) = \mathbf{w}'\mathbf{x} = \sum_{i=1}^{n} w_i x_i$, where the bias term is ignored but can easily be accommodated, the regularization operator in classic SVR is the squared ℓ_2-norm of the normal vector $\mathbf{w} \in \Re^n$; i.e., $P[f] \equiv \|\mathbf{w}\|_2^2 = \sum_{i=1}^{n} w_i^2$.

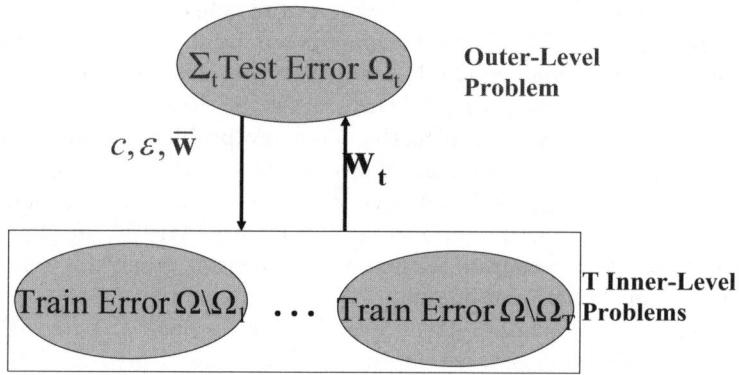

Fig. 1. SVR Model Selection as a Stackelberg Game

The classic SVR approach has two hyper-parameters, the regularization constant C and the tube width ε, that are typically selected by cross validation based on the mean square error (MSE) or mean absolute deviation (MAD) measured on the out-of-sample data. In what follows, we focus on the latter and introduce additional parameters for feature selection and improved regularization and control. We partition the ℓ data points into T disjoint partitions, Ω_t for $t = 1, \ldots, T$, such that $\bigcup_{t=1}^{T} \Omega_t = \Omega$. Let $\overline{\Omega}_t \equiv \Omega \setminus \Omega_t$ be the subset of the data other than those in group Ω_t. The sets $\overline{\Omega}_t$ are called *training sets* while the sets Ω_t are called the *validation sets*. We denote $\overline{\mathcal{N}}_t$ and $\mathcal{N}_t$ to be their index sets respectively. For simplicity, we will ignore the bias term, b, but the method can easily be generalized to accommodate it. In a fairly general formulation in which we list only the essential constraints, the model selection bilevel program is to find the parameters ε, C, and $\mathbf{w}^t$ for $t = 1, \cdots, T$, and also the bounds $\underline{\mathbf{w}}$ and $\overline{\mathbf{w}}$ in order to

$$
\begin{aligned}
\underset{C,\varepsilon,\mathbf{w}^t,\underline{\mathbf{w}},\overline{\mathbf{w}}}{\text{minimize}} \quad & \frac{1}{T} \sum_{t=1}^{T} \frac{1}{|\mathcal{N}_t|} \sum_{i \in \mathcal{N}_t} |\mathbf{x}_i' \mathbf{w}^t - y_i| \\
\end{aligned}
\tag{1}
$$

$$
\text{subject to} \quad \varepsilon, C, \geq 0, \quad \underline{\mathbf{w}} \leq \overline{\mathbf{w}},
$$

$$
\begin{aligned}
\text{and for} \quad & t = 1, \ldots, T, \\
& \mathbf{w}^t \in \underset{\underline{\mathbf{w}} \leq \mathbf{w} \leq \overline{\mathbf{w}}}{\arg \min} \left\{ C \sum_{j \in \mathcal{N}_t} \max(|\mathbf{x}_j' \mathbf{w} - y_j| - \varepsilon, 0) + \frac{1}{2} \| \mathbf{w} \|_2^2 \right\},
\end{aligned}
\tag{2}
$$

where the argmin in the last constraint denotes the set of optimal solutions to the convex optimization problem (2) in the variable $\mathbf{w}$ for given hyper-parameters ε, C, $\mathbf{w}_0$, $\underline{\mathbf{w}}$, and $\overline{\mathbf{w}}$. Problem 1 is called the first-level or outer-level problem. Problem (2) is referred to as the the second-level or inner-level problem. The bilevel program is equivalent to the Stackelberg game shown in figure 1. The bilevel programming approach has no difficulty handling the additional hyper-parameters and other convex constraints (such as prescribed upper bounds on these parameters) because it is based on constrained optimization methodology.

The parameters, $\underline{\mathbf{w}}$ and $\overline{\mathbf{w}}$, are related to feature selection and regularization. The bound constraints $\underline{\mathbf{w}} \leq \mathbf{w} \leq \overline{\mathbf{w}}$ enforce the fact that the weights on each descriptor must fall in a range for all of the cross-validated solutions. This effectively constrains the capacity of each of the functions, leading to an increased likelihood of improving the generalization performance. It also forces all the subsets to use the same descriptors, a form of variable selection. This effect can be enhanced by adopting the one-norm, which forces $\mathbf{w}$ to be sparse. The box constraints will ensure that consistent but not necessarily identical sets will be used across the folds. This represents a fundamentally new way to do feature selection, embedding it within cross validation for model selection.

Note that the loss functions used in the first level and second level—to measure errors—need not match. For the inner-level optimization, we adopt the ε-insensitive loss function because it produces robust solutions that are sparse in the dual space. But typically, ε-insensitive loss functions are not employed in the outer cross-validation objective; so here we use mean absolute deviation (as an example). Variations of the bilevel program (1) abound, and these can all be treated by the general technique described next, suitably extended/modified/specialized to handle the particular formulations. For instance, we may want to impose some restrictions on the bounds $\underline{\mathbf{w}}$ and $\overline{\mathbf{w}}$ to reflect some *a priori* knowledge on the desired support vector $\mathbf{w}$. In particular, we use $-\underline{\mathbf{w}} = \overline{\mathbf{w}} \geq 0$ in Section 5 to restrict the search for the weights to square boxes that are symmetric with respect to the origin. Similarly, to facilitate comparison with grid search, we restrict C and ε to be within prescribed upper bounds.

3.1   Bilevel Problems as MPECs

The bilevel optimization problem (1) determines all of the model parameters via the minimization of the outer objective function. Collecting all the weight vectors across the folds, $\mathbf{w}^t$, column-wise into the matrix W for compactness, the cross-validation error measured as mean average deviation across all the folds is

$$\Theta(W) = \frac{1}{T} \sum_{t=1}^{T} \frac{1}{|\mathcal{N}_t|} \sum_{i \in \mathcal{N}_t} | \mathbf{x}_i' \mathbf{w}^t - y_i |, \tag{3}$$

and is subject to the simple restrictions on these parameters, and most importantly, to the additional inner-level optimality requirement of each $\mathbf{w}^t$ for $t = 1, \ldots, T$. To solve (1), we rewrite the inner-level optimization problem (2) by introducing additional slack variables, $\boldsymbol{\xi}^t \geq 0$ within the t-th fold as follows: for given ε, C, $\underline{\mathbf{w}}$, and $\overline{\mathbf{w}}$,

$$\begin{aligned} \underset{\mathbf{w}^t, \boldsymbol{\xi}^t}{\text{minimize}} \quad & C \sum_{j \in \mathcal{N}_t} \xi_j^t + \frac{1}{2} \| \mathbf{w}^t \|_2^2 \\ \text{subject to} \quad & \underline{\mathbf{w}} \leq \mathbf{w}^t \leq \overline{\mathbf{w}}, \\ & \left. \begin{array}{l} \xi_j^t \geq \mathbf{x}_j' \mathbf{w}^t - y_j - \varepsilon \\ \xi_j^t \geq y_j - \mathbf{x}_j' \mathbf{w}^t - \varepsilon \\ \xi_j^t \geq 0 \end{array} \right\} \quad j \in \overline{\mathcal{N}}_t, \end{aligned} \tag{4}$$

which is easily seen to be a convex quadratic program in the variables $\mathbf{w}^t$ and $\boldsymbol{\xi}^t$. By letting $\boldsymbol{\gamma}^{t,\pm}$ be the multipliers of the bound constraints, $\underline{\mathbf{w}} \leq \mathbf{w} \leq \overline{\mathbf{w}}$, respectively, and $\alpha_j^{t,\pm}$ be the multipliers of the constraints $\xi_j^t \geq \mathbf{x}_j' \mathbf{w}^t - y_j - \varepsilon$ and $\xi_j^t \geq y_j - \mathbf{x}_j' \mathbf{w}^t - \varepsilon$, respectively, we obtain the Karush-Tucker-Tucker optimality conditions of (4) as the following linear complementarity problem in the variables $\mathbf{w}^t$, $\boldsymbol{\gamma}^{t,\pm}$, $\boldsymbol{\alpha}_j^{t,\pm}$, and $\boldsymbol{\xi}_j^t$:

$$\begin{aligned} & 0 \leq \boldsymbol{\gamma}^{t,-} \perp \mathbf{w}^t - \underline{\mathbf{w}} \geq 0, \\ & 0 \leq \boldsymbol{\gamma}^{t,+} \perp \overline{\mathbf{w}} - \mathbf{w}^t \geq 0, \\ & \left. \begin{array}{l} 0 \leq \alpha_j^{t,-} \perp \mathbf{x}_j' \mathbf{w}^t - y_j + \varepsilon + \xi_j^t \geq 0 \\ 0 \leq \alpha_j^{t,+} \perp y_j - \mathbf{x}_j' \mathbf{w}^t + \varepsilon + \xi_j^t \geq 0 \\ 0 \leq \xi_j^t \quad \perp C - \alpha_j^{t,+} - \alpha_j^{t,-} \geq 0 \end{array} \right\} \quad \forall j \in \overline{\mathcal{N}}_t, \\ & 0 = \mathbf{w}^t + \sum_{j \in \overline{\mathcal{N}}_t} (\alpha_j^{t,+} - \alpha_j^{t,-}) \mathbf{x}_j + \boldsymbol{\gamma}^{t,+} - \boldsymbol{\gamma}^{t,-}, \end{aligned} \tag{5}$$

where $a \perp b$ means $a'b = 0$. The orthogonality conditions in (5) express the well-known complementary slackness properties in the optimality conditions of the inner-level (parametric) quadratic program. All the conditions (5) represent the Karush-Kuhn-Tucker conditions. The overall two-level regression problem is therefore

$$\text{minimize} \quad \frac{1}{T} \sum_{t=1}^{T} \frac{1}{|\mathcal{N}_t|} \sum_{i \in \mathcal{N}_t} z_i^t$$

subject to $\varepsilon, C, \geq 0, \quad \underline{\mathbf{w}} \leq \overline{\mathbf{w}},$

and $\quad$ for all $t = 1, \ldots, T$

$$-z_i^t \leq \mathbf{x}_i'\mathbf{w}^t - y_i \leq z_i^t, \qquad \forall\, i \in \mathcal{N}_t,$$

$$\left.\begin{array}{l} 0 \leq \alpha_j^{t,-} \perp \mathbf{x}_j'\mathbf{w}^t - y_j + \varepsilon + \xi_j^t \geq 0 \\[6pt] 0 \leq \alpha_j^{t,+} \perp y_j - \mathbf{x}_j'\mathbf{w}^t + \varepsilon + \xi_j^t \geq 0 \\[6pt] 0 \leq \xi_j^t \quad \perp C - \alpha_j^{t,+} - \alpha_j^{t,-} \geq 0 \end{array}\right\} \forall j \in \overline{\mathcal{N}}_t, \qquad (6)$$

$$0 \leq \boldsymbol{\gamma}^{t,-} \perp \mathbf{w}^t - \underline{\mathbf{w}} \geq 0,$$

$$0 \leq \boldsymbol{\gamma}^{t,+} \perp \overline{\mathbf{w}} - \mathbf{w}^t \geq 0,$$

$$0 = \mathbf{w}^t + \sum_{j \in \overline{\mathcal{N}}_t} (\alpha_j^{t,+} - \alpha_j^{t,-})\mathbf{x}_j + \boldsymbol{\gamma}^{t,+} - \boldsymbol{\gamma}^{t,-}.$$

The most noteworthy feature of the above optimization problem is the complementarity conditions in the constraints, making the problem an instance of a linear program with linear complementarity constraints (sometimes called an LPEC). The discussion in the remainder of this paper focuses on this case.

4   Alternative Bilevel Optimization Methods

The bilevel cross-validation model described above searches the continuous domain of hyper-parameters as opposed to classical cross validation via grid search, which relies on the discretization of the domain. In this section, we describe two alternative methods for solving the model. We also describe the details of the classical grid search approach.

The difficulty in solving the LPEC reformulation (6) of the bilevel optimization problem (1) stems from the linear complementarity constraints formed from the optimality conditions of the inner problem (5); all of the other constraints and the objective are linear. It is well recognized that a straightforward solution using the LPEC formulation is not appropriate because of the complementarity constraints, which give rise to both theoretical and computational anomalies that require special attention. Among various proposals to deal with these constraints, two are particularly effective for finding a local solution: one is to relax the complementarity constraints and retain the relaxations in the constraints. The other proposal is via a penalty approach that allows the violation of these constraints but penalizes the violation by adding a penalty term in the objective function of (6). There are extensive studies of both treatments, including detailed convergence analyses and numerical experiments on realistic applications and random problems. Some references are [9,10,11,6] and [4]. In this work, we experiment with both approaches.

4.1   A Relaxed NLP Reformulation

Exploiting the LPEC structure, the first solution method that is implemented in our experiments for solving (6) employs a relaxation of the complementarity constraint. In the relaxed complementarity formulation, we let **tol** > 0 be a prescribed tolerance of the complementarity conditions. Consider the relaxed formulation of (6):

$$\text{minimize} \quad \frac{1}{T} \sum_{t=1}^{T} \frac{1}{|\mathcal{N}_t|} \sum_{i \in \Omega_t} z_i^t$$

$$\text{subject to } \varepsilon, C \geq 0, \quad \underline{\mathbf{w}} \leq \overline{\mathbf{w}},$$

$$\text{and} \qquad \text{for all } t = 1, \dots, T$$

$$-z_i^t \leq \mathbf{x}_i' \mathbf{w}^t - y_i \leq z_i^t, \qquad\qquad \forall\, i \in \mathcal{N}_t$$

$$\left. \begin{aligned} 0 \leq \alpha_j^{t,-} \perp_{\textbf{tol}} \mathbf{x}_j' \mathbf{w}^t - y_j + \varepsilon + \xi_j^t \geq 0 \\[4pt] 0 \leq \alpha_j^{t,+} \perp_{\textbf{tol}} y_j - \mathbf{x}_j' \mathbf{w}^t + \varepsilon + \xi_j^t \geq 0 \\[4pt] 0 \leq \xi_j^t \quad \perp_{\textbf{tol}} C - \alpha_j^{t,+} - \alpha_j^{t,-} \geq 0 \end{aligned} \right\} \; \forall j \in \overline{\mathcal{N}}_t \qquad (7)$$

$$0 \leq \boldsymbol{\gamma}^{t,-} \perp_{\textbf{tol}} \mathbf{w}^t - \underline{\mathbf{w}} \geq 0,$$

$$0 \leq \boldsymbol{\gamma}^{t,+} \perp_{\textbf{tol}} \overline{\mathbf{w}} - \mathbf{w}^t \geq 0,$$

$$0 = \mathbf{w}^t + \sum_{j \in \overline{\mathcal{N}}_t} (\alpha_j^{t,+} - \alpha_j^{t,-}) \mathbf{x}_j + \boldsymbol{\gamma}^{t,+} - \boldsymbol{\gamma}^{t,-},$$

where $a \perp_{\textbf{tol}} b$ means $a'b \leq \textbf{tol}$. The latter formulation constitutes the *relaxed bilevel support-vector regression problem* that we employ to determine the hyper-parameters C, ε, $\underline{\mathbf{w}}$ and $\overline{\mathbf{w}}$; the computed parameters are then used to define the desired support-vector model for data analysis.

The relaxed complementary slackness is a novel feature that aims at enlarging the search region of the desired regression model; the relaxation corresponds to *inexact cross validation* whose accuracy is dictated by the prescribed scalar, **tol**. This reaffirms an advantage of the bilevel approach mentioned earlier, namely, it adds flexibility to the model selection process by allowing early termination of cross validation, and yet not sacrificing the quality of the out-of-sample errors.

The above NLP remains a non-convex optimization problem; thus, finding a global optimal solution is hard, but the state-of-the-art general-purpose NLP solvers such as FILTER (see [12] and [13]) and SNOPT (see [14]) are capable of computing good-quality feasible solutions. These solvers are available on the NEOS server – an internet server that allows remote users to utilize professionally implemented state-of-the-art optimization algorithms. To solve a given problem, the user first specifies the problem in an algebraic language, such as AMPL or GAMS, and submits the code as a job to NEOS. Upon receipt, NEOS assigns a number and password to the job, and places it in a queue. The remote solver unpacks, processes the problem, and sends the results back to the user.

The nonlinear programming solver, FILTER, was chosen to solve our problems. We also experimented with SNOPT but as reported in [1], we found FILTER to work better overall. FILTER is a sequential quadratic programming (SQP) based method, which is a Newton-type method for solving problems with nonlinear objectives and nonlinear constraints. The method solves a sequence of approximate convex quadratic programming subproblems. FILTER implements a SQP algorithm using a trust-region approach with a "filter" to enforce global convergence [12]. It terminates either when a Karush-Kuhn-Tucker point is found within a specified tolerance or no further step can be processed (possibly due to the infeasibility of a subproblem).

4.2 Penalty Reformulation

Another approach to solving the problem (6) is the penalty reformulation. Penalty and augmented Lagrangian methods have been widely applied to solving LPECs and MPECs, for instance, by [15]. These methods typically require solving an unconstrained optimization problem. In contrast, penalty methods penalize only the complementarity constraints in the objective by means of a penalty function.

Consider the LPEC, (6), resulting from the reformulation of the bilevel regression problem. Define S_t, for $t = 1, \ldots, T$, to be the constraint set within the t-th fold, without the complementarity constraints:

$$
S_t := \left\{ \begin{array}{c} z^t, \boldsymbol{\alpha}^{t,\pm}, \boldsymbol{\xi}^t, \\ \boldsymbol{\gamma}^{t,\pm}, \mathbf{r}^t, s^t \end{array} \middle| \begin{array}{l} -z_i^t \leq \mathbf{x}_i' \mathbf{w}^t - y_i \leq z_i^t, \quad \forall i \in \mathcal{N}_t, \\[4pt] \left. \begin{array}{l} \mathbf{x}_j' \mathbf{w}^t - y_j + \varepsilon + \xi_j^t \geq 0 \\[3pt] y_j - \mathbf{x}_j' \mathbf{w}^t + \varepsilon + \xi_j^t \geq 0 \\[3pt] C - \alpha_j^{t,+} - \alpha_j^{t,-} \geq 0 \end{array} \right\} \forall j \in \overline{\mathcal{N}}_t, \\[18pt] \underline{\mathbf{w}} \leq \mathbf{w}^t \leq \overline{\mathbf{w}}, \\[4pt] 0 = \mathbf{w}^t + \displaystyle\sum_{j \in \overline{\mathcal{N}}_t} (\alpha_j^{t,+} - \alpha_j^{t,-}) \mathbf{x}_j + \boldsymbol{\gamma}^{t,+} - \boldsymbol{\gamma}^{t,-}, \\[12pt] \mathbf{w}^t = \mathbf{r}^t - \mathbb{1}\, s^t, \\[4pt] z^t, \boldsymbol{\alpha}^{t,\pm}, \boldsymbol{\xi}^t, \boldsymbol{\gamma}^{t,\pm}, \mathbf{r}^t, s^t \geq 0. \end{array} \right\}, \quad (8)
$$

where we rewrite the weight vector, $\mathbf{w}^t$, within each fold as $\mathbf{w}^t = \mathbf{r}^t - \mathbb{1}\, s^t$, with $\mathbf{r}^t$, $s^t \geq 0$ and $\mathbb{1}$ denotes a vector of ones of appropriate dimension. Also, let S_0 be defined as the set of constraints on the outer-level variables:

$$
S_0 := \left\{ \begin{array}{c} C, \varepsilon, \\ \overline{\mathbf{w}}, \underline{\mathbf{w}}, \mathbf{w}_0 \end{array} \middle| \begin{array}{c} C, \varepsilon, \mathbf{w}_0, \overline{\mathbf{w}}, \underline{\mathbf{w}} \geq 0 \\ \underline{\mathbf{w}} \leq \overline{\mathbf{w}} \end{array} \right\}. \quad (9)
$$

Then, the overall constraint set for the LPEC (6), without the complementarity constraints is defined as $S_{\mathsf{LP}} := \bigcup_{t=0}^{T} S_t$. Let all the variables in (8) and (9) be collected into the vector $\boldsymbol{\zeta} \geq 0$.

In the penalty reformulation, all the complementarity constraints of the form $a \perp b$ in (6) are moved into the objective via the penalty function, $\phi(a, b)$. This effectively converts the LPEC (6) into a penalty problem of minimizing some, possibly non-smooth, objective function on a *polyhedral set*. Typical penalty functions include the differentiable quadratic penalty term, $\phi(a, b) = a'b$, and the non-smooth piecewise-linear penalty term, $\phi(a, b) = \min(a, b)$. In this paper, we consider the quadratic penalty. The penalty term, which is a product of the complementarity terms is

$$\phi(\boldsymbol{\zeta}) = \sum_{t=1}^{T} \left(\overbrace{\frac{1}{2}\|\mathbf{w}^t\|_2^2 + C \sum_{j \in \mathcal{N}_t} \xi_j^t}^{\Theta_p^t} \atop \begin{array}{l} +\frac{1}{2}\sum_{i \in \mathcal{N}_t}\sum_{j \in \mathcal{N}_t}(\alpha_i^{t,+} - \alpha_i^{t,-})(\alpha_j^{t,+} - \alpha_j^{t,-})\mathbf{x}_i'\mathbf{x}_j \\ +\varepsilon \sum_{j \in \mathcal{N}_t}(\alpha_j^{t,+} + \alpha_j^{t,-}) + \sum_{j \in \mathcal{N}_t} y_j(\alpha_j^{t,+} - \alpha_j^{t,-}) \\ \underbrace{-\overline{\mathbf{w}}'\boldsymbol{\gamma}^{t,+} + \underline{\mathbf{w}}'\boldsymbol{\gamma}^{t,-}}_{-\Theta_d^t} \end{array} \right). \tag{10}$$

When all the hyper-parameters are fixed, the first two terms in the quadratic penalty constitute the primal objective, Θ_p^t, while the last five terms constitute the negative of the dual objective, Θ_d^t, for support vector regression in the t-th fold. Consequently, the penalty function is a combination of T differences between the primal and dual objectives of the regression problem in each fold. Thus,

$$\phi(\boldsymbol{\zeta}) = \sum_{t=1}^{T} \left(\Theta_p^t(\boldsymbol{\zeta}_p^t) - \Theta_d^t(\boldsymbol{\zeta}_d^t) \right),$$

where $\boldsymbol{\zeta}_p^t \equiv (\mathbf{w}^t, \boldsymbol{\xi}^t)$, the vector of primal variables in the t-th primal problem and $\boldsymbol{\zeta}_d^t \equiv (\boldsymbol{\alpha}^{t,\pm}, \boldsymbol{\gamma}^{t,\pm})$, the vector of dual variables in the t-th dual problem. However, the penalty function also contains the hyper-parameters, C, ε and $\overline{\mathbf{w}}$ as variables, rendering $\phi(\boldsymbol{\zeta})$ non-convex. Recalling that the linear cross-validation objective was denoted by Θ, we define the penalized objective: $P(\boldsymbol{\zeta}; \mu) = \Theta(\boldsymbol{\zeta}) + \mu\,\phi(\boldsymbol{\zeta})$, and the penalized problem, $PF(\mu)$, is

$$\min_{\boldsymbol{\zeta}} \quad P(\boldsymbol{\zeta}; \mu)$$
$$\text{subject to } \boldsymbol{\zeta} \in S_{\mathsf{LP}}. \tag{11}$$

This penalized problem has some very nice properties that have been extensively studied. First, we know that finite values of μ can be used, since local solutions of LPEC, as defined by strong stationarity, correspond to stationarity points of $PF(\mu)$. The point $\boldsymbol{\zeta}^*$ is a stationary point of $PF(\mu)$ if and only if there exists a Lagrangian multiplier vector $\boldsymbol{\rho}^*$, such that $(\boldsymbol{\zeta}^*, \boldsymbol{\rho}^*)$ is a KKT point of $PF(\mu)$. In general, KKT points do not exist for LPECs. An alternative local optimality condition, strong stationarity of the LPEC, means that ζ^* solves an LP formed by fixing the LPEC complementarity conditions appropriately. See Definition 2.2. [9] for precise details on strong stationarity. Finiteness ensures that the penalty parameter can be set to reasonable values, contrasting with other approaches in which the penalty problem only solve the original problem in the limit.

Theorem 1 (Finite penalty parameter). *[[9], Theorem 5.2] Suppose that ζ^* is a strongly stationary point of (6), then for all μ sufficiently large, there exists a Lagrangian multiplier vector ρ^*, such that (ζ^*, ρ^*) is a KKT point of $PF(\mu)$ (11).*

It is perhaps not surprising to note that the zero penalty corresponds to a point where the primal and dual objectives are equal in (4.2). These strongly stationary solutions correspond to solutions of (11) with $\phi(\zeta) = 0$, i.e., a zero penalty. The quadratic program, $PF(\mu)$, is non-convex, since the penalty term is not positive definite. Continuous optimization algorithms will not necessarily find a global solution of $PF(\mu)$. But we do know know that local solutions of $PF(\mu)$ that are feasible for the LPEC are also local optimal for the LPEC.

Theorem 2 (Complementary $PF(\mu)$ solution solves LPEC). *[[9], Theorem 5.2] Suppose ζ^* is a stationary point of $PF(\mu)$ (11) and $\phi(\zeta^*) = 0$. Then ζ^* is a strongly stationary for (6).*

One approach to solving exact penalty formulations like (11) is the successive linearization algorithm, where a sequence of problems with a linearized objective,

$$\Theta(\zeta - \zeta^k) + \mu \nabla\phi(\zeta^k)'(\zeta - \zeta^k), \tag{12}$$

is solved to generate the next iterate. We now describe the *Successive Linearization Algorithm for Model Selection* (SLAMS).

4.3   Successive Linearization Algorithm for Model Selection

The QP, (11), can be solved using the Frank-Wolfe method of [10] which simply involves solving a sequence of LPs until either a global minimum or some locally stationary solution of (6) is reached. In practice, a sufficiently large value of μ will lead to the penalty term vanishing from the penalized objective, $P(\zeta^*; \mu)$. In such cases, the locally optimal solution to (11) will also be feasible and locally optimal to the LPEC (6).

Algorithm 1 gives the details of SLAMS. In Step 2, the notation *arg vertex min* indicates that ζ^k is a vertex solution of the LP in Step 2. The step size in Step 4 has a simple closed form solution since a quadratic objective subject to bounds constraints is minimized. The objective has the form $f(\lambda) = a\lambda^2 + b\lambda$, so the optimal solution is either 0, 1 or $\frac{-b}{2a}$, depending on which value yields the smallest objective. SLAMS converges to a solution of the finite penalty problem (11). SLAMS is a special case of the Frank-Wolfe algorithm and a convergence

---

Algorithm 1. Successive linearization algorithm for model selection

---

Fix $\mu > 0$.

1. *Initialization*:
   Start with an initial point, $\zeta^0 \in S_{\mathsf{LP}}$.
2. *Solve Linearized Problem*:
   Generate an intermediate iterate, $\bar{\zeta}^k$, from the previous iterate, ζ^k, by solving the linearized penalty problem, $\bar{\zeta}^k \in \arg \underset{\zeta \in S_{\mathsf{LP}}}{\text{vertex min}} \nabla_\zeta P(\zeta^k; \mu)' (\zeta - \zeta^k)$.
3. *Termination Condition*:
   Stop if the minimum principle holds, i.e., if $\nabla_\zeta P(\zeta^k; \mu)' (\bar{\zeta}^k - \zeta^k) = 0$.
4. *Compute Step Size*:
   Compute step length $\lambda \in \underset{0 \le \lambda \le 1}{\arg \min} P\left((1 - \lambda) \zeta^k + \lambda \bar{\zeta}^k; \mu \right)$, and get the next iterate, $\zeta^{k+1} = (1 - \lambda) \zeta^k + \lambda \bar{\zeta}^k$.

---

proof of the Frank-Wolfe algorithm with no assumptions on the convexity of $P(\zeta^j, \mu)$ can be found in [11], thus we offer the convergence result without proof.

Theorem 3 (Convergence of SLAMS). *[[11]] Algorithm 1 terminates at ζ^k that satisfies the minimum principle necessary optimality condition of $PF(\mu)$: $\nabla_\zeta P(\zeta^k; \mu)'(\zeta - \zeta^k) \ge 0$ for all $\zeta \in S_{\mathsf{LP}}$, or each accumulation $\bar{\zeta}$ of the sequence $\{\zeta^k\}$ satisfies the minimum principle.*

Furthermore, for the case where SLAMS generates a complementary solution, SLAMS finds a strongly stationary solution of the LPEC.

Theorem 4 (SLAMS solves LPEC). *Let ζ^k be the sequence generated by SLAMS that accumulates to $\bar{\zeta}$. If $\phi(\bar{\zeta}) = 0$, then ζ is strongly stationary for LPEC (6).*

Proof. For notational convenience let the set $S_{\mathsf{LP}} = \{\zeta \,|\, \mathbf{A}\zeta \ge \mathbf{b}\}$, with an appropriate matrix, $\mathbf{A}$, and vector, $\mathbf{b}$. We first show that $\bar{\zeta}$ is a KKT point of the problem

$$\min_\zeta \nabla_\zeta P(\zeta; \mu)$$

$$\text{s.t. } \mathbf{A}\zeta \ge \mathbf{b}.$$

We know that $\bar{\zeta}$ satisfies $\mathbf{A}\bar{\zeta} \ge \mathbf{b}$ since ζ^k is feasible at the k-th iteration. By Theorem 3 above, $\bar{\zeta}$ satisfies the minimum principle; thus, we know the system of equations

$$\nabla_\zeta P(\bar{\zeta}; \mu)'(\zeta - \bar{\zeta}^k) < 0, \quad \zeta \in S_{\mathsf{LP}},$$

has no solution for any $\zeta \in S_{\mathsf{LP}}$. Equivalently, if $I = \{i | A_i \bar{\zeta} = \mathbf{b}_i\}$, then

$$P(\bar{\zeta}; \mu)'(\zeta - \bar{\zeta}) < 0, \quad A_i \zeta \ge 0, \ i \in I,$$

has no solution. By Farkas' Lemma, there exists $\bar{\mathbf{u}}$ such that

$$\nabla_\zeta P(\bar{\zeta}; \mu) - \sum_{i \in I} \bar{\mathbf{u}}_i A_i = 0, \quad \bar{\mathbf{u}} \ge 0.$$

Thus $(\bar{\zeta}, \bar{\mathbf{u}})$ is a KKT point of $PF(\mu)$ and $\bar{\zeta}$ is a stationary point of $PF(\mu)$. By Theorem 2, $\bar{\zeta}$ is also a strongly stationary point of LPEC (6).

4.4   Early Stopping

Typically, in many machine learning applications, emphasis is placed on generalization and scalability. Consequently, inexact solutions are preferred to globally optimal solutions as they can be obtained cheaply and tend to perform reasonably well. Noting that, at each iteration, the algorithm is working to minimize the LPEC objective as well as the complementarity penalty, one alternative to speeding up termination at the expense of the objective is to stop as soon as *complementarity is reached*. Thus, as soon as an iterate produces a solution that is feasible to the LPEC, (6), the algorithm is terminated. We call this approach *Successive Linearization Algorithm for Model Selection with Early Stopping* (EZ-SLAMS). This is similar to the well-known machine learning concept of early stopping, except that the criterion used for termination is based on the status of the complementarity constraints i.e., feasibility to the LPEC. We adapt the finite termination result in [11] to prove that EZ-SLAMS terminates finitely for the case when complementary solutions exist, which is precisely the case of interest here. Note that the proof relies upon the fact that S_{LP} is polyhedral with no straight lines going to infinity in both directions.

Theorem 5 (Finite termination of EZ-SLAMS). *Let ζ^k be the sequence generated by SLAMS that accumulates to $\bar{\zeta}$. If $\phi(\bar{\zeta}) = 0$, then EZ-SLAM terminates at an LPEC (6) feasible solution ζ^k in finitely many iterations.*

Proof. Let $\mathcal{V}$ be the finite subset of vertices of S_{LP} that constitutes the vertices $\{\bar{\mathbf{v}}^k\}$ generated by SLAMS. Then,

$$\{\zeta^k\} \in \text{convex hull}\{\zeta^0 \cup \mathcal{V}\},$$
$$\bar{\zeta} \in \text{convex hull}\{\zeta^0 \cup \mathcal{V}\}.$$

If $\bar{\zeta} \in \mathcal{V}$, we are done. If not, then for some $\zeta \in S_{\mathsf{LP}}$, $\mathbf{v} \in \mathcal{V}$ and $\lambda \in (0, 1)$,

$$\bar{\zeta} = (1 - \lambda)\zeta + \lambda \mathbf{v}.$$

For notational convenience define an appropriate matrix M and vector b such that $0 = \phi(\bar{\zeta}) = \bar{\zeta}'(M\bar{\zeta} + q)$. We know $\bar{\zeta} \geq 0$ and $M\bar{\zeta} + q \geq 0$. Hence,

$$\mathbf{v}_i = 0, \text{ or } M_i\mathbf{v} + q_i = 0.$$

Thus, $\mathbf{v}$ is feasible for LPEC (6).

The results comparing SLAMS to EZ-SLAMS are reported in Sections 6 and 7. It is interesting to note that there is always a significant decrease in running time with typically no significant degradation in generalization performance when early stopping is employed.

4.5 Grid Search

In classical cross-validation, parameter selection is performed by discretizing the parameter space into a grid and searching for the combination of parameters that minimizes the validation error (which corresponds to the upper level objective in the bilevel problem). This is typically followed by a local search for fine-tuning the parameters. Typical discretizations are logarithmic grids of base 2 or 10 on the parameters. In the case of the classic SVR, cross validation is simply a search on a two-dimensional grid of C and ε.

This approach, however, is not directly applicable to the current problem formulation because, in addition to C and ε, we also have to determine $\overline{\mathbf{w}}$, and this poses a significant combinatorial problem. In the case of k-fold cross validation of n-dimensional data, if each parameter takes d discrete values, cross validation would involve solving roughly $O(kd^{n+2})$ problems, a number that grows to intractability very quickly. To counter the combinatorial difficulty, we implement the following heuristic procedures:

- Perform a two-dimensional grid search on the unconstrained (classic) SVR problem to determine C and ε. We call this the *unconstrained grid search* (Unc. Grid). A coarse grid with values of 0.1, 1 and 10 for C, and 0.01, 0.1 and 1 for ε was chosen.
- Perform an n-dimensional grid search to determine the features of $\overline{\mathbf{w}}$ using C and ε obtained from the previous step. Only two distinct choices for each feature of $\overline{\mathbf{w}}$ are considered: 0, to test if the feature is redundant, and some large value that would not impede the choice of an appropriate feature weight, otherwise. Cross validation under these settings would involve solving roughly $O(3.2^N)$ problems; this number is already impractical and necessitates the heuristic. We label this step the *constrained grid search* (Con. Grid).
- For data sets with more than 10 features, *recursive feature elimination* [16] is used to rank the features and the 10 largest features are chosen, then constrained grid search is performed.

5 Experimental Design

Our experiments aim to address several issues. The experiments were designed to compare the successive linearization approaches (with and without early stopping) to the classical grid search method with regard to generalization and running time. The data sets used for these experiments consist of randomly generated synthetic data sets and real world chemoinformatics (QSAR) data.

5.1 Synthetic Data

Data sets of different dimensionalities, training sizes and noise models were generated. The dimensionalities i.e., number of features considered were $n = 10$, 15 and 25, among which, only $n_r = 7$, 10 and 16 features respectively, were relevant. We trained on sets of $\ell = 30$, 60, 90, 120 and 150 points using 3-fold cross

Table 1. The Chemoinformatics (QSAR) data sets

Data set	# Obs.	# Train	# Test	# Vars.	# Vars. (stdized)	# Vars. (postPCA)
AQUASOL	197	100	97	640	149	25
B/B BARRIER (BBB)	62	60	2	694	569	25
CANCER	46	40	6	769	362	25
CHOLECYSTOKININ (CCK)	66	60	6	626	350	25

validation and tested on a hold-out set of a further $1,000$ points. Two different noise models were considered: Laplacian and Gaussian. For each combination of feature size, training set size and noise model, 5 trials were conducted and the test errors were averaged. In this subsection, we assume the following notation: $U(a,b)$ represents the uniform distribution on $[a,b]$, $N(\mu,\sigma)$ represents the normal distribution with probability density function $\frac{1}{\sqrt{2\pi}\sigma}\exp\left(-\frac{(x-\mu)^2}{2\sigma^2}\right)$, and $L(\mu,b)$ represents the Laplacian distribution with the probability density function $\frac{1}{2b}\exp\left(-\frac{|x-\mu|}{b}\right)$.

For each data set, the data, $\mathbf{w}_{\text{REAL}}$ and labels were generated as follows. For each point, 20% of the features were drawn from $U(-1,1)$, 20% were drawn from $U(-2.5, 2.5)$, another 20% from $U(-5,5)$, and the last 40% from $U(-3.75, 3.75)$. Each feature of the regression hyperplane $\mathbf{w}_{\text{REAL}}$ was drawn from $U(-1,1)$ and the smallest $n - n_r$ features were set to 0 and considered irrelevant. Once the training data and $\mathbf{w}_{\text{REAL}}$ were generated, the noise-free regression labels were computed as $y_i = \mathbf{x}_i'\mathbf{w}_{\text{REAL}}$. Note that these labels now depend only on the relevant features. Depending on the chosen noise model, noise drawn from $N(0, 0.4\sigma_{\mathbf{y}})$ or $L(0, \frac{0.4\sigma_{\mathbf{y}}}{\sqrt{2}})$ was added to the labels, where $\sigma_{\mathbf{y}}$ is the standard deviation of the noise-less training labels.

5.2   Real-World QSAR Data

We examined four real-world regression chemoinformatics data sets: Aquasol, Blood/Brain Barrier (BBB), Cancer, and Cholecystokinin (CCK), previously studied in [17]. The goal is to create Quantitative Structure Activity Relationship (QSAR) models to predict bioactivities typically using the supplied descriptors as part of a drug design process. The data is scaled and preprocessed to reduce the dimensionality. As was done in [17], we standardize the data at each dimension and eliminate the uninformative variables that have values outside of ± 4 standard deviations range. Next, we perform principle components analysis (PCA), and use the top 25 principal components as descriptors. The training and hold out set sizes and the dimensionalities of the final data sets are shown in Table 1. For each of the training sets, 5-fold cross validation is optimized using bilevel programming. The results are averaged over 20 runs.

The LPs within each iterate in both SLA approaches were solved with CPLEX. The penalty parameter was uniformly set to $\mu = 10^3$ and never resulted in complementarity failure at termination. The hyper-parameters were bounded as $0.1 \leq C \leq 10$ and $0.01 \leq \varepsilon \leq 1$ so as to be consistent with the hyper-parameter ranges used in grid search. All computational times are reported in seconds.

5.3   Post-processing

The outputs from the bilevel approach and grid search yield the bound $\overline{\mathbf{w}}$ and the parameters C and ε. With these, we solve a constrained support vector problem on all the data points:

$$\text{minimize}\quad C\sum_{i=1}^{\ell}\max(\,|\,\mathbf{x}_i'\mathbf{w} - y_i\,| - \varepsilon, 0\,) + \frac{1}{2}\,\|\,\mathbf{w}\,\|_2^2$$
$$\text{subject to}\quad -\overline{\mathbf{w}} \leq \mathbf{w} \leq \overline{\mathbf{w}}$$

to obtain the vector of model weights $\widehat{\mathbf{w}}$, which is used in computing the generalization errors on the hold-out data:

$$\text{MAD} \equiv \frac{1}{1000}\sum_{(\mathbf{x},y)\ \text{hold-out}}|\,\mathbf{x}'\widehat{\mathbf{w}} - y\,|$$

and

$$\text{MSE} \equiv \frac{1}{1000}\sum_{(\mathbf{x},y)\ \text{hold-out}}(\,\mathbf{x}'\widehat{\mathbf{w}} - y\,)^2.$$

The computation times, in seconds, for the different algorithms were also recorded.

6   Computational Results: Synthetic Data

In the following sections, constrained (abbreviated con.) methods refer to the bilevel models that have the box constraint $-\overline{\mathbf{w}} \leq \mathbf{w} \leq \overline{\mathbf{w}}$, while unconstrained (abbreviated unc.) methods refer to the bilevel models without the box constraint. In this section, we compare the performance of several different methods on synthetic data sets.

Five methods are compared: unconstrained and constrained grid search (Unc. Grid and Con. Grid), constrained SLAMS (SLAMS), constrained SLAMS with early stopping (EZ-SLAMS) and constrained FILTER based sequential quadratic programming (Filter SQP).

There are in total 15 sets of problems being solved; each set corresponds to a given dimensionality ($n = 10, 15$ or 25) and a number of training points ($\ell = 30, 60, \ldots 150$). For each set of problems, 5 methods (as described above) were employed. For each method, 10 random instances of the same problem are solved, 5 with Gaussian noise and 5 with Laplacian noise. The averaged results for the 10, 15, and 25-d data sets are shown in Tables 2, 3 and 4 respectively. Each table shows the results for increasing sizes of the training sets for a fixed

Table 2. 10-d synthetic data with Laplacian and Gaussian noise under 3-fold cross validation. Results that are significantly better or worse are tagged ✓ or ✗ respectively.

Method	Objective	Time (sec.)	MAD	MSE
30 pts				
Unc. Grid	1.385 ± 0.323	$5.2\pm\ 0.5$	1.376	2.973
Con. Grid	1.220 ± 0.276	635.3 ± 59.1	1.391	3.044
Filter (SQP)	1.065 ± 0.265	$18.7\pm\ 2.2$	**1.284 ✓**	**2.583 ✓**
Slams	1.183 ± 0.217	$2.4\pm\ 0.9$	1.320	2.746
EZ-Slams	1.418 ± 0.291	$0.6\pm\ 0.1$	1.308	2.684
60 pts				
Unc. Grid	1.200 ± 0.254	$5.9\pm\ 0.5$	1.208	2.324
Con. Grid	1.143 ± 0.245	709.2 ± 55.5	1.232	2.418
Filter (SQP)	1.099 ± 0.181	$23.3\pm\ 4.4$	1.213	2.328
Slams	1.191 ± 0.206	$3.7\pm\ 2.4$	1.186	2.239
EZ-Slams	1.232 ± 0.208	$1.3\pm\ 0.3$	1.186	2.238
90 pts				
Unc. Grid	1.151 ± 0.195	$7.2\pm\ 0.5$	1.180	2.215
Con. Grid	1.108 ± 0.192	789.8 ± 51.7	**1.163 ✓**	**2.154 ✓**
Filter (SQP)	1.069 ± 0.182	39.6 ± 14.1	**1.155 ✓**	**2.129 ✓**
Slams	1.188 ± 0.190	$5.8\pm\ 2.6$	**1.158 ✓**	**2.140 ✓**
EZ-Slams	1.206 ± 0.197	$2.7\pm\ 0.8$	**1.159 ✓**	**2.139 ✓**
120 pts				
Unc. Grid	1.124 ± 0.193	$7.0\pm\ 0.1$	1.144	2.087
Con. Grid	1.095 ± 0.199	704.3 ± 15.6	1.144	2.085
Filter (SQP)	1.037 ± 0.187	$30.2\pm\ 7.8$	1.161	2.152
Slams	1.116 ± 0.193	15.6 ± 15.3	1.141	2.082
EZ-Slams	1.137 ± 0.191	$4.2\pm\ 1.1$	1.143	2.089
150 pts				
Unc. Grid	1.091 ± 0.161	$8.2\pm\ 0.3$	1.147	2.098
Con. Grid	1.068 ± 0.154	$725.1\pm\ 2.7$	1.142	2.081
Filter (SQP)	1.029 ± 0.171	$40.6\pm\ 5.9$	1.150	2.110
Slams	1.103 ± 0.173	$20.1\pm\ 5.5$	1.136	2.063
EZ-Slams	1.110 ± 0.172	$7.4\pm\ 1.1$	1.136	2.062

dimensionality. The criteria used for comparing the various methods are validation error (cross-validation objective), test error (generalization error measured as MAD or MSE on the 1000-point hold-out test set) and computation time (in seconds). For MAD and MSE, the results in bold refer to those that are significantly different than those of the unconstrained grid as measured by a two-sided t-test with significance of 0.1. The results that are significantly better and worse are tagged with a check (✓) or a cross (✗) respectively.

From an optimization perspective, the bilevel programming methods consistently tend to outperform the grid search approaches significantly. The objective values found by the bilevel methods, especially FILTER, are much smaller than those found by their grid-search counterparts. Of all the methods, FILTER finds a lower objective most often. The coarse grid size and feature elimination heuristics used in the grid search cause it to find relatively poor objective values.

Table 3. 15-d synthetic data with Laplacian and Gaussian noise under 3-fold cross validation. Results that are significantly better or worse are tagged ✓ or ✗ respectively.

Method	Objective	Time (sec.)	MAD	MSE
30 pts				
UNC. GRID	1.995 ± 0.421	9.1 ± 9.3	1.726	4.871
CON. GRID	1.659 ± 0.312	735.8 ± 92.5	1.854	5.828
FILTER (SQP)	1.116 ± 0.163	28.7 ± 7.3	1.753	5.004
SLAMS	1.497 ± 0.258	5.0 ± 1.3	1.675	4.596
EZ-SLAMS	1.991 ± 0.374	0.9 ± 0.2	1.697	4.716
60 pts				
UNC. GRID	1.613 ± 0.257	7.3 ± 1.3	1.584	4.147
CON. GRID	1.520 ± 0.265	793.5 ± 83.1	1.589	4.254
FILTER (SQP)	1.298 ± 0.238	52.6 ± 36.4	1.511	3.874
SLAMS	1.565 ± 0.203	8.3 ± 3.5	1.504	3.820
EZ-SLAMS	1.673 ± 0.224	2.3 ± 0.3	1.498	3.807
90 pts				
UNC. GRID	1.553 ± 0.261	8.2 ± 0.5	1.445	3.553
CON. GRID	1.575 ± 0.421	866.2 ± 67.0	1.551	4.124
FILTER (SQP)	1.333 ± 0.254	64.7 ± 12.9	**1.407** ✓	**3.398** ✓
SLAMS	1.476 ± 0.182	16.3 ± 6.3	**1.411** ✓	**3.398** ✓
EZ-SLAMS	1.524 ± 0.197	3.8 ± 0.9	1.412	**3.404** ✓
120 pts				
UNC. GRID	1.481 ± 0.240	7.5 ± 0.0	1.396	3.350
CON. GRID	1.432 ± 0.171	697.9 ± 2.2	1.395	3.333
FILTER (SQP)	1.321 ± 0.168	57.5 ± 11.6	1.388	3.324
SLAMS	1.419 ± 0.166	32.6 ± 18.6	1.375	**3.273** ✓
EZ-SLAMS	1.474 ± 0.181	6.2 ± 0.8	1.379	3.291
150 pts				
UNC. GRID	1.448 ± 0.264	8.7 ± 0.1	1.362	3.221
CON. GRID	1.408 ± 0.232	723.2 ± 2.0	1.376	3.268
FILTER (SQP)	1.333 ± 0.204	85.7 ± 23.6	1.371	3.240
SLAMS	1.436 ± 0.217	41.8 ± 17.5	1.360	3.214
EZ-SLAMS	1.459 ± 0.216	10.1 ± 1.8	1.359	3.206

The reported times provide a rough idea of the computational effort of each algorithm. As noted above, the computation times for the NEOS solver, FILTER, includes transmission, and waiting times as well as solve times. For grid search methods, smart restart techniques were used to gain a considerable increase in speed. However, for Con. Grid, even these techniques cannot prevent the running time from becoming impractical as the problem size grows. While the computation times of FILTER are both much less than that of Con. Grid, it is the SLA approaches that really dominate. The efficiency of the SLA approaches is vastly computationally superior to both grid search and FILTER.

The bilevel approach is much more computationally efficient than grid search on the fully parameterized problems. The results, for FILTER, are relatively efficient and very acceptable when considering that they include miscellaneous times for solution by NEOS. It is reasonable to expect that a FILTER implementation on

Table 4. 25-d synthetic data with Laplacian and Gaussian noise under 3-fold cross validation. Results that are significantly better or worse are tagged ✓ or ✗ respectively.

Method	Objective	Time (sec.)	MAD	MSE
30 pts				
Unc. Grid	3.413 ± 0.537	5.7 ± 0.0	2.915	13.968
Con. Grid	2.636 ± 0.566	628.5 ± 0.5	**3.687** ✗	**22.065** ✗
filter (SQP)	1.087 ± 0.292	18.0 ± 2.7	2.916	13.881
Slams	1.684 ± 0.716	7.9 ± 2.4	2.962	14.607
ez-Slams	3.100 ± 0.818	1.5 ± 0.2	2.894	13.838
60 pts				
Unc. Grid	2.375 ± 0.535	6.2 ± 0.0	2.321	8.976
Con. Grid	2.751 ± 0.653	660.9 ± 1.6	**3.212** ✗	**16.734** ✗
filter (SQP)	1.467 ± 0.271	53.1 ± 11.1	2.282	8.664
Slams	2.065 ± 0.469	15.3 ± 7.1	2.305	8.855
ez-Slams	2.362 ± 0.441	3.1 ± 0.4	2.312	8.894
90 pts				
Unc. Grid	2.256 ± 0.363	7.0 ± 0.0	2.161	7.932
Con. Grid	2.927 ± 0.663	674.7 ± 1.0	**3.117** ✗	**15.863** ✗
filter (SQP)	1.641 ± 0.252	86.0 ± 15.5	**2.098** ✓	**7.528** ✓
Slams	2.149 ± 0.304	29.3 ± 12.2	2.119	7.711
ez-Slams	2.328 ± 0.400	6.3 ± 1.2	2.131	7.803
120 pts				
Unc. Grid	2.147 ± 0.343	8.4 ± 0.0	2.089	7.505
Con. Grid	2.910 ± 0.603	696.7 ± 1.5	**3.124** ✗	**15.966** ✗
Slams	2.156 ± 0.433	45.6 ± 16.4	**2.028** ✓	**7.121** ✓
ez-Slams	2.226 ± 0.461	10.3 ± 1.5	**2.034** ✓	**7.154** ✓
150 pts				
Unc. Grid	2.186 ± 0.383	9.9 ± 0.1	1.969	6.717
Con. Grid	2.759 ± 0.515	721.1 ± 1.7	**2.870** ✗	**13.771** ✗
Slams	2.069 ± 0.368	63.5 ± 30.5	1.949	6.636
ez-Slams	2.134 ± 0.380	14.2 ± 2.5	**1.947** ✓	6.619

a local machine (instead of over the internet) would require significantly less computation times, which could bring it even closer to the times of Unc. Grid or the SLA methods. The filter approach does have a drawback, in that is that it tends to struggle as the problem size increases. For the synthetic data, filter failed to solve 10 problems each from the 25d data sets with 120 and 150 points and these runs have been left out of Table 4.

Of course, in machine learning, an important measure of performance is generalization error. These problems were generated with irrelevant variables; presumably, appropriate choices of the symmetric box parameters in the bilevel problem could improve generalization. (This topic is worth further investigation but is beyond the scope of this paper.) Compared to classic SVR optimized with Unc. Grid, filter and the SLA approaches yield solutions that are better or comparable to the test problems and never significantly worse. In contrast, the generalization performance of Con. Grid steadily degrades as problem size and dimensionality grow.

Table 5. Results for QSAR data under 5-fold cross validation. Results that are significantly better or worse are tagged ✓ or ✗ respectively.

Method	Objective	Time (sec.)	MAD	MSE
Aquasol				
Unc. Grid	0.719 ± 0.101	$17.1\pm\ \ 0.4$	0.644	0.912
Con. Grid	0.778 ± 0.094	$1395.9\pm\ \ 3.5$	**0.849 ✗**	**1.605 ✗**
filter (SQP)	0.574 ± 0.083	1253.0 ± 533.7	0.676	0.972
Slams	0.670 ± 0.092	$137.8\pm\ 52.0$	0.647	0.911
ez-Slams	0.710 ± 0.088	$19.1\pm\ \ 3.3$	0.643	0.907
Blood/Brain Barrier				
Unc. Grid	0.364 ± 0.048	$13.4\pm\ \ 1.9$	0.314	0.229
Con. Grid	0.463 ± 0.081	1285.7 ± 155.3	**0.733 ✗**	**0.856 ✗**
filter (SQP)	0.204 ± 0.043	572.7 ± 339.5	0.338	0.214
Slams	0.363 ± 0.042	$17.1\pm\ \ 9.8$	0.312	0.231
ez-Slams	0.370 ± 0.042	$8.0\pm\ \ 1.6$	0.315	0.235
Cancer				
Unc. Grid	0.489 ± 0.032	$10.3\pm\ \ 0.9$	0.502	0.472
Con. Grid	0.477 ± 0.065	$1035.3\pm\ \ 1.5$	**0.611 ✗**	**0.653 ✗**
filter (SQP)	0.313 ± 0.064	$180.8\pm\ 64.3$	0.454	0.340
Slams	0.476 ± 0.086	$25.5\pm\ \ 9.2$	0.481	**0.336 ✓**
ez-Slams	0.567 ± 0.096	$5.2\pm\ \ 1.1$	0.483	**0.341 ✓**
Cholecystokinin				
Unc. Grid	0.798 ± 0.055	$12.0\pm\ \ 0.4$	1.006	1.625
Con. Grid	0.783 ± 0.071	$1157.6\pm\ \ 1.8$	**1.280 ✗**	**2.483 ✗**
filter (SQP)	0.543 ± 0.063	542.3 ± 211.5	0.981	1.520
Slams	0.881 ± 0.108	$35.1\pm\ 20.4$	**1.235 ✗**	**2.584 ✗**
ez-Slams	0.941 ± 0.092	$9.1\pm\ \ 1.3$	**1.217 ✗**	**2.571 ✗**

Finally, the SLA approaches that employ early stopping tend to generalize very similarly to the SLA approaches that do not stop early. The objective is usually worse but generalization is frequently comparable. This is a very important discovery because it suggests that allowing the SLA approaches to iterate to termination is very expensive, and it is without any corresponding improvement in the cross-validation objective or the generalization performance. The early stopping method, EZ-SLAMS is clearly competitive with the classical Unc. Grid approach with respect to validation and generalization; their main advantage is their efficiency even when handling many hyper-parameters (which Unc. Grid is unable to do).

7   Computational Results: QSAR Data

Table 5 shows the average results for the QSAR data. After the data is preprocessed, we randomly partition the data into 20 different training and testing sets. For each of the training sets, 5-fold cross validation is optimized using bilevel programming. The results are averaged over the 20 runs. We report results for the same 5 methods as those used for synthetic data. The parameter settings used in the grid searches, FILTER and SLA approaches and the statistics reported are the same as those used for the synthetic data.

Again, as with the synthetic data, FILTER finds solutions with the smallest cross-validated training errors or equivalently objective value. SLAMS also finds good quality solutions except for the clearly suboptimal solution found on the Cholecystokinin data. The SLA method have very good computational times. However, computation times for FILTER are not competitive with the SLA methods or Unc. Grid. Unsurprisingly, constrained grid search has the worst computation time. The difficulty of the underlying bilevel optimization problem is underscored by the fact that the greedy Con. Grid search in Section 4.5 sometimes fails to find a better solution than the unconstrained grid search. The constrained search drops important variables that cause it to have bad generalization.

In terms of test set error, FILTER performs the best. SLA also performs quite well on all data sets except on Cholecysotkinin, where the SLA get trapped in poor local minima. However, on the remaining data sets, the SLA approaches generalize very well and tend to be competitive with Unc. Grid with regard to execution time. The best running times, however, are produced by the early stopping based SLA approaches, which SLAM the door on all other approaches computationally while maintaining as of good generalization performance as SLAMS.

8 Discussion

We showed how the widely used model selection technique of cross validation (for support vector regression) could be formulated as a bilevel programming problem; the formulation is more flexible and can deal with many more hyper-parameters than the typical grid search strategy which quickly becomes intractable as the hyper-parameter space increases. The proposed bilevel problem is converted to an instance of a linear program with equilibrium constraints (LPEC). This class of problems is difficult to solve due to the non-convexity created by the complementarity constraints introduced in the reformulation. A major outstanding question has always been the development of efficient algorithms for LPECs and bilevel programs. To this end, we proposed two approaches to solve the LPEC: a relaxed NLP-based approach which was solved using the off-the-shelf, SQP-based, NLP solver, FILTER and a exact penalty-based approach which was solved using a finite successive linearization algorithm.

Our preliminary computational results indicate that general purpose SQP solvers can tractably find high-quality solutions that generalize well. The computation times of the FILTER solver are especially impressive considering the fact that they are obtained via internet connections and shared resources. Generalization results on random data show that FILTER yields are comparable, if not better than current methods. Interestingly, SLAMS typically finds worse solutions than FILTER in terms of the objective (cross-validated error) but with very comparable generalization. The best estimate of the generalization error to be optimized in the bilevel program remains an open question. The SLAMS algorithm computationally outperforms classical grid search and the FILTER solver especially as the number of hyper-parameters and data points grows. We have demonstrated scalability to high dimensional data sets containing up to thousands points (results not reported here).

The computational speed of NLP- or the SLA-based approaches can be improved by taking advantage of the structure inherent in bilevel problems arising from machine learning applications. Machine learning problems, especially support vector machines, are highly structured, and yield elegant and sparse solutions, a fact that several decomposition algorithms such as sequential minimal optimization target. Despite the non-convexity of the LPECs, bilevel programs for machine learning problems retain the structure inherent in the original machine learning problems. In addition, the variables in these LPECs tend to decouple, for example, in cross validation, the variables may be decoupled along the folds. This suggests that applying decomposition or cutting-plane methods to bilevel approaches can make them even more efficient. An avenue for future research is developing decomposition-based or cutting-plane algorithms that can train on data sets containing tens of thousands of points or more.

While support vector regression was chosen as the machine learning problem to demonstrate the potency of the bilevel approach, the methodology can be extended to several machine learning problems including classification, semi-supervised learning, multi-task learning, missing value imputation, and novelty detection. Some of these formulations have been presented in [2], while others remain open problems. Aside from discriminative methods, bilevel programming can also be applied to generative methods such as Bayesian techniques. Furthermore, the ability to optimize a large number of parameters allows one to consider new forms of models, loss functions and regularization.

Another pressing question, however, arises from a serious limitation of the formulation presented herein: the model can only handle linear data sets. Classical machine learning addresses this problem by means of the kernel trick. It was shown in [2] that the kernel trick can be incorporated into a generic bilevel model for cross validation. The flexibility of the bilevel approach means that one can even incorporate input-space feature selection into the kernelized bilevel model. This type of bilevel program can be reformulated as an instance of a mathematical program with equilibrium constraints (MPEC). The MPECs arising from kernelized bilevel machine learning problems tend to have several diverse sources of non-convexity because they have nonlinear complementarity constraints; this leads to very challenging mathematical programs and an equally challenging opening for future pursuits in this field.

References

1. Bennett, K., Hu, J., Ji, X., Kunapuli, G., Pang, J.: Model selection via bilevel optimization. In: International Joint Conference on Neural Networks (IJCNN 2006), pp. 1922–1929 (2006)
2. Kunapuli, G., Bennett, K., Hu, J., Pang, J.: Bilevel model selection for support vector machines. In: Hansen, P., Pardolos, P. (eds.) CRM Proceedings and Lecture Notes. American Mathematical Society (in press, 2008)

3. Bracken, J., McGill, J.: Mathematical programs with optimization problems in the constraints, vol. 21, pp. 37–44 (1973)
4. Luo, Z., Pang, J., Ralph, D.: Mathematical Programs With Equilibrium Constraints. Cambridge University Press, Cambridge (1996)
5. Facchinei, F., Pang, J.: Finite-Dimensional Variational Inequalities and Complementarity Problems. Springer, New York (2003)
6. Outrata, J., Kocvara, M., Zowe, J.: Nonsmooth Approach to Optimization Problems with Equilibrium Constraints: Theory, Applications and Numerical Results. Kluwer Academic Publishers, Dordrecht (1998)
7. Dempe, S.: Foundations of Bilevel Programming. Kluwer Academic Publishers, Dordrecht (2002)
8. Dempe, S.: Annotated bibliography on bilevel programming and mathematical programs with equilibrium constraints. Optimization 52, 333–359 (2003)
9. Ralph, D., Wright, S.: Some properties of regularization and penalization schemes for mpecs. Optimization Methods and Software 19, 527–556 (2004)
10. Mangasarian, O.: Misclassification minimization. Journal of Global Optimization 5, 309–323 (1994)
11. Bennett, K.P., Mangasarian, O.L.: Bilinear separation of two sets in n-space. Computational Optimization and Applications 2, 207–227 (1993)
12. Fletcher, R., Leyffer, S.: Nonlinear programming without a penalty function. Mathematical Programming 91, 239–269 (2002)
13. Fletcher, R., Leyffer, S.: User manual for filtersqp Tech. Report NA/181, Department of Mathematics, University of Dundee (1999),
    http://www-unix.mcs.anl.gov/leyffer/papers/SQP_manual.pdf
14. Gill, P., Murray, W., Saunders, M.: User's guide for snopt version 6: A fortran package for large-scale nonlinear programming (2002)
15. Huang, X., Yang, X., Teo, K.: Partial augmented lagrangian method and mathematical programs with complementarity constraints. Journal of Global Optimization 35, 235–254 (2006)
16. Guyon, I., Elisseeff, A.: An introduction to variable and feature selection. Journal of Machine Learning Research 3, 1157–1182 (2003)
17. Demiriz, A., Bennett, K., Breneman, C., Embrecht, M.: Support vector regression methods in cheminformatics. Computer Science and Statistics 33 (2001)

Bayesian Ying Yang System, Best Harmony Learning, and Gaussian Manifold Based Family

Lei Xu

Department of Computer Science and Engineering,
The Chinese University of Hong Kong
lxu@cse.cuhk.edu.hk

Abstract. Two intelligent abilities and three inverse problems are re-elaborated from a probability theory based two pathway perspective, with challenges of statistical learning and efforts towards the challenges overviewed. Then, a detailed introduction is provided on the Bayesian Ying-Yang (BYY) harmony learning. Proposed firstly in (Xu,1995) and systematically developed in the past decade, this approach consists of a two pathway featured BYY system as a general framework for unifying a number of typical learning models, and a best Ying-Yang harmony principle as a general theory for parameter learning and model selection. The BYY harmony learning leads to not only a criterion that outperforms typical model selection criteria in a two-phase implementation, but also model selection made automatically during parameter learning for several typical learning tasks, with computing cost saved significantly. In addition to introducing the fundamentals, several typical learning approaches are also systematically compared and re-elaborated from the BYY harmony learning perspective. Moreover, a further brief is made on the features and applications of a particular family called Gaussian manifold based BYY systems.

1 Introduction

1.1 Two Intelligent Abilities and Three Inverse Problems

An intelligent system, which could be an individual or a collection of men, animals, robots, agents, and other intelligent bodies, survives in its world with needs of two types of intelligent abilities. As illustrated by Fig.1, implemented by a top-down or outbound pathway, Type-I consists of abilities of discovering the knowledge about its world, including not only understanding ability to explain its world but also motoring ability to track the changes in its world. The knowledge is obtained either from pieces of uncertain evidences (or called samples) about the world or from certain existing authorized sources (e.g., textbooks) that were obtained from samples in past. Therefore, Type-I abilities are actually obtained via processes that we usually call *learning*, during which an intelligent system gradually senses its world from samples and modifies itself to adapt the world. This learning task aims at common features or regularities among an ensemble of uncertain evidences (or called samples) from the world.

J.M. Zurada et al. (Eds.): WCCI 2008 Plenary/Invited Lectures, LNCS 5050, pp. 48–78, 2008.

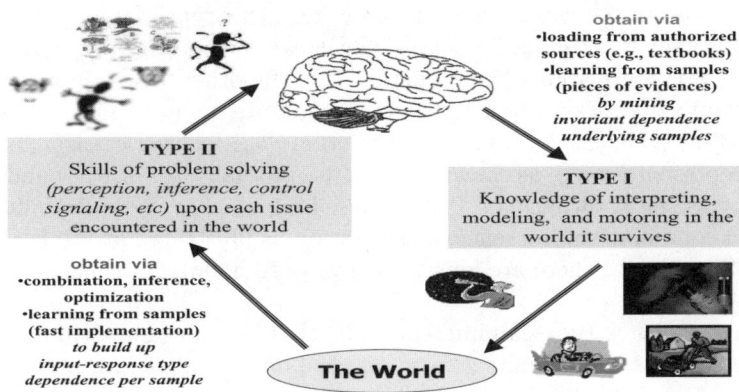

Fig. 1. Two types of intelligent ability and how to get the abilities

On the other hand, implemented by a bottom-up or inbound pathway, Type-II consists of problem solving skills, ranging from perceiving events that are encountered to producing signals that activate the outbound pathway. These skills can be roughly classified into two categories. One is made via evidence combination, inference, optimization, based on a priori knowledge of Type-I. The other is developing a fast implementing device (or called problem solver) for those often encountered events that usually need a rapid response. Specifically, the problem solver is developed via learning from samples either based on the existing Type-I knowledge or in help of a teacher who teaches a desired response as each sample comes (e.g., in supervised pattern recognition, function approximation, control system, ..., etc). This learning task is featured by aiming at the dependence of input-response type per one or several samples encountered.

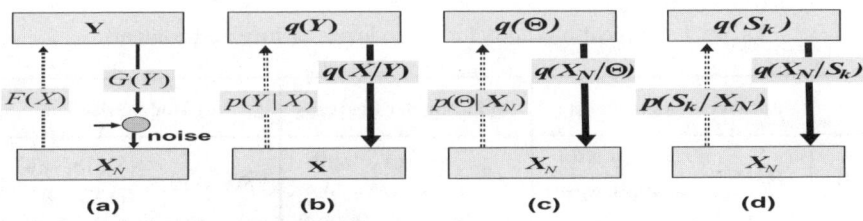

Fig. 2. Three levels of inverse problems

Insights on how two types of of intelligent abilities can be further observed from three levels of inverse problems, illustrated in Fig.2. Provided with an observation x that can be regarded as either generated from an inner representation y or a consequence from a cause y via a given mapping $G : y \rightarrow x$, the Type-II ability makes an inverse inference $x \rightarrow y$, as shown in Fig.2(a). When $G : y \rightarrow x$ is one-to-one, its inverse one-to-one mapping $F : x \rightarrow y$ is analytically solvable.

Generally, it is not so simple due to uncertainties. One type of uncertainties is incurred externally by observation noises, which can be described by a distribution $q(x|y, \theta_{x|y})$ for a probabilistic mapping $y \rightarrow x$. The other type origins internally from a mapping $G : y \rightarrow x$ of many-to-one or infinite many to one, which can be considered by $q(x|y, \theta_{x|y})$ plus a distribution $q(y|\theta_y)$ for every reasonable cause or inner representation y, as shown in Fig.2(b). Actually, $q(x|y, \theta_{x|y})$ and $q(y|\theta_y)$ jointly act as the knowledge of Type I, based on which a Type-II ability is obtained via combination, inference, optimization as illustrated at the left-bottom in Fig.1. Specifically, there are four typical ways to handle it, as listed in the 1st column of Tab. 1.

The first choice is Bayesian inference (BI) that provides a distribution $p(y|x)$ for a probabilistic inverse map $x \rightarrow y$ via combining evidences from $q(x|y, \theta_{x|y})$ and $q(y|\theta_y)$ in a normalized way, which involves an integral with a computational complexity that is usually too high to be practical. The difficulty is tackled by seeking a most probable mapping $x \rightarrow y$ in a sense of the largest probability $p(y|x)$, called the maximum Bayes (MB) or MAximum Posteriori (MAP). It further degenerates into $y^* = arg\max_y q(x|y, \theta_{x|y})$ when there is no knowledge about $q(y|\theta_y)$. In some cases, making maximization may also be computationally expensive. Instead, the last choice is to Learn a Parametric Distribution (LPD) $p(y|x, \theta_{y|x})$ by which an inverse mapping $x \rightarrow y$ can be fast implemented. To get this $p(y|x, \theta_{y|x})$, we need its structure pre-specified and then learn the parameter set $\theta_{y|x}$ from samples either based on $q(x|y, \theta_{x|y})$ and $q(y|\theta_y)$ or in help of a teacher who teaches a desired response to each sample. Actually, this LPD is a special case of the following second type of inverse problems.

The second level of inverse problems considers the situations that $q(x|y, \theta_{x|y})$ and $q(y|\theta_y)$ are unknown but provided with their parametric structures. As illustrated in Fig.2(c), the scenario becomes that we have a set of samples $\mathcal{X}_N = \{x_t\}_{t=1}^N$ from a map $\Theta \rightarrow \mathcal{X}_N$, and the task is getting an inverse mapping $\mathcal{X}_N \rightarrow \Theta$, usually referred by the term *estimation* or *parameter learning* for Θ.

Table 1. Typical methods for three levels of inverse problems

	(a) Inverse Inference on y	(b) Parameter Learning	(c) Model Selection
BI	$p(y\|x) = \dfrac{q(x\|y,\theta_{x\|y})q(y\|\theta_y)}{q(x\|\theta)}$ $q(x\|\theta) = \int q(x\|y,\theta_{x\|y})q(y\|\theta_y)dy$	$p(\theta\|X_N) = \dfrac{q(X_N\|\theta)q(\theta)}{q(X_N\|S)}$ $q(X_N\|S) = \int q(X_N\|\theta)q(\theta)d\theta$	$p(k\|X_N) = \dfrac{q(X_N\|S_k)q(k)}{q(X_N\|\aleph)}$ $q(X_N\|\aleph) = \sum_k q(X_N\|S_k)q(k)$
MB	$max_y[q(x\|y,\theta_{x\|y})q(y\|\theta_y)]$	$max_\theta[q(X_N\|\theta)q(\theta)]$	$max_k[q(X_N\|S_k)q(k)]$
ML	$max_y\, q(x\|y,\theta_{x\|y})$	$max_\theta\, q(X_N\|\theta)$	$max_k\, q(X_N\|S_k)$
LPD	$p(y\|x,\theta_{y\|x})$	$p(\theta\|X_N)$	$p(k\|X_N)$
BI for Bayesian Inference,	MB for Maximum Bayes or called Maximum Posteriori (MAP)		
ML for Maximum Likelihood or Marginal Likelihood,	LPD for Learned Parametric Distribution		

This Θ consists of $\theta_{x|y}, \theta_y$, as well as $\theta_{y|x}$ (if the above LPD is considered to-gether). There could be different directions to pursuit an inverse mapping $\mathcal{X}_N \rightarrow \Theta$. One most widely studied one is that similar to Fig.2(b), with uncertainties considered by two distributions $q(\mathcal{X}_N|\Theta)$ and $q(\Theta)$. Usually, $q(\mathcal{X}_N|\Theta)$ is described by $q(\mathcal{X}_N|\Theta) = \int q(\mathcal{X}_N|\mathcal{Y}_N, \theta_{x|y})q(\mathcal{Y}_N|\theta_y)d\mathcal{Y}_N$. When the samples in $\mathcal{X}_N$ are inde-pendently and identically distributed (i.i.d.), we have $q(\mathcal{X}_N|\Theta) = \prod_{t=1}^{N} q(x_t|\Theta)$ with $q(x_t|\Theta)$ given by the choice BI of the 1st column in Tab. 1.

Based on $q(\mathcal{X}_N|\Theta)$ and $q(\Theta)$, again there are four ways for getting an inverse mapping $\mathcal{X}_N \rightarrow \Theta$, as shown in the 2nd column of Table 1. The simplest and most widely studied one is the maximum likelihood (ML) learning $\max_\Theta q(\mathcal{X}_N|\Theta)$. With a priori distribution $q(\Theta)$ in consideration, we are lead to the choice MB of the 2nd column in Table 1, i.e., $\max_\Theta[q(\mathcal{X}_N|\Theta)q(\Theta)]$, on which extensive studies have been made under different names [25,32,42], and are collectively referred in term of Bayesian school. The challenge is how to get an appropriate $q(\Theta)$, which needs a priori knowledge that we may not have. Related efforts also include those made under *Tikhonov regularization* [40,26] or regularization approaches. Conceptually, we may also consider the BI choice in the 2nd column of Table 1 for a probabilistic inverse mapping by a distribution $p(\Theta|\mathcal{X}_N)$, while it encounters an integral over Θ.

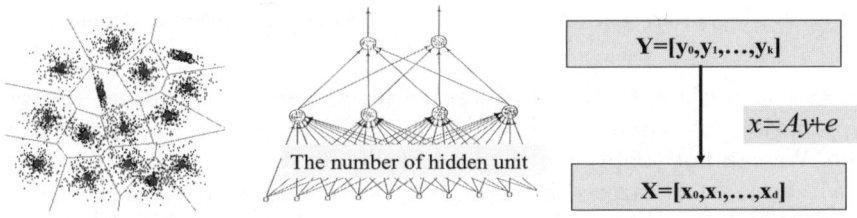

Fig. 3. A combination of a series of individual simple structures

Being too difficult to compute except some special cases, this integral over Θ is encountered not just as above but also in the 3rd column of Table 1. An alternative is using a particularly designed parametric structure in place of $p(\Theta|\mathcal{X}_N)$, i.e., the choice LPD in the 2nd column of Table 1. Moreover, even in implementing the ML learning, we have to handle either a summation or a numerical integral over y for getting $q(x|\Theta)$ (see the choice BI of the 1st column in Table 1), which also involves a huge computing cost except special cases. Instead, the choice LPD in the 1st column of Table 1 is considered via a particularly designed parametric structure $p(y|x, \theta_{y|x})$. Studies on learning either or both of $p(y|x, \theta_{y|x})$ and $p(\Theta|\mathcal{X}_N)$ jointly with the *parameter learning* for Θ have been made in the Helmholtz free energy based learning [15,11], BYY Kullback learning [64], and BYY harmony learning [64,47]. Detailed discussions are referred to Sec.3.2.

Until now, we assume that the parametric structures of $q(x|y, \theta_{x|y})$ and $q(y|\theta_y)$, as well as of $p(y|x, \theta_{y|x})$ are provided in advance. In fact, we do not know how to pre-specify these structures. Usually, we consider a family of infinite many structures $\{S_\mathbf{k}(\Theta_\mathbf{k})\}$ via combining a set of individual simple structures (or simply called units) via a simple combination scheme, as shown in Fig.3. Every unit can be simply one point, one dimension in a linear space, or one simple computing unit. The types of the basic units and the combination scheme jointly act as a seed or meta structure ℵ that grows into a family $\{S_\mathbf{k}(\Theta_\mathbf{k})\}$ with each $S_\mathbf{k}$ sharing a same configuration but in different scales, each of which is labeled by a scale parameter $\mathbf{k}$ in term of one integer or a set of integers. That is, each specific $\mathbf{k}$ corresponds to one candidate model with a specific complexity. We can enumerate each candidate via enumerating [1] $\mathbf{k}$.

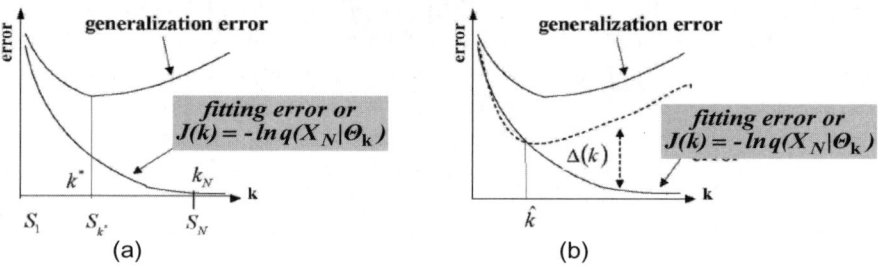

Fig. 4. Model selection : fitting performance vs generalization performance

As shown in Fig.2(d), the third level of inverse problems considers selecting an appropriate $\mathbf{k}^*$ based on $\mathcal{X}_N = \{x_t\}_{t=1}^N$ only, usually referred as *model selection*. We can not simply use the best likelihood value as a measure to guide this selection. As illustrated in Fig.4(a), $J(k) = -\max_\Theta lnq(\mathcal{X}_N|\Theta)$ will keep decreasing as k increases and reaches zero at a value k_N that is usually much larger than the appropriate one, as long as the size N is finite. Though a $S_\mathbf{k}(\Theta_\mathbf{k})$ with $\mathbf{k}^* \prec \mathbf{k}$ can get a low value $J(\mathbf{k})$ and thus $\mathcal{X}_N$ got well described, it has a poor generalization performance (i.e., performing poorly on new samples with the same regularity underlying $\mathcal{X}_N$). This is also called *over-fitting* problem.

1.2   Efforts Towards Challenges

In the past 30 or 40 years, several learning principles or theories have been proposed and studied for an appropriate $J(\mathbf{k})$, roughly along three directions.

Those measures summarized in Table 1 are featured by the most probable principle based on probability theory. The efforts of the first direction can be summarized under this principle. As discussed above, the ML choice of the 2nd column in Table 1 can not serve as $J(\mathbf{k})$. Studies on the BI choice of the 2nd column, i.e., $J(\mathbf{k}) = -\max_\Theta[q(\mathcal{X}_N|\Theta)q(\Theta)]$, have been made under the name of *minimum message length* (MML)[42]. It can provide an improved performance

---

[1] We say that $\mathbf{k}_1$ proceeds $\mathbf{k}_2$ or $\mathbf{k}_1 \prec \mathbf{k}_2$ if $S_{\mathbf{k}_1}$ is a part (or called a substructure) of $S_{\mathbf{k}_2}$. When $\mathbf{k}$ consists of only one integer, $\mathbf{k}_1 \prec \mathbf{k}_2$ becomes simply $\mathbf{k}_1 < \mathbf{k}_2$.

over $J(\mathbf{k}) = -\max_\Theta q(\mathcal{X}_N|\Theta)$ but is sensitive to whether an appropriate $q(\Theta)$ is pre-specified, which is difficult. Studies on the BI choice of the 3rd column in Table 1 have also been conducted widely in the literature. Usually assuming that $q(\mathbf{k})$ is equal for every $\mathbf{k}$, we are lead to the ML (marginal likelihood) choice of the 3rd column, i.e., $J(\mathbf{k}) = -\ln q(\mathcal{X}_N|S_k)$, by which the effect of $q(\Theta)$ has been integrated out. However, the integral over Θ is difficult to compute and thus is approximately tackled by turning it into the following format:

$$J(k) = -\max_\Theta \ln q(\mathcal{X}_N|\Theta) + \Delta(\mathbf{k}), \tag{1}$$

where the term $\Delta(\mathbf{k})$ is resulted from a rough approximation such that it is computable. Differences on $q(\Theta)$ and on methods for approximating the integral result in different specific forms. Typical efforts include those under the names of *Bayesian Information Criterion* [34,23], Bayes Factors [21], the evidence or the marginal likelihood [22], etc. The *Akaike Information Criterion (AIC)* can also be obtained as a special case though it was orginally derived from a different perspective [1,2].

The second direction follows the well known principle of Ockham Razor, i.e., seeking a most economic model that represents $\mathcal{X}_N$. It is implemented via mimizing a two part coding length. One is for encoding the residuals or errors incurred by the model in representing $\mathcal{X}_N$, which actually corresponds to the first term in eq.(1). The other is for encoding the model itself, which actually corresponds to the second term in eq.(1). Different specific forms maybe obtained due to differences on what measure is used for the length and on how to evaluate the measure, which is usually difficult, especially for the second part coding. Studies have been made under the names of *minimum message length* (MML)[42], *minimum description length* (MDL) [29], best information transfer, etc. After this or that type of approximation, the resulted criteria turn out closely related to or even same as those obtained along the above first direction.

Another direction is towards estimating the generalization performance directly. One typical approach is called *cross-validation* (CV). $\mathcal{X}_N$ is randomly and evenly divided into $D_i, i = 1, \cdots, m$ parts, each D_i is used to measure the performance of $S_\mathbf{k}$ with its $\Theta_\mathbf{k}$ determined from the rest samples in $\mathcal{X}_N$ after taking D_i away. Then we use the average performance measures of m times as an estimation of $J(\mathbf{k})$ [39,30]. One other approach is using the VC dimension based learning theory [41] to estimate a bound of generalization performance via theoretical analysis. A rough bound can be obtained for some special cases, e.g., a Gaussian mixture [44]. Generally, such a bound is difficult to get because it is very difficult to estimate the VC dimension of a learning model.

Even with a $J(\mathbf{k})$ available, evaluating its optimal values invovles a discrete optimization nested with a series of implementations of parameter learning for a best $\Theta_\mathbf{k}^*$ at each $\mathbf{k}$. The task usually incurs a huge computing cost, while many practical applications demand that learning is made adaptively upon each sample comes. Moreover, the parameter learning performance deteriorates rapidly as $\mathbf{k}$ increases, which makes the value of $J(\mathbf{k})$ evaluated unreliably. Efforts have been made on tackling this challenge along two directions. One is featured by incremental algorithms that attempts to incorporate as much as possible what learned as **k** increases step by step, focusing on learning newly added parameters. Such an incremental implementation can save computing costs in certain extent. However, parameter learning has to be made by enumerating the values of **k**, and computing costs are still very high. Also, it usually leads to suboptimal performance because not only those newly added parameters but also the old parameter set $\Theta_{\mathbf{k}}$ have to be re-learned. Another type of efforts has been made on a widely encountered category of structures that consists of individual substructures, e.g., a Gaussian mixture that consists of several Gaussian components. A local error criterion is used to check whether a new sample x belongs to each substructure. If x is regarded as not belonging to anyone of substructures, an additional substructure is added to accommodate this new x. This incremental implementation is much faster. However, the local evaluating nature makes it very easy to be trapped into a poor performance, except for some special cases that $\mathcal{X}_N = \{x_t\}_{t=1}^N$ come from substructures that are well separated.

The other direction consists of learning algorithms that start with **k** at a large value and decrease **k** step by step, with extra parameters discarded and the remaining parameter updated. These algorithms are further classified into two types. One is featured by decreasing **k** step by step, based on evaluating the value of $J(\mathbf{k})$ at each **k**. The other is called automatic model selection, with extra structural parts removed automatically during parameter learning. One early effort is Rival Penalized Competitive Learning (RPCL) [65,48] for a structure that consists of k individual substructures. With k initially given a value larger enough, a coming sample x is allocated to one of the k substructures via competition, and the winner adapts this sample by a little bit, while the rival (i.e., the second winner) is de-learned a little bit to reduce a duplicated allocation. This rival penalized mechanism will discard those extra substructures, making model selection automatically during learning. Various extensions have been made in the past one decade and half. Readers are referred to a recent encyclopedia paper [48].

1.3   Two-Pathway Approaches and the Scope of This Paper

RPCL learning was heuristically proposed in lack of theoretical guide. Proposed firstly in [64] and systematically developed in the past decade [47,49], the Bayesian Ying-Yang (BYY) harmony learning acts as a general statistical theory that guides various learning tasks with model selection achieved automatically during parameter learning, which is featured by using a Bayesian Ying-Yang (BYY) system to model an intelligent system and three level of inverse problems shown in Fig.1 and Fig.2.

The two-pathway idea has been adopted in the literature of modelling a perception system for decades. One early example is the adaptive resonance theory developed in the 1970s [14], featured by a resonance between bottom-up input and top-down expectation in help of a mechanism motivated from a cognitive science view. Efforts have been further made on multi-layer net featured two-pathway approaches, e.g., under the least mean square error based auto-association [6], the LMSER self-organization [66]. However, these early studies were neither motivated nor targeted at a probability theory based perspective as shown in Fig.2.

In addition to those approaches discussed in Table 1, studies on a probabilistic two-path way perspective include the BYY learning, the Helmholtz free energy based learning or *Helmholtz machine* [15,11], variational approximation methods [20,19]. Motivated differently, these approaches share certain common features and also have different properties. Firstly proposed in 1995 [64,56,50,51,47] and developed in the past decade, BYY learning not only acts as a general framework for a unified perspective on these approaches as well as the approaches in Table 1, but also provides a new theory for model selection on a finite size of samples, both on deriving a criterion that outperforms typical model selection criteria in a two-phase implementation, and on developing learning algorithms for several typical learning tasks with an appropriate model scale obtained automatically during parameter learning, while with computing cost saved significantly.

In the rest of this paper, Section 2 introduces the fundamentals of Bayesian Ying Yang system and best harmony learning theory, the implementable structures for Yang machine and the distributed log-quadratic inner structures for Ying machine. In Section 3, relations and differences of a number of existing typical learning approaches are rather systematically compared and re-elaborated from the perspective of BYY learning under the principles of best harmony versus best matching. Finally, a further introduction is made on a particular family of BYY systems featured with Gaussian manifolds as components.

2   Bayesian Ying-Yang Learning

2.1   Bayesian Ying-Yang System and Best Harmony Learning

As shown in Fig.5, a unified scenario of Fig.2 is considered by regarding that the observation set $\mathbf{X} = \{x\}$ are generated via a top-down path from its inner representation $\mathbf{R} = \{\mathbf{Y}, \Theta\}$. Given a system architecture, the parameter set Θ collectively represents the underlying structure of $\mathbf{X}$, while one element $y \in \mathbf{Y}$ is the corresponding inner representation of one element $x \in \mathbf{X}$. A mapping $\mathbf{R} \to \mathbf{X}$ and an inverse mapping $\mathbf{X} \to \mathbf{R}$ are jointly considered via the joint distribution of $\mathbf{X}$ and $\mathbf{R}$ in two types of Bayesian decomposition shown at the right-bottom of Fig.5. In a compliment to the famous ancient Ying-Yang philosophy, the decomposition of $p(\mathbf{X}, \mathbf{R})$ coincides the Yang concept with a visible domain $p(\mathbf{X})$ for a Yang space and a forward pathway by $p(\mathbf{R}|\mathbf{X})$ as a Yang pathway. Thus, $p(\mathbf{X}, \mathbf{R})$ is called Yang machine. Similarly, $q(\mathbf{X}, \mathbf{R})$ is called Ying machine with an invisible domain $q(\mathbf{R})$ for a Ying space and a backward pathway by $q(\mathbf{X}|\mathbf{R})$ as a Ying pathway. Such a Ying-Yang pair is called *Bayesian Ying-Yang (BYY) system*.

As shown in Fig.5, the system is further divided into two layers. The front layer is actually the one shown in Fig.2(b), with a parametric Ying-Yang pair at the left-bottom of Fig.5, which consists of four components with each associated

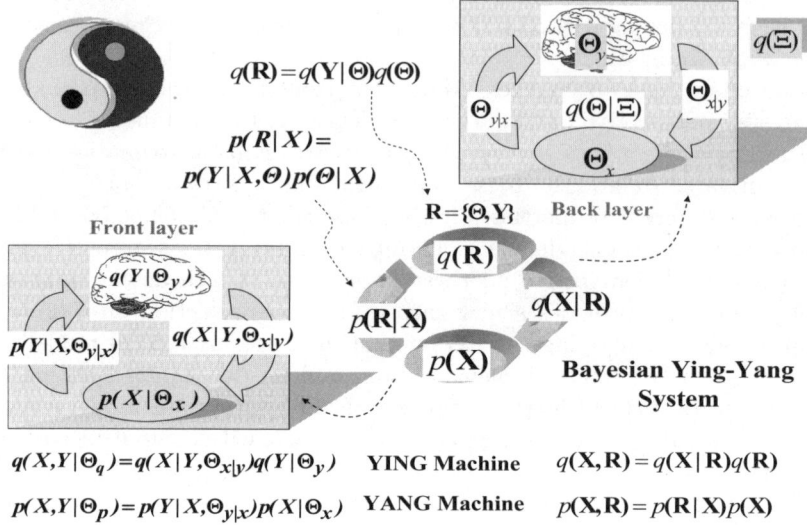

Fig. 5. Bayesian Ying-Yang System

with a subset of parameters $\Theta = \{\Theta_p, \Theta_q\}$, where $\Theta_p = \{\theta_{y|x}, \theta_x\}$ and $\Theta_q = \{\theta_y, \theta_{x|y}\}$. This Θ is accommodated on the back layer with a priori structure $q(\Theta|\Xi_q)$ to back up the front layer, the back layer may be modulated by a meta knowledge from a meta layer $q(\Xi)$. Correspondingly, an inference on Θ is given by $p(\Theta|\mathbf{X}, \Xi_p)$ that integrates information from both the front layer and the meta layer. Putting together, we have

$$q(\mathbf{X}, \mathbf{R}) = q(\mathbf{X}|\mathbf{Y}, \theta_{x|y})q(\mathbf{Y}|\theta_y)q(\Theta|\Xi_q),$$
$$p(\mathbf{X}, \mathbf{R}) = p(\Theta|\mathbf{X}, \Xi_p)p(\mathbf{Y}|\mathbf{X}, \theta_{y|x})p(\mathbf{X}|\theta_x). \qquad (2)$$

The external input is only a set of samples $\mathcal{X}_N = \{x_t\}_{t=1}^N$ of $\mathbf{X} = \{x\}$, based on which we form an estimate of $p(\mathbf{X}|\theta_x)$ either directly or with a unknown sclar parameter $\theta_x = h$, as shown in Tab.2. Based on this very limited knowledge, the goal of building up the entire system is too ambitious to pursuit. We need to further specify certain structures of $p(\mathbf{X}, \mathbf{R})$ and $q(\mathbf{X}, \mathbf{R})$. Summarized in Tab 2 are typical scenarios of both $p(\mathbf{X}, \mathbf{R})$ and $q(\mathbf{X}, \mathbf{R})$, and further details will be introduced in the subsequent two subsections.

Similar to the discussions made at the end of Sec.1.1, the Ying Yang system is also featured by a given meta structure $\aleph$ that grows into a family $\{S_\mathbf{k}(\Theta_\mathbf{k})\}$ with each $S_\mathbf{k}$ sharing a same configuration but in different scales of $\mathbf{k}$. The meta structure $\aleph$ consists $\aleph_q, \aleph_p$ for the Ying machine and the Yang machine respectively, from which we get the structures of $q(\mathbf{X}, \mathbf{Y}|\Theta_q)$ and $p(\mathbf{X}, \mathbf{Y}|\Theta_p)$ in different scales. Though it is difficult to precisely define, the scale $\mathbf{k}$ of an entire system is featured by the scale or complexity for representing R, which is roughly regarded as consisting of the scale $\mathbf{k}_Y$ for representing Y and the number n_f of free parameters in Θ.

As shown in Tab.2, different structures of the Ying machine $q(\mathbf{X}, \mathbf{Y}|\Theta_q)$ are considered to accommodate the world knowledge and different types of dependences encountered in various learning tasks. First, an expression format is needed for each inner representation $\mathbf{Y}$. It has four typical choices as shown in Tab.2. The general case is the last one, i.e., $\mathbf{Y} = \{\mathbf{Y}_v, \mathbf{L}\}$ with $\mathbf{Y}_v = \{y_v\}$, $\mathbf{L} = \{\ell\}$. Each ℓ takes a finite number of integers to denote one of several labels for tasks of pattern classification, choice decision, and clustering analyses, etc., while each y_v is a vector that acts as an inner coding or cause for observations. Moreover, $q(\mathbf{Y}_v|\theta_y)$ describes the structure dependence among a set of values that $\mathbf{Y}_v$ may take. Second, $q(\mathbf{X}|\mathbf{Y}_v, \theta_{x|y})$ describes the knowledge about the dependence relation from inner representation to observation. Third, in addition to these structures, the knowledge is also represented by Θ jointly, which is confined by a background knowledge via a priori structure $q(\Theta|\Xi)$ with a unknown parameter set Ξ_q. Some choices are shown in Tab.2.

As to the Yang machine $p(\mathbf{X}, \mathbf{Y}|\Theta_p)$, we already have the above discussed input $p(\mathbf{X}|\theta_x)$. Similar to the case of Fig.2(b), the structures of $p(\mathbf{Y}|\mathbf{X}, \theta_{y|x}) = p(\mathbf{Y}_v|\mathbf{X}, L, \theta_{y|x})p(L|\mathbf{X}, \theta_{y|x})$ not only make a fast implementation of a desired problem solving but also act as an inverse role of the Ying machine $q(\mathbf{X}, \mathbf{Y}|\Theta_q)$. If we are also provided with the structure of $p(\Theta|\mathbf{X}, \Xi_p)$, what still remains unknown consists of $\mathbf{k}$ and $\Xi = \{\Xi_q, \Xi_p\}$. An analogy of this Ying Yang system to the ancient Ying-Yang philosophy motivates to determine the unknowns under a best harmony principle, which is mathematically implemented by maximizing the following harmony measure

$$\max_{\{\mathbf{k}, \Xi\}} H(p\|q, \mathbf{k}, \Xi), \ H(p\|q, \mathbf{k}, \Xi) = \int p(\mathbf{R}|\mathbf{X})p(\mathbf{X}) \ln [q(\mathbf{X}|\mathbf{R})q(\mathbf{R})]d\mathbf{X}d\mathbf{R}$$

$$= \int p(\Theta|\mathbf{X}, \Xi)H_f(\mathbf{X}, \Theta, \mathbf{k}, \Xi)d\Theta,$$

$$H_f(\mathbf{X}, \Theta, \mathbf{k}, \Xi) = \sum_L p(L|\mathbf{X}, \theta_{y|x})H_f(\mathbf{X}, L, \Theta, \mathbf{k}, \Xi), \qquad (3)$$

$$H_f(\mathbf{X}, L, \Theta, \mathbf{k}, \Xi) = \int p(\mathbf{Y}_v|\mathbf{X}, L, \theta_{y|x})p(\mathbf{X}|\theta_x) \times$$
$$\times \ln [q(\mathbf{X}|\mathbf{Y}_v, L, \theta_{x|y})q(\mathbf{Y}_v|L, \theta_y)q(L|\theta_L)q(\Theta|\Xi_q)]d\mathbf{Y}_v.$$

On one hand, maximizing $H(p\|q)$ forces $q(\mathbf{X}|\mathbf{R})q(\mathbf{R})$ to match $p(\mathbf{R}|\mathbf{X})p(\mathbf{X})$. Due to the constraints on the given Ying and Yang structures, a perfect matching $p(\mathbf{R}|\mathbf{X})p(\mathbf{X}) = q(\mathbf{X}|\mathbf{R})q(\mathbf{R})$ may not be really reached but still be approached as possible as it can. At this equality, $H(p\|q)$ becomes the negative entropy that describes the complexity of system. Further maximizing $H(p\|q)$ with $\mathbf{k}, \Xi$ is actually minimizing the complexity of system, which provides a model selection ability on $\mathbf{k}$. Such an ability can also be observed from other perspectives, with details referred to [52,49].

The first difficulty we encounter is where to get the structure of $p(\Theta|\mathbf{X}, \Xi_p)$ that specifies a probabilistic inverse mapping $\mathcal{X}_N \to \Theta$ shown in Fig.2(c). One possibility is the BI choice in the second column of Tab.1 or written as the choice B for $p(\Theta|\mathbf{X}, \Xi_p)$ in Tab.2. As previously discussed, it usually involves a difficult computation for not only an integral over Θ but also an integral over Y. To avoid this difficulty, we usually consider the choice A and choice C in Tab.2.

Table 2. Typical scenarios of $q(\mathbf{X},\mathbf{R}) = q(\mathbf{X}|\mathbf{Y},\theta_{x|y})q(\mathbf{Y}|\theta_y)q(\Theta|\Xi_q)$ and $p(\mathbf{X},\mathbf{R}) = p(\Theta|\mathbf{X},\Xi_p)p(\mathbf{Y}|\mathbf{X},\theta_{y|x})p(\mathbf{X}|\theta_x)$

$q(\mathbf{X},\mathbf{R})$	$q(\mathbf{Y}|\boldsymbol{\theta_y})$	$q(\mathbf{X}|\mathbf{Y},\boldsymbol{\theta_{x|y}})$	$q(\theta\mid\Xi)=$ $\prod_{\xi\in\{x|y,x,y|x,y\}}q(\theta_\xi\mid\Xi_\xi)$
Choice 1	$q(L\mid\Theta_L), L=\{\ell\}$ $\ell=1,\cdots,k$	$q(\mathbf{X}\mid L,\Theta_L)$	Ignored
Choice 2	$q(Y\mid\Theta_y), Y=\{y\}$ real $y=[y_1,\cdots,y_m]^T$	$G(X\mid g(Y,\theta_{X|Y}),\Sigma_{X|Y})$ for real X	Non-informative
Choice 3	$q(Y\mid\Theta_y), Y=\{y\}$ binary $y=[y_1,\cdots,y_m]^T$	$q(X\mid Y,\theta_{X|Y})$ both X,Y are binary	$q(\Theta)\propto\dfrac{1}{\sum_{t=1}^N q(u_t\mid\Theta)}$
Choice 4	Hybrid $q(Y,L\mid\Theta_y)=q(Y\mid L,\theta_y)q(L\mid\Theta_L)$	Hybrid $q(\mathbf{X}\mid Y,L,\theta_{X|Y,L})$	Parametric $q(\theta\mid\Xi)$

	$p(X\mid\theta_x)$	$p(L\mid X,\theta_{y|x})$
Choice A	$\delta(X-X_N)$	Free of structure
Choice B	Sample-based $p_h(\mathbf{X})=\prod_{t=1}^N G(x\mid x_t,h^2 I)$	Bayesian structure $p(L\mid X,\Theta_L)=\dfrac{q(L\mid\theta_y)q(X\mid L,\theta_{x|y},\theta_y)}{\sum_L q(L\mid\theta_y)q(X\mid L,\theta_{x|y},\theta_y)}$ $q(X\mid L,\Theta_L)=\int q(X\mid Y,\theta_{x|y})q(Y\mid L,\theta_y)dY$
Choice C	Batch-based $p_h(\mathbf{X})=\prod_{t=1}^N p_h(x_t)$ $p_h(x_t)=\frac{1}{N}\sum_{t=1}^N G(x\mid x_t,h^2 I)$	$\pi_L(X,\Theta_p)=c_L+B_L\text{vec}(X)+\beta_L\text{vec}(X)^T Q_L\text{vec}(X)$ $p(L\mid X,\Theta_L)=\dfrac{e^{\pi_L(X,\Theta_L)}}{\sum_L e^{\pi_L(X,\Theta_L)}}$, $\Theta_L=\{B_L,Q_L,c_L\}$ could be (1) consisting of free unknown parameters, (2) either Q_L is the Hessian of $\ln[q(L\mid\theta_y)q(X\mid L,\theta_{x|y},\theta_y)]$ with respect to $\text{vec}(X)$ or a part of Q_L is the counterpart part of the Hessian, while the rest is unknown parameters.

	$p(\theta\mid X_N,\Xi)$	$p(Y\mid X,L,\theta_{y|x})$
Choice A	*Free of structure*	*Free of structure*
Choice B	Bayesian structure $p(\theta\mid X_N)=\dfrac{q(X_N\mid\theta)q(\theta\mid\Xi)}{\int q(X_N\mid\theta)q(\theta\mid\Xi)d\theta}$	Bayesian structure $p(y\mid x,L,\theta_{y|x})=\dfrac{q(x\mid y,L,\theta_{x|y})q(y,L\mid\theta_y)}{\int q(x\mid y,L,\theta_{x|y})q(y\mid L,\theta_y)dy}$
Choice C	*Ying machine induced* $G(\text{vec}(\theta)\mid\mu(X_N,\Theta),\Sigma(X_N,\Theta))$ • $\mu(X_N,\Theta)$ is either $\theta^*(X_N,\Xi)=$ $\text{argmax}_\theta H_f(X_N,\theta,k,\Xi)$ or a prespecified linear or nonlinear function of X_N with unknown parameters Ξ. • $\Sigma^{-1}(X_N,\Theta)$ is either the Hessian of $H_f(X_N,\theta,k,\Xi)$ with respect to $\text{vec}(\theta)$ or a simplfied approximation of this Hessian.	*Ying machine induced* *For Y consisting of real variables* $G(\text{vec}(Y)\mid\mu(X,\phi_{\mu,L}),\Sigma(Y,\phi_{\Sigma,L}))$ • $\mu(X,\phi_{\mu,L})$ is a prespecified function of X with unknown parameters $\phi_{\mu,L}$. • $\Sigma^{-1}(Y,\phi_{\Sigma,L})$ is either the Hessian of $\ln[q(X\mid Y,L,\theta_{x|y})q(Y\mid L,\theta_y)]$ with respect to $\text{vec}(Y)$ or a simpilfied approximation of this Hessian. *For Y consisting of binary variables* $p(Y\mid X,L,\theta_{y|x})$ is in a given structure based on $\eta_L(X,\Theta_p)=c_{b,L}+B_{b,L}\text{vec}(X)+\text{vec}(X)^T Q_{b,L}\text{vec}(X)$ similar to the Choice(C) of $p(L\mid X,\theta_{y|x})$.

First, we consider Choice A, i.e., a $p(\Theta|\mathbf{X}, \Xi_p)$ free of structure. Maximizing $H(p\|q)$ with respect to such a $p(\Theta|\mathbf{X}, \Xi_p)$ leads to

$$p(\Theta|\mathbf{X}, \Xi_p) = \delta(\Theta - \Theta^*), \ \ \Theta^* = \max_{\Theta} H_f(\mathbf{X}, \Theta, \mathbf{k}, \Xi). \tag{4}$$

That is, the problem becomes seeking a best harmony between the front layer Ying-Yang pair. But it also incurs a problem. With $p(\mathbf{X}|\theta_x)$ given empirically from $\mathcal{X}_N$, the mapping to Θ^* from $\mathcal{X}_N$ of random samples is probabilistic. However, $\delta(\Theta - \Theta^*)$ can not take this uncertainty in consideration. Actually, $\Theta^*(\mathcal{X}_N)$ by eq.(4) only takes over the information of the first order statistics from the Ying machine. In other words, maximizing $H(p\|q)$ with respect to a free $p(\Theta|\mathbf{X}, \Xi_p)$ can only make a best Ying Yang harmony in term of the first order statistics.

This uncertainty is considered by $p(\Theta|\mathbf{X}, \Xi_p)$ in the Choice B or Choice C such that a best Ying Yang harmony in term of not only the first order statistics but also the statistics of the second order or higher. To be detailed in the next subsection, the approximation will make $H(p\|q, \mathbf{k}, \Xi)$ in eq.(3) approximately turned into the following format:

$$H(p\|q, \mathbf{k}, \Xi) = H_f(\mathcal{X}_N, \Theta^*, \mathbf{k}, \Xi) + \Delta(\Theta^*, \mathbf{k}, \Xi), \tag{5}$$

where Θ^* and $H_f(\mathcal{X}_N, \Theta, \mathbf{k}, \Xi)$ are given in eq.(4), and $\Delta(\Theta^*, \mathbf{k}, \Xi)$ either involves no integral over Θ or an integral over a subset of Θ that is analytically solvable. If the meta parmeters Ξ is given, we can directly maximize the above $H(p\|q, \mathbf{k}, \Xi)$ to select $\mathbf{k}$. If the meta parameters Ξ is unknown, we need to make $\max_{\Xi} H(p\|q, \mathbf{k}, \Xi)$ too. Actually, getting Θ^* by eq.(4) depends on Ξ. In other words, the process of seeking an approriate Ξ^* is coupled with finding Θ^*. In general, we can estimate Ξ^* and Θ^* jointly by iterating the following two steps:

$$\Theta \ step: \Theta^{(t+1)} = \Theta^{(t)} + \eta \nabla_{\Theta} H_f(\mathcal{X}_N, \Theta, \mathbf{k}, \Xi^{(t)})_{\Theta = \Theta^{(t)}}, \tag{6}$$

$$or \ \ \ \Theta^{(t+1)} = arg \max_{\Theta} H_f(\mathcal{X}_N, \Theta, \mathbf{k}, \Xi^{(t)}) \text{ if it is analytically solvable,}$$

$$\Xi \ step: \Xi^{(t+1)} = \Xi^{(t)} + \eta \nabla_{\Xi \ at \ \Xi^{(t)}} [H_f(\mathcal{X}_N, \Theta^{(t+1)}, \mathbf{k}, \Xi) + \Delta(\Theta^{(t+1)}, \mathbf{k}, \Xi)],$$

which starts with an initialization $\Theta^{(0)}$ and $\Xi^{(0)}$ and reaches a convergence.

In a summary, the best Ying Yang harmony by maximizing $H(p\|q, \mathbf{k}, \Xi)$ is made via the following two stage implementation :

$$Stage \ I: \ \text{get } \Xi^*, \Theta^* \text{ by eq.(6) for } \forall \mathbf{k} \in \mathcal{K}, \ \mathcal{K} \text{ is a set of values of } \mathbf{k}; \tag{7}$$

$$Stage \ II: \mathbf{k}^* = arg \min_{\mathbf{k} \in \mathcal{K}} J(\mathbf{k}), J(\mathbf{k}) = -H_f(\mathcal{X}_N, \Theta^*, \mathbf{k}, \Xi^*) + \Delta(\Theta^*, \mathbf{k}, \Xi^*).$$

As mentioned previously, the scale $\mathbf{k}$ of a BYY system is contributed from two parts. One is featured by $\mathbf{k}_Y$ for representing Y and the rest is featured by the number n_f of free parameters in Θ. The degrees of difficulty for estimating the two parts are quite different. When $q(\mathbf{Y}|\theta_y)$ is in a so called scale reducible structure, an appropriate $\mathbf{k}_Y$ will be determined automatically during parameter learning on Ξ^* and Θ^* by eq.(6), with $\mathbf{k}$ initialized at one big enough value. The details are referred to Sec.2.3. Interestingly, the model selection problem in many typical learning tasks [49,52] can be reformulated into a BYY system for selecting merely this $\mathbf{k}_Y$ part. This favorable feature makes both parameter learning for Ξ^*, Θ^* and model selection for $\mathbf{k}_Y$ implemented simultaneously by only implementing eq.(6), which can significantly reduce the computational cost that is needed in a two stage implementation by eq.(7). However, the performance of this automatic model selection will deteriorate as the sample size N reduces. In such a case, we can implement both the stages in eq.(7) with a computational cost similar to those conventional two stage implementations of typical model selection criteria. Still, eq.(7) will provide an improvement over those typical criteria since the contribution by $\mathbf{k}_Y$ has been addressed more accurately, though the contribution featured by the number n_f of free parameters in Θ is roughly estimated in a way similar to those typical criteria.

2.2　Yang Machine: Implementable Scenarios

With $p(\mathbf{X}|\theta_x)$ given empirically from $\mathcal{X}_N$, i.e., Choice A in Tab. 2, it follows from eq.(3) that we further have

$$H_f(\mathcal{X}_N, \Theta, \mathbf{k}, \Xi) = \sum_L p(L|\mathcal{X}_N, \theta_{y|x}) H_f(\mathcal{X}_N, L, \Theta, \mathbf{k}, \Xi),$$

$$H_f(\mathcal{X}_N, L, \Theta, \mathbf{k}, \Xi) = \int p(\mathbf{Y}_v|\mathcal{X}_N, L, \theta_{y|x}) \mathcal{L}(\mathcal{X}_N, L, \mathbf{Y}, \Theta_q) d\mathbf{Y}_v - Z(\Theta|\Xi_q),$$

$$\mathcal{L}(\mathcal{X}_N, L, \mathbf{Y}_v, \Theta_q) = \ln[q(\mathcal{X}_N|\mathbf{Y}_v, L, \theta_{x|y})q(\mathbf{Y}_v|L, \theta_y)q(L|\theta_L)],$$

$$Z(\Theta|\Xi_q) = -\ln q(\Theta|\Xi_q), \quad \Theta_q = \{\theta_{x|y}, \theta_y, \theta_L\}. \tag{8}$$

There still remains an integral over $\mathbf{Y}_v$, which is handled differently according to the choices of $p(\mathbf{Y}_v|\mathbf{X}, L, \theta_{y|x})$ in Tab.2, whenever there is no confusion, y_v is denoted by y for simplicity. For the Choice A, maximizing $H(p\|q)$ with respect to a $p(\mathbf{Y}_v|\mathbf{X}, L, \theta_{y|x})$ free of structure leads to

$$p(\mathbf{Y}_v|\mathbf{X}, L, \theta_{y|x}) = \delta(\mathbf{Y}_v - \mathbf{Y}^*_{vL}(\Theta_q)), \quad \mathbf{Y}^*_{vL}(\Theta_q) = \max_{\mathbf{Y}_v} \mathcal{L}(\mathcal{X}_N, L, \mathbf{Y}_v, \Theta_q),$$

$$H_f(\mathcal{X}_N, L, \Theta, \mathbf{k}, \Xi) = \mathcal{L}(\mathcal{X}_N, L, \mathbf{Y}^*_{vL}(\Theta_q), \Theta_q) - Z(\Theta|\Xi_q). \tag{9}$$

The computational difficulty incurred by the integral over $\mathbf{Y}_v$ has been avoided. But it also incurs two problems. First, the above $\mathbf{Y}^*_{vL}(\Theta_q)$ may not have a differentiable expression with respect to Θ_q, or even no analytical expression. Thus, a gradient based algorithm for $\max_\Theta H(p\|q, \Theta)$ can not take the relation $\mathbf{Y}^*_{vL}(\Theta_q)$ in consideration, which makes learning fragile to local optimal performance. Second, the mapping from a set $\mathcal{X}_N$ of random samples to the inner representations is probabilistic while $\delta(\mathbf{Y}_v - \mathbf{Y}^*_{vL}(\Theta_q))$ can not take this uncertainty in consideration, since it only takes over the information of the first order statistics from the Ying machine. Similar to the discussion after eq.(4), considered in eq.(9) is a best Ying Yang harmony only in term of the first order statistics. It is improved by $p(\mathbf{Y}_v|\mathbf{X}, L, \theta_{y|x})$ in the Choice C of Tab.2 such that a best Ying Yang harmony in the front layer via approximately considering the second order statistics.

Considering a Taylor expansion of $Q(\xi)$ around $\xi^* = \max_\xi Q(\xi)$ up to the second order and noticing $\nabla_\xi Q(\xi) = 0$ at $\xi = \xi^*$, we approximately have

$$\int p(\xi)Q(\xi)d\xi \approx Q(\xi^*) + \frac{1}{2}Tr[\Sigma H_Q(\xi^*)], \quad \Sigma = \int p(\xi)(\xi - \xi^*)(\xi - \xi^*)^T d\xi, \tag{10}$$

where the Hessian matrix $H_Q(\xi) = \partial^2 Q(\xi)/\partial\xi\partial\xi^T$ is negative definite in a neighborhood of ξ^*. Moreover, $q(\xi) = e^{-Q(\xi)} / \int e^{-Q(\xi)}d\xi$ defines a distribution. If we use $G(\xi|\mu, \Sigma)$ to approximate $q(\xi)$, the solution is $\mu = \xi^*$, $\Sigma = H_Q^{-1}(\xi^*)$ [45].

With $\mathbf{Y}_v$ as ξ and $\mathcal{L}(\mathcal{X}_N, L, \mathbf{Y}_v, \Theta_q)$ as $Q(\xi)$, it follows from eq.(8) and eq.(10) that we approximately have

$$H_f(\mathcal{X}_N, L, \Theta, \mathbf{k}, \Xi) \approx \mathcal{L}(\mathcal{X}_N, L, \mathbf{Y}_{vL}^*(\Theta_q), \Theta_q) - 0.5d_{\mathbf{k}_Y}(L, \Theta_q) - Z(\Theta),$$
$$d_{\mathbf{k}_Y}(L, \Theta) = Tr[\Sigma_L(\mathbf{Y}_{vL}^*, \Theta)H_L(\mathbf{Y}_v, \Theta_q)]_{\mathbf{Y}_v = \mathbf{Y}_{vL}^*(\Theta_q)},$$
$$H_L(\mathbf{Y}_v, \Theta_q) = -\frac{\partial^2 \mathcal{L}(\mathcal{X}_N, L, \mathbf{Y}_v, \Theta_q)}{\partial\mathbf{Y}_v\partial\mathbf{Y}_v^T},$$
$$\Sigma_L(\mathbf{Y}_{vL}^*, \theta_{y|x}) = \int[\mathbf{Y}_v - \mathbf{Y}_{vL}^*(\Theta_q)][\mathbf{Y}_v - \mathbf{Y}_{vL}^*(\Theta_q)]^T p(\mathbf{Y}_v|\mathcal{X}_N, L, \theta_{y|x})d\mathbf{Y}_v.$$

Following the discussion after eq.(10), see the Choice C in Tab.2, we consider

$$p(\mathbf{Y}_v|\mathcal{X}_N, L, \theta_{y|x}) = G(\mathbf{Y}_v|\mu(\mathcal{X}_N, \phi_{\mu,L}), \Sigma(\mathcal{X}_N, \phi_{\Sigma,L})). \tag{11}$$

Let $\mu(\mathcal{X}_N, \phi_{\mu,L}) = \mathbf{Y}_{vL}^*(\Theta_q)$ and $\Sigma(\mathcal{X}_N, \phi_{\Sigma,L}) = H_L^{-1}(\mathbf{Y}_{vL}^*(\Theta_q), \Theta_q)$, we have

$$\Sigma_L(\mathbf{Y}_{vL}^*, \theta_{y|x}) = H_L^{-1}(\mathbf{Y}_v^*(\Theta_q), \Theta_q), \ d_{\mathbf{k}_Y}(\Theta) = Tr[I_Y] = d_Y, \tag{12}$$

where d_Y is the dimension of $\mathbf{Y}_v$. Comparing with eq.(9), we can find that the only difference is this integer d_Y. This term is useful in Stage II of eq.(7) for making model selection on $\mathbf{k}$. However, the two problems mentioned after eq.(9) largely remain. Alternatively, we let $\Sigma(\mathcal{X}_N, \phi_{\Sigma,L}) = H_L^{-1}(\mathbf{Y}_v^*(\Theta_q), \Theta_q)$ but leave $\mu(\mathcal{X}_N, \phi_{\mu,L})$ to be a parametric function (e.g., a linear function or nonlinear function) with a unknown set $\phi_{\mu,L}$. Let $\mathbf{Y}_v - \mathbf{Y}_{vL}^* = \mathbf{Y}_v - \mu(\mathcal{X}_N, \phi_{\mu,L}) + \mu(\mathcal{X}_N, \phi_{\mu,L}) - \mathbf{Y}_{vL}^*$, we have

$$\Sigma(\mathbf{Y}_{vL}^*, \theta_{y|x}) = H_L^{-1}(\mathbf{Y}_{vL}^*, \Theta_q) + e(\mathbf{Y}_{vL}^*, \Theta)e^T(\mathbf{Y}_{vL}^*, \Theta),$$
$$d_{\mathbf{k}_Y}(L, \Theta) = d_Y + e^T(\mathbf{Y}_{vL}^*, \Theta)H_L(\mathbf{Y}_{vL}^*, \Theta_q)e(\mathbf{Y}_{vL}^*, \Theta),$$
$$\mathbf{Y}_{vL}^* = \mathbf{Y}_{vL}^*(\Theta_q), \ e(\mathbf{Y}_{vL}^*, \Theta) = \mu(\mathcal{X}_N, \phi_{\mu,L}) - \mathbf{Y}_{vL}^*. \tag{13}$$

In addition to d_Y, the second term in $d_{\mathbf{k}_Y}(L, \Theta)$ takes uncertainties in consideration. This term is updated via Θ_q and will gradually disappear as learning converges and $e(\mathbf{Y}_{vL}^*, \Theta)$ tends to 0. Even without an analytical expression for $\mathbf{Y}_v^*(\Theta_q)$, we can get a value $\mathbf{Y}_{vL}^*$ via $\max_\Theta H(p\|q, \Theta)$ and then update Θ with the above $d_{\mathbf{k}_Y}(L, \Theta)$ in effect by certain extent. If $\mathbf{Y}_{vL}^*(\Theta_q)$ is obtained in a differentiable expression, a further improvement can be obtained via further taking $\nabla_{\Theta_q}\mathbf{Y}_{vL}^*(\Theta_q)$ in consideration through the chain rule.

When $\mathbf{Y}_v$ consists of vectors in binary variables. The above approach does not apply because we can not use eq.(10). In these cases, the integral over $\mathbf{Y}_v$ becomes summation, which can be computed but usually with a high computational complexity. Still, we can consider $p(\mathbf{Y}_v|\mathcal{X}_N, L, \theta_{y|x})$ in a parametric structure to facilitate the computation. An example is listed in Tab.2 and details will be further discussed in Sec.2.3.

We continue to proceed beyond the case of eq.(4) by considering $p(\Theta|\mathcal{X})$ in the Choice C of Tab.2 for a best Ying Yang harmony with not only $\Theta^*(\mathcal{X}_N)$ but also its corresponding second order statistics in consideration. With $p(\mathbf{X}|\theta_x)$ given empirically from $\mathcal{X}_N$, i.e., Choice A in Tab. 2, from eq.(3) we have $\int p(\Theta|\mathcal{X}_N, \Xi) H_f(\mathcal{X}_N, \Theta, \mathbf{k}, \Xi) d\Theta$. Regarding Θ as ξ and $H_f(\mathcal{X}_N, \Theta, \mathbf{k}, \Xi)$ as $Q(\xi)$, it follows again from eq.(10) that we approximately get eq.(5), that is

$$H(p\|q, \mathbf{k}, \Xi) = H_f(\mathcal{X}_N, \Theta^*, \mathbf{k}, \Xi) + \Delta(\Theta^*, \mathbf{k}, \Xi), \quad \Theta^* = \max_{\Theta} H_f(\mathcal{X}_N, \Theta, \mathbf{k}, \Xi),$$

$$\Delta(\Theta^*, \mathbf{k}, \Xi) = -0.5 d_{\mathbf{k}}, \quad d_{\mathbf{k}} = Tr[\Sigma(\Theta^*) H_H(\Theta^*)], \tag{14}$$

$$\Sigma(\Theta^*) = \int (\Theta - \Theta^*)(\Theta - \Theta^*)^T p(\Theta|\mathcal{X}_N) d\Theta, \quad H_H(\Theta) = -\frac{\partial^2 H_f(\mathcal{X}_N, \Theta, \mathbf{k}, \Xi)}{\partial\Theta\partial\Theta^T}.$$

Further let $p(\Theta|\mathcal{X}_N) = G(\Theta|\Theta^*, H_H^{-1}(\Theta^*))$, we get that $d_{\mathbf{k}} = Tr[I]$ is the number n_f of free parameters in Θ [46,47,51]. It follows from eq.(14) that both Θ^* and $H_H(\Theta^*)$ depend $\mathcal{X}_N, \Xi$. Let $p(\Theta|\mathcal{X}_N) = G(\Theta|\mu(\mathcal{X}_N, \Xi), H_H^{-1}(\Theta^*))$ with $\mu(\mathcal{X}_N, \Xi)$ in a parametric function of $\mathcal{X}_N, \Xi$, similar to eq.(13) we also get

$$\Delta(\Theta^*, \mathbf{k}, \Xi) = -0.5 d_{\mathbf{k}},$$
$$d_{\mathbf{k}} = n_f + (\mu(\mathcal{X}_N, \Xi) - \Theta^*)^T H_H(\Theta^*)(\mu(\mathcal{X}_N, \Xi) - \Theta^*). \tag{15}$$

Revising the direction of thinking, put $p(\Theta|\mathcal{X}_N) = G(\Theta|\mu(\mathcal{X}_N, \Xi), H_H^{-1}(\Theta^*))$ into eq.(3) we can also consider

$$H(p\|q, \mathbf{k}, \Xi) = \int G(\Theta|\mu(\mathcal{X}_N, \Xi), H_H^{-1}(\Theta^*)) H_f^-(\mathcal{X}_N, \Theta, \mathbf{k}, \Xi) d\Theta + \Delta(\Theta^*, \mathbf{k}, \Xi),$$

$$\Delta(\Theta^*, \mathbf{k}, \Xi) = \int G(\Theta|\mu(\mathcal{X}_N, \Xi), H_H^{-1}(\Theta^*)) \ln q(\Theta|\Xi_q) d\Theta,$$

$$H_f^-(\mathcal{X}_N, \Theta, \mathbf{k}, \Xi) = \sum_L \int p(L|\mathcal{X}_N, \theta_{y|x}) p(\mathbf{Y}_v|\mathcal{X}_N, L, \theta_{y|x}) \mathcal{L}(\mathcal{X}_N, L, \mathbf{Y}_v, \Theta_q) d\mathbf{Y}_v,$$

where $\mathcal{L}(\mathcal{X}_N, L, \mathbf{Y}_v, \Theta_q)$ is still given by eq.(8). There may be two ways to handle the integral in the first term. One is considering several typical structures on which the integral can be handled, with details referred to Sec.2.3. The other way is similar to eq.(10). Considering a Taylor expansion of $Q(\xi)$ around μ up to the second order, we approximately have

$$\int G(\xi|\mu, \Sigma) Q(\xi) d\xi \approx Q(\xi)_{\xi=\mu} + \frac{1}{2} Tr[\Sigma \partial^2 Q(\xi)/\partial\xi\partial\xi^T]_{\xi=\mu}. \tag{16}$$

With $G(\Theta|\mu(\mathcal{X}_N, \Xi), H_H^{-1}(\Theta^*))$ as ξ and $H_f^-(\mathcal{X}_N, \Theta, \mathbf{k}, \Xi)$ as $Q(\xi)$, we get

$$\int G(\Theta|\mu(\mathcal{X}_N, \Xi), H_H^{-1}(\Theta^*)) H_f^-(\Theta, \mathbf{k}, \Xi) d\Theta = H_f^-(\mathcal{X}_N, \mu(\mathcal{X}_N, \Xi), \mathbf{k}, \Xi)$$

$$+ \frac{1}{2} Tr[H_H^{-1}(\Theta^*)\{\partial^2 H_f^-(\Theta, \mathbf{k}, \Xi)/\partial\Theta\partial\Theta^T\}]_{\Theta=\mu(\mathcal{X}_N, \Xi)}. \tag{17}$$

For the second term of $H(p\|q, \mathbf{k}, \Xi)$, i.e., $\Delta(\Theta^*, \mathbf{k}, \Xi)$, we may handle it by either observing the specific structure of $q(\Theta|\Xi_q)$ or using eq.(16). By the latter, we reach $H(p\|q, \mathbf{k}, \Xi)$ in eq.(14) again but with

$$\Delta(\Theta^*, \mathbf{k}, \Xi) = -0.5 d_{\mathbf{k}}, \quad d_{\mathbf{k}} = Tr[H_H^{-1}(\Theta^*) H_H(\mu(\mathcal{X}_N, \Xi))], \tag{18}$$

which becomes $d_{\mathbf{k}} = n_f$ when $\mu(\mathcal{X}_N, \Xi)$ reaches Θ^*.

Generally, it is difficult to consider $p(\Theta|\mathcal{X})$ given by a Bayesian structure, i.e., the Choices B in Tab.2. In some cases, we may divide the set Θ into two parts Θ' and Θ'' and assume that $p(\Theta|\mathcal{X}) = p(\Theta'|\mathcal{X})p(\Theta'|\mathcal{X})$ and $q(\Theta|\Xi_q) = q(\Theta'|\Xi_q')q(\Theta''|\Xi_q'')$. For one part Θ'', we may handle $\int p(\Theta''|\mathcal{X})\ln q(\Theta''|\Xi_q)d\Theta''$ analytically. Then, the remining parts are handled as before.

The last but not least, we further proceed to the case that $p(\mathbf{X}|\theta_x) = p_h(\mathbf{X})$ is given by Choice B in Tab. 2, with an extra unknown input h in consideration. Let $\mathbf{X}$ to be replaced by $\mathbf{X}, h$ and notice that only $\mathbf{X}$ relates to h, we have

$$p(\mathbf{X}, h) = p(\mathbf{X}|h)p(h), \ p(\mathbf{X}|h) = p_h(\mathbf{X}), \ p(\mathbf{R}|\mathbf{X}, h) = p(\mathbf{R}|\mathbf{X}),$$
$$q(\mathbf{X}, h|\mathbf{R}) = q(h|\mathbf{X}, \mathbf{R})q(\mathbf{X}|\mathbf{R}), \ q(h|\mathbf{X}, \mathbf{R}) = q(h|\mathbf{X}), \tag{19}$$

with $q(\mathbf{R})$ remains unchanged. Put it into eq.(3), we have

$$H(p\|q, \mathbf{k}, \Xi) = \int p(h)p(\Theta|\mathbf{X}, \Xi)H_f(\mathbf{X}, \Theta, h, \mathbf{k}, \Xi)d\Theta d\mathbf{X} dh \tag{20}$$

Maximizing $H(p\|q, \mathbf{k}, \Xi)$ with $p(h)$ free of constraint leads to

$$p(h) = \delta(h - h^*), \ h^* = arg \max_h H_h(p\|q, \mathbf{k}, \Xi),$$
$$H_h(p\|q, \mathbf{k}, \Xi) = \int p(\Theta|\mathbf{X}, \Xi)H_f(\mathbf{X}, \Theta, h, \mathbf{k}, \Xi)d\Theta. \tag{21}$$

Equivalently, h^* is also given by $h^* = arg \max_h H_f(\mathbf{X}, \Theta, h, \mathbf{k}, \Xi)$. Moreover, it further follows from eq.(16) that we have

$$H_f(\mathbf{X}, \Theta, h, \mathbf{k}, \Xi) = H_f(\mathcal{X}_N, \Theta, \mathbf{k}, \Xi) + 0.5h^2 Tr[\Sigma(\mathcal{X}_N)] - Z(h),$$
$$Z(h) = -\ln q(h|\mathcal{X}_N), \ \Sigma(\mathbf{X}) = \frac{\partial^2 \sum_L p(L|\mathcal{X}_N, \theta_{y|x})\ln q(\mathbf{X}|\mathbf{Y}_v, L, \theta_{x|y})}{\partial \mathbf{X}\partial \mathbf{X}^T}.$$

An example of $q(h|\mathcal{X}_N)$ is obtained from $p_h(\mathbf{X})$ by eq.(29) in the next subsection.

Therefore, we can modify the two stage implementation by eq.(6) and eq.(7) with all the appearances of Θ replaced by $\{\Theta, h\}$ and $H_f(\mathcal{X}_N, \Theta, \mathbf{k}, \Xi)$ by $H_f(\mathcal{X}_N, \Theta, h, \mathbf{k}, \Xi)$. Recalling the discussion after eq.(7), the performance of automatic model selection on $\mathbf{k}_Y$ by eq.(6) will deteriorate as the sample size N reduces. With an approriate h^* learned together with Θ^*, a considerable improvement can be obtained to reduce this deterioration, as verified by experiments [35].

2.3   Ying Machine : Distributed Log-Quadratic Inner Structures

We proceed to typical scenarios of $q(\mathbf{X}, \mathbf{R})$. To accommodate the world knowledge and the dependences underlying $\mathcal{X}_N$ appropriately, there are several issues to be considered, consisting of inner representation with a wide coverage of typical learning tasks, computational feasibility, scalability of complicated problems, and structural scale reducibility that does not impede automatic model selection.

For many learning tasks, these issues are collectively considered in a class of structures for Ying machine, namely distributed log-quadratic inner structures. As already discussed in Sec.2.1, we consider a distributed inner representation

$\mathbf{Y} = \{\mathbf{Y}_v, \mathbf{L}\}$, i.e., vector based inner representations $\mathbf{Y}_v = \{y_v\}$ are distributively and collaboratively described in a collection $\mathbf{L} = \{\ell\}$ with the help of $q(\mathcal{X}_N | \mathbf{Y}_v, \mathbf{L}, \theta_{x|y})$ and $q(\mathbf{Y}_v | \theta_y)$ in the following log-quadratic structures :

$$\begin{aligned}\mathcal{L}(\mathcal{X}_N, \mathbf{L}, \mathbf{Y}, \Theta_q) &= \ln\left[q(\mathcal{X}_N | \mathbf{Y}_v, \mathbf{L}, \theta_{x|y})q(\mathbf{Y}_v | \mathbf{L}, \theta_y)q(\mathbf{L})\right] \\ &= Tr[\Phi_{\mathcal{X}_N}(\Theta_{\mathbf{L}}^q)\mathbf{Y}_v\mathbf{Y}_v^T + \Psi_{\mathcal{X}_N}(\Theta_{\mathbf{L}}^q)\mathbf{Y}_v + \phi_{\mathcal{X}_N}(\Theta_{\mathbf{L}}^q)] + \ln q(\mathbf{L}),\end{aligned} \quad (22)$$

where $\Phi_{\mathcal{X}_N}(\Theta_{\mathbf{L}}^q), \Psi_{\mathcal{X}_N}(\Theta_{\mathbf{L}}^q), \phi_{\mathcal{X}_N}(\Theta_{\mathbf{L}}^q)$ are in given expressions. Eq.(8) becomes

$$H_f(\mathcal{X}_N, \Theta, \mathbf{k}, \Xi) = \sum_{\mathbf{L}} p(\mathbf{L}|\mathcal{X}_N, \theta_{y|x})\ln q(\mathbf{L}) + \sum_{\mathbf{L}} p(\mathbf{L}|\mathcal{X}_N, \theta_{y|x})Tr[\phi_{\mathcal{X}_N}(\Theta_{\mathbf{L}}^q)] +$$

$$\sum_{\mathbf{L}} p(\mathbf{L}|\mathcal{X}_N, \theta_{y|x})Tr[\Psi_{\mathcal{X}_N}(\Theta_{\mathbf{L}}^q)\mu(\mathcal{X}_N, \theta_{y|x}) + \Phi_{\mathcal{X}_N}(\Theta_{\mathbf{L}}^q)\Gamma_{\mathbf{L}}(\mathcal{X}_N, \theta_{y|x})] - Z(\Theta),$$

$$\mu(\mathcal{X}_N, \theta_{y|x}) = \int \mathbf{Y}_v p(\mathbf{Y}_v | \mathcal{X}_N, \mathbf{L}, \theta_{y|x})d\mathbf{Y}_v,$$

$$\Gamma_{\mathbf{L}}(\mathcal{X}_N, \theta_{y|x}) = \int \mathbf{Y}_v\mathbf{Y}_v^T p(\mathbf{Y}_v | \mathcal{X}_N, \mathbf{L}, \theta_{y|x})d\mathbf{Y}_v. \quad (23)$$

As a result, the integral over $\mathbf{Y}_v$ for $H_f(\mathcal{X}_N, \Theta, \mathbf{k}, \Xi)$ has been turned into the integrals for getting the first order and second order statistics of $p(\mathbf{Y}_v | \mathcal{X}_N, \mathbf{L}, \theta_{y|x})$. Let $p(\mathbf{Y}_v | \mathcal{X}_N, \mathbf{L}, \theta_{y|x}) = G(\mathbf{Y}_v | \mathbf{Y}_{vL}^*(\Theta_q), H_L^{-1}(\mathbf{Y}_{vL}^*(\Theta_q)))$,

Table 3. Typical structures of $q(y) = q(y|\theta_y)$ and $q(x|y) = q(x|y, \theta_{x|y}) = G(x|\mu_y, \Sigma_y)$

y	$\{l\}$	$\{y\}$	$\{y,l\}$
$q(y|\theta_y)$	$q(l) = a_l \geq 0, \sum_{l=1}^{k} a_l = 1$	$q(y) = \prod_{j=1}^{m} q(y^{(j)})$	$q(y,l) = q(y|l)a_l, q(y|l) = \prod_{j=1}^{m_l} q(y^{(j)}|l)$

(a)

Case	(1) Gaussian	(2) Bernoulli	(3) $y^{(j)} = \{y_g^{(j)}, i^{(j)}\}$	(4) Mixture
$q(y^{(j)})$	$G(y^{(j)}|0, \lambda^{(j)})$	$q_j^{y^{(j)}}(1-q_j)^{1-y^{(j)}}$	$G(y_g^{(j)}|v^{(j,i)}, \lambda^{(j,i)})\beta^{(j,i)}$	$\sum_{i=1}^{\kappa^{(j)}} \beta^{(j,i)} q^{(i)}(y_g^{(j)})$
$q(y^{(j)}|l)$	$G(y^{(j)}|0, \lambda_l^{(j)})$	$q_{lj}^{y^{(j)}}(1-q_{lj})^{1-y^{(j)}}$	$G(y_g^{(j)}|v_l^{(j,i)}, \lambda_l^{(j,i)})\beta_l^{(j,i)}$	$\sum_{i=1}^{\kappa^{(j)}} \beta_l^{(j,i)} q_l^{(i)}(y_g^{(j)})$

Note: for Type (3) we have $q(y^{(j)}) = q(y_g^{(j)}, i^{(j)}|l) = q(y_g^{(j)}|i^{(j)}, l)q(i^{(j)}|l)$ with

$$q(y_g^{(j)}|i^{(j)}, l) = G(y_g^{(j)}|v_l^{(j,i)}, \lambda_l^{(j,i)}), \quad q(i^{(j)}|l) = \beta_l^{(j,i)} \geq 0, \sum_{i=1}^{\kappa^{(j)}} \beta_l^{(j,i)} = 1.$$

(b)

Type	A	B	C		Case	(1)	(2)	(3)	(4)
y	$\{l\}$	$\{y\}$	$\{y,l\}$	$q(x|y, \theta_{x|y}) = G(x|\mu_y, \Sigma_y)$	Noises relates to	none of $\{y,l\}$	only $\{l\}$	only $\{y\}$	both $\{y,l\}$
μ_y	μ_l	$Ay + \mu$	$A_l y + \mu_l$		Σ_y	Σ	Σ_l	Σ_y	$\Sigma_{y,l}$

(c)

we have $\mu(\mathcal{X}_N, \theta_{y|x}) = \mathbf{Y}_{vL}^*(\Theta_q)$ and $\Gamma_L(\mathcal{X}_N, \theta_{y|x}) = H_L^{-1}(\mathbf{Y}_{vL}^*(\Theta_q) + \mathbf{Y}_{vL}^*(\Theta_q)\mathbf{Y}_{vL}^{*T}(\Theta_q)$ and thus it returns back to the situation same as that in eq.(11) and eq.(12). From eq.(11) with $\Sigma(\mathcal{X}_N, \phi_{\Sigma,L}) = H_L^{-1}(\mathbf{Y}_{vL}^*(\Theta_q), \Theta_q)$, we also have

$$\mu(\mathcal{X}_N, \theta_{y|x}) = \mu(\mathcal{X}_N, \phi_{\mu,L}),$$
$$\Gamma_L(\mathcal{X}_N, \theta_{y|x}) = H_L^{-1}(\mathbf{Y}^*(\Theta_q), \Theta_q) + \mu(\mathcal{X}_N, \phi_{\mu,L})\mu^T(\mathcal{X}_N, \phi_{\mu,L}), \quad (24)$$

and thus encounter a situation equivalent to that by eq.(11) and eq.(13).

Beyond those cases based on eq.(10), eq.(23) is also applicable to the cases that $\mathbf{Y}_v$ consists of binary or discrete variables. It is applicable to any structures in the log-quadratic form by eq.(22) or in this form approximately. Beyond eq.(11), we consider $p(\mathbf{Y}_v|\mathcal{X}_N, \mathbf{L}, \theta_{y|x})$ in a structure that the integrals for $\mu(\mathcal{X}_N, \theta_{y|x})$ and $\Gamma_L(\mathcal{X}_N, \theta_{y|x})$ in eq.(23) can be solved analytically or computed efficiently.

Table 4. Typical parametric structures of $p(y|x, \theta_{y|x})$

y	$\{l\}$	$\{y_v\}$	$\{y_v, l\}$
$p(y\|x, \theta_{y\|x})$	$p(l\|x) = \alpha_l(x) \geq 0,\ \alpha_l(x) = \varsigma_l(x)\Big/ \sum_{l=1}^{k} \varsigma_l(x)$	$p(y_v\|x)$	$p(y_v, l\|x) = p(y_v\|x, l)\alpha_l(x)$

(A)

Case	(1) Gaussian	(2) Bernoulli	(3) $y = [y_g, \{i^\theta\}]$
$p(y\|x)$	$G(y\|\zeta(x), \Gamma)$	$\prod_{j=1}^{m} \varsigma_j^{y^\theta}(x)[1 - \varsigma_j(x)]^{1-y^\theta}$	$G(y_g\|\varsigma(x), \Gamma)\prod_{j=1}^{m}\beta(y_g^{(j)})$
$p(y\|x, l)$	$G(y\|\zeta_l(x), \Gamma_l)$	$\prod_{j=1}^{m_l} \varsigma_{j,l}^{y^\theta}(x)[1 - \varsigma_{j,l}(x)]^{1-y^\theta}$	$G(y_g\|\varsigma_l(x), \Gamma_l)\prod_{j=1}^{m_l}\beta_l(y_g^{(j)})$

For case (3), $p(y \mid x, l) = p(y_g, \{i^\theta\} \mid x, l) = p(y_g \mid x, l)\prod_{j=1}^{m_l} p(i^{(j)} \mid y_g^{(j)}, l)$

$p(y_g \mid x, l) = G(y_g \mid \varsigma_l(x), \Gamma_l),\ p(i^{(j)} \mid y_g^{(j)}, x, l) = p(i^{(j)} \mid y_g^{(j)}, l) = \beta_l(y_g^{(j)})$

Γ, Γ_l given by eq.(24) in the i.i.d. special case.

$$\beta_l(y_g^{(j)}) = exp(\pi_l^{(j,i)})\Big/ \Sigma_{i=1}^{\kappa_l^{(j)}} exp(\pi_l^{(j,i)})$$

(a) Posteriori $\pi_l^{(j,i)} = ln[G(y_g^{(j)} \mid v_l^{(j,i)}, \lambda_l^{(j,i)})\beta_l^{(j,i)}]$	(b) $\pi_l^{(j,i)}$ or $\beta_l(y_g^{(j)})$ is a free constant	(c) Parametric $\pi_l^{(j,i)} = b_0 + b_1 y_g^{(j)} + b_2 y_g^{(j)2}$

(B)

$\varsigma_l(x) = [\varsigma_{1,l}(x), \ldots, \varsigma_{m_l,l}(x)]^T$, $\varsigma(x) = \varsigma_l(x)$ at $k=1$, $o_{j,l}(x) = \beta_l x^T H_{j,l} x + w_{j,l}^T x + c_{j,l}$, see eq.(27)		
(1) Gaussian	**(2) Bernoulli**	**(3) Multinomial $\{l\}$**
$\varsigma_l(x) = o_l$ with $U_{j,l} = 0, \forall j$	$\varsigma_{j,l}(x) = s(o_{j,l}), 0 \leq s(r) \leq 1$ is a sigmoid function, e.g., $(1+e^{-r})^{-1}$	$\varsigma_{j,l}(x) = s(o_{j,l})$ with $m_l = 1, \forall l$ $0 \leq s(r)$ is monotonic, e.g., e^{-r}

(C)

Moreover, instead of specifying the entire $p(\mathbf{Y}_v|\mathcal{X}_N, \mathbf{L}, \theta_{y|x})$, we can merely design $\mu(\mathcal{X}_N, \theta_{y|x})$ and $\Gamma_{\mathbf{L}}(\mathcal{X}_N, \theta_{y|x})$ in certain pre-specified parametric functions that are analytically computable.

The above discussions apply to the general scenarios that there maybe also temporal or even graphical dependence among that elements of $\mathbf{X} = \{x\}$. Correspondingly, we consider certain structure among the elements of $\mathbf{Y} = \{y_v, \ell\}$ to accommodate this dependence. E.g., further details about temporal dependences are referred to [59,50]. To get further insights, here we focus on the cases that the elements of $\mathbf{X} = \{x\}$ are independently and identically distributed (i.i.d.), and thus the elements of $\mathbf{Y} = \{y_v, \ell\}$ are i.i.d. That is, we consider

$$q(\mathbf{Y}|\theta_y) = \prod q(y_v, \ell|\theta_y), \ \ q(y_v, \ell|\theta_y) = q(y_v|\ell, \theta_{y,\ell})q(\ell),$$
$$p(\mathbf{Y}|\mathbf{X}, \theta_{y|x}) = \prod p(y_v, \ell|x, \theta_{y|x}), \ \ q(\mathbf{X}|\mathbf{Y}, \theta_{x|y}) = \prod q(x|y_v, \ell, \theta_{x|y}), \quad (25)$$

with $p(y_v, \ell|x, \theta_{y|x}) = p(y_v|x, \ell, \theta_{y|x,\ell})p(\ell|x, \theta_{\ell|x})$.

Shown in Tab. 3 and Tab.4 are several typical structures, covering several typical learning tasks [49]. All of them satisfy the format by eq.(23), except the case (d) for $q(y|\theta_y)$. Even for this exceptional case as well as other structures that fail to satisfy the format by eq.(23), we can still approximately use a Taylor expansion of $\ln[q(x|y, \theta_{x|y})q(y|\theta_y)]$ or $\ln q(y^{(j)}|\ell)$ with respect to y up to the second order. Readers are also referred to [53,46,47] for various other choices.

In the i.i.d. cases by eq.(25) and with the structures given in Tabs. 3 & 4, we can get a further insight on eq.(23) via a more detailed expression. Considering $q(y|\theta_y)$ at its Case (1) & Case (2) in Tab.3(b), we write eq.(23) into

$$H_f(\mathcal{X}_N, \Theta, \mathbf{k}, \Xi) = \sum_t \sum_\ell p(\ell|x_t, \theta_{y|x})[\ln \alpha_\ell + \phi(x_t, \Theta_q^\ell)] - Z(\Theta) + \quad (26)$$

$$\sum_t \sum_\ell p(\ell|x_t, \theta_{y|x})Tr\{\Psi(x_t, \Theta_q^\ell)\mu_\ell(x_t) + \Phi(x_t, \Theta_q^\ell)[\Gamma_\ell(x_t) + \mu_\ell(x_t)\mu_\ell^T(x_t)]\},$$

$$\mu_\ell(x) = \int yp(y|x, \ell)dy, \ \ \Gamma_\ell(x) = \int (y - \mu_\ell(x))(y - \mu_\ell(x))^T p(y|x, \ell)dy,$$

The analytical expressions for $\mu_\ell(x)$ and $\Gamma_\ell(x)$, as well as the corresponding $\phi(x_t, \Theta_q^\ell)$, $\Psi(x_t, \Theta_q^\ell)$, and $\Phi(x_t, \Theta_q^\ell)$ are given in Tab.5.

Moreover, $p(\ell|x_t, \theta_{y|x})$ is given by Tab.4(a) with $\zeta_\ell(x)$ in the choice (3) of Tab.4(c), which is a simplification of $p(\mathbf{L}|\mathcal{X}_N, \theta_{y|x})$ in the choice C(B) of Tab.2. Also, we can get the simplified counterpart of the choice C of Tab.2 as follows

$$p(\ell|x_t, \theta_{y|x}) = e^{-o_\ell(x_t)}/\sum_{j=1}^k e^{-o_j(x_t)}, \ \ o_\ell(x) = \beta x^T H_\ell x + b_\ell^T x + c_\ell,$$
$$H_\ell = \partial^2 \ln \int q(x|y_v, \ell, \theta_{x|y})q(y_v|\ell, \theta_y)dy_v/\partial x \partial x^T. \quad (27)$$

Furthermore, we are ready to elaborate the issue of structural scale reducibility, mentioned several times previously but without a further interpretation yet. As discussed after eq.(3), maximizing $H(p\|q, \mathbf{k}, \Xi)$ will push $\mathbf{k}$ as least as possible. Similarly, $\max_{\Theta,h} H_f(\mathcal{X}_N, \Theta, h, \mathbf{k}, \Xi)$ will push the entropy of $q(\mathcal{X}_N|\mathbf{Y}_v,\mathbf{L},\theta_{x|y})\, q(\mathbf{Y}_v|\mathbf{L},\theta_y)q(\mathbf{L})$ as least as possible. One part of this entropy is contributed from the representation scale $\mathbf{k}_Y$ of $\mathbf{Y} = \{\mathbf{Y}_v,\mathbf{L}\}$. Interestingly, each integer of $\mathbf{k}_Y$ is associated with one or several parameters in Θ_q, and the least complexity nature will push these parameters towards 0 if the corresponding integer represents a redundant scale part. We say that $q(\mathcal{X}_N|\mathbf{Y}_v,\mathbf{L},\theta_{x|y})q(\mathbf{Y}_v|\mathbf{L},\theta_y)q(\mathbf{L})$ has scale reducibility if there is no constraint to impede or block these parameters to be pushed towards 0.

For the structures in eq.(25), $\mathbf{k}_Y$ consists of k and $\{m_\ell\}_{\ell=1}^k$. We observe that pushing one $q(\ell) = \alpha_\ell$ towards zero is equivalent to reducing the scale k to $k-1$, with all the corresponding structures discarded in effect. Moreover, pushing the variance of $q(y^{(j)}|\ell)$ towards 0 means that those of the j-th dimension can be discarded. We say that $q(\ell)$ is scale reducible if no constraint prevents $q(\ell)$ to become 0 for every ℓ, and that $q(y_v|\ell,\theta_{y,\ell})$ is scale reducible if no constraint prevents $q(y^{(j)}|\ell,\theta_{y,\ell})$ to become zero for every j,ℓ. Thus, $q(y|\theta_y)$ is scale reducible when both $q(\ell)$ and $q(y_v|\ell,\theta_{y,\ell})$ are scale reducible. If there are extra parts of structure was allocated in a scale reducible $q(y|\theta_y)$, $\max_{\Theta,h} H_f(\mathcal{X}_N,\Theta,h,\mathbf{k},\Xi)$ will drive those extra parameters towards zero. i.e., automatic model selection on $\mathbf{k}_Y$ is made during parameter learning on Ξ^*,Θ^* by eq.(6). Taking $H_f(\mathcal{X}_N,\Theta,\mathbf{k},\Xi)$ in eq.(26) as an example, maximizing the term $\sum_t\sum_\ell p(\ell|x_t,\theta_{y|x})\ln\alpha_\ell$ of $H_f(\mathcal{X}_N,\Theta,\mathbf{k},\Xi)$ becomes equivalently maximizing $N\sum_\ell \alpha_\ell \ln\alpha_\ell$ with $\alpha_\ell = \sum_t p(\ell|x_t,\theta_{y|x})/N$, which tends to push α_ℓ towards zero when it is extra.

The last, we consider the details of $q(\Theta|\Xi)$ in Tab.2. The simplest case is the choice 1, i.e., ignoring its role simply by letting $Z(\Theta) = -\ln q(\Theta) = 0$. Moreover,

Table 5. Major terms of $H_f(\mathcal{X}_N,\Theta,\mathbf{k},\Xi)$ with $p(y|x,\ell)$ in Tab. 4, $q(x|y,\theta_{x|y})$ and $q(y|\theta_y)$ at Case (1) & Case (2) in Tab. 3

$p(y\mid x,l)$	$G(y\mid\zeta_l(x),\Sigma_l)$	$\prod_{j=1}^{m_l}\varsigma_{j,l}^{y^{(j)}}(x)[1-\varsigma_{j,l}(x)]^{1-y^{(j)}}$
$\mu_l(x)$	$\zeta_l(x)$	$\varsigma_l(x)=[\varsigma_{1,l}(x),\dots,\varsigma_{m_l,l}(x)]^{\mathrm{T}}$
$\Sigma_l(x)$	Σ_l	$\mathbf{diag}\,[\varsigma_{1,l}(x)-\varsigma_{1,l}^2(x),\dots,\varsigma_{m_l,l}(x)-\varsigma_{m_l,l}^2(x)]^{\mathrm{T}}$

(b)

	$q(x\mid y,l)=G(x\mid A_ly+\mu_l,\Sigma_l)$	
$q(y\mid l)$	Gaussian $G(y\mid 0,\Lambda_l)$	Bernoulli $\prod_{j=1}^{m_l}q_{l,j}^{y^{(j)}}(1-q_{l,j})^{1-y^{(j)}}$
$\Phi(x,\theta_q^l)$	$-0.5(\Lambda_l^{-1}+\Sigma_{A,l}^{-1})$	$-0.5\Sigma_{A,l}^{-1}$
$\Psi(x,\theta_q^l)$	$e_l(x)$	$[\ln\dfrac{q_{l,1}}{1-q_{l,1}},\dots,\ln\dfrac{q_{l,m_l}}{1-q_{l,m_l}}]^{\mathrm{T}}+e_l(x)$
$-\phi(x,\theta_q^l)$	$-0.5[D_{\Sigma_l}(x)+\ln|\Lambda_l|+m_l\ln(2\pi)]$	$-0.5D_{\Sigma_l}(x)+\sum_{j=1}^{m_l}(1-q_{l,j})$
$e_l(x)=(x-\mu_l)\Sigma_l^{-1}A_l,\ \ \Sigma_{A,l}^{-1}=A_l^T\Sigma_l^{-1}A_l,\ \ D_{\Sigma_l}(x)=(x-\mu_l)\Sigma_l^{-1}(x-\mu_l)+\ln|\Sigma_l|+d\ln(2\pi)$		

(a)

we may further divided each Θ_ξ into independent groups $q(\Theta_\xi) = \prod_j q(\Theta_\xi^{(j)})$, and let $q(\Theta_\xi^{(j)})$ in the following choices [25,21,32]:

(1) The simplest way is to consider the so called improper priori, e.g., a uniform improper priori

$$q(\Theta_\xi^{(j)}) \propto constant, \tag{28}$$

for real a parameter. For a scalar $\gamma > 0$, we consider $q(\gamma)$ by a Jeffrey improper priori $q(\gamma) \propto 1/\gamma$. When Σ is a $d \times d$ covariance matrix, a Jeffreys improper priori is $q(\Sigma) \propto 1/|\Sigma|^{0.5(d+1)}$.

(2) For different parameters, we consider different specific distribution. For examples, a gamma distribution for a scalar $\gamma > 0$, an inverted Wishart distribution for Σ of a $d \times d$ covariance matrix, a Beta/Dirichlet distribution for the proportional parameters $\{\alpha_\ell\}$, and a uniform improper priori by eq.(28) or a Gaussian distribution for real parameters in a vector or matrix.

(3) Another general way for getting an improper priori is the choice 3 in Tab.2, which was developed in [46,47,51]. Here we explain its rationale. Given a density $p(u|\theta)$, we have $\int p(u|\theta)du = 1$ that does not depend on θ for an infinite size of samples. This is no longer true for $s = \sum_{t=1}^N p(u_t|\theta)$ by considering a finite sample size of samples $\{u_t\}_{t=1}^N$. This s actually varies with θ and imposes an implicit distribution θ. Considering a priori $q(\theta) \propto \frac{1}{s}$ can balance off this unnecessary bias. E.g., for h in eq.(19) we have

$$q(h) \propto [\sum_{t=1}^N \sum_{\tau=1}^N G(x_t|x_\tau, h^2 I)/N]^{-1}. \tag{29}$$

Other examples are referred to [46,47,51], e.g., eqn.(11) for $q(\Theta)$ in [47].

3 Best Harmony vs Best Matching: Relations to Others

3.1 Special Cases: Relations to Existing Approaches

It is interesting to further observe how the best harmony learning degenerates as a BYY system degenerates to a conventional model $q(\mathbf{X}|\Theta)$. We consider $\mathbf{R} = \{\Theta\}$ without an inner representation part $\mathbf{Y}$, which leads us back to Fig.2(c), and simplifies $H(p\|q) = H(p\|q, \mathbf{k}, \Xi)$ in eq.(3) into

$$H(p\|q) = \int p(\Theta|\mathbf{X})p(\mathbf{X}) \ln [q(\mathbf{X}|\Theta)q(\Theta)]d\mathbf{X}d\Theta. \tag{30}$$

For a $p(\Theta|\mathbf{X})$ free of structure and $p(\mathbf{X})$ of the choice (A) in Tab.2, maximizing $H(p\|q)$ with respect to $p(\Theta|\mathbf{X})$ leads to the MB type Bayesian learning in Tab.1, i.e., $\max_\Theta \ln [q(\mathcal{X}_N|\Theta)q(\Theta)]$, while $J(\mathbf{k})$ in eq.(7) becomes

$$\mathbf{k}^* = arg \min_{\mathbf{k}} J(\mathbf{k}), J(\mathbf{k}) = - \max_\Theta \ln [q(\mathcal{X}_N|\Theta)q(\Theta)] + 0.5d_{\mathbf{k}}, \tag{31}$$

which is a Bayesian learning based extension of AIC. For a non-informative $q(\Theta)$, it further degenerates to exactly AIC [1,2]. Moreover, for a general case with $p(\mathbf{X}, h)$ by eq.(19), it follows from eq.(22) that eq.(30) is extended into

$$H(p\|q) = \int p(h)p(\Theta|\mathcal{X}_N)H_h(p\|q,\Theta)d\Theta \approx \max_{\Theta,h} H_h(p\|q,\Theta) - 0.5d_{\mathbf{k}},$$
$$H_h(p\|q,\Theta) = \ln[q(\mathcal{X}_N|\Theta)q(\Theta)] + 0.5h^2 Tr[\Sigma(\mathcal{X}_N)] - Z(h),$$
$$Z(h) = -\ln q(h|\mathcal{X}_N), \Sigma(\mathbf{X}) = \partial^2 \ln q(\mathbf{X}|\Theta)/\partial\mathbf{X}\partial\mathbf{X}^T. \tag{32}$$

With $p(\Theta|\mathbf{X})$ in a given structure, the BYY harmony learning is different from the conventional Bayesian learning. E.g., we consider $p(\Theta|\mathbf{X})$ with the BI structure in Tab.1 and rewrite eq.(30) into

$$H(p\|q) = \int p(\Theta|\mathbf{X})p(\mathbf{X})\ln p(\Theta|\mathbf{X})d\mathbf{X}\Theta + \int p(\mathbf{X})\ln q(\mathcal{X}|S)d\mathbf{X}. \tag{33}$$

Particularly, for $p(\mathbf{X})$ of the choice (A) in Tab.2, it further becomes

$$H(p\|q) = \int p(\Theta|\mathcal{X}_N)\ln p(\Theta|\mathcal{X}_N)d\Theta + \ln q(\mathcal{X}_N|S). \tag{34}$$

The maximization of its second term is exactly the MI (marginal likelihood) choice in Tab.1. As already discussed in Section 1, it has been previously studied under various names [34,23,21,22]. The first term in eq.(34) is the negative entropy of $p(\Theta|\mathcal{X}_N)$ and its maximization is seeking an inverse inference $\mathcal{X}_N \to \Theta$ with a least uncertainty. More generally, it follows from eq.(3) that we get an extension of eq.(34) as follows:

$$H(p\|q) = \int p(\mathbf{R}|\mathcal{X}_N)\ln p(\mathbf{R}|\mathcal{X}_N)d\mathbf{R} + \ln q(\mathcal{X}_N|S),$$
$$p(\mathbf{R}|\mathcal{X}) = q(\mathcal{X}|\mathbf{R})q(\mathbf{R})/q(\mathcal{X}|S), \quad q(\mathcal{X}|S) = \int q(\mathcal{X}|\mathbf{R})q(\mathbf{R})d\mathbf{R}. \tag{35}$$

Even generally, we also let S to be included in the inner representation $\mathbf{R}$, and get a further generalization of eq.(30) as follows:

$$H(p\|q) = \sum_S p(S|\mathbf{X})p(\mathbf{X})\ln[q(\mathbf{X}|S)q(S)]. \tag{36}$$

When $p(S|\mathbf{X})$ is free of structure, maximizing $H(p\|q)$ with respect to $p(S|\mathbf{X})$ leads to $\max_S \ln[q(\mathcal{X}_N|S)q(S)]$ for model selection, i.e., the BI choice in Tab.1. In the special case that $q(S)$ is equal for each candidate S, it further degenerates to $\max_S \ln q(\mathcal{X}_N|S)$, i.e., the ML choice in Tab.1. Also, a generalized counterpart of eq.(34) becomes

$$H(p\|q) = \sum_S p(S|\mathcal{X}_N)\ln p(S|\mathcal{X}_N) + \ln q(\mathcal{X}_N), \quad q(\mathcal{X}_N) = \sum_S q(\mathcal{X}_N|S)q(S).$$

3.2   Best Harmony Versus Best Matching

For a BYY system, in addition to making the best harmony learning by eq.(3), an alternative has also been proposed and studied in [64] under the name of

Bayesian Kullback Ying Yang (BKYY) learning that performs the following *best matching* principle:

$$\min KL(p\|q), \ KL(p\|q) = \int p(\mathbf{R}|\mathbf{X})p(\mathbf{X}) \ln \frac{p(\mathbf{R}|\mathbf{X})p(\mathbf{X})}{q(\mathbf{X}|\mathbf{R})q(\mathbf{R})} d\mathbf{X}d\mathbf{R} \qquad (37)$$

$$= \int p(\Theta|\mathbf{X})\{\int p(\mathbf{Y}|\mathbf{X}, \theta_{y|x})p(\mathbf{X}) \ln \frac{p(\Theta|\mathbf{X})p(\mathbf{Y}|\mathbf{X}, \theta_{y|x})p(\mathbf{X})}{q(\mathbf{X}|\mathbf{Y}, \theta_{x|y})q(\mathbf{Y}|\theta_y)q(\Theta)} d\mathbf{X}d\mathbf{Y}\}d\Theta,$$

which reaches to the best matching $KL(p\|q) = 0$ at $p(\mathbf{R}|\mathbf{X})p(\mathbf{X}) = q(\mathbf{X}|\mathbf{R})q(\mathbf{R})$.

As a BYY system degenerates to a conventional model $q(\mathbf{X}|\Theta)$, the above eq.(37) is simplified into the following counterpart of eq.(30):

$$\min KL(p\|q), \ KL(p\|q) = \int p(\Theta|\mathbf{X})p(\mathbf{X}) \ln \frac{p(\Theta|\mathbf{X})p(\mathbf{X})}{q(\mathbf{X}|\Theta)q(\Theta)} d\mathbf{X}d\Theta. \qquad (38)$$

When $p(\Theta|\mathbf{X})$ is free of structure, minimizing $KL(p\|q)$ with respect to $p(\Theta|\mathbf{X})$ leads to $p(\Theta|\mathbf{X}) = q(\mathbf{X}|\Theta)q(\Theta)/q(\mathcal{X}|S)$ and $q(\mathcal{X}|S) = \int q(\mathbf{X}|\Theta)q(\Theta)\mu(d\Theta)$. As a result, eq.(38) becomes

$$\min KL(p\|q), \ KL(p\|q) = \int p(\mathbf{X}) \ln \left[p(\mathbf{X})/q(\mathcal{X}|S) \right] d\mathbf{X}, \qquad (39)$$

where $p(\mathbf{X})$ is an input irrelevant to $q(\mathcal{X}|S)$ and $q(\Theta)$. For $p(\mathbf{X})$ of the choice (A) in Tab.2, eq.(39) further becomes equivalent to the MI (marginal likelihood) choice in Tab.1. For a general case with $p(\mathbf{X}, h)$ by eq.(19), eq.(39) provides its data smoothing version with not only Θ but also h learned.

Alternatively we may also consider $\min_{q(\Theta)} KL(p\|q)$ when $q(\Theta)$ is a free of constraint, which leads to $q(\Theta) = p(\Theta|\mathbf{X})$ and

$$\min KL(p\|q), \ KL(p\|q) = \int p(\Theta|\mathbf{X})p(\mathbf{X}) \ln \left[p(\mathbf{X})/q(\mathbf{X}|\Theta) \right] d\mathbf{X}d\Theta. \qquad (40)$$

When $p(\mathbf{X})$ is an input irrelevant to $q(\mathbf{X}|\Theta)$, it is equivalent to

$$\max \int p(\Theta|\mathbf{X})p(\mathbf{X}) \ln q(\mathbf{X}|\Theta) d\mathbf{X}d\Theta, \qquad (41)$$

which further becomes $\max \int p(\Theta|\mathcal{X}_N) \ln q(\mathcal{X}_N|\Theta)d\Theta$ for $p(\mathbf{X})$ of the choice (A) in Tab.2. Its maximization with a structural free $p(\Theta|\mathbf{X})$ leads to the classical ML learning again. Moreover, in help of eq.(10), we are again lead to eq.(31), i.e., the ML learning based AIC [1,2].

Next, we return to eq.(37) with its inner representation $\mathbf{Y}$ in consideration. When $p(\mathbf{Y}|\mathbf{X}, \theta_{y|x})$ is free, $\min_{p(\mathbf{Y}|\mathbf{X},\theta_{y|x})} KL(p\|q)$ leads to eq.(38) again with

$$p(\mathbf{Y}|\mathbf{X}, \theta_{y|x}) = q(\mathbf{X}|\mathbf{Y}, \theta_{x|y})q(\mathbf{Y}|\theta_y)/q(\mathbf{X}|\Theta),$$
$$q(\mathbf{X}|\Theta) = \int q(\mathbf{X}|\mathbf{Y}, \theta_{x|y})q(\mathbf{Y}|\theta_y)d\mathbf{Y}. \qquad (42)$$

In other words, we can integrate over the effect of inner representation $\mathbf{Y}$ to get $q(\mathbf{X}|\Theta)$ and then handle it by eq.(38).

On the other hand, $\min_{q(\Theta)} KL(p\|q)$ with a free $q(\Theta)$ results in $q(\Theta) = p(\Theta|\mathbf{X})$ and also

$$\min KL(p\|q) = \int p(\Theta|\mathbf{X})KL(p\|q, \Theta)d\Theta \geq \min_f KL(p\|q, \Theta).$$

$$KL(p\|q,\Theta) = \int p(\mathbf{Y}|\mathbf{X},\theta_{y|x})p(\mathbf{X})\ln\frac{p(\mathbf{Y}|\mathbf{X},\theta_{y|x})p(\mathbf{X})}{q(\mathbf{X}|\mathbf{Y},\theta_{x|y})q(\mathbf{Y}|\theta_y)}d\mathbf{X}d\mathbf{Y}. \quad (43)$$

This $\min_f KL(p\|q,\Theta)$ was originally proposed in 1995 under the name Bayesian Kullback Ying Yang (BKYY) learning [64]. From $\min_{p(\mathbf{Y}|\mathbf{X},\theta_{y|x})} KL(p\|q,\Theta)$, we are lead to the above discussed eq.(42) and eq.(41) again.

The difference between the best Ying Yang matching by eq.(37) and the best Ying Yang harmony learning by eq.(3) can be better understood from the following relation:

$$KL(p\|q) = \int p(\mathbf{R}|\mathbf{X})p(\mathbf{X})\ln[p(\mathbf{R}|\mathbf{X})p(\mathbf{X})]d\mathbf{X}d\mathbf{R} - H(p\|q). \quad (44)$$

In addition to maximizing $H(p\|q)$, minimizing $KL(p\|q)$ also includes minimizing the first term that is the negative entropy of the Yang representation, which cancels out the least complexity nature that was discussed after eq.(3). Therefore, the best Ying Yang harmony learning by eq.(3) considerably outperforms the best Ying Yang matching learning by eq.(37) for learning tasks that need model selection, as already verified by a number of experimental comparisons and applications [35].

3.3 BKYY Learning, Helmholtz Machine, and Variational Approach

Aiming at avoiding the integral $q(\mathbf{X}|\Theta) = \int q(\mathbf{X}|\mathbf{Y},\theta_{x|y})q(\mathbf{Y}|\theta_y)d\mathbf{Y}$ for the ML learning, maximizing the likelihood function is suggested to be replaced by maximizing one of its lower bound via the *Helmholtz free energy or called variational free energy* [11,24], by which $\max_\Theta q(\mathbf{X}|\Theta)$ is replaced by maximizing

$$\begin{aligned}
F &= -\int p(\mathbf{Y}|\mathcal{X}_N,\theta_{y|x})\ln[p(\mathbf{Y}|\mathcal{X}_N,\theta_{y|x})/q(\mathcal{X}_N|\mathbf{Y},\Theta)q(\mathbf{Y}|\theta_y)]d\mathbf{Y} \\
&= -\int p(\mathbf{Y}|\mathcal{X}_N,\theta_{y|x})\ln\frac{p(\mathbf{Y}|\mathcal{X}_N,\theta_{y|x})}{q(\mathbf{Y}|\mathcal{X}_N,\Theta)}d\mathbf{Y} + \ln q(\mathcal{X}_N|\Theta) \le \ln q(\mathcal{X}_N|\Theta), \\
q(\mathbf{Y}|\mathcal{X}_N,\Theta) &= q(\mathcal{X}_N|\mathbf{Y},\theta_{x|y})q(\mathbf{Y}|\theta_y)/q(\mathcal{X}_N|\Theta). \quad (45)
\end{aligned}$$

Instead of computing $q(\mathcal{X}_N|\Theta)$ and $q(\mathbf{Y}|\mathcal{X}_N,\Theta)$, a parametric model is considered for $p(\mathbf{Y}|\mathcal{X}_N,\theta_{y|x})$, and learning is made for determining the unknown parameters $\theta_{y|x}$ together with Θ via maximizing F.

In fact, maximizing F by eq.(45) is equivalent to $\min_f KL(p\|q,\Theta)$ by eq.(43) with $p(\mathbf{X})$ in the choice (A) of Tab.2. In other words, the two approaches coincide in this situation, though they were motivated from two different perspectives. Maximizing F by eq.(45) directly aims at approximating the ML learning on $q(\mathcal{X}_N|\Theta)$, with an approximation gap to trade off computational efficiency via a pre-specified parametric $p(\mathbf{Y}|\mathcal{X}_N,\theta_{y|x})$. This gap disappears if $p(\mathbf{Y}|\mathcal{X}_N,\theta_{y|x})$ is able to reach the posteriori $q(\mathbf{Y}|\mathcal{X}_N,\Theta)$. Instead, minimizing $KL(p\|q,\Theta)$ by eq.(43) is not motivated from a purpose of approximating the ML learning though it was also shown in [64] that $\min_{p(\mathbf{Y}|\mathbf{X},\theta_{y|x})} KL(p\|q,\Theta)$ for a $p(\mathbf{Y}|\mathbf{X},\theta_{y|x})$ free of constaint makes $\min_f KL(p\|q,\Theta)$ become the ML learning with $p(\mathbf{X})$ in the choice (A) of Tab.2. The motivation is determining all the unknowns in the Ying-Yang pair to make the pair best matched. Beyond becoming equivalent to the ML learning and approximating the ML learning, studies on $\min_f KL(p\|q, \Theta)$ by eq.(43) covers not only extensions to the general case with $p(\mathbf{X}, h)$ by eq.(19), but also the problems of minimizing $KL(p\|q, \Theta)$ with respect to a free $q(\mathbf{X}|\mathbf{Y}, \theta_{x|y})$, which leads to

$$\min \int p(\mathbf{Y}|\Theta_p) \ln \frac{p(\mathbf{Y}|\Theta_p)}{q(\mathbf{Y}|\theta_y)} \mu(d\mathbf{Y}), \ \ p(\mathbf{Y}|\Theta_p) = \int p(\mathbf{Y}|\mathbf{X}, \theta_{y|x}) p(\mathbf{X}) \mu(d\mathbf{X}). \ (46)$$

If $q(\mathbf{Y}|\theta_y)$ is independent among its components and $p(\mathbf{Y}|\mathbf{X}, \theta_{y|x})$ has a postlinear structure, eq.(46) becomes equivalent to the minimum mutual information (MMI) base ICA learning [3]. The details are referred to [64,56,51,52].

In the past decade, extensive studies have also been made under the name of variational approximation methods [20,19], which further put the basic idea of the *Helmholtz free energy* [11,24] in a general framework of approximation methods rooting from techniques in the calculus of variations and with a wide variety of uses [33]. The key idea is turning a complex problem into a simpler one, featured by a decoupling of the degrees of freedom in the original problem. This decoupling is achieved via an expansion of the problem to include additional parameters (called variational parameters), in help of convex duality [31]. The variational approximation method revisits the *Helmholtz free energy* approach under the formulation of probability theory, in a sense that $p(\mathbf{Y}|\mathcal{X}_N, \theta_{y|x})$ is used as an additional parameter to turn the problem of the integral $q(\mathbf{X}|\Theta) = \int q(\mathbf{X}|\mathbf{Y}, \theta_{x|y}) q(\mathbf{Y}|\theta_y) d\mathbf{Y}$ into eq.(45).

3.4   A Relationship Map

A summary of the BYY learning related approaches is provided in Fig.6 under the principles of best harmony versus best matching, as well as their relations to typical learning approaches.

The common part of all the approaches is the shadowed center area, featured by using a probabilistic model to best match a data set $\mathcal{X}_N$ via determining three levels of its unknowns. The first two levels are the ML learning for unknown parameter learning and model selection shown in the ML row of Tab.1, which has been widely studied from various perspective as previously discussed in Sec. 1 [34,23,21,22]. The third level is evaluating or selecting an appropriate meta structure $\aleph$ via $q(\mathcal{X}_N|\aleph)$, i.e., the second term in eq.(37), for which few studies have been made yet but deserve to explore.

Outbound from this shadowed center we have two directions. One is to the left-side. Priori probabilities are taken in consideration for determining three levels of its unknowns. The first two levels are the MB choices for parameter learning and model selection in Tab.1. As discussed in Sec. 1, studies have made under the name of Bayesian learning or Bayesian approach [25,32], as well as MML [42]. The third level is again evaluating an appropriate meta structure $\aleph$ via $q(\mathcal{X}_N|\aleph)q(\aleph)$ with a priori $q(\aleph)$ in consideration. Moving forward even left,

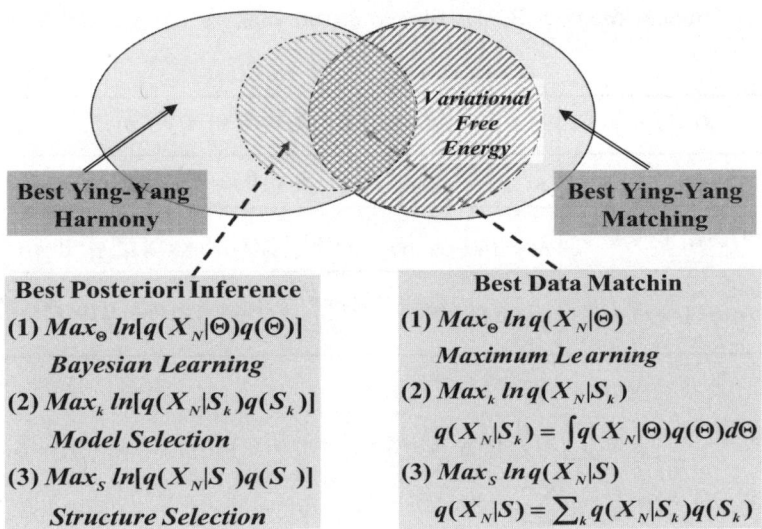

Fig. 6. Best harmony, Best matching, and Typical learning approaches

we are lead to those areas of the best Ying Yang harmony learning by eq.(3), which includes but goes beyond the areas of the ML and MB approaches, as already discussed in Sec.3.1.

The second direction goes the right-side, the domain of the best Ying Yang matching by eq.(37). Out of the shadowed center, we enter the common area shared with the approach of *variational free energy* or the Helmholtz machine [11,24]. Moving right still, we proceed beyond and lead to a number of other cases, as already discussed in Sec.3.2.

In addition to the mathematical relation by eq.(44), the difference between the best Ying Yang matching and the best Ying Yang harmony can also be understood from a best information transfer perspective and a projection geometry perspective. The details are referred to Section II(C) and Section III of [51], respectively. Other discussions on relations and differences are further referred to several recent papers [47,46,49].

4   Gaussian Manifold Based Systems, Typical Applications, and Concluding Remarks

One common structural feature shared by those structures in Tab.3 & Tab.4 is that each of them actually describes samples in the space of x via a number of Gaussian manifolds in certain organization. Shown in Tab.6 are three examples of mixtures of Gaussian manifolds, by combining $q(x|y) = q(x|y, \theta_{x|y}) = G(x|A_\ell y + \mu_\ell, \Sigma_\ell)$ with three different types of $q(y) = q(y|\theta_y)$. Taking the LFA case as an example, it follows from eq.(11), eq.(12), and eq.(13) that we have

Table 6. Gaussian mixture (GM), Binary factor analysis (BFA), and Local factor analysis (LFA)

$q(y\|\theta_y)$	$q(x\|y,l,\theta_{x\|y,l}) = G(x\|A_l y + \mu_l, \Sigma_l)$
$q(l) = \alpha_l \geq 0$	At $A_\ell = 0$ we get **GM** $\quad q(x\|\theta) = \sum_{\ell=1}^{k} \alpha_\ell G(x\|\mu_\ell, \Sigma_\ell)$
$q(y) = \prod_{j=1}^{m} q_j^{y^{(j)}} (1-q_j)^{1-y^{(j)}}$	At k=1 we get **BFA** $\quad q(x\|\theta) = \sum_y G(x\|Ay+\mu,\Sigma) \prod_{j=1}^{m} q_j^{y^{(j)}} (1-q_j)^{1-y^{(j)}}$
$q(y,l) = \alpha_l \prod_{j=1}^{m_l} G(y^{(j)}\|0,\lambda_l^{(j)})$	we get **LFA** $\quad q(x\|\theta) = \sum_{\ell=1}^{k} \alpha_\ell \int G(x\|A_l\bar{y}+\mu_l,\Sigma_l) \prod_{j=1}^{m_l} G(y^{(j)}\|0,\lambda_l^{(j)}) dy$

$$H_f(\mathcal{X}_N, \Theta, \mathbf{k}, \Xi) = \sum_t H_f^{(t)}(\mathcal{X}_N, \Theta, h, \mathbf{k}, \Xi) + \ln q(\Theta|\Xi) + \ln q(h|\mathcal{X}_N),$$

$$H_f^{(t)}(\mathcal{X}_N, \Theta, h, \mathbf{k}, \Xi) = \sum_\ell p(\ell|x_t) \ln \left[G(x|A_\ell y(x_t) + \mu_\ell, \Sigma_\ell) G(y(x_t)|0, \Lambda_\ell) \alpha_\ell \right]$$

$$-0.5 \sum_\ell p(\ell|x_t) \{ m_\ell + [y(x_t) - y*(x_t)]^T [\Lambda_\ell^{-1} + A_\ell^T \Sigma_\ell^{-1} A_\ell][y(x_t) - y*(x_t)] \},$$

$$-0.5 h^2 \sum_\ell p(\ell|x_t) Tr[\Sigma_\ell^{-1}], \quad \Theta = \{ A_\ell^T A_\ell = I, \mu_\ell, \Sigma_\ell, \Lambda_\ell, \alpha_\ell, W_\ell, w_\ell, b_\ell, c_\ell \}_{\ell=1}^{k},$$

$$y(x) = W_\ell x + w_\ell, \quad y*(x) = [\Lambda_\ell^{-1} + A_\ell^T \Sigma_\ell^{-1} A_\ell]^{-1} A_\ell^T \Sigma_\ell^{-1}(x - \mu_\ell), \qquad (47)$$

$$p(\ell|x_t, \theta_{y|x}) = \frac{e^{-o_\ell(x_t)}}{\sum_{j=1}^{k} e^{-o_j(x_t)}}, \quad o_\ell(x) = \beta x^T [\Sigma_\ell + A_\ell \Lambda_\ell A_\ell^T]^{-1} x + b_\ell^T x + c_\ell.$$

where $p(\ell|x_t, \theta_{y|x})$ is given by eq.(27) with $H_\ell = \Sigma_\ell + A_\ell \Lambda_\ell A_\ell^T$, and $q(h|\mathcal{X}_N)$ is given by eq.(29). From the above $H_f(\mathcal{X}_N, \Theta, \mathbf{k}, \Xi)$, we can develop a gradient based adaptive algorithm to implement learning by eq.(6). Moreover, we can also get $J(\mathbf{k})$ for Stage II in eq.(7)(e.g., see eqn.(86) in [46]).

Using Gaussian manifold based systems, computational feasibility is guaranteed by the fact that the integral over y is analytically solvable, scalability of complicated problems is obtained via increasing the number of Gaussian manifolds in consideration, and coverage of typical learning tasks is achieved via different ways that organize Gaussian manifolds. Moreover, the scale $\mathbf{k}_Y$ of Gaussian manifold based systems is simply featured by the variance of a Gaussian variable in Y and the probability of a discrete variable in Y, which ensures scale reducibility of $q(\mathbf{Y}|\theta_y)$. Due to these natures, Gaussian manifold based systems have been applied to various tasks. Several examples are listed below:

- Cluster analysis, Gaussian mixture, and mixture of shape-structures (including lines, planes, curves, surfaces, and even complicated shapes) [63,60,57,56,55,46].
- Factor analysis (FA) and local FA, including PCA, subspace analysis and local subspaces, etc [57,36,35,18,17].

- Independent subspace analysis, including independence components analysis (ICA), binary factor analysis (BFA), nonGaussian factor analysis (NFA), and LMSER, as well as three layer net [56,54,51,53,4].
- Independent state space analysis, including temporal factor analysis (TFA), independent hidden Markov model (HMM), temporal LMSER, and variants [61,59,56,50].
- Combination of multiple inference, including multiple classifier combination, RBF nets, mixture of experts, etc [62,60,46].

Proposed firstly in 1995 [64], the BYY harmony learning has been systematically developed in the past decade. Studies have demonstrated the feasibility of using BYY system as a general framework for unifying a number of typical learning models and a promising direction of adopting best Ying-Yang harmony as a general theory for parameter learning and model selection. The BYY harmony learning leads to not only a criterion that outperforms existing typical model selection criteria in a two-phase implementation, but also automatic model selection during parameter learning with computing cost saved significantly. Readers are referred to [47,46,49] for a tutorial and recent systematic overviews and also to some earlier papers for a similar purpose [58,51,52]. Moreover, readers are referred to [61,59,50,56] for the studies on the BYY harmony learning with temporal dependences taken in consideration.

Acknowledgement. The work described in this paper was fully supported by a Hong Kon RGC grant Project No: CUHK4173/06E).

References

1. Akaike, H.: A new look at the statistical model identification. IEEE Tr. Automatic Control 19, 714–723 (1974)
2. Akaike, H.: Likelihood of a model and information criteria. Journal of Econometrics 16, 3–14 (1981)
3. Amari, S., Cichocki, A., Yang, H.: A new learning algorithm for blind signal separation. In: Advances in NIPS, vol. 8, pp. 757–763. MIT Press, Cambridge (1996)
4. An, Y.J., et al.: A Comparative Investigation on Model Selection in Independent Factor Analysis. J. Mathematical Modelling and Algorithms 5, 447–473 (2006)
5. Barndorff-Nielson, O.E.: Methods of Information and Exponential Families. Wiley, Chichester (1978)
6. Bourlard, H., Kamp, Y.: Auto-association by multilayer Perceptrons and singular value decomposition. Biological Cybernetics 59, 291–294 (1988)
7. Bozdogan, H.: Model Selection and Akaike's Information Criterion: The general theory and its analytical extension. Psychometrika 52, 345–370 (1987)
8. Bozdogan, H., Ramirez, D.E.: FACAIC: Model selection algorithm for the orthogonal factor model using AIC and FACAIC. Psychometrika 53(3), 407–415 (1988)
9. Brown, L.: Fundamentals of Statistical Exponential Families. Institute of Mathematical Statistics, Hayward, CA (1986)
10. Cavanaugh, J.E.: Unifying the derivations for the Akaike and corrected Akaike information criteria. Statistics & Probability Letters 33, 201–208 (1997)

11. Dayan, P., Hinton, G.E., Neal, R.M., Zemel, R.S.: The Helmholtz machine. Neural Computation 7(5), 889–904 (1995)
12. Gilks, W.R., Richardson, S., Spiegelhakter, D.J.: Markov Chain Monte carlo in Practice. Chapman and Hall, London (1996)
13. Girosi, F., et al.: Regularization theory and neural architectures. Neural Computation 7, 219–269 (1995)
14. Grossberg, S.: Adaptive patten classification and universal recording: I&II. Biological Cybernetics 23, 187–202 (1976)
15. Hinton, G.E., Zemel, R.S.: Autoencoders, minimum description length and Helmholtz free energy. In: Advances in NIPS, vol. 6, pp. 3–10 (1994)
16. Hinton, G.E., Dayan, P., Frey, B.J., Neal, R.N.: The wake-sleep algorithm for unsupervised learning neural networks. Science 268, 1158–1160 (1995)
17. Hu, X.L., Xu, L.: A Comparative Study on Selection of Cluster Number and Local Subspace Dimension in the Mixture PCA Models. In: Wang, J., Yi, Z., Żurada, J.M., Lu, B.-L., Yin, H. (eds.) ISNN 2006. LNCS, vol. 3971, pp. 1214–1221. Springer, Heidelberg (2006)
18. Hu, X.L., Xu, L.: A comparative investigation on subspace dimension determination. Neural Networks 17, 1051–1059 (2004)
19. Jaakkola, T.S.: Tutoiral on variational approximation methods. In: Opper, Saad (eds.) Advanced Mean Field methods: Theory & Pratice, pp. 129–160. MIT Press, Cambridge (2001)
20. Jordan, M., Ghahramani, Z., Jaakkola, T., Saul, L.: Introduction to variational methods for graphical models. Machine Learning 37, 183–233 (1999)
21. Kass, R.E., Raftery, A.E.: Bayes Factors. Journal of the American Statistical Association 90, 773–795 (1995)
22. MacKay, D.J.C.: Information Theory, Inference, and Learning Algorithms. Cambridge University Press, Cambridge (2003)
23. Neath, A.A., Cavanaugh, J.E.: Regression and time series model selection using variants of the Schwarz information criterion. Communications in Statistics A 26, 559–580 (1997)
24. Neal, R., Hinton, G.E.: A view of the EM algorithm that justifies incremental, sparse, and other variants. In: Jordan, M.I. (ed.) Learning in graphical models, pp. 355–368. MIT Press, Cambridge (1999)
25. Press, S.J.: Bayesian statistics: principles, models, and applications. Factors. John Wiley & Sons, Inc., Chichester (1989)
26. Poggio, T., Girosi, F.: Networks for approximation and learning. Proc. of IEEE 78, 1481–1497 (1990)
27. Redner, R.A., Walker, H.F.: Mixture densities, maximum likelihood, and the EM algorithm. SIAM Review 26, 195–239 (1984)
28. Rissanen, J.: Stochastic complexity and modeling. Annals of Statistics 14(3), 1080–1100 (1986)
29. Rissanen, J.: Stochastic Complexity in Statistical Inquiry. World Scientific, Singapore (1989)
30. Rivals, I., Personnaz, L.: On Cross Validation for Model Selection. Neural Computation 11, 863–870 (1999)
31. Rockafellar, R.: Convex Analysis. Princeton University Press, Princeton (1972)
32. Ruanaidh, O., Joseph, J.K.: Numerical Bayesian methods applied to signal processing. Springer, New York (1996)
33. Rustagi, J.: Variational Method in Statistics. Academic Press, New York (1976)
34. Schwarz, G.: Estimating the dimension of a model. Annals of Statistics 6, 461–464 (1978)

35. Shi, L.: Bayesian Ying-Yang harmony learning for local factor analysis: a comparative investigation. In: Tizhoosh, Ventresca (eds.) Oppositional Concepts in Computational Intelligence (Studies in CI). Springer, Heidelberg (2008)
36. Shi, L., Xu, L.: Local Factor Analysis with Automatic Model Selection: A Comparative Study and Digits Recognition Application. In: Kollias, S., Stafylopatis, A., Duch, W., Oja, E. (eds.) ICANN 2006. LNCS, vol. 4132, pp. 260–269. Springer, Heidelberg (2006)
37. Stone, M.: Cross-validatory choice and assessment of statistical prediction. J. Royal Statistical Society B 36, 111–147 (1974)
38. Stone, M.: An asymptotic equivalence of choice of model by cross-validation and Akaike's criterion. J. Royal Statistical Society B 39(1), 44–47 (1977)
39. Stone, M.: Cross-validation: A review. Math. Operat. Statist. 9, 127–140 (1978)
40. Tikhonov, A.N., Arsenin, V.Y.: Solutions of Ill-posed Problems. Winston and Sons (1977)
41. Vapnik, V.N.: The Nature Of Statistical Learning Theory. Springer, Heidelberg (1995)
42. Wallace, C.S., Boulton, D.M.: An information measure for classification. Computer Journal 11, 185–194 (1968)
43. Wallace, C.S., Freeman, P.R.: Estimation and inference by compact coding. J. of the Royal Statistical Society 49(3), 240–265 (1987)
44. Wang, L., Feng, J.: Learning Gaussian mixture models by structural risk minimization. In: Proc. ICMLC 2005, August 19-21, Guangzhou, China, pp. 4858–4863 (2005)
45. Xu, L.: Machine learning problems from optimization perspective. Journal of Global Optimization (to appear, 2008)
46. Xu, L.: A unified perspective and new results on RHT computing, mixture based learning, and multi-learner based problem solving. Pattern Recognition 40, 2129–2153 (2007)
47. Xu, L.: Bayesian Ying Yang learning. Scholarpedia 2(3), 1809 (2007), http://scholarpedia.org/article/BayesianYingYangLearning
48. Xu, L.: Rival penalized competitive learning. Scholarpedia 2(8), 1810 (2007), http://scholarpedia.org/article/RivalPenalizedCompetitiveLearning
49. Xu, L.: A trend on regularization and model selection in statistical learning: a Bayesian Ying Yang learning perspective. In: Duch, W., Mandziuk, J. (eds.) Challenges for Computational Intelligence, pp. 365–406. Springer, Heidelberg (2007)
50. Xu, L.: Temporal BYY encoding, Markovian state spaces, and space dimension determination. IEEE Tr. Neural Networks 15, 1276–1295 (2004)
51. Xu, L.: Advances on BYY harmony learning: information theoretic perspective, generalized projection geometry, and independent factor auto-determination. IEEE Tr. Neural Networks 15, 885–902 (2004)
52. Xu, L.: Bayesian Ying Yang learning (I) & (II). In: Zhong, Liu (eds.) Intelligent Technologies for Information Analysis, pp. 615–706. Springer, Heidelberg (2004)
53. Xu, L.: BI-directional BYY learning for mining structures with projected polyhedra and topological map. In: Lin, Smale, Poggio, Liau (eds.) Proc. of FDM 2004: Foundations of Data Mining, Brighton, UK, pp. 5–18 (2004)
54. Xu, L.: BYY learning, regularized implementation, and model selection on modular networks with One hidden layer of binary units. Neurocomputing 51, 227–301 (2003)
55. Xu, L.: Data smoothing regularization, multi-sets-learning, and problem solving strategies. Neural Networks 15(5-6), 817–825 (2003)

56. Xu, L.: Independent component analysis and extensions with noise and time: a Bayesian Ying-Yang learning perspective. Neural Information Processing Letters and Reviews 1, 1–52 (2003)
57. Xu, L.: BYY harmony learning, structural RPCL, and topological self-organizing on unsupervised and supervised mixture models. Neural Networks 15, 1125–1151 (2002)
58. Xu, L.: Bayesian Ying Yang harmony learning. In: Arbib, M.A. (ed.) The Handbook of Brain Theory and Neural Networks, pp. 1231–1237. The MIT Press, Cambridge (2002)
59. Xu, L.: BYY harmony learning, independent state space and generalized APT financial analyses. IEEE Tr. Neural Networks 12, 822–849 (2001)
60. Xu, L.: Best harmony, unified RPCL and automated model selection for unsupervised and supervised learning on Gaussian mixtures, ME-RBF models and three-layer nets. Intl J. Neural Systems 11, 3–69 (2001)
61. Xu, L.: Temporal BYY learning for state space approach, hidden Markov model and blind source separation. IEEE Tr. on Signal Processing 48, 2132–2144 (2000)
62. Xu, L.: RBF nets, mixture experts, and Bayesian Ying-Yang learning. Neurocomputing 19(1-3), 223–257 (1998)
63. Xu, L.: Bayesian Ying-Yang machine, clustering and number of clusters. Pattern Recognition Letters 18(11-13), 1167–1178 (1997)
64. Xu, L.: Bayesian-Kullback coupled YING-YANG machines: unified learnings and new results on vector quantization. In: Proc. ICONIP 1995, Beijing, October 30-November 3, pp. 977–988 (1995)
65. Xu, L., Krzyzak, A., Oja, E.: Rival penalized competitive learning for clustering analysis, RBF net and curve detection. IEEE Tr. on Neural Networks 4, 636–649 (Its early version on In: Proc. of 11th ICPR92. vol.I, pp. 672–675 (1992& 1993))
66. Xu, L.: Least mean square error reconstruction for self-organizing neural-nets. Neural Networks 6, 627–648 (1993) (Its early version on Proc. IJCNN 1991 Singapore. pp. 2363–2373 (1991& 1993))

The Berlin Brain-Computer Interface*

Benjamin Blankertz[1,2,**], Michael Tangermann[1], Florin Popescu[2],
Matthias Krauledat[1], Siamac Fazli[2], Márton Dónaczy[2],
Gabriel Curio[3], and Klaus-Robert Müller[1,2]

[1] Berlin Institute of Technology, Machine Learning Laboratory, Berlin, Germany
blanker@cs.tu-berlin.de
[2] Fraunhofer FIRST (IDA), Berlin, Germany
[3] Campus Benjamin Franklin, Charité University Medicine Berlin, Germany

1 Introduction

The Berlin Brain-Computer Interface (BBCI) uses a machine learning approach to extract subject-specific patterns from high-dimensional EEG-features optimized for revealing the user's mental state. Classical BCI application are brain actuated tools for patients such as prostheses (see Section 4.1) or mental text entry systems ([2] and see [3,4,5,6] for an overview on BCI). In these applications the BBCI uses natural motor competences of the users and specifically tailored pattern recognition algorithms for detecting the user's intent. But beyond rehabilitation, there is a wide range of possible applications in which BCI technology is used to monitor other mental states, often even covert ones (see also [7] in the fMRI realm). While this field is still largely unexplored, two examples from our studies are exemplified in Section 4.3 and 4.4.

1.1 The Machine Learning Approach

The advent of machine learning (ML) in the field of BCI has led to significant advances in real-time EEG analysis. While early EEG-BCI efforts required neurofeedback training on the part of the user that lasted on the order of days, in ML-based systems it suffices to collect examples of EEG signals in a so-called *calibration measurement* during which the user is cued to perform repeatedly anyone of a small set of mental tasks. This data is used to adapt the system to the specific brain signals of each user (*machine training*). This step of adaption seems to be instrumental for effective BCI performance due to a large inter-subject variability with respect to the brain signals ([8]). After this preparation step, which is very short compared to the subject training in the operant conditioning approach ([9,10]), the feedback application can start. Here, the users can actually transfer information through their brain activity and control applications. In this phase, the system is composed of the classifier that discriminates between different mental states and the control logic that translates the classifier output into control signals, e.g., cursor position or selection from an alphabet.

---

* This paper is a copy of the manuscript submitted to appear as [1].
** Corresponding author.

J.M. Zurada et al. (Eds.): WCCI 2008 Plenary/Invited Lectures, LNCS 5050, pp. 79–101, 2008.
© Springer-Verlag Berlin Heidelberg 2008

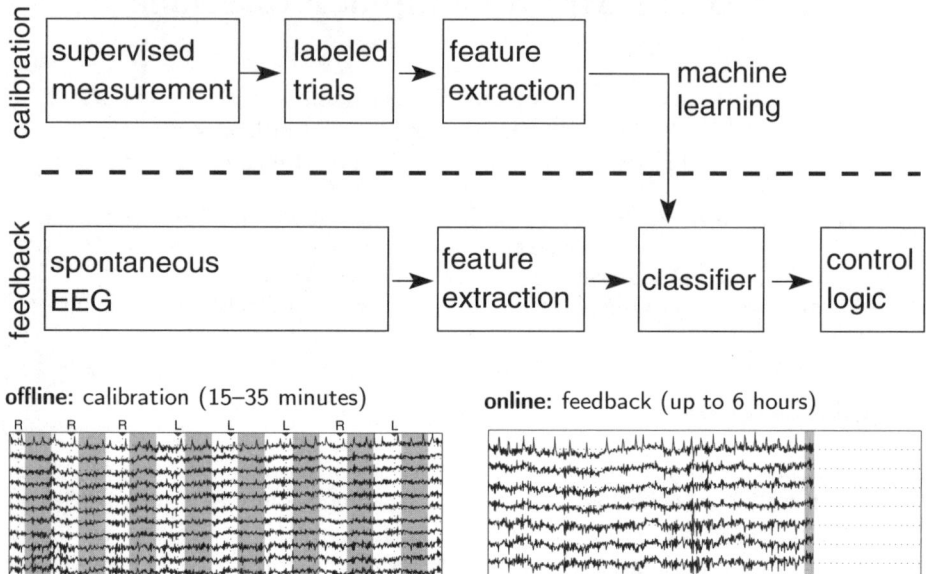

Fig. 1. Overview of the machine-learning-based BCI system. The system runs in two phases. In the calibration phase, we instruct the subjects to perform certain tasks and collect short segments of labeled EEG (trials). We train the classifier based on these examples. In the feedback phase, we take sliding windows from continuous stream of EEG; the classifier outputs a real value that quantifies the likeliness of class membership; we run a feedback application that takes the output of the classifier as input. Finally the subject receives the feedback on the screen as, e.g., cursor control.

An overview of the whole process in an ML-based BCI is sketched in Fig. 1. Note that in alternative applications of BCI technology (see Section 4.3 and 4.4), the calibration may need novel nonstandard paradigms, as the sought-after mental states (like lack of concentration, specific emotions, workload) might be difficult to induce in a controlled manner.

1.2   Neurophysiological Features

Readiness Potential. Event-related potentials (ERPs) are transient brain responses that are time-locked to some event. This event may be an external sensory stimulus or an internal state signal, associated with the execution of a motor, cognitive, or psychophysiologic task. Due to simultaneous activity of many sources in the brain, ERPs are typically not visible in single trials (i.e., the segment of EEG related to *one* event) of raw EEG. For investigating ERPs, EEG is acquired during many repetitions of the event of interest. Then short segments (called epochs or trials) are cut out from the continuous EEG signals around each event and are averaged across epochs to reduce event-unrelated background activity. In BCI applications based on ERPs, the challenge is to detect ERPs in single trials.

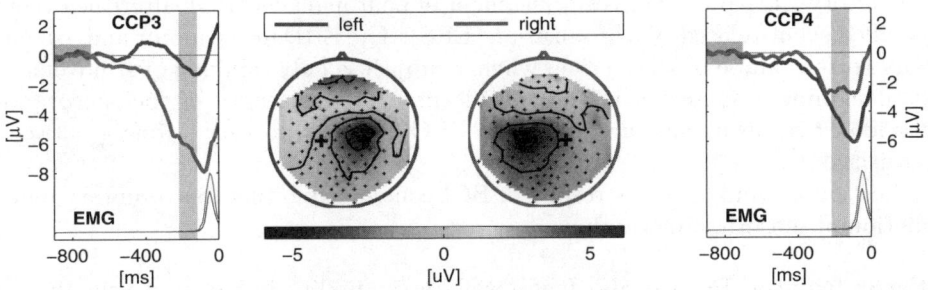

Fig. 2. Response *averaged* event-related potentials (ERPs) of a right-handed subject in a left vs. right hand finger tapping experiment (N =275 resp. 283 trials per class). Finger movements were executed in a self-paced manner, i.e., without any external cue, using an approximate inter-trial interval of 2 seconds. The two scalp plots show the topographical mapping of scalp potentials averaged within the interval -220 to -120 ms relative to keypress (time interval vertically shaded in the ERP plots; initial horizontal shading indicates the baseline period). Larger crosses indicate the position of the electrodes CCP3 and CCP4 for which the ERP time course is shown in the subplots at both sides. For comparison time courses of EMG activity for left and right finger movements are added. EMG activity starts after -120 ms and reaches a peak of 70 µV at -50 ms. The readiness potential is clearly visible, a predominantly contralateral negativation starting about 600 ms before movement and raising approximately until EMG onset.

The *readiness potential* (RP, or Bereitschaftspotential) is an ERP that reflects the intention to move a limb, and therefore precedes the physical (muscular) initiation of movements. In the EEG it can be observed as a pronounced cortical negativation with a focus in the corresponding motor area. In hand movements the RP is focussed in the central area contralateral to the performing hand, cf. [11,12,13] and references therein for an overview. See Fig. 2 for an illustration. Section 4.2 shows an application of BCI technology using the readiness potential. Further details about our BCI-related studies involving RP can be found in [14,8,15,16].

Sensorimotor Rhythms. Apart from transient components, EEG comprises rhythmic activity located over various areas. Most of these rhythms are so-called idle rhythms, which are generated by large populations of neurons in the respective cortex that fire in rhythmical synchrony when they are not engaged in a specific task. Over motor and sensorimotor areas in most subjects oscillations with a fundamental frequency between 9 and 13 Hz can be observed, the so called µ-rhythm. Due to its comb-shape, the µ-rhythm is composed of several harmonics, i.e., components of double and sometimes also triple the fundamental frequency ([17]) with a fixed phase synchronization, cf. [18]. These sensorimotor rhythms (SMRs) are attenuated when engagement with the respective limb takes place. As this effect is due to loss of synchrony in the neural populations, it is termed event-related desynchronization (ERD), see [19]. The increase of oscillatory EEG (i.e., the reestablishment of neuronal synchrony after the event) is called event-related synchronization (ERS). The ERD in the motor and/or sensory cortex can be observed even when a subject is only thinking of a movement or imagining a sensation in the specific limb. The strength of the sensorimotor idle rhythms as measured by scalp EEG is known to vary strongly between subjects.

Section 3.1 and 3.2 show results of BCI control exploiting the voluntary modulation of sensorimotor rhythm.

Error-Related Potentials. It is a well-known finding in human psychophysics that a subject's recognition of having committed a response error is accompagnied by specific EEG variations that can be observed in (averaged) ERPs (e.g. [20]). The ERP after an error trial is characterized by two components: a negative wave called error negativity (N_E) [21] (or error-related negativity (ERN, [22])) and a following broader positive peak labeled as error positivity (P_E), [20]. It has been demonstrated that the P_E is more specific to errors while the N_E can also be observed in correct trials, cf. [20], [23]. Although both amplitude and latency depend on the specific task, the N_E occurs delayed and less intense in correct trials than in error trials. The N_E is also elicited by negative feedback ([24]) and by error observation ([25]). Furthermore [26] investigated error-related potentials in response to errors that are made by an interface in human-computer interaction.

Section 3.3 investigates the detectability of error-related potentials after erroneous BCI feedback, which gives a perspective of the potential use in BCI systems as a 'second-pass' response verification.

2   Processing and Machine Learning Techniques

Due to the simlutaneous activity of many sources in the brain and additional influence by noise the detection of relevant components of brain activity in single trials as required for BCIs is a data analytical challenge. One approach to compensate for the missing opportunity to average across trials is to record brain activity from many sensors and to exploit the multi-variateness of the acquired signals, i.e., to average across space in an intelligent way. Raw EEG scalp potentials are known to be associated with a large spatial scale owing to volumne conduction ([27]). Accordingly all EEG channels are highly correlated and powerful spatial filters are required to extract localized information with a good signal to noise ratio (see also the motivation for the need of spatial filtering in [28]).

In the case of detecting ERPs, such as RP or error-related potentials, the extraction of features from one source is mostly done by linear processing methods. In this case the spatial filtering can be acomplished implicitly in the classification step (interchangability of linear processing steps). For the detection of modulations of SMRs, the processing is non-linear (e.g. calculation of band power). In this case, the prior application of spatial filtering is extremely beneficial. The methods used for BCIs range from simple fixed filters like Laplacians ([29]), and data driven unsupervised techniques like independent component analysis (ICA) [30] or model based approaches ([31]) to data driven supervised techniques like common spatial patterns analysis (CSP) [28].

In this Section we summarize the two techniques that we consider most important for classifying multi-variate EEG signals, CSP and regularized linear discriminant analysis. For a more complete and detailed review of signal processing and pattern recognition techniques see [32,33,8].

2.1 Common Spatial Patterns Analysis

The CSP technique (see [34]) allows to determine spatial filters that maximize the variance of signals of one condition and at the same time minimize the variance of signals of another condition. Since variance of band-pass filtered signals is equal to band-power, CSP filters are well suited to detect amplitude modulations of sensorimotor rhythms (see Section 1.2) and consequently to discriminate mental states that are characterized by ERD/ERS effects. As such it has been well used in BCI systems ([35,14]) where CSP filters are calculated individually for each subject on the data of a calibration measurement.

The CSP technique decomposes multichannel EEG signals in the sensor space. The number of spatial filters equals the number of channels of the original data. Only few filters have properties that make them favorable of classification. The discriminative value of a CSP filter is quantified by its generalized eigenvalue. This eigenvalue is relative to the sum of the variances in both conditions. An eigenvalue of 0.9 for class 1 means an average ratio of 9:1 of variances during condition 1 and 2. See Fig. 3 for an illustration of CSP filtering.

For details on the technique of CSP analysis and its extensions we refer to ([28,36,37,38,39]).

2.2 Regularized Linear Classification

For known Gaussian distributions with the same covariance matrix for all classes, it can be shown that Linear Discriminant Analysis (LDA) is the optimal classifier in the sense that it minimizes the risk of misclassification for new samples drawn from the same distributions ([40]). Note that LDA is equivalent to Fisher Discriminant and Least Squares Regression ([40]). For EEG classification the assumption of Gaussianity can be achieved rather well by appropriate preprocessing of the data. But the mean and covariance matrix of the distributions have to be estimated from the data, since the true distributions are not known. Especially for high-dimensional data with few trials the estimation of the covariance matrix is very imprecise, because the number of unknown parameters is quadratic in the number of dimensions. In the estimation of covariance matrices this leads to a systematic error: Large eigenvalues of the original covariance matrix are estimated too large, and small eigenvalues are estimated too small, see Fig. 4. This error in the estimation degrades classification performance (and invalidates the optimality statement for LDA). A common remedy for the systemtic bias, is shrinkage of the estimated covariance matrices (e.g. [41]):

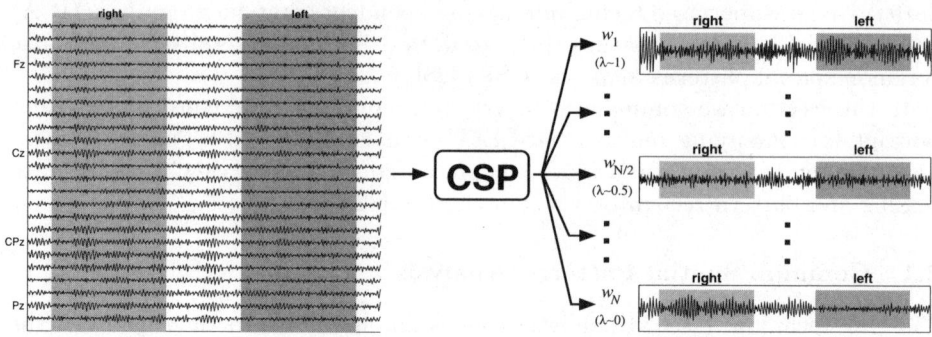

Fig. 3. The input the CSP analysis are (band-pass filtered) multi-channel EEG signals which are recorded for two conditions (here 'left' and 'right' hand motor imagery. The reults of CSP analysis is a sequence of spatial filters. The number of filters (here N) is equal to the number of EEG channels. When these filters are applied to the continuous EEG signals, the (average) relative variance in the two conditions is given by the eigenvalues. An eigenvalue near 1 results in large variance of signals of condition 1 and an eigenvalue near 0 reults in small variance for condition 1. Most eigenvalues are near 0.5 such that the corresponding filters do not contribute to the discrimination.

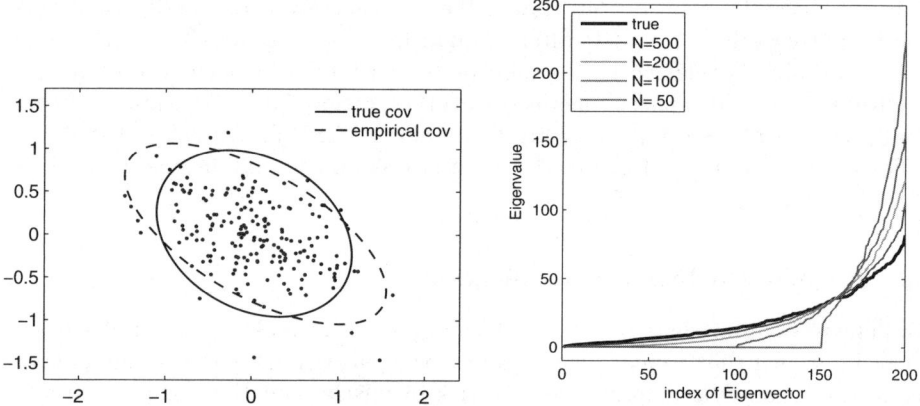

Fig. 4. *Left:* Data points drawn from a Gaussian distribution (gray dots; $d = 200$ dimensions) with true covariance matrix indicated by an ellipsoid in solid line, and estimated covariance matrix in dashed line. *Right:* Eigenvalue spectrum of a given covariance matrix (bold line) and eigenvalue spectra of covarinace matrices estimated from a finite number of samples drawn ($N = 50, 100, 200, 500$) from a corresponding Gaussian distribution.

The estimator of the covariance matrix $\hat{\Sigma}$ is replaced by

$$\tilde{\Sigma} = (1 - \gamma)\hat{\Sigma} + \gamma\lambda\mathbf{I}$$

for a $\gamma \in [0, 1]$ and λ defined as average eigenvalue $\mathrm{trace}(\hat{\Sigma})/d$ with d being the dimensionality of the feature space and $\mathbf{I}$ being the identity matrix.. Then the following holds. Since $\hat{\Sigma}$ is positive semi-definite we can have an eigenvalue decomposition $\hat{\Sigma} = \mathbf{VDV}^\top$ with orthonormal $\mathbf{V}$ and diagonal $\mathbf{D}$. Due to the orthogonality of $\mathbf{V}$ we get

$$\tilde{\Sigma} = (1-\gamma)\mathbf{VDV}^\top + \gamma\lambda\mathbf{I} = (1-\gamma)\mathbf{VDV}^\top + \gamma\lambda\mathbf{VIV}^\top = \mathbf{V}\left((1-\gamma)\mathbf{D} + \gamma\lambda\mathbf{I}\right)\mathbf{V}^\top$$

as eigenvalue decomposition of $\tilde{\Sigma}$. That means

- $\tilde{\Sigma}$ and $\hat{\Sigma}$ have the same Eigenvectors (columns of $\mathbf{V}$)
- extreme eigenvalues (large or small) are modified (shrunk or elongated) towards the average λ.
- $\gamma = 0$ yields unregularized LDA, $\gamma = 1$ assumes spherical covariance matrices.

Using LDA with such modified covariance matrix is termed regularized LDA. The parameter γ needs to be estimated from training data, e.g. by cross validation.

3 BBCI Control Using Motor Paradigms

3.1 High Information Transfer Rates

In order to preserve ecological validity (i.e., the correspondence between intention and control effect) we let the users perform motor tasks for applications like cursor movements. For paralyzed patients the control task is to attempt movements (e.g., left hand or right hand or foot), other subjects are instructed to perform kinesthetically imagined movements ([42]) or quasi-movements ([43]).

As a test application of the performance of our BBCI system we implemented a 1D cursor control. One of the two fields on the left and right edge of the screen was highlighted as target at the beginning of a trial, see Fig. 5. The cursor was initially at the center of the screen and started moving according to the BBCI classifier output about half a second after the indication of the target. The trial ended when the cursor touched one of the two fields. That field was then colored green or red, depending on whether or not it was the correct target. After a short period the next target cue was presented (see [8,44] for more details).

The aim of our first feedback study was to explore the limits of possible information transfer rates (ITRs) in BCI systems not relying on user training or evoked potentials. The ITR derived in Shannon's information theory can be used to quantify the information content, which is conveyed through a noisy (i.e., error introducing) channel. In BCI context:

$$\text{bitrate}(p, N) = \left(p\log_2(p) + (1-p)\log_2\left(\frac{1-p}{N-1}\right) + \log_2(N)\right) \qquad (1)$$

where p is the accuracy of the subject in making decisions between N targets, e.g., in the feedback explained above, $N = 2$ and p is the accuracy of hitting the correct bars. To include the speed of decision into the performance measure:

$$\text{ITR [bits/min]} = \frac{\#\text{ of decisions}}{\text{duration in minutes}} \cdot \text{bitrate}(p, N) \qquad (2)$$

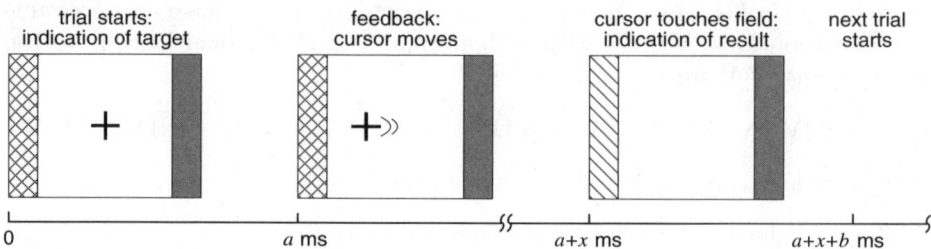

Fig. 5. Course of a feedback trial. The target cue (field with crosshatch) is indicated for a ms, where a is chosen individual according to the capabilities of the user. Then the cursor starts moving according to the BCI classifier until it touches one of the two fields at the edge of the screen. The duration depends on the performance and is therefore different in each trial (x ms). The touched field is colored green or red according to whether its was the correct target or not (for this black and white reproduction, the field is hatched with diagonal lines). After b ms, the next trial starts, where b is chosen indivudally for the subject.

In this form, the ITR takes different average trial durations (i.e., the speed of decisions) and different number of classes into account. Therefore, it is often used as a performance measure of BCI systems ([45]). Note, that it gives reasonable results only if some assumptions on the distribution of error are met, see [46].

The subjects of the study ([8,14]) were 6 staff members, most of which had performed feedback with earlier versions of the BBCI system before. (Later, the study was extended by 4 further subjects, see [44]). First the parameters of preprocessing were selected and a classifier was trained based on a calibration measurement individually for each subject. Then feedback was switched on and further parameters of the feedback were adjusted according to the subject's request.

For one subject, no significant discrimination between the mental imagery conditions was found, see [44] for an analysis of that specific case. The other five subjects performed 8 runs of 25 cursor control trials as explained above. Table 1 shows the performance result in accuracy (percentage of trials in which the subject hit the indicated target) and as ITR (see above). As a test of practical usability, subject al operated a simple text entry system based on BBCI cursor control. In a free spelling mode, he spelled 3 German sentences with a total of 135 characters in 30 minutes, which is a spelling speed of 4.5 letters per minutes. Note that the subject corrected all errors using the deletion symbol. For details, see [47]. Recently, using the novel mental text entry system Hex-o-Spell which was developed in cooperation with the Human-Computer Interaction Group at the University of Glasgow, the same subject achieved a spelling speed of more than 7 letters per minute, cf. [2,48].

3.2   Good Performance without Subject Training

The goal of our second feedback study was to investigate for what proportion of naive subjects our system could provide successful feedback in the very first

Table 1. Results of a feedback study with 6 healthy subjects (identification code in the first column). From the three classes used in the calibration measurement the two chosen for feedback are indicated in second column (L: left hand, R: right hand, F: right foot). The accuracies obtained online in cursor control are given in column 3. The average duration ± standard deviation of the feedback trials is provided in column 4 (duration from cue presentation to target hit). Subjects are sorted according to feedback accuracy. Columns 5 and 6 report the information transfer rates (ITR) measured in bits per minute as obtained by Shannon's formula, cf. (1). Here the complete duration of each run was taken into account, i.e., also the inter-trial breaks from target hit to the presentation of the next cue. The column *overall ITR* (oITR) reports the average ITR of all runs (of 25 trials each), while column *peak ITR* (pITR) reports the peak ITR of all runs.

subject	classes	accuracy [%]	duration [s]	oITR [b/m]	pITR [b/m]
al	LF	98.0 ± 4.3	2.0 ± 0.9	24.4	35.4
ay	LR	95.0 ± 3.3	1.8 ± 0.8	22.6	31.5
av	LF	90.5 ± 10.2	3.5 ± 2.9	9.0	24.5
aa	LR	88.5 ± 8.1	1.5 ± 0.4	17.4	37.1
aw	RF	80.5 ± 5.8	2.6 ± 1.5	5.9	11.0
mean		90.5 ± 7.6	2.3 ± 0.8	15.9	27.9

session ([49]). The design of this study was similar to the one described above. But here the subjects were 14 individuals who never performed in a BCI experiment before. Furthermore the parameters of the feedback have been fixed beforehand for all subjects to conservative values.

For one subject no distinguishable classes were identified. The other 13 subjects performed feedback: 1 near chance level, 3 with 70-80%, 6 with 80-90% and 3 with 90-100% hits. The results of all feedbacks runs are shown in Fig. 6.

This clearly shows that a machine learning based approach to BCI such as the BBCI is able to let BCI novices perform well from the first session. Note that in all BCI studies – independent of whether machine learning is used or not – non-performing subjects are encountered (e.g. [50]). It is an open problem how to alleviate this issue.

3.3 Automatic Response Verification

An elegant approach to cope with BCI misclassifications is a response checking mechanism that is based on the subject's brain signals themselves. This approach was first explored in [51] in an offline analysis of BCI feedback data. A simple amplitude threshold criterium for the detection of error-related potentials was used to demonstrate the potential use of the approach. Several studies have shown the possibility to detect error-related potentials in choice reaction tasks ([52,16,53]) with more advanced pattern recognition algorithms. The results taken together give a clear indication that a response verification might be a worthwhile add-on to BCIs in the following sense of a two-pass system. We

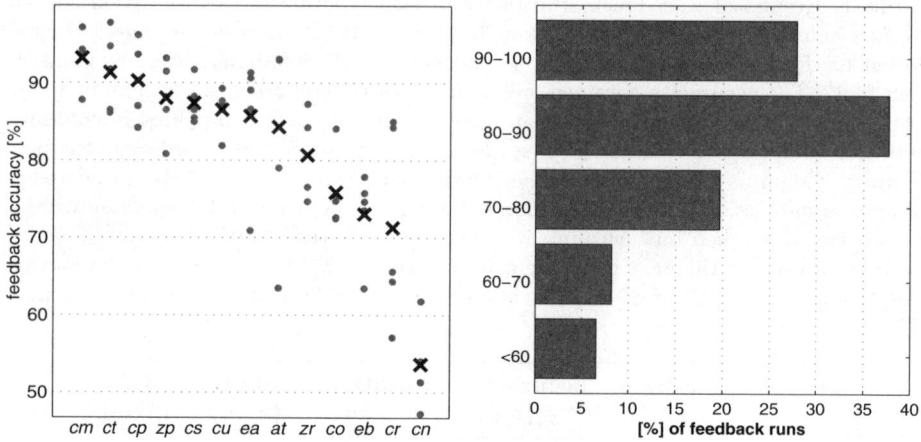

Fig. 6. *Left:* Feedback accuracy of all runs (gray dots) and intra-subject averages (black crosses). *Right:* Histogram of accuracies obtained in BBCI-controlled cursor movement task in all feedback runs of the study.

call the original classification of the BCI feedback first-pass. Then in the second-pass, the interval after the response feedback is subjected to the error potential detector. If that indicates that the user perceived the feedback as an error, the decision is rejected[1] . Surprisingly, so far no online BCI application with error-detection was reported. Nevertheless, further important evidence was provided in [54,26] by showing the detecability of potentials elicited by interaction errors in a simulated BCI. But due to the discrete feedback with fixed timing used in that study, it remains open how the situation would be in a continuous cursor control feedback where an upcoming error might be anticipated by the users by predictions about the cursor movement (e.g., no classical phasic error-related component might be elicited when the cursor starts moving slowly towards the wrong field).

Fig. 7 shows the ERPs for correct and erroneous feedback trials with respect to time point $t = 0$ when the cursor enters either the correct or the wrong field (for the design of the feedback, see Fig. 5). In this subject the error-related positivity as well as the error-related negativity is clearly visible at fronto-central and parieto-central scalp position. In other subjects often only the positive component was observed. It can be speculated that the shorter negative component is obscured by the jitter on the time point of error recognition owing to the feedback paradigm (see remark above). This issue is subject of an ongoing investigation.

In order to quantify the potential gain of an automatic error rejection, we calculate the bitrate of a two-pass BCI system as outlined above. Let tp be the

---

[1] In binary decisions the outcome could even be reverted. But practically it was observed that such a strategy leads to less improvement if the error detection itself is also error prone ([54]).

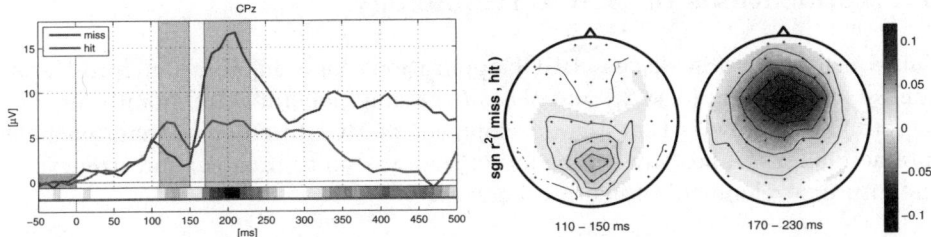

Fig. 7. *Left:* ERPs for correct and erroneous feedback trials. *Right:* Topography of signed r^2 values for the time intervals of error-related negativity (110 to 150 ms) and error-related positivity (170 to 230 ms).

rate of true positives (erroneous trials, classified as errors) and tn the rate of true negatives (correct trials, classified as correct). Then we can calculated the bitrate of a system that rejects trials which were classified as errors in the following way ([54]):

$$r_{\text{accepted}} = p\, tn + (1-p)(1-tp) \qquad \text{rate of accepted trials}$$

$$p_{\text{accepted}} = p\frac{tn}{r_{\text{accepted}}} \qquad \text{accuracy on accepted trials}$$

$$\text{bitrate}_{\text{rv}}(p, tp, tn, N) = \text{bitrate}(p_{\text{accepted}}, N) \tag{3}$$

Fig. 8 shows the improvement in ITR that would have been achieved by using the response verification with rejecting decision for trials which were classified as erroneous. The relative gain obtained through response verification is 80 % on average for the worse performing subjects and 25 % for better performing subjects.

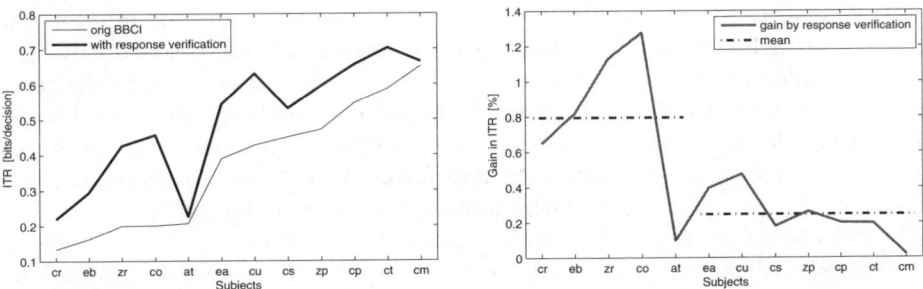

Fig. 8. *Left:* Bitrates (Eq. (1)) of original BCI classification (thin line) and calculated bitrates (Eq. (3)) for the case that trials are rejected which are classified as errors by the error-potential detector (thick line). Only those subjects are taken into this investigation who committed at least 20 error and had above chance performance. *Right:* Relative gain obtained through response verification. The mean for the worse performing subjects is 80 % and the mean for better performing subjects is 25 %.

4 Applications of BBCI Technology

Subsequently we will discuss BBCI applications for rehabilitation (prosthetic control and spelling [3,48,2]) and *beyond* (gaming, mental state monitoring [55, 56] etc.). Our view is that the development of BCI to enhance man machine interaction for the healthy will be an important step to broaden and strengthen the future development of neurotechnology.

4.1 Prosthetic Control

Motor-intention based BCI offers the possibility of a direct and intuitive control modality for persons disabled by high-cervical spinal cord injury, i.e., tetraplegics, whose control of all limbs is severely impaired. The advantage of this type of BCI over other interface modalities is that by directly translating movement intention into a command to a prosthesis, the link between cortical activity related to motor control of the arm and physical action is restored, thereby offering a possible rehabilitation function, as well as enhanced motivation factor for daily use. Testing of this concept is the main idea driving the Brain2Robot project (see Acknowledgement). However, two important challenges must be fully met before non-invasive, EEG based motor imagery BCI can be practically used by the disabled.

One such challenge is the cumbersome nature of standard EEG set-up, involving application of gel, limited recording time, and subsequent removal of the set-up, which involves washing the hair. It is unlikely that disabled persons, in need of BCI technology for greater autonomy, would adopt such a system. Meanwhile, short of any invasive or minimally invasive recording modality, the only available option is the use of so called 'dry' electrodes, i.e. not requiring the use of conductive gel or other liquids in such a way that electrode application and removal takes place in a matter of minutes. We have developed such technology (a 'dry cap') and tested it for motor-imagery based BCI [57]. The cap required about 5 minutes for set-up and exhibited an average of 70 % of the information transfer rate achieved for the same subjects with respect to a standard EEG 'gel cap', the difference being most likely attributed to the use of 6 electrodes used in the dry cap vs. 64 electrodes used in the gel cap. Although the locations of the 6 electrodes were chosen judiciously (by analyzing which electrode positions in the gel cap were most important, as expected 3 electrodes over each cortical motor area), some performance degradation was unavoidable and necessary – a full 64 electrode dry cap would also be cumbersome.

Another challenge for EEG-BCI control of prosthetics is inherent safety. This is of paramount importance, whether the prosthetic controlled is an orthosis (a worn mechanical device which augments the function of a set of joints) or a robot (which may move the paralysed arm or be near the body but unattached to it, as in the case of Brain2Robot), or even a neuroprosthesis, i.e. a system which electrically activates muscles in the user's arm or peripheral neurons which innervate these muscles. Specifically, the BCI interface should not output spurious or unintended action commands to the prosthetic device, as these could cause

injuries, or even in the case in which the probability of injury is low and secondary safety 'escape commands' are incorporated, it may (reasonably) cause
fear in the otherwise immobile user and therefore discourage him or her from
continuing to use the system. Therefore we have looked at necessary enhancements to commonly used 'BCI feedback' control which could incorporate the
use of a 'rest' or 'idle' state, i.e. a continuous output of the classifier which not
only outputs a command related to a trained brain state (say, imagination of
left hand movement) but a 'do nothing' command related to a state in which the
user performs daily activities unrelated to motor imagination (a 'rest' or 'idle'
state) and in which the prosthetic should do nothing. Thus we have begun to
look at the trade-off between speed of BCI (information transmission rate or
ITR) and safety (false positive rate) achievable by incorporating a 'control' law,
which is a differential equation whose inputs are continuous outputs of the clas-
sifer, in our case a quadratic-type classifier, and whose output is the command
to the prosthetic ([58]). It remains to be seen how much each particular subject,
whose 'standard' BCI performance varies greatly, must trade reduced speed for
increased safety.

A final implicit goal of all BCI research is to improve the maximally achievable
ITR for each type of brain imaging modality. In the case of EEG the ITR is
seems to be limited to about 1 decision every 2 seconds ([44], fastest subject
performed at an average speed of 1 binary decision every 1.7 s) despite intensive
research effort to improve it. In the case of Brain2Robot further information
about the desired endpoint of arm movement is obtained by 3D tracking of gaze
– eye movement and focus being normally intact in the tetraplegic population,
and the achievable ITR is sufficient, since it lies in the range of the frequency of
discrete reaching movements of the hand. However, competing issues of cognitive
load, safety and achievable dexterity can only be assessed by testing BCI for
prosthetic control with the intended user group while paying attention to the
level of disability and motor-related EEG patterns in each subject, as both are
likely to vary significantly.

4.2   Time-Critical Applications: Prediction of Upcoming Movements

In time-critical control situations, BCI technology might provide early detection
of reactive movements based on preparatory signals for the reduction of the
time span between the generation of an intention (or reactive movements) and
the onset of the intended technical operation (e.g. in driver-assisted measures for
vehicle safety). Through detection of particularly early readiness potentials (see
Section 1.2) which reflect the mental preparation of movements, control actions
can be prepared or initiated before the actual movement and thus we intend to
decode these signals in a very timely and accurate manner.

In order to explore the prospective value of BCI for such applications, we
conducted a two alternative forced choice experiment (d2-test), in which the
subject had to respond as fast as possible with a left or right index finger key
press, see [59]. Fig. 9 (left) compares the readiness potentials in such reactive
finger movements with those in selfpaced finger movements ($t = 0$ for key press).

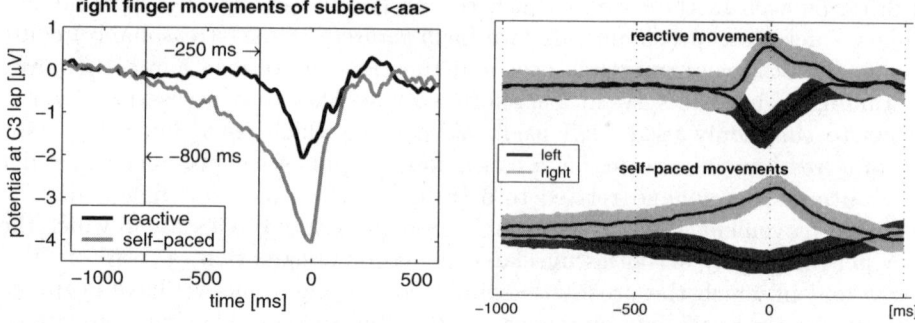

Fig. 9. *Left:* Averaged readiness potential in spontaneous selfpaced (grey) and reactive (dark) finger movements (with $t = 0$ at key press) for one subject. *Right:* Distribution of the continuous classifier output in both experimental settings.

Fig. 9 (right) shows the traces of continuous classifier output for reactive (upper subplot) and selfpaced (lower subplot) finger movements. As expected, the discrimination between upcoming left vs. right finger movements is better possible for the self-paced movements at an *early* stage, but towards the time point of key press performance is similar. In particular, 100 ms before the keypress even for movements in fast reations, a separation becomes substantial. The discriminability already at this point in time confirms the potential value of BCI technology for time-critical applications. For more details and classification results, we refer the interested reader to [59].

4.3   Neuro Usability

In the development of many new products or in the improvement of existing products, usability studies play an important role. They are performed in order to measure to what degree a product meets the intended purpose with regard to the aspects effectiveness, efficiency and user satisfaction. A further goal is to quantify the joy of use. While effectiveness can be quantified quite objectively, e.g., in terms of task completion, the other aspects are more intricate to assess. Even psychic variables consciously unaccessible to the persons themselves might be involved. Furthermore, in usability studies it is of interest to perform an effortless continuous acquistion of usability parameters whilst not requiring any action on the side of the subject as this might interfer with the task at hand. For these reasons, BCI technology could become a crucial tool for usability studies in the future.

We exemplify the potential benefit of BCI technology in one example ([55]). Here, usability of new car features is quantified by the mental workload of the car driver. In the case of a device that uses fancy man-machine interface technology, the producer should demonstrate that it does not distract the driver from the traffic (mental workload is not increased when the feature is used). In case of a tool for which the manufacturer claims it relieves the driver from workload (e.g., automatic distance control), this effect should be demonstrated as objectively as possible.

Since there is no ground truth available on the cognitive workload to which the driver is exposed, we designed a study[2] in which additional workload was induced in a controlled manner. For details, please refer to [55]. EEG was acquired from 12 male and 5 female subjects while driving on a highway at a speed of 100 km/h (primary task). Second, the subjects had an auditory reaction task: one of two buttons mounted on the left and right index finger had to be hit every 7.5 s according to a given vocal prompt. For the tertiary task, two different conditions have been used. (a) mental calculation; (b) following one of two simultaneously broadcast voice recordings. In a first a calibration phase, the developed BBCI workload detector was adapted to the individual driver. After that, the system was able to predict the cognitive workload of the driver online. This information was used in the test phase to switch off the auditory reaction task, when high workload was detected ('mitigation').

As a result of the mitigation strategy, the average reaction time in the test phase was on average 100 ms faster than in the (un-mitigated) calibration phase ([55]). Since in total the workload during the two phases has been equal, it can be conjectured that the average reactivity was the same. Thus, the difference in reaction times can only be explained by the fact that the workload detector switched off the reaction task during periods of reduced reactivity.

Note, that the high intersubject variabiltiy, which is a challenge for many BCI applications comes as an advantage here: for neuro-usability studies, top subjects (with respect to the detectability of relevant EEG components) of a study can be selected according to the appropriateness of their brain signals.

Beyond the neuro usability aspect of the study, one could speculate that such devices might be incorporated in future cars in order to reduce distractions (e.g., navigation system is switched off during periods of high workload) to a minimum when the drivers' brain is already over-loaded by other demands during potentially hazardous situations.

4.4 Mental State Monitoring

When aiming to optimize the design of user interfaces or, more general, of a work flow, the mental state of a user during the task execution can provide useful information. This information can not only be exploited for the improvement of BCI applications, but also for improving industrial production environments, the user interface of cars and for many other applications. Examples of these mental states are the levels of arousal, fatigue, emotion, workload or other variables whose brain activity correlates (at least partially) are amenable to measurement. The improvement of suboptimal user interfaces reduces the number of critical mental states of the operators. Thus it can lead to an increase in production yield, less errors and accidents, and avoids frustration of the users.

Typically, information collected about the mental states of interest is exploited in an offline analysis of the data and leads to a re-design of the task or the interface. In addition, it might be desirable that a method for mental state

---

[2] This study was performed in cooperation with the Daimler AG. For further information, please refer to [55].

monitoring can be applied online during the execution of a task. Traditional methods for capturing mental states and user ratings are questionnaires, video surveillance of the task, or the analysis of errors made by the operator. However questionnaires are of limited use for precisely assessing the information of interest as the delivered answers are often distorted by subjectiveness. Questionnaires cannot determine the quantities of interest in real-time (during the execution of the task) but only in retrospect; moreover, they are intrusive i.e. they interfere with the task. Even the monitoring of eye blinks or eye movements only allows for an indirect access to the user's mental state. Although the monitoring of a user's errors is a more direct measure, it detects critical changes of the user state post-hoc only. Neither is the anticipation of an error possible, nor can suitable countermeasures be taken to avoid it.

As a new approach we propose the use of EEG signals for mental state monitoring and combine it with BBCI classfication methods for data analysis. With this approach the brain signals of interest can be isolated from background activity as in BCI systems; this combination allows for the non-intrusive evaluation of mental states in real-time and on a single-trial basis such that an online system with feedback can be build.

In a pilot study ([56]) we evaluated the use of EEG signals for arousal monitoring. The experimental setting simulates a security surveillance system where the sustained concentration ability of the user in a rather boring task is crucial. As in BCI, the system had to be calibrated to the individual user in order to recognize and predict mental states, correlated with attention, task involvement or a high or low number of errors of the subject respectively.

Experimental Setup for Attention Monitoring. In this study a subject was seated approx. 1 m in front of a computer screen that displayed different stimuli in a forced choice setting. She was asked to respond quickly to stimuli by pressing keys of a keyboard with either the left or right index finger; recording was done with a 128 channel EEG at 100 Hz. The subject had to rate several hundred x-ray images of luggage objects as either dangerous or harmless by a key press after each presentation. The experiment was designed as an oddball paradigm where the number of the harmless objects was much larger than that of the dangerous objects. The terms standard and deviant will subsequently be used for the two conditions. One trial was usually performed within 0.5 seconds after the cue presentation.

The subject was asked to perform 10 blocks of 200 trials each. Due to the monotonous nature of the task and the long duration of the experiment, the subject was expected to show a fading level of arousal which results in worse concentration and the generation of more and more erroneous decisions during later blocks.

For the offline analysis of the collected EEG signals, the following steps were applied. After exclusion of channels with bad impedances a spatial Laplace filter was applied and the band power features from 8-13 Hz were computed on 2 s windows. The resulting band power values of all channels were concatenated into a final vector. As the subject's correct and erroneous decisions were known,

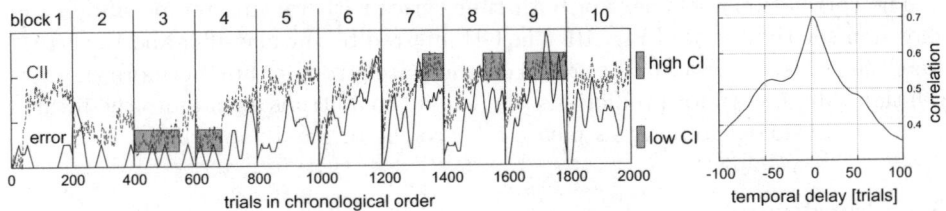

Fig. 10. *Left*: Comparison of the concentration insufficiency index (CII, dotted curve) and the error index for the subject. The error index (the true performed errors smoothed over time) reflects the inverse of the arousal of the subject. *Right*: Correlation coefficient between the CII (returned by the classifier) and the true performance for different time shifts. Highest correlation is around a zero time shift as expected. Please note that the CII has an increased correlation with the error even before the error appears.

a supervised LDA classifier was trained on the data. The classification error of this procedure was estimated by a cross-validation scheme that left out a whole block of 200 trials during each fold for testing. As the number of folds was determined by the number of experimental blocks it varied slightly from subject to subject.

Results. The erroneous decisions taken by a subject were recorded and smoothed in order to form a measure for the arousal. This measure is further referred to as *error index* and reflects the ability of the subject to concentrate and fulfill the security task. To enhance the contrast of the discrimination analysis, two thresholds were introduced for the error index and set after visual inspection. Extreme trials outside these thresholds defined two sets of trials with a rather high rsp. a low value. The EEG data of the trials were labeled as *sufficiently concentrated* or *insufficiently concentrated* depending on these thresholds for later analysis. Fig. 10 shows the error index. The subject did perform nearly error-free during the first blocks but then showed increasing errors beginning with block 4. However, as the blocks were separated by short breaks, the subject could regain attention at the beginning of each new block at least for a small number of trials. The trials of high and low error index formed the training data for teaching a classifier to discriminate mental states of insufficient arousal based on single trial EEG data.

A so-called Concentration Insufficiency Index (CII) of a block was generated by an LDA classifier that had been trained off-line on the labeled training data of the remaining blocks. The classifier output (CII) of each trial is plotted in Fig. 10 together with the corresponding error index. It can be observed that the calculated CII mirrors the error index for most blocks. More precisely the CII mimics the error increase inside each block and in blocks 3 and 4 it can anticipate the increase of later blocks, i.e. out-of-sample. For those later blocks the CII reveals that the subject could not recover its full arousal during the breaks. Instead it shows a short-time arousal for the time immediately after a break, but the CII accumulates over time.

The correlation coefficient of both time series with varying temporal delay is shown in the right plot of Fig. 10. The CII inferred by the classifier and the errors that the subject had actually produced correlate strongly. Furthermore the correlation is high even for predictions that are up to 50 trials ahead into the future.

For a physiological analysis please refer to the original paper [56].

5   Conclusion

The chapter provides a brief overview on the Berlin Brain-Computer Interface. We would like to emphasize that the use of modern machine learning tools – as put forward by the BBCI group – is pivotal for a successful and high ITR operation of a BCI from the first session [49,44]. Note that due to space limitations the chapter can only discuss general principles of signal processing and machine learning for BCI; for details ample references are provided (see also [3]). Our main emphasis was to discuss the wealth of applications of neurotechnology beyond rehabilitation. While BCI is an established tool for opening a communication channel for the severely disabled ([60,61,62,63,64], its potential as an instrument for enhancing man-machine interaction is underestimated. The use of BCI technology as a direct channel additional to existing means to communicate opens applications in mental state monitoring [55,56], gaming [65,66], virtual environment navigation[67], vehicle safety [55], rapid image viewing [68] and enhanced user modeling. To date only proofs of concept and first steps have been given that still need to move a long way to innovative products, but already the attention monitoring and neuro usability applications outlined in Section 4.3 and 4.4 show the usefulness of neurotechnology for the monitoring of complex cognitive mental states. With our novel technique at hand, we can make direct use of mental state monitoring information to enable Human-Machine Interaction to exhibit adaptive anticipatory behaviour.

To ultimately succeed in these promising applications the BCI field needs to proceed in multiple aspects: (a) improvement of EEG technology beyond gel electrodes and (e.g. [57]) towards cheap and portable devices, (b) understanding of the BCI-illiterates phenomenon, (c) improved and more robust signal processing and machine learning methods, (d) higher ITRs for non-invasive devices and finally (e) the development of compelling industrial applications also outside the realm of rehabilitation.

Acknowledgement

The studies were partly supported by the *Bundesministerium für Bildung und Forschung* (BMBF), FKZ 01IBE01A/B, by the German Science Foundation (DFG, contract MU 987/3-1), by the European Union's Marie Curie Excellence Team project MEXT-CT-2004-014194, entitled 'Brain2Robot' and by their IST Programme under the PASCAL Network of Excellence, IST-2002-506778. This publication only reflects the authors' views. We thank our coauthors for allowing us to use published material from [49,8,59,57,55,56].

References

1. Blankertz, B., Tangermann, M., Popescu, F., Krauledat, M., Fazli, S., Dónaczy, M., Curio, G., Müller, K.R.: The berlin brain-computer interface. In: Graimann, B., Pfurtscheller, G. (eds.) Non-Invasive and Invasive Brain-Computer Interfaces. Springer, Heidelberg (2008) The Frontiers Collection (in review, 2008)
2. Blankertz, B., Krauledat, M., Dornhege, G., Williamson, J., Murray-Smith, R., Müller, K.R.: A Note on Brain Actuated Spelling with the Berlin Brain-Computer Interface. In: Stephanidis, C. (ed.) UAHCI 2007 (Part II). LNCS, vol. 4555, pp. 759–768. Springer, Heidelberg (2007)
3. Dornhege, G., del, R., Millán, J., Hinterberger, T., McFarland, D., Müller, K.R. (eds.): Toward Brain-Computer Interfacing. MIT Press, Cambridge (2007)
4. Wolpaw, J.R., Birbaumer, N., McFarland, D.J., Pfurtscheller, G., Vaughan, T.M.: Brain-computer interfaces for communication and control. Clin. Neurophysiol. 113(6), 767–791 (2002)
5. Allison, B., Wolpaw, E., Wolpaw, J.R.: Brain-computer interface systems: progress and prospects. Expert Rev. Med. Devices 4(4), 463–474 (2007)
6. Pfurtscheller, G., Neuper, C., Birbaumer, N.: Human Brain-Computer Interface. In: Riehle, A., Vaadia, E. (eds.) Motor Cortex in Voluntary Movements, pp. 367–401. CRC Press, New York (2005)
7. Haynes, J., Sakai, K., Rees, G., Gilbert, S., Frith, C.: Reading hidden intentions in the human brain. Current Biology 17, 323–328 (2007)
8. Blankertz, B., Dornhege, G., Lemm, S., Krauledat, M., Curio, G., Müller, K.R.: The Berlin Brain-Computer Interface: Machine learning based detection of user specific brain states. J. Universal Computer Sci. 12(6), 581–607 (2006)
9. Elbert, T., Rockstroh, B., Lutzenberger, W., Birbaumer, N.: Biofeedback of slow cortical potentials. I. Electroencephalogr Clin. Neurophysiol. 48, 293–301 (1980)
10. Birbaumer, N., Kübler, A., Ghanayim, N., Hinterberger, T., Perelmouter, J., Kaiser, J., Iversen, I., Kotchoubey, B., Neumann, N., Flor, H.: The though translation device (TTD) for completly paralyzed patients. IEEE Trans. Rehab. Eng. 8(2), 190–193 (2000)
11. Kornhuber, H.H., Deecke, L.: Hirnpotentialänderungen bei Willkürbewegungen und passiven Bewegungen des Menschen: Bereitschaftspotential und reafferente Potentiale. Pflügers Arch 284, 1–17 (1965)
12. Lang, W., Lang, M., Uhl, F., Koska, C., Kornhuber, A., Deecke, L.: Negative cortical DC shifts preceding and accompanying simultaneous and sequential movements. Exp. Brain Res. 74(1), 99–104 (1989)
13. Cui, R.Q., Huter, D., Lang, W., Deecke, L.: Neuroimage of voluntary movement: topography of the Bereitschaftspotential, a 64-channel DC current source density study. Neuroimage 9(1), 124–134 (1999)
14. Blankertz, B., Dornhege, G., Krauledat, M., Müller, K.R., Kunzmann, V., Losch, F., Curio, G.: The Berlin Brain-Computer Interface: EEG-based communication without subject training. IEEE Trans. Neural Sys. Rehab. Eng. 14(2), 147–152 (2006)
15. Blankertz, B., Dornhege, G., Krauledat, M., Kunzmann, V., Losch, F., Curio, G., Müller, K.R.: The berlin brain-computer interface: Machine-learning based detection of user specific brain states. In: Dornhege, G., del, R., Millán, J., Hinterberger, T., McFarland, D., Müller, K.R. (eds.) Toward Brain-Computer Interfacing, pp. 85–101. MIT press, Cambridge (2007)

16. Blankertz, B., Dornhege, G., Schäfer, C., Krepki, R., Kohlmorgen, J., Müller, K.R., Kunzmann, V., Losch, F., Curio, G.: Boosting bit rates and error detection for the classification of fast-paced motor commands based on single-trial EEG analysis. IEEE Trans. Neural Sys. Rehab. Eng. 11(2), 127–131 (2003)

17. Krusienski, D., Schalk, G., McFarland, D.J., Wolpaw, J.: A mu-rhythm matched filter for continuous control of a brain-computer interface. IEEE Trans. Biomed. Eng. 54(2), 273–280 (2007)

18. Nikulin, V.V., Brismar, T.: Phase synchronization between alpha and beta oscillations in the human electroencephalogram. Neuroscience 137, 647–657 (2006)

19. Pfurtscheller, G., Lopes da Silva, F.: Event-related EEG/MEG synchronization and desynchronization: basic principles. Clin. Neurophysiol. 110(11), 1842–1857 (1999)

20. Falkenstein, M., Hoormann, J., Christ, S., Hohnsbein, J.: ERP components on reaction errors and their functional significance: a tutorial. Biol. Psychol. 51(2-3), 87–107 (2000)

21. Falkenstein, M., Hohnsbein, J., Hoormann, J., Blanke, L.: Effects of errors in choice reaction tasks on the ERP under focused and divided attention. In: Brunia, C., Gaillard, A., Kok, A. (eds.) Psychophysiological Brain Research, pp. 192–195. Tilburg University Press, Tilburg (1990)

22. Gehring, W., Coles, M., Meyer, D., Donchin, E.: The error-related negativity: an event-related brain potential accompanying errors. Psychophysiology 27, 34 (1993)

23. Vidal, F., Hasbroucq, T., Grapperon, J., Bonnet, M.: Is the 'error negativity' specific to errors? Biological Psychology 51, 109–128 (2000)

24. Miltner, W.H.R., Braun, C.H., Coles, M.G.H.: Event-related brain potentials following incorrect feedback in a time-estimation task: Evidence for a 'generic' neural system for error-detection. J. Cogn. Neurosci. 9, 788–798 (1997)

25. Schie, H.T.v., Mars, R.B., Coles, M.G., Bekkering, H.: Modulation of activity in medial frontal and motor cortices during error observation. Nat. Neurosci. 7(5), 549–554 (2004)

26. Ferrez, P., Millán, J.: Error-related eeg potentials generated during simulated brain-computer interaction. IEEE Trans. Biomed. Eng. (accepted)

27. Nunez, P.L., Srinivasan, R., Westdorp, A.F., Wijesinghe, R.S., Tucker, D.M., Silberstein, R.B., Cadusch, P.J.: EEG coherency I: statistics, reference electrode, volume conduction, Laplacians, cortical imaging, and interpretation at multiple scales. Electroencephalogr Clin. Neurophysiol. 103(5), 499–515 (1997)

28. Blankertz, B., Tomioka, R., Lemm, S., Kawanabe, M., Müller, K.R.: Optimizing spatial filters for robust EEG single-trial analysis. IEEE Signal Proc. Magazine 25(1), 41–56 (2008)

29. McFarland, D.J., McCane, L.M., David, S.V., Wolpaw, J.R.: Spatial filter selection for EEG-based communication. Electroencephalogr Clin. Neurophysiol. 103, 386–394 (1997)

30. Hill, N., Lal, T.N., Tangermann, M., Hinterberger, T., Widman, G., Elger, C.E., Schölkopf, B., Birbaumer, N.: Classifying event-related desynchronization in EEG, ECoG and MEG signals. In: Dornhege, G., del, R., Millán, J., Hinterberger, T., McFarland, D., Müller, K.R. (eds.) Toward Brain-Computer Interfacing, pp. 235–260. MIT press, Cambridge (2007)

31. Grosse-Wentrup, M., Gramann, K., Buss, M.: Adaptive spatial filters with predefined region of interest for EEG based brain-computer-interfaces. In: Schölkopf, B., Platt, J., Hoffman, T. (eds.) Advances in Neural Information Processing Systems 19, pp. 537–544 (2007)

32. Dornhege, G., Krauledat, M., Müller, K.R., Blankertz, B.: General signal processing and machine learning tools for BCI. In: Dornhege, G., del, R., Millán, J., Hinterberger, T., McFarland, D., Müller, K.R. (eds.) Toward Brain-Computer Interfacing, pp. 207–233. MIT Press, Cambridge (2007)
33. Parra, L.C., Spence, C.D., Gerson, A.D., Sajda, P.: Recipes for the linear analysis of EEG. NeuroImage 28(2), 326–341 (2005)
34. Fukunaga, K.: Introduction to Statistical Pattern Recognition, 2nd edn. Academic Press, San Diego (1990)
35. Guger, C., Ramoser, H., Pfurtscheller, G.: Real-time EEG analysis with subject-specific spatial patterns for a Brain Computer Interface (BCI). IEEE Trans. Neural Sys. Rehab. Eng. 8(4), 447–456 (2000)
36. Ramoser, H., Müller-Gerking, J., Pfurtscheller, G.: Optimal spatial filtering of single trial EEG during imagined hand movement. IEEE Trans. Rehab. Eng. 8(4), 441–446 (2000)
37. Lemm, S., Blankertz, B., Curio, G., Müller, K.R.: Spatio-spectral filters for improving classification of single trial EEG. IEEE Trans. Biomed. Eng. 52(9), 1541–1548 (2005)
38. Dornhege, G., Blankertz, B., Krauledat, M., Losch, F., Curio, G., Müller, K.R.: Optimizing spatio-temporal filters for improving brain-computer interfacing. In: Advances in Neural Inf. Proc. Systems (NIPS 2005), vol. 18, pp. 315–322. MIT Press, Cambridge (2006)
39. Tomioka, R., Aihara, K., Müller, K.R.: Logistic regression for single trial EEG classification. In: Schölkopf, B., Platt, J., Hoffman, T. (eds.) Advances in Neural Information Processing Systems 19, pp. 1377–1384. MIT Press, Cambridge (2007)
40. Duda, R.O., Hart, P.E., Stork, D.G.: Pattern Classification, 2nd edn. Wiley & Sons, Chichester (2001)
41. Friedman, J.H.: Regularized discriminant analysis. J Amer. Statist. Assoc. 84(405), 165–175 (1989)
42. Neuper, C., Scherer, R., Reiner, M., Pfurtscheller, G.: Imagery of motor actions: Differential effects of kinesthetic and visual-motor mode of imagery in single-trial EEG. Brain Res. Cogn. Brain Res. 25(3), 668–677 (2005)
43. Nikulin, V.V., Hohlefeld, F.U., Jacobs, A.M., Curio, G.: Quasi-movements: A novel motor-cognitive phenomenon. In: Neuropsychologia (in press, 2008)
44. Blankertz, B., Dornhege, G., Krauledat, M., Müller, K.R., Curio, G.: The non-invasive Berlin Brain-Computer Interface: Fast acquisition of effective performance in untrained subjects. NeuroImage 37(2), 539–550 (2007)
45. Wolpaw, J.R., McFarland, D.J., Vaughan, T.M.: Brain-computer interface research at the Wadsworth Center. IEEE Trans. Rehab. Eng. 8(2), 222–226 (2000)
46. Schlögl, A., Kronegg, J., Huggins, J., Mason, S.G.: Evaluation Criteria for BCI Research. In: Dornhege, G., del, R., Millán, J., Hinterberger, T., McFarland, D., Müller, K.R. (eds.) Towards Brain-Computer Interfacing, pp. 297–312. MIT press, Cambridge (2007)
47. Dornhege, G.: Increasing Information Transfer Rates for Brain-Computer Interfacing. PhD thesis, University of Potsdam (2006)
48. Müller, K.R., Blankertz, B.: Toward noninvasive brain-computer interfaces. IEEE Signal Proc. Magazine 23(5), 125–128 (2006)
49. Blankertz, B., Losch, F., Krauledat, M., Dornhege, G., Curio, G., Müller, K.R.: The Berlin Brain-Computer Interface: Accurate performance from first-session in BCI-naive subjects. IEEE Trans. Biomed. Eng. (accepted, 2008)

50. Kübler, A., Müller, K.R.: An introduction to brain computer interfacing. In: Dornhege, G., del, R., Millán, J., Hinterberger, T., McFarland, D., Müller, K.R. (eds.) Toward Brain-Computer Interfacing, pp. 1–25. MIT press, Cambridge (2007)
51. Schalk, G., Wolpaw, J.R., McFarland, D.J., Pfurtscheller, G.: EEG-based communication: presence of an error potential. Clin. Neurophysiol 111, 2138–2144 (2000)
52. Blankertz, B., Schäfer, C., Dornhege, G., Curio, G.: Single Trial Detection of EEG Error Potentials: A Tool for Increasing BCI Transmission Rates. In: Dorronsoro, J.R. (ed.) ICANN 2002. LNCS, vol. 2415, pp. 1137–1143. Springer, Heidelberg (2002)
53. Parra, L., Spence, C., Gerson, A., Sajda, P.: Response error correction - a demonstration of improved human-machine performance using real-time EEG monitoring. IEEE Trans. Neural. Sys. Rehab. Eng. 11(2), 173–177 (2003)
54. Ferrez, P., Millán, J.: You are wrong! – automatic detection of interaction errors from brain waves. In: 19th International Joint Conference on Artificial Intelligence, pp. 1413–1418 (2005)
55. Kohlmorgen, J., Dornhege, G., Braun, M., Blankertz, B., Müller, K.R., Curio, G., Hagemann, K., Bruns, A., Schrauf, M., Kincses, W.: Improving human performance in a real operating environment through real-time mental workload detection. In: Dornhege, G., del, R., Millán, J., Hinterberger, T., McFarland, D., Müller, K.R. (eds.) Toward Brain-Computer Interfacing, pp. 409–422. MIT press, Cambridge (2007)
56. Müller, K.R., Tangermann, M., Dornhege, G., Krauledat, M., Curio, G., Blankertz, B.: Machine learning for real-time single-trial EEG-analysis: From brain-computer interfacing to mental state monitoring. J Neurosci. Methods 167(1), 82–90 (2008)
57. Popescu, F., Fazli, S., Badower, Y., Blankertz, B., Müller, K.R.: Single trial classification of motor imagination using 6 dry EEG electrodes. PLoS ONE 2(7) (2007)
58. Fazli, S., Dónaczy, M., Kawanabe, M., Popescu, F.: Asynchronous, adaptive BCI using movement imagination training and rest-state inference. In: IASTED's Proceedings on Artificial Intelligence and Applications, pp. 85–90 (2008)
59. Krauledat, M., Dornhege, G., Blankertz, B., Curio, G., Müller, K.R.: The Berlin brain-computer interface for rapid response. Biomed. Tech. 49(1), 61–62 (2004)
60. Kübler, A., Kotchoubey, B., Kaiser, J., Wolpaw, J., Birbaumer, N.: Brain-computer communication: Unlocking the locked in. Psychol. Bull 127(3), 358–375 (2001)
61. Kübler, A., Nijboer, F., Mellinger, J., Vaughan, T.M., Pawelzik, H., Schalk, G., McFarland, D.J., Birbaumer, N., Wolpaw, J.R.: Patients with ALS can use sensorimotor rhythms to operate a brain-computer interface. Neurology 64(10), 1775–1777 (2005)
62. Birbaumer, N., Cohen, L.: Brain-computer interfaces: communication and restoration of movement in paralysis. J. Physiol. 579, 621–636 (2007)
63. Birbaumer, N., Weber, C., Neuper, C., Buch, E., Haapen, K., Cohen, L.: Physiological regulation of thinking: brain-computer interface (BCI) research. Prog. Brain Res. 159, 369–391 (2006)
64. Hochberg, L., Serruya, M., Friehs, G., Mukand, J., Saleh, M., Caplan, A., Branner, A., Chen, D., Penn, R., Donoghue, J.: Neuronal ensemble control of prosthetic devices by a human with tetraplegia. Nature 442(7099), 164–171 (2006)

65. Krepki, R., Blankertz, B., Curio, G., Müller, K.R.: The Berlin Brain-Computer Interface (BBCI): towards a new communication channel for online control in gaming applications. Journal of Multimedia Tools and Applications 33(1), 73–90 (2007)
66. Krepki, R., Curio, G., Blankertz, B., Müller, K.R.: Berlin brain-computer interface - the hci communication channel for discovery. Int. J. Hum. Comp. Studies 65, 460–477 (2007) Special Issue on Ambient Intelligence
67. Leeb, R., Lee, F., Keinrath, C., Scherer, R., Bischof, H., Pfurtscheller, G.: Brain-computer communication: motivation, aim, and impact of exploring a virtual apartment. IEEE Trans. Neural Sys. Rehab. Eng. 15(4), 473–482 (2007)
68. Gerson, A., Parra, L., Sajda, P.: Cortically coupled computer vision for rapid image search. IEEE Trans. Neural. Sys. Rehab. Eng. 14(2), 174–179 (2006)

Basic Scheme of Neuroinformatics Platform: XooNIps

Shiro Usui and Yoshihiro Okumura

RIKEN, Brain Science Institute, 2-1 Hirosawa, Wako, Saitama 351-0198, Japan
usuishiro@riken.jp,okumura@brain.riken.go.jp

Abstract. To promote international cooperation in the new field of Neuroinformatics (NI), the Neuroinformatics Japan Center at RIKEN Brain Science Institute (BSI) has been established in 2005 as the Japan-Node (J-Node) for coordination with the International Neuroinformatics Coordinating Facility. The Laboratory for Neuroinformatics was established in 2002 at RIKEN BSI, and created the NI base-platform "XooNIps" following the concepts and experience acquired from the Visiome Platform, which was developed under the project of the Neuroinformatics Research in Vision. XooNIps features better scalability, extensibility, and customizability to operate under various site policies supporting different databases and portals. Utilizing XooNIps, eight J-Node platforms have been developed by each platform committee which were selected from active research areas in Japan. XooNIps contributes not only in NI field but in diverse areas such as library repositories and university research resources.

Keywords: Neuroinformatics, Database, Platform, XooNIps, INCF, Japan-Node.

1 Introduction

A pressing need for a concerted international effort to help researchers to understand brain mechanisms and functions using information technology was documented in the report on Neuroinformatics (NI) from the Global Science Forum Neuroinformatics Working Group of the Organization for Economic Cooperation and Development (OECD) [1]. International Neuroinformatics Coordinating Facility (INCF) has then been established at Karolinska to promote international interdisciplinary cooperation in NI [2].

NI combines neuroscience and informatics researches to develop and apply advanced tools and approaches essential for a major advancement in understanding the structure and function of the brain. NI undertakes the challenge of developing mathematical models, databases, data analysis, and tools necessary for establishing such NI platforms. The major emphasis of the NI platform is the organization of neuroscience data and knowledge-based contents to facilitate the development of computational models and tools.

J.M. Zurada et al. (Eds.): WCCI 2008 Plenary/Invited Lectures, LNCS 5050, pp. 102–116, 2008.

Responding to this need to promote international cooperation in the new field of NI, the Neuroinformatics Japan Center (NIJC) at RIKEN Brain Science Institute (BSI) has been established in 2005 as the Japan-Node (J-Node) [3] in coordination with INCF. The Laboratory for Neuroinformatics (NI team), established in 2002 as a team of the Advanced Technology Development Group at RIKEN BSI, created the NI base-platform "XooNIps" [4] following the concepts of the Visiome Platform (VP) [5], which was developed under the Neuroinformatics Research in Vision (NRV) project [6]. NRV was a pioneering NI project initiated in 1999 and completed in March 2004 with the primary aim of building the foundation of NI research in Japan. It focused on the visual system to promote experimental, theoretical and technical research as a pilot study on NI and was made available to the public. Since then, VP has been improving its contents continuously to answer the needs of the users and re-released as a new site in January 2007 [7] with subsequent improvements introduced by XooNIps. In other words, VP was a foundation for the development of the base-platform which offers reduced costs while at the same time speeds-up the development of new platforms with flexible management style providing a framework for accumulating, sharing and making public resources which were once a difficult task. XooNIps is also designed for developing databases in different research fields through customization of the option menu; digital resources are stored according to their respective categories, each associated with their related metadata. It features high scalability, extensibility, and customizability to operate under various site policies. It can also contribute not only in NI field but in such diverse areas as library repositories and university research resources.

2   XooNIps

The scientific results from the brain and neuroscience researches include not only published papers, but also other various electronic resources such as experimental data, mathematical models, simulation programs, simulation results, measuring methods, URLs, etc. It is crucial to house, develop, share and disseminate such resources for the further understanding of the brain as a system.

There is a need to develop such framework of an open data system like VP [5,6,7] not only in the field of vision science but also in other research fields as well. Following this basis, as it presently requires enormous effort and expenditures to continue building various databases concomitantly in the brain scientific research field, there is a need for a common base-platform which reduces the costs for the development and management of such databases, offering a flexible management style which simplifies the addition and installation of specialized functions and accepts the respective policy in the particular research fields.

Similar digital archive systems are available such as EPrints [8] and DSpace [9]. EPrints is a flexible platform for building high value repositories. DSpace is an open source solution for accessing, managing, and preserving scholarly works. However, these software aim basically to construct a digital archive platform for institutional repositories. For the management of NI-Platform (NI-PF) systems, the demands for more customizable system are escalated as the aforementioned systems do not permit to exchange information among researchers, to add extended functions required in a respective field, or to customize the management policy.

To solve these issues, the NI base-platform XooNIps has been developed, based on the concept and experience from previous research on VP. It is a web based system to share multifarious electronic data such as files, URLs, books, papers as well as related metadata.

2.1   Main View

Figure 1 shows the main view of XooNIps and its main features are briefly explained bellow.

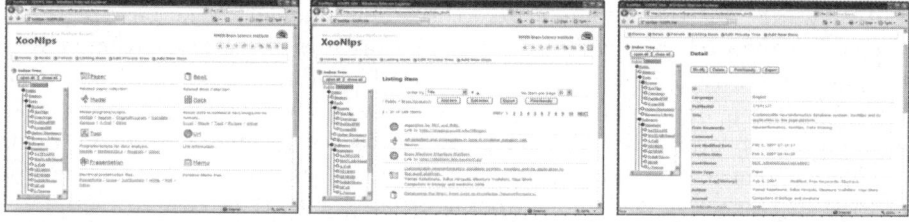

Fig. 1. Main Views of XooNIps. From the left top page, listing items, detail information.

Index Tree: For a better view of the inter-relationship among research resources, XooNIps provides a tree-structured keyword index as in Figure 1. User can open and close each tree node and browse the listed contents by clicking a chosen keyword.

Item Types: XooNIps categorizes digital resources according to item types as in Figure 1. The item types are provided as extension modules of the content management system XOOPS [10]. There are required metadata associated with the digital resources as well as formats for submission, modification, browsing, printing and searching. XooNIps has 12 kinds of item type in the basic model. A particular feature is that the site manager can determine the preferred item types to combine during the development of the platform. Another advantage is that new item types can be created, for example, "Article" was developed by Media Center at Keio University, which uses MODS metadata scheme.

Finding Items and Accessing Detailed Information: To find items, the following three types of item search method are available. The first is a direct search by browsing the index tree as described in "Index Tree". Second is a "Keyword search" which search items matching the specified keywords. The third is an "Advanced search" where the item types and metadata fields are specified. The results of these search methods are displayed as a list of items with 20 records by default (Figure 1, Center). Users can access detailed information of each item by clicking the item from the list (Figure 1, Right).

2.2 System Outline

To operate a database smoothly, XooNIps offers three levels of accessible area and five user authorities. This enables a series of workflows such as making items public, sharing items within a group and so on. Figure 2 illustrates the basic features and system outline of XooNIps.

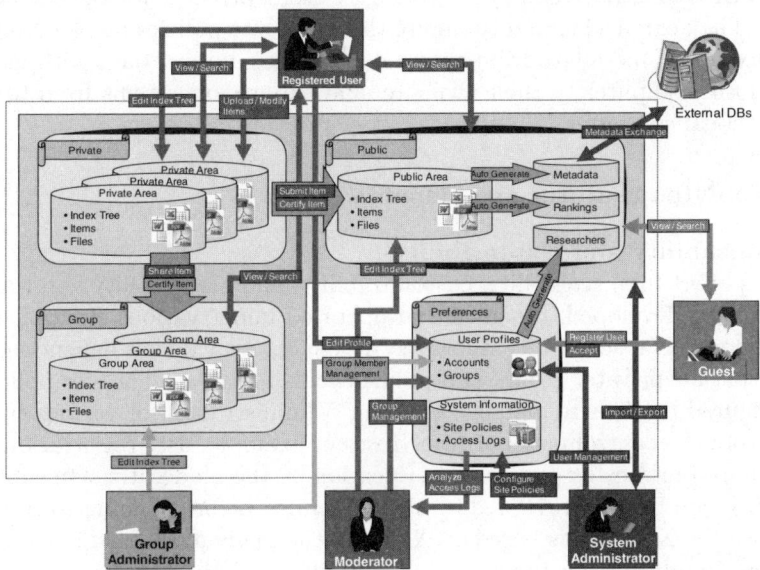

Fig. 2. System Outline of XooNIps

The three levels of accessible areas are: Public, Group and Private. The Public area is accessible to everyone. It can be accepted by the moderator to release the item in public. The Group area is reserved for its group members and is shared for common usage. It is available to search or read only for the members of the group. It can be accepted by the group moderator to share the items. The Private area is a personal space for registered users exclusively.

XooNIps has five types of inclusive relationship user's authorities named: Guest, Registered User, Group Administrator, Moderator, and System Administrator.

Guest is a limited non-logged in user. They can search and browse public items, and request for an account.

Registered User is a typical logged in users. They can register items in the private domain assigned only to a registered user, share items in a domain assigned only to a group after belonging to the group, submit items to the public domain, search and browse items registered in public, private and group shared domains, and edit Index Tree in their private domains.

Group Administrator is a special user's privilege for the group domain management. They can manage membership of their own group, edit the index in shared domains by their group, and approve or reject items to share in his/her group domain.

Moderator is a special user's privilege for quality control of registered users and publication items. They can approve a new user account, edit Index Tree in the public domain, create a new researcher group, watch their own site access logs for usage statistics, and approve or reject items to open in the public domain.

System Administrator is the strongest user's privilege for the site administration. They can decide and configure their own site policies on XooNIps, elicit the collection of metadata from external databases, import data with metadata from a local computer to their own site, and export some items from their own site to a local computer.

2.3   XooNIps Features and Benefits

Customizability and Extensibility

XooNIps offers high scalability, customizability, and extensibility and thus platforms can be developed, designed and operated under various site policies and environment. The scalability allows to design its own policy depending on its application for private, group, and public accessible areas. The site policies can be configured by the site manager (System Administrator) for various site operations from the control panel of XooNIps. For example, user registration can be accepted automatically, or require verification by the moderator. On submitting an item, it can be peer-reviewed by the moderator before opening to the public. And since the XooNIps is based on XOOPS, the many available modules can be combined to add further functions for customization and extension.

Quality control (Peer-review system)

By default, user registration requires verification by the moderator. This may avoid the registration of malicious users. When registering items in the Public or Group domain, the moderator or group administrator can verify the contents to be released (peer-reviewed) beforehand. Therefore, it is possible to maintain the quality of users and items. Among these workflows, the event notice function is effective to encourage the operation from one user to another. The user can receive this notice by a private message or e-mail. Consequently, it brings workflow efficiency.

Communicate with External Systems

For the communication with external databases, XooNIps supports two protocols: Open Archives Initiative Protocol for Metadata Harvesting (OAI-PMH) [11] and XML-RPC.

OAI-PMH Metadata exchange: XooNIps utilizes OAI-PMH to support interoperability with other XooNIps platforms. The OAI-PMH provides an application independent interoperability framework based on metadata harvesting. There are two classes of participants in the OAI-PMH framework. First is Data Providers, which administer systems that support the OAI-PMH as a means of exposing metadata. Second is Service Providers, which use metadata harvested via the OAI-PMH as a basis for building value-added services. The Data Providers supporting metadata schemes for XooNIps are OAI-DC [12], JuNii and JuNii2 [13]. The OAI-DC is based on Dublin Core simple metadata format. It is an essential requirement for using the OAI-PMH. The JuNii and JuNii2 are another metadata format for Japanese Institutional Repositories proposed by the National Institute of Informatics. Since the metadata from a given database will be propagated to other XooNIps or other Data Providers (such as EPrints or DSpace) in conformity with the OAI-PMH, users can search keywords simultaneously across different databases on the Internet.

XML-RPC Web API: Due to the fact that XooNIps is a web-application software, all operations require web-browsers such as Internet Explorer. This causes inconvenience to exchange data between the local machine and the XooNIps. To solve this issue, XooNIps provides a XML-RPC Web API that works over the Internet, which is an implementation of a Remote Procedure Calling protocol. This function allows users to operate the XooNIps from an external application with a more user friendly interface. A software which utilizes this function is being developed by the Laboratory for Neuroinformatics at RIKEN BSI, called "Concierge" [14,15].

Accessibility and Incentive rewarding

XooNIps provides an automatic fill-in function for the basic item types such as Paper or Book to save time for inputting metadata information. For the Paper item, the user only needs to provide the PubMed ID to extract the information from the PubMed database, which is automatically filled in its respective metadata fields. For the Book item, it extracts information from the Amazon database from the ISBN (ASIN), which are also filled in its respective metadata fields.

For making a file-attached item public, users are required to input the copyright information and utilization condition for the file. XooNIps offers a framework for selecting pre-defined terms and conditions, such as "All rights reserved" which is an unrestricted description, and "Some rights reserved" licensed by the Creative Commons [16] In the latter, users can select among the following conditions: "Attribution", "Attribution Share Alike", "Attribution No Derivatives", "Attribution Non-commercial", "Attribution Non-commercial Share Alike" and "Attribution Non-commercial No Derivatives".

For the analysis of site access, the moderator can download and/or visualize the event log categorized by the respective events or time period.

To encourage the motivation for the users, XooNIps provides a ranking function and users introducing function. Ranking is a function in such that the user can recognize registered users' activities and the site, indicating the "newly-arrived item", "most popular read item", "most popular downloaded item", "user who prepares the most released item", and "the keywords which were searched the most". Referring to this incentive rewards function, registered users can grasp the popularity of the item they registered and it motivates registered users to make an improvement to register the items. User introduction is a function wherein the researchers can highlight their achievements and personal history.

2.4   System Architecture

XooNIps has been developed as an extended module of XOOPS. The principal appealing point is that it reduces the cost for constructing and managing a specific platform since it is an open source software. It is possible to construct varied kinds of sites by combining the existing modules of XOOPS. Furthermore, the site layout can easily be substituted by replacing the theme of XOOPS. That is, XooNIps is an extended module of XOOPS which is an OS independent web application written in PHP. Installation merely requires a copy of the XooNIps modules to be placed in the XOOPS modules directory, which becomes visible for activation in the administration menu. As shown in Figure 3, XooNIps architecture comprises three main components which are user interface, web application, and database. Users can access XooNIps through web browsers. Web application components include a web server, XOOPS, and modules. The XooNIps module is at the same functional level as other XOOPS general modules. In addition, MySQL is used as a relational database backend for storing data from XOOPS and its general modules including XooNIps.

The recommended software environments of XooNIps are:

1. XOOPS: XOOPS 2.0.16a-JP
2. Web server: Apache 2.0+
3. MySQL: 4.1+ with InnoDB
4. PHP: 5.1+ (required extensions: xml, zlib, gd, fileinfo, mbstring)
5. External commands: wvText, pdftotext, xlhtml, ppthtml (These are used to create a search index of the file content.)
6. Web browser: Microsoft Internet Explorer 6+, Mozilla FireFox 1.5+, Apple Safari 2.0+

XooNIps is an open source development project, released under the GPL (GNU General Public License) [17], therefore all contributions are warmly welcomed and appreciated. The current version 3.31 has been released in November 2007. The software is available for download at the official site http://xoonips. sourceforge.jp/

3   XooNIps Based VP and Other Application

VP has been improving, managing, collecting and registering the contents continuously to answer the needs for the users since opened in public under the VP committee. To respond to user's needs, VP has been modified based on XooNIps and re-released as a new site since January 2007 with the modification of the improvement of index tree, registering of contents, and enhancement of the coordination with Japan-Node under INCF (Figure 4).

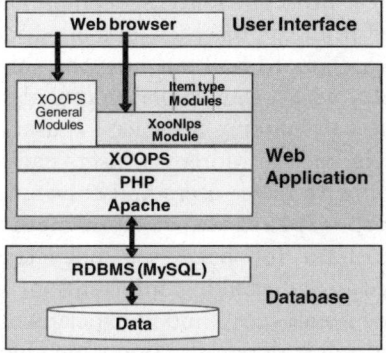

Fig. 3. XooNIps architecture. XooNIps has been developed as a module of XOOPS. XooNIps architecture comprises three main components: User Interface, Web Application and Database. XooNIps is an OS-independent system. It is written in PHP script language.

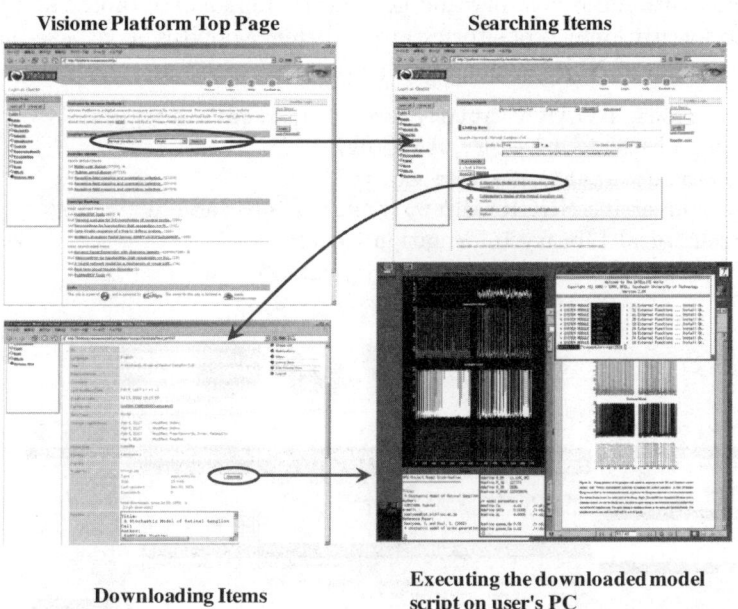

Fig. 4. Visiome Platform Top Page, Downloading items, Searching items, Executing the downloaded model script on user's PC

VP is being developed to answer a critical need for a database to assist in the exploration of complex visual functions. It is designed for a web-based database where published references, mathematical models, experimental data, analytical tools and many other resources can be archived in files (in zip, lzh or other compression formats) including any formats of model, data or stimulus with files of explanatory figures, program sources, readme and other related files so that they can be accessed from the Internet, uploaded, downloaded, and tested. The platform allows researchers to find out how the submitted models work or compare their own results with other experimental data. It also allows users to improve existing models by making it easier for users to integrate their new hypothesis into the existing model. Moreover, users can export/import items to the database to be shared with other users and colleagues.

VP is accessible at `http://platform.visiome.neuroinf.jp/` as illustrated in Figure 4. At the left of the top page, the index tree section in which the main area displays updates and ranking information, and the search results. VP currently has 10 basic item types; model, experimental data, stimulus, tool, URL, presentation, paper, book, demonstration which includes the item of movie collection and binder. It contains a total of more than 3,000 contents.

3.1 The Other Applications of XooNIps

The flexibility of XooNIps enables users to utilize not only as a NI platform but also as an institutional repository and laboratory/personal database system and now diverse institutional repositories are utilizing XooNIps as a base-platform under the support of NIJC such as for example:

– Keio University, KOARA (`http://koara.lib.keio.ac.jp/`),
– Asahikawa Medical University, AMCoR
  (`http://amcor.asahikawa-med.ac.jp/`),
– Saitama University, SUCRA (`http://sucra.saitama-u.ac.jp/`),
– Nara Prefectural Library Information Center
  (`http://www2.library.pref.nara.jp/nlmc/`),

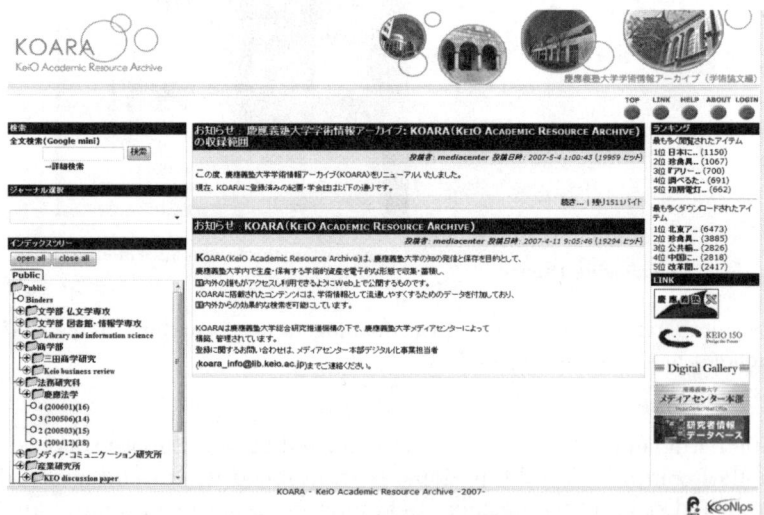

Fig. 5. The top page of KOARA (Keio Academic Resource Archive)

- Nara Cultural Research Center (http://repository.nabunken.go.jp/),
- Sapporo Medical University, Kansei Gakuin University and Waseda University (in planning).

4   Platforms under Japan-Node

4.1   Japan-Node

NIJC organizes Japanese activities in NI research and also participates in the international coalition of INCF (Figure 6). NIJC supports researchers developing and maintaining neuroscience databases, provides a portal for these databases and NI, and is designing the infrastructure for Japanese NI. It is also developing database technologies, and facilitates cooperation and distribution of the information stored in those databases. NIJC thus provides links and smooth integrations of the NI-PFs (J-Node Portal), supports the NI platforms, and supplies and supports NI tools and systems such as the base-platform XooNIps in cooperation with BSI NI team. Based on the basic features and function of VP, the eight platforms described below have been developed by each Platform Committee under the J-Node utilizing XooNIps (Figure 7). The registered users can freely download its contents and upload their own contents by a simple procedure with the approval of a Platform Committee. These platforms are accessible at the J-Node site. We here introduce each PF in brief.

4.2   Access Statistics of the Japan-Node Portal

The average number of accesses to the J-Node site has significantly increased because of the simplification of registration and search for the data of users'

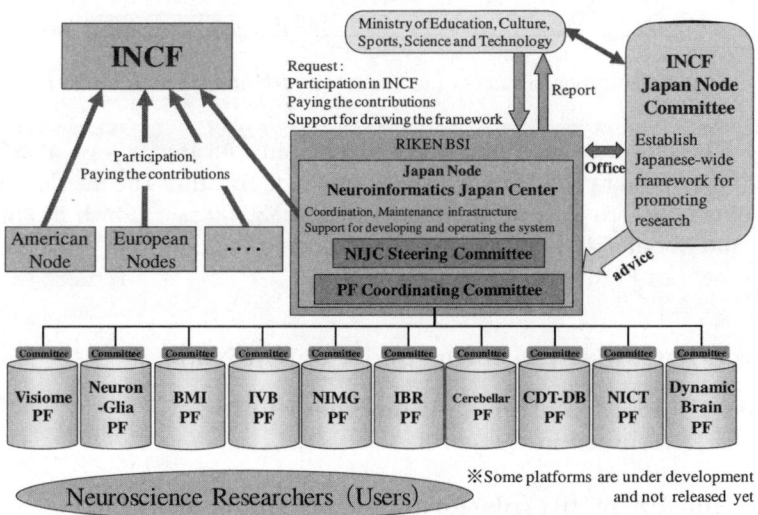

Fig. 6. Scheme of INCF Neuroinformatics Japan Node and Platforms

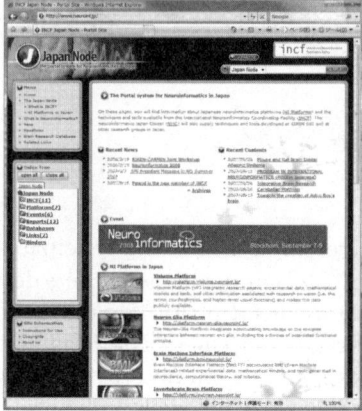

Fig. 7. Top page of Japan-Node Portal

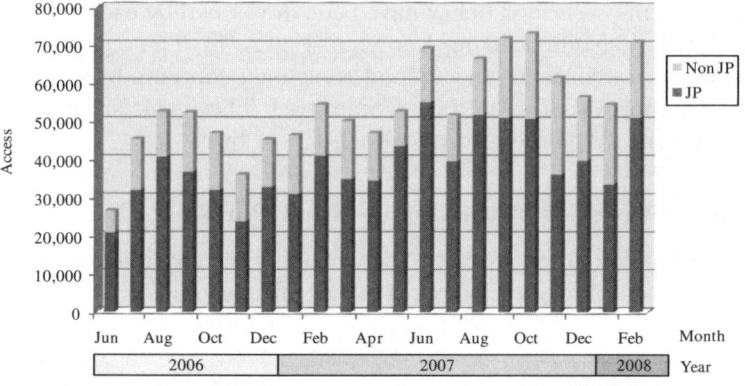

Fig. 8. Number of access per month in J-Node since June 2006

interests. Therefore, it has being utilized not only in Japan, but also internationally. The contents shall progressively enhance to fulfill the user's need. The number of accesses to J-Node site has averaged 53,900 per month in Japan and globally, and about 1,132,000 times in total during the period of June 2006 - Feb 2008 as shown in Figure 8.

4.3   Overview of NI-Platforms under Japan-Node

Visiome PF http://platform.visiome.neuroinf.jp/
Refer to the description in chapter 3.

Neuron-Gila PF http://platform.neuron-glia.neuroinf.jp/
The objective of Neuron-Gila Platform (NGP) is to provide a platform on which experimental as well as theoretical neuroscientists can share new findings and

ideas. Model descriptions (mathematical and theoretical) of new findings and new ideas regarding the properties of neurons, glia cells, and neuron-glia networks are accumulated on the platform. The Platform offers two functions: 1.Publicize & archive models and related data and tools regarding Neuron-Gila functions. 2. Provide workspace for sharing models, data, tools and personal notes privately among group members of registered users.

Invertebrate Brain (IVB) PF http://platform.invbrain.neuroinf.jp/

Developing the platform to collect and share experimental data, mathematical models and research tools about invertebrate brains, neurons and behaviors is the goal. Main contents and applications are galleries of invertebrate brains and nervous systems, confocal serial images of neurons, models of the 3D neural structure reconstructed from confocal images, research tools, bibliographic information about the invertebrate brain, and information about the academic community. It allows for original image data of invertebrate neurons to be collected on the site.

Cerebellar PF http://platform.cerebellum.neuroinf.jp/

It is a digital research archive for cerebellar research. Available resources include mini-reviews of contemporary cerebellar research, and a list of papers and mathematical models for cerebellar operation. It provides a history of cerebellar research and the basic concepts of the cerebellar structure and function, references and images, experimental data for modeling, source codes of neural network models, and other tools for study of the cerebellum. It can be downloaded and use of its contents is free and also one can upload their own contents on the platform by a simple procedure with the approval of the platform committee.

Brain-Machine Interface (BMI) PF http://platform.bmi.neuroinf.jp/

It is a database covering the research fields of the brain-machine related neuroscience, computational theory, robotics, etc. The aim of this platform is to provide organically linked information about BMI to researchers of the field inside and outsideJapan. BMI-PF contains the following contents: database of papers associated with BMI, physiological data of brain activity and muscle activity, programs of computational theory and algorithms, experimental data and demonstration of robotics, and database of research sites around the world.

Integrative Brain Research (IBR) PF
http://www.togo-nou.nips.ac.jp/

The Integrative Brain Research project is a grant group of neuroscientists funded by a Japanese Ministry and includes about 300 Pls. The group consists of five subgroups: Integrative Brain Research, System study on higher brain functions, Elucidation of neural network function in the brain, Molecular Brain Science, and Research on Patho- mechanisms. And to encourage interdisciplinary interaction, the Database Committee of the grant group is actively working on establishing a network of neuroscientists. The Committee maintains three major programs;

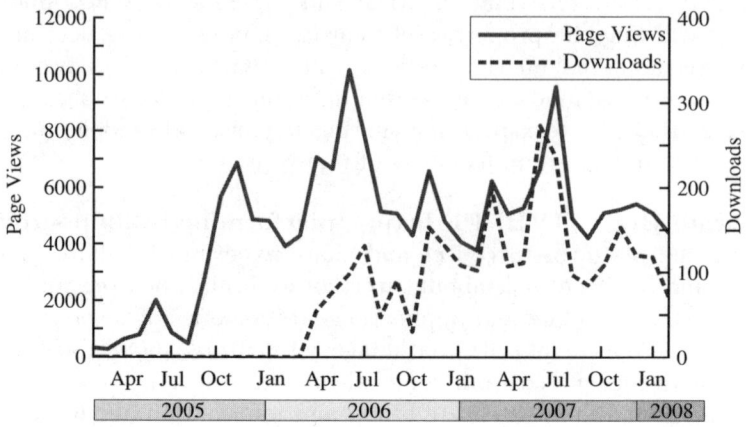

Fig. 9. Number of XooNIps downloaded per month for the years 2005-2008

(1) Neuroscientist database, (2) Neuroscientist SNS (social networking service), and (3) Mouse phenotype database.

Neuroimaging (NIMG) PF http://platform.nimg.neuroinf.jp/
NIMG-PF committee is developing an extended module for visualization of XooNIps contents to display 3D-brain images, to allow searching papers that include activations at the locations specified by pointing on the images. Separately, free software called sBrain is registered for standalone use. It has functions of visualization and searching as well as simulations of neural activation.

Neuroinformatics Common Tools (NICT) PF
http://platform.nict.neuroinf.jp/
Sharing common base technology for neuroscience promotes studies for not only theoretical neuroscientists but also experimental neuroscientists. NI-TECH PF aims to share mathematical theories, analytical tools and neuroinformatics (NI) supporting tools. It includes software tools such as XooNIps, Concierge, SATELLITE[18], and Samurai-Graph[19]. In addition, via this platform, it is expected that collaborated studies can be rapidly and seamlessly conducted on the Internet.

Cerebellar Development Transcriptome (CDT-DB) Database
http://www.cdtdb.brain.riken.jp/
This database is developed independently by Labratory for Molecular Neurogenesis at BSI as one of the database under J-Node. Mouse cerebellum develops through a series of cellular events all within the first three weeks after birth. It studies the entire transcription system (transriptome) responsible for cerebellar development by analyzing the temporal gene expression with fluorescent differential display. GeneChip and microarray analyzes the spatial cellular gene

expression with in situ hybridization. For data mining, annotation expression data basically follows according to the GENE ONTOLOGY (GO) term and systematizes them in a platform searchable by keywords and expression patterns as well as by their combinations. In addition, it is linked to the data of many relevant database websites for easy access to additional information.

5 Conclusion

Since VP answers growing demands of intelligent organization of various types of information, the access to contents from a given research field through such a platform is increasing. Following the VP, a new base-platform system for NI, the XooNIps, has been developed. In addition to the XooNIps database function, XOOPS general modules expanded the platform into a community site to interact with members and visitors. This tool will fulfill the next generation researchers' need to comprehended better their achievements. The future vision for the enhancement of the Identity Management is the consolidation of user's information using LDAP or PKI. The simulation environment alters from an accumulation to a creation. Its ongoing potential will increase with Data Grid Management System SRB (Storage Resource Broker) [20].

The average number of accesses to the J-Node site has significantly increased as in Figure 9 since its contents were enhanced to fulfill the user's convenience. A XooNIps advantage is that it can develop a platform very easily because it is designed as a customizable foundation for a database system. XooNIps gets large numbers of monthly downloads of approximately equal numbers of Japanese and international users. The next step is to allow a seamless exchange of data between these tools in order to build an integrative Neuroinformatics Research Environment for neuroscience communities .

Acknowledgments

We acknowledge Dr. Kazutsuna Yamaji at NII and Mr.Taniguchi at Broadleaf who worked with us for the development of the base-platform XooNIps. We also appreciate to all the platform committee members under the J-Node for the collaborations and member of the NI team, especially Mr. Kamiji for helping us to prepare this manuscript.

References

1. Report on Neuroinformatics from the Global Science Forum Neuroinformatics Working Group of the Organization for Economic Co-operation and Development (June 2002)
2. INCF, http://www.incf.org/
3. Neuroinformatics Japan-Node, http://www.neuroinf.jp/
4. XooNIps official Site, http://xoonips.sourceforge.jp/

5. Usui, S., et al.: Visiome environment:enterprise solution for neuroinformatics in vision science. Neurocomputing 58-60, 1097–1101 (2004)
6. Usui, S.: Visiome: Neuroinformatics Research in Vision Project. Neural Networks 16(9), 1293–1300 (2003)
7. Visiome Platform, http://platform.visiome.neuroinf.org/
8. EPrints.org, http://www.eprints.org/
9. DSpace Foundation, http://www.dspace.org/
10. XOOPS official site, http://www.xoops.org/
11. Open Archives Initiative Protocol for Metadata Harvesting, http://www.openarchives.org/pmh/
12. Simple Dublin core XML format, http://standards-catalogue.ukoln.ac.uk/index/OAI_DC
13. JuNii2, http://ju.nii.ac.jp/
14. Usui, S., et al.: Concierge: Personal database software for managing digital research resources frontiers. In: NEUROINFORMATICS, http://frontiersin.org/neuroscience/
15. Concierge official site, http://concierge.sourceforge.jp/
16. Creative Commons, http://creativecommons.org/
17. GNU General Public License, http://www.gnu.org/copyleft/gpl.html
18. SATELLITE official site, http://stellite.sourceforge.jp/
19. Samurai Graph, http://samurai-graph.sourceforge.jp/
20. Storage Resource Broker, http://www.sdsc.edu/srb/index.php/Main_Page

Collaborative Architectures of Fuzzy Modeling

Witold Pedrycz

[1] Department of Electrical & Computer Engineering
University of Alberta, Edmonton, T6R 2G7 Canada and
Systems Research Institute Polish Academy of Sciences, Warsaw, Poland
pedrycz@ee.ualberta.ca

Abstract. There are evident and profoundly articulated needs to deal with distributed sources of data (such as e.g., sensors and sensor networks, web sites, distributed databases). While recognizing limited accessibility of such data at a global level (which could be associated with technical constraints and/or privacy issues) and fully acknowledging benefits and potentials of collaborative processing, we introduce a concept of Collaborative Computational Intelligence (CI), and collaborative fuzzy models, in particular. Collaboration is realized in different ways by engaging a host of bidirectional interactions between all local processing sites (models) or by proceeding with unidirectional communication in which we establish some mechanisms of developing experience consistency of fuzzy modeling. We offer a coherent taxonomy of various schemes of interaction which in the sequel implies a certain development of a suite of algorithms. In this setting, we highlight a pivotal role of granular information in the establishing of the mechanisms of interaction. In the realm of collaborative fuzzy models and fuzzy modeling we elaborate on the concept of knowledge sharing. We also bring forward a concept of experience–consistent fuzzy system identification showing how fuzzy models built on a basis of limited data can benefit from taking advantage of the past experience conveyed in the form of previously constructed fuzzy models. Proceeding with a more detailed algorithmic framework, we elaborate on the key design issues concerning fuzzy rule-based systems which constitute a dominant category of fuzzy models. Collaboration invokes some mechanisms of aggregation and reconciliation of local findings. We emphasize that the resulting findings such as specific components of models can be quantified in terms of type-2 fuzzy sets – a pursuit which offers an interesting motivation behind this higher type of fuzzy sets

Keywords: distributed Computational Intelligence, fuzzy sets, fuzzy models, information granules, collaboration.

1 Introductory Comments

In the realm of intelligent systems we can witness an ongoing growth of interest in distributed systems whose components (say, nodes, agents, databases, robots) operate in a collaborative fashion. We envision numerous collaborative structures of multi-agent topologies. There is a great deal of methodological and algorithmic pursuits as

J.M. Zurada et al. (Eds.): WCCI 2008 Plenary/Invited Lectures, LNCS 5050, pp. 117–139, 2008.
© Springer-Verlag Berlin Heidelberg 2008

well a wave of application-oriented developments cf. [[3][17][19] Given the nature of the problem tackled by such systems where we commonly encounter nodes (agents) operating quite independently at various levels of specificity, it is very likely that the effectiveness of the overall system depends heavily upon a way in which the agents collaborate and effectively exchange their findings [1][7].

With this regard, given their essential abilities to tackle information granularity, fuzzy sets offer an important avenue to realize a variety of schemes of interaction (communication) in multi-agent systems where various findings obtained locally (viz. at the level of individual agents) are represented in the form of information granules [19][20][21][26][27] rather than plain numeric entities. The communication is realized at the far abstract level given the issues of data security and privacy as well as encountering related technical constraints which prevent us from moving around the masses of numeric data (as e.g., encountered in wireless sensor networks or swarms of robots).

In fuzzy information processing and fuzzy modeling, in particular, not too much has been said about their distributed processing schemes. While there has been a wealth of methodological and algorithmic developments in fuzzy modeling, the subject of distributed and collaborative fuzzy models has not been investigated in great detail. For instance, a lot has been said about rule-based fuzzy models of the form "if x is A_i then $y = f_i(x, a_i)$, $i=1, 2, \ldots, c$ where A_i are fuzzy sets defined in the multidimensional input space and f_i denotes a local model endowed with some parameters (a_i). What if we encounter individual data sites D[1], D[2], ..., D[P] for which such models have to be constructed? Not only they have to be formed on a basis of locally available data D[ii], ii =1, 2, ..., P but they need to collaborate and exchange their findings, reconcile eventual differences and collectively develop fuzzy constructs. What is visible though, is that the communication dwells on *knowledge* rather than *data*. In communication of this nature, we witness a process of knowledge sharing. Formally, the underlying knowledge residing at data site "ii" and being shared between the individual sites can be concisely described as $\mathbf{K}$[ii]. For instance, for the rule based-systems, the shared knowledge assumes the form $\mathbf{K}$[ii] = {A_i[ii], i=1, 2, ..., c}where A_i[ii] are the information granules (fuzzy sets) formed at D[ii]. The knowledge of these fuzzy sets is communicated to all other data sites. We may have another format of $\mathbf{K}$[ii] being a more comprehensive version of knowledge sharing which concerns now both the information granules and the local models, that is $\mathbf{K}$[ii] = {A_i[ii], f_i[ii], a_i[ii]}. A schematic, high-end visualization of such machinery of knowledge sharing is presented in Figure 1.

There is a vast array of mechanisms of interaction between individual components of the distributed system. The two of them deserve careful attention. In the first one all nodes operate locally while collaboration can be established between any two of them, see Figure 2 (a). The linkages between the nodes are bidirectional; knowledge sharing is realized in both directions. Node "ii" benefits from findings (knowledge) shared with it by node "jj" and vice versa: knowledge acquired and conveyed by node "ii" helps to carry out processing (modeling) completed at node "ii". The scenario illustrated in Figure 2 (b) exhibits a highly asymmetric behavior: a single node benefits from various sources of knowledge (models) already formed at some other nodes thus

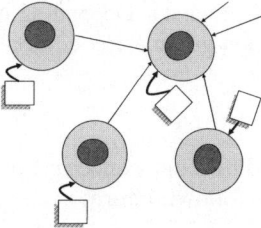

Fig. 1. An overview of a multi-agent distributed system; each agent develops and evolves on a basis of experimental data that are locally available and communicates with other agents establishing some interaction at the global level. Distinguished are the computing core (dark center) and the communication layer of the nodes (surrounding light color region) present in the distributed systems.

augmenting its performance along with the locally available data. This mode of interaction can be referred to as a formation of experience-consistent fuzzy models as the obtained model is not only formed with the aid of some locally available data but, what is more important, takes advantage of some previous experience captured by the already constructed fuzzy models. Note that in this case of interaction, the connections are unidirectional, namely one data site becomes affected by some other models.

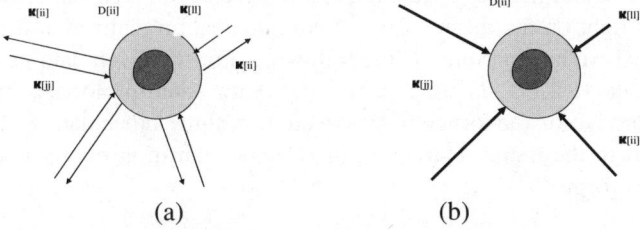

(a)   (b)

Fig. 2. Examples of modes of interaction in a distributed system: (a) collaboration between any two nodes of the system, and (b) experience consistent fuzzy modeling with highly asymmetric interaction

There is also a great deal of in-between collaboration scenarios in which the nodes could engage in some selective interaction where strength of interaction itself could vary quite substantially from node to node. There is also no need to have a fully connected network of nodes. In case of bidirectional links they need not be symmetric. There could be stronger impact exerted by node "jj" on node "ii" while a far weaker connection could be established for interaction realized in the opposite direction.

Referring to fuzzy rule-based systems we mentioned earlier, the experience consistent fuzzy model is formed by considering data D and the knowledge accumulated in the form of the models which now becomes available. Schematically we denote this accumulated knowledge **K** as follows

$$\mathbf{K} = \{ \ D, \ \{ \ \mathbf{K}[ii] \ , ii=1, 2, ..., P \} \ \} \tag{1}$$

where D is the locally available data set. Depending on the details of interaction, we could assume that the information granules (condition parts of the rules) are also utilized and communicated

$$\mathbf{K} = \{ \, D, \; \{ \, \mathbf{K} \, [ii] = \{A_i[ii], i=1, 2, ..., c\} \, , ii=1, 2, ..., P\} \, \} \qquad (2)$$

Alternatively, a transfer of knowledge concerns both conditions and conclusions of the rules, that is $\mathbf{K}$ assumes the following form

$$\mathbf{K} = \{ \, D, \; \{ \, \mathbf{K} \, [ii] = \mathbf{K}[ii] = \{A_i[ii], f_i[ii], a_i[ii]\} \, , ii=1, 2, ..., P\} \, \} \qquad (3)$$

2 Fuzzy Clustering, Information Granules and Communication Mechanisms

In this study, fuzzy clustering is considered as an algorithmic vehicle of information granulation. To make the ensuing discussion strongly focused and make sure that it well links with the detailed algorithmic considerations, we use here the Fuzzy C-Means (FCM) clustering method [2][9][10]. One may note that the presented concepts are far more general and may invoke the use of any other method fuzzy clustering. The FCM algorithm is well documented in the literature and the reader is referred to it with regard to the computational details. What is of interest in our considerations is an observation about an interesting relationship between prototypes and partition matrices as it sheds light on the mechanisms of communication realized at the level of information. What we have in mind is the following. For the given data $\{x_1, x_2, ..., x_N\}$ $x_k \in \mathbf{D} \subset \mathbf{R}^n$, k=1, 2, ..., N we are provided with some prototypes $v_1, v_2, ..., v_c$ which are reflective of the structure discovered in some other data $\mathbf{F} \subset \mathbf{R}^n$ These prototypes lead to the induced structure in $\mathbf{D}$ whose partition matrix is computed in the well-known form

$$u_{ik} = \frac{1}{\sum\limits_{j=1}^{c} \left(\dfrac{\| x_k - v_i \|}{\| x_k - v_j \|} \right)^{2/(m-1)}} \qquad (4)$$

and

$$v_i = \frac{\sum\limits_{i=1}^{c} u_{ik}^m x_k}{\sum\limits_{i=1}^{c} u_{ik}^m} \qquad (5)$$

i =1, 2,...,c; k=1, 2, ..., N. The fuzzification coefficient "m" assumes values greater than 1. The patterns (data) x_k are treated as vectors in $\mathbf{R}^n$ and the distance between two elements in this space $\| . \|$ is typically realized as a weighted Euclidean distance.

More specifically, for any two data $\mathbf{x}$ and $\mathbf{y}$ in $\mathbf{R}^n$, we have $\|\mathbf{x}-\mathbf{y}\|^2 = \sum_{j=1}^{n} \dfrac{(x_j - y_j)^2}{\sigma_j^2}$

where σ_j^2 is the variance of the j-th coordinate (variable) of the feature space.

Apparently the communication of the structural findings is realized at the level of information granules which in this case are represented in the form of prototypes. The calculations of the partition matrix (4) are realized assuming that $\mathbf{D}$ and $\mathbf{F}$ are the same. While this could be the case, we can generalize (4) by admitting that the calculations involve the attributes which belong to the intersection of $\mathbf{D}$ and $\mathbf{F}$ that means that the partition matrix is calculated as

$$u_{ik} = \frac{1}{\sum_{j=1}^{c} \left(\dfrac{\| \mathbf{x}_k - \mathbf{v}_i \|_{\mathbf{F} \cap \mathbf{D}}}{\| \mathbf{x}_k - \mathbf{v}_j \|_{\mathbf{F} \cap \mathbf{D}}} \right)^{2/(m-1)}} \qquad (6)$$

The symbol used above, $\| \ \|_{\mathbf{F} \cap \mathbf{D}}$, underlines that the distance is computed using the features which are shared across $\mathbf{D}$ and $\mathbf{F}$. Evidently an empty intersection or the intersection comprising only a very few variables makes the communication infeasible. The results of computations at $\mathbf{D}$ which give rise to the locally available partition matrix can be communicated outside the local node in the form of prototypes.

The expressions (4) and (5) realize a pair of one-to-one transformations between partition matrices and prototypes considered as the generic mechanism of communication between clustering operating at the level of locally available data.

3   Collaborative Clustering

The communication of knowledge involves a structure $\mathbf{K}[ii]$ which embraces a collection of information granules – fuzzy clusters. Considering that such clusters have been constructed with the use of the FCM algorithm, they are fully characterized by their prototypes and partition matrices. As a matter of fact, these two characterizations are equivalent in the sense highlighted in the previous section. The prototypes and partition matrices are the two possible communication vehicles between the data sites. Given the fact that the data sites concern different data sets, sharing knowledge about the partition matrices is not feasible at all. The prototypes, on the other hand, form a viable alternative to establish this communication. Communicating a limited number of prototypes is also highly attractive since no significant overhead is built in this manner. As the FCM optimization focuses on the partition matrices as one of its components to be adjusted throughout collaboration, we introduce a concept of so-called *induced* partition matrices. Consider the ii-th data site. The prototypes produced at the jj-th data site $\mathbf{v}_1[jj], \mathbf{v}_2[jj],\ldots, \mathbf{v}_c[jj]$ are communicated to the ii-th data site. Given this collection of the prototypes, we induce a partition matrix over the data site D[ii]. Denote it by $U^\sim[ii|jj]$ where the two indexes (ii and jj) point at data sites taking part in this interaction. Its entries are determined in a standard way encountered in FCM computing [2], that is

$$u^\sim_{ik}[ii|jj] = \cfrac{1}{\displaystyle\sum_{j=1}^{c}\left(\cfrac{\|\,\mathbf{x}_k\,[ii] - \mathbf{v}_i\,[jj]\,\|}{\|\,\mathbf{x}_k\,[ii] - \mathbf{v}_j\,[jj]\,\|}\right)^{2}} \tag{7}$$

i=1, 2,…, c; k=1, 2,…,N[ii] and $\mathbf{x}_k \in$ D[ii]. Refer also to Figure 3 which highlights the essence of this mechanism of the collaboration by showing how the communication links have been established.

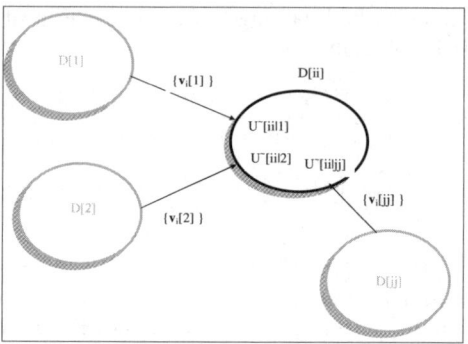

Fig. 3. Data sites and communication realized through passing prototypes and the consecutive generation of the induced partition matrices U$^\sim$[ii|jj]

Proceeding with all other data sites, D[1], …, D[ii-1], D[ii+1],…, D[P], we end up with P-1 induced partition matrices, U$^\sim$[ii|1], U$^\sim$[ii|2],…., U$^\sim$[ii|ii-1], U$^\sim$[ii|ii+1],…., U$^\sim$[ii|P]. The minimization of differences between the U[ii] and U$^\sim$[ii|jj] is used to establish some collaborative activities occurring between the data sites. At the ii-th site, the clustering is guided by the augmented objective function assuming the following form

$$Q[ii] = \sum_{k=1}^{N[ii]}\sum_{i=1}^{c} u^{2}_{ik}[ii]\|\mathbf{x}_k - \mathbf{v}_i\|^{2} + \beta\sum_{\substack{jj=1\\jj\neq ii}}^{P}\sum_{k=1}^{N[ii]}\sum_{i=1}^{c}(u_{ik}[ii] - u^\sim_{ik}[ii\,|\,jj])^{2}\,d^{2}_{ik} \tag{8}$$

where β is a certain nonnegative number. The objective function Q[ii] consists of two components. The first one is nothing but a standard sum of weighted distances between the patterns in D[ii] and their prototypes, $d^{2}_{ik} = \|\,\mathbf{x}_k - \mathbf{v}_i[ii]\,\|^{2}$. In this sense, it is just the objective function encountered in the standard FCM being applied to D[ii] with the fuzzification coefficient m = 2.0. The second component reflects an impact coming from the structures formed at all remaining data sites. The distance between the optimized partition matrix and the induced partition matrices is to be minimized – this requirement is captured by this part of the objective function (8). The scaling coefficient β strikes a balance between the optimization guided by the structure in D[ii] and the already developed structures available at the remaining sites. The value of β implies a certain level of intensity of collaboration; the higher its value, the stronger the collaboration. For β = 0 no collaboration occurs and the problem reduces to the collection of "P" independently run clustering tasks being confined to the corresponding data sites.

In brief, the problem of collaborative clustering can be defined as follows:

> Given a finite number of disjoint data sites with patterns defined in the same feature space, develop a scheme of collective development and reconciliation of a fundamental cluster structure across the sites that it is based upon exchange and communication of local findings where the communication needs to be realized at some level of information *granularity*. The development of the structures at the local level exploits the communicated findings in an *active* manner through minimization of the corresponding objective function augmented by the structural findings developed outside the individual data site. We also allow for retention of key individual (specific) findings that are essential (unique) for the corresponding data site.

We can offer another important and visible category of applications which deal with wireless sensor networks. In such networks, we envision a collection of randomly scattered sensors whose communication is established on *ad hoc* basis. Each node (sensor) collects the data available in its neighborhood and realizes their processing leading to the determination of the *local* characteristics of data (say, formulated as a collection of clusters being observed at this particular local level of the given sensor). At the same time it is recognized that the local processing could benefit from some collective activities established between the sensors. This need for a *global* and collective style of processing is motivated by a limited amount of data available locally and a need to establish a global view at the data collected by the overall network. Each sensor formulates a very limited and localized perception of the environment that has to be augmented by local findings formed by other sensors.

There are essential differences between the proposed approach and the concept which has been encountered in the literature under the umbrella of distributed clustering, cf. [4][5][6][8][12][13][14]16]. In distributed clustering forming some interesting extension of clustering [11][18] it is assumed that the clusters are the same across all data sites. In particular, an assumed mixture model studied there assumes that at each data site there are exactly the same clusters being modeled by Gaussian distributions $N(\mathbf{m}_i, \Sigma_i)$ described by some mean vectors $\mathbf{m}_i$ and covariance matrices Σ_i and put together in the form of some linear combination, cf. [17]. More specifically, we encounter the relationship to be in the form $\sum_{i=1}^{c} \lambda_{ji} N(\mathbf{m}_i, \Sigma_i)$, j=1, 2, ,,,., P where the values of the mixing parameters λ_{ji} are potentially unique for each data site. In contrast, in this study no specific assumptions are being made. The only assumption which is being made here concerns the same granularity of the findings (viz. number of clusters at each data site). As a result, the structure at each data site makes an attempt to reconcile differences however retains and quantifies those that are of particular relevance to the given data site. In the sequel, they are expressed in the form of the fuzzy sets of prototypes or when it comes to membership degrees arise in the format of type-2 fuzzy sets.

Similarly the concept of cluster ensemble, which is present in the literature, is based upon different concepts. Cluster ensemble methods differ in two main ways, that is the way the generic clustering procedure is developed and a way in which the results are combined [23]. Topchy et. al. [24] proposed a consensus function based on informative-theoretic principles and generalized mutual information, in particular. A different consensus function was developed in [6] which is based on some voting/merging method providing a pairwise iterative scheme of combination. Strehl and Ghosh [23] proposed three different ensemble clustering models based on a certain consensus method. All of them use various hypergraph operations to construct the solution. An interesting clustering proposal has been put forward by Wiswedel and Berthold [25].

4   The General Flow of Collaborative Processing

The essence of collaborative clustering pertains to the development of structures at individual data sites on the basis of effective communication of the findings obtained at the level of the individual data sites. There are two phases, namely an optimization of the structures at the individual sites and an interaction between them when exchanging the findings. They intertwine so that these two phases occur in a fixed sequence. A general view of the processing along with its main phases is included in Figure 4.

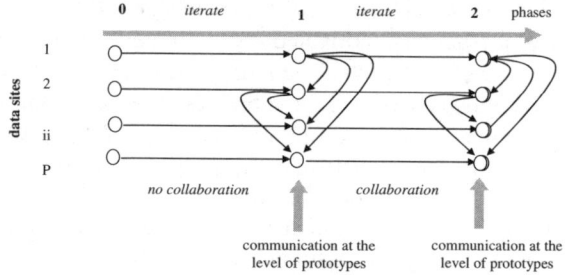

Fig. 4. A general functional view at the processing realized in collaborative clustering

Initially, the FCM algorithm is run independently at each data site (which happens without any communication). After FCM has been terminated at each site, processing stops and the data sites communicate their findings. As already stressed, this communication needs to be realized at some level of information granularity. The effectiveness of the interaction depends on the way in which one data site "talks" to others in terms of what has been discovered so far. Once communication has been established and the nodes are informed about structural findings at other sites, each site proceeds with its optimization pursuits by focusing on the local data while taking into consideration the findings communicated by other data sites. These optimization processes are run independently from each other. Once all of them have declared termination of computing, they are ready to engage in the communication phase. Again they communicate the findings and set up new conditions for the next phase of the FCM optimization. The pair of optimization and communication processes is referred to as a collaboration phase. The overall collaboration takes a finite number of collaboration phases (phases, for short), which terminates once no further significant change in the revealed structure is reported.

As has become clear from this high-end description of the collaboration, there are two important components crucial to the overall process. First, we have to specify a way of communicating and representing findings at some level of granularity (let us recall that we are not allowed to communicate at the level of individual data but have to establish communication at the higher level of abstraction by engaging the exchange of the granular constructs). Second, we have to come up with an augmented objective function whose minimization embraces both the structures at the local level of the individual data sites and reconciles them with the structures communicated by other data sites.

The overall scheme of the collaborative clustering is outlined as follows.

Given: data sites D[1], D[2], ..., D[P]

Choose the number of clusters (c) to be looked for in the collaborative clustering, set up some termination criterion of the FCM, and establish a level of collaboration (interaction) by choosing some nonnegative value of β.

Initial phase   Carry out clustering (FCM) for each data site producing a collection of prototypes $\{v_i[ii]\}$, i=1,2,...,c for each data site.

Collaboration
Iterate {successive phases of collaboration}

Communicate the results about the structure determined at each data site.
For each data site (ii)
{
Minimize (8) at each data site by iteratively proceeding with the iterative calculations of the partition matrix and the prototypes, that is

$$u_{rs}[ii] = \frac{1}{\displaystyle\sum_{j=1}^{c} \frac{d_{rs}^2}{d_{js}^2}}\left[1 - \sum_{j=1}^{c}\frac{\beta\displaystyle\sum_{\substack{jj=1\\jj\neq ii}}^{P} u_{js}^{\sim}[ii\,|\,jj]}{[1+\beta(P-1)]}\right] + \frac{\beta\displaystyle\sum_{\substack{jj=1\\jj\neq ii}}^{P} u_{rs}^{\sim}[ii\,|\,jj]}{[1+\beta(P-1)]} \qquad (9)$$

and

$$v_{rt}[ii] = \frac{\displaystyle\sum_{k=1}^{N[ii]} u_{rk}^2[ii]x_{kt} + \beta\displaystyle\sum_{\substack{jj=1\\jj\neq ii}}^{P}\sum_{k=1}^{N[ii]} (u_{rk}[ii]-u_{rk}^{\sim}[ii\,|\,jj])^2 x_{kt}}{\displaystyle\sum_{k=1}^{N[ii]} u_{rk}^2[ii] + \beta\displaystyle\sum_{\substack{jj=1\\jj\neq ii}}^{P}\sum_{k=1}^{N[ii]} (u_{rk}[ii]-u_{rk}^{\sim}[ii\,|\,jj])^2} \qquad (10)$$

$r=1,2,...,c; t =1, 2, ..., n; s =1, 2, ..., N[ii]$
    } for data site

until termination condition of the collaboration activities has been satisfied.

5 Evaluation of the Quality of Collaboration: Forming a Compromise between Global and Local Characteristics of Data

The evaluation of the quality of the results of collaboration between the data sites requires a careful assessment. As there are partition matrices associated to each of the D[ii]'s, one could think of computing distance between them and treat it as a measure of quality of the ongoing process. While the idea sounds convincing, its realization requires more attention. We should stress the fact that a direct comparison of two partition matrices could not be feasible as we may not have a direct correspondence between their rows (respective clusters). This is a well-known problem identified in the literature, cf. [15]. To get around this shortcoming, we use the concept of proximity and proximity matrix induced by a given partition matrix. Let us recall that for any partition matrix $U = [u_{ik}]$, i=1,2,.., c, k=1,2, ...,m, an induced proximity matrix, that is Prox = [prox(k,l)], k, l=1,2,...,m, comes with entries which satisfy the following properties

(a) symmetry $prox (k_1, k_2) = prox(k_2, k_1)$

(b) reflexitivity $prox(k_1, k_1) = 1.0$

Interestingly enough, here we do not require transitivity (which, albeit nice to have, is always difficult to achieve in practice). The proximity values are based on the corresponding membership degrees occurring in the partition matrix

$$prox(k_1, k_2) = \sum_{i=1}^{c} \min(u_{ik_1}, u_{ik_2}) \qquad (11)$$

It is worth noting that the proximity matrix is more abstract in this form than the original partition matrix it is based upon. It "abstracts" the clusters themselves and this is what we need in this construct. Given the proximity matrix, we cannot "retrieve" the original entries of the partition matrix it was generated from.

Let us consider now the ii-th data site with its partition matrix U[ii] and the induced partition matrices $U^\sim[ii|jj]$, jj =1, 2, ..., ii-1, ii+1, ... , P. To quantify the consistency between the structure revealed at the ii-th data site with those existing at remaining sites by computing the following expression

$$W[ii] = \frac{1}{(N^2[ii]/2)} \sum_{\substack{jj=1 \\ jj \neq ii}}^{P} \left\| Prox(U[ii]) - Prox(U^\sim[ii | jj]) \right\| \qquad (12)$$

More specifically, we consider that the distance between the corresponding proximity matrices is realized in the form of the Hamming distance. In other words, we have

$$\| \text{Prox}(U[ii]) - \text{Prox}(U^{\sim}[ii \mid jj]) \| =$$

$$\sum_{k_1=1}^{N[ii]} \sum_{k_2>k_1}^{N[ii]} | \text{prox}(k_1, k_2)[ii] - \text{prox}(k_1, k_2)^{\sim}[ii \mid jj] | \qquad (13)$$

where $\text{prox}(k_1, k_2)[ii]$ denotes the (k_1, k_2)- entry of the proximity matrix $U[ii]$. Similarly, $\text{prox}(k_1, k_2)^{\sim}[ii|jj]$ is the corresponding (k_1, k_2) entry of the proximity matrix produced by the induced partition matrix $U^{\sim}[ii|jj]$. In a nutshell, rather than working at the level of comparing the individual partition matrices (which requires knowledge of the explicit correspondence between the rows of the partition matrices), we generate their corresponding proximity matrices that allows us to carry out comparison at this more abstract level. Next summing up the values of $W[ii]$ over all data sites, we arrive at the global level of consistency of the structure discovered collectively through the collaboration

$$W = W[1] + W[2] + \ldots + W[P] \qquad (14)$$

The lower the value of W, the higher is the consistency between the "P" structures. Likewise the values of W being reported during successive phases of the collaboration can serve as a sound indicator as to the progress and quality of the collaborative process and serve as a suitable termination criterion; refer to Table 1. In particular, when tracing the successive values of W, one could stop the collaboration once no further changes in the values of W are reported. The use of the above consistency measure is also essential when gauging the intensity of collaboration and adjusting its level through changes of β. Let us recall that this parameter shows up in the minimized objective function and shows how much other data sites impact the formation of the clusters at the given site. Higher values of β imply stronger collaborative linkages established between the sites. By reporting the values of W treated as a function of β, that is $W = W(\beta)$, we can experimentally optimize the intensity of collaboration. One may anticipate that while for low values of β no collaboration occurs and the values of W tend to be high, large values of β might lead to competition and subsequently the values of $W(\beta)$ may tend to be high. Under some conditions, no convergence of the collaboration process could be reported. There might be some regions of optimal values of β. Obviously, the optimal level (intensity) of collaboration depends upon a number of parameters of the collaborative clustering, in particular the number of clusters and the number of data sites involved in the collaboration. It could also depend upon the data themselves.

6 Fuzzy Sets of Type-2 in the Quantification of the Effect of Collaboration

As we have underlined, it is also advantageous to assess the quality of the results by evaluating their consistency and expressing a level of differences. Here the quantification of results completed in terms of type-2 fuzzy set constitutes an interesting alternative or prototypes being treated as granular constructs, which is fuzzy sets rather than plain numeric entities. Type-2 fuzzy sets are granular constructs that generalize fuzzy sets in the sense that their membership functions do no assume numeric membership grades but instead of them we encounter fuzzy sets defined in the unit interval. Interestingly, type-2 fuzzy sets have been discussed in various settings however very little was said about a determination of their membership functions. In collaborative clustering we estimate the membership function on a basis of a collection of membership grades available in different partition matrices. To be more specific, let us revisit what becomes known about cluster membership of pattern $\mathbf{x}$ in D[ii] given the available results of collaborative clustering. The membership in the i-th cluster is computed using the prototypes of D[ii] and is denoted as $u=u_i$. The prototypes optimized for the jj-th data site, jj =1, 2, ..., ii-1, ii+1, ..., P give rise to the membership of $\mathbf{x}$ to the same i-th cluster. Denote them by $z_1, z_2, ..., z_{P-1}$. All in all, we obtain a collection of membership grades which are now captured in a form of type-2 fuzzy set. The corresponding membership function is determined by solving a certain optimization problem [21][22] which realizes an idea which could be referred to as a principle of *justifiable* granularity. We consider triangular fuzzy set as one of the simplest versions of the membership functions. It is also legitimate in the context of this application given that we operate in presence of limited experimental evidence. The modal value of the fuzzy set is the membership value obtained with the use of the prototypes present at D is equal to "u". Consider now the values of z_i that are lower than u, $z_i < u$ We use them in the formation of the left-hand side of the linear portion of the membership function, refer to Figure 5.

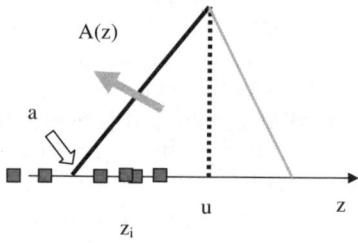

Fig. 5. Computation of a membership function of fuzzy set of type-2; note that is order to maximize the performance index, we rotate the linear segment of the membership function around the modal value of the fuzzy set. Small dark boxes denote available experimental data. The same estimation procedure applies to the right-hand side of the fuzzy set.

There are two requirements guiding the design of the fuzzy set, namely
(a) maximize the experimental evidence of the fuzzy set; this implies that we tend to "cover" as many numeric data as possible, viz. the coverage has to be made as high as possible. Graphically, in the optimization of this requirement, we rotate the linear segment up (clockwise) as illustrated in Figure 4. Normally, the sum of the membership grades $A(z_i)$,

$$\sum_i A(z_i)$$ where A is the linear membership function to be optimized

with respect to its slope and z_i is located to the left to the modal value (u) has to be maximized

(b) Simultaneously, we would like to make the fuzzy set as specific as possible so that is comes with some well defined semantics. This requirement is met by making the support of A as small as possible, that is $\min_a |u - a|$

To accommodate the two conflicting requirements, we have to combine (a) and (b) into a form of a single scalar index which in turn becomes maximized. Two alternatives could be sought, say

$$\max_{a \neq u} \frac{\sum_i A(z_i)}{|u - a|} \tag{15}$$

or

$$\sum_i (1 - A(z_i))(u - a) \tag{16}$$

The linearly decreasing portion of the membership function positioned at the right-hand side of the modal value (u) is optimized in the same manner. We exclude a trivial solution of a = u in which case the fuzzy set of type-2 collapses to a type-1 fuzzy sets (with numeric values of membership function). We can use this construct in the formation of granular prototypes and fuzzy sets of type-2.

7 Hierarchical Clusters of Clusters

In the previous collaboration strategy, we have assumed that the collaborating data sites exchange their findings (prototypes) at the same level of granularity (there is the same number of clusters c[ii] across all collaborating parties). One can envision a different architecture and the underlying strategy of reconciling findings at the local level. This brings the concept of *clusters of clusters*. The essence of the method is that the structural findings formed at the lowest level are reconciled in the form of structure that is common to all local data sites. The prototypes at each D[ii] are considered together and clustered into "cc" clusters formed at the higher level. In the sequel, the resulting partition matrix is used to convey information about the behavior of the original prototypes when being confronted with structural findings (prototypes) at other data sites. More specifically, using the partition matrix U formed at the higher level of this hierarchy, we form some relevancy index $\gamma(U)$ to quantify the impact on any of the prototypes coming from the data site. The index which applies to each column of U associates the ith prototype at data site D[ii] with $\gamma_i(U)[ii]$ which articulates how much identity this prototypes retains when confronted with the data structure obtained at other data sites. The index is included in the modified objective function used to cluster data at the ii-th data site

$$Q = \sum_{i=1}^{c[ii]} \sum_{x_k \in X[ii]} \gamma_i(U)[ii] \| x_k - v_i[ii] \|^2 \tag{17}$$

The formation of the clusters of clusters is an interactive process: we start with the development of structure individually at D[ii], cluster the obtained prototypes and use the relevancy index to minimize the modified objective function as shown above. The clusters formed in this way are again clustered at the higher level of the hierarchy. This leads to the new values of the relevance index and the process iterates until it stabilizes. The number of clusters "cc" assumed at the higher level plays an important role as a measure to express the intensity of reconciliation of the individual findings. Strong interaction becomes realized when we consider only a few clusters. In case when $cc = c[1] + c[2]+... + c[P]$ there is no interaction at all (each prototypes retains its identity) and the values of $\gamma_i(U)[ii]$ are all equal to 1 not affecting the form of the objective function and thus not changing the prototypes. The strength of the structural interaction controlled by the values of the number of clusters "cc" may affect the dynamics of collaboration with the likelihood that its lower values associated with stronger collaboration may imply eventual instability.

8   Experience Consistent Fuzzy Models: A Concept

In this modeling scenario, it becomes advantageous not only consider currently available data but also actively exploit previously obtained findings. Such observations bring us to the following formulation of the problem:

> Given some experimental data, construct a model which is *consistent* with the findings (models) produced for some previously available data. Owing to the existing requirements such as data privacy or data security of data as well as some other technical limitations, in the construction of the model an access to these previous data is not available however we can take advantage of the knowledge of the parameters of the existing models.

Considering the need to achieve a certain desired consistency of the proposed model with the previous findings, we refer to the development of such models as *experience-based* or *experience-consistent* fuzzy modeling.

When dealing with experience-consistent models, we may encounter a number of essential constraints which imply a way in which the underlying processing can be realized. For instance, it is common that the currently available data are quite limited in terms of its size (which implies a limited evidence of the data set) while the previously available data sets could be substantially larger meaning that relying on the models formed in the past could be beneficial for the development of the current model. There is also another reason in which the experience –driven component plays a pivotal role. The data set D could be quite small and affected by a high level of noise – in this case it becomes highly legitimate to seriously consider any additional experimental evidence available around.

In the realization of the consistent-oriented modeling, we consider the following scenario. Given is a data set D using which we intent to construct a fuzzy rule-based model. There is a collection of data sets $D_1, D_2, ..., D_P$. For each of them developed is an individual fuzzy model. Those local models are available when seeking consistency with the fuzzy models formed for D_{ii}, $ii=1, 2, ..., P$. At the same time, it is worth stressing that the data sets themselves are not available to any processing and modeling realized at the level of D.

The underlying architectural details of the rule-based model considered in this study are as follows. For each data site D and D_{ii}, we consider the rules with local regression models assuming the form

Data D

$$\text{-if } \mathbf{x} \text{ is } B_i \text{ then } y = \mathbf{a}_i^T \mathbf{x} \qquad (18)$$

where $\mathbf{x} \in \mathbf{R}^{n+1}$ and B_i are fuzzy sets defined in the n-dimensional input space, i=1, 2,..., c. The local regression model standing in the i-th rule is a linear regression function described by a certain vector of parameters $\mathbf{a}_i$. More specifically, the n-dimensional vector of the original input variables is augmented by a constant input so we have $\mathbf{x} = [x_1 \ x_2 \ ... \ x_n \ 1]^T$ and $\mathbf{a} = [a_1 \ a_2 \ ... \ a_n \ a_0]^T$ where a_0 stands for a bias term that translates the original hyperplane.

The same number of rules (c) is encountered at all other data sites, $D_1, D_2, ..., D_P$. The format of the rules is the same as for D, that is for the ii-th data sited D_{ii} we have

$$\text{-if } \mathbf{x} \text{ is } B_i[ii] \text{ then } y = \mathbf{a}_i[ii]^T \mathbf{x} \qquad (19)$$

As before the fuzzy sets in the condition part of the i-th rule are denoted by $B_i[ii]$ while the parameters of the local model are denoted by $\mathbf{a}_i[ii]$. The index in the square brackets refers to the specific data site, that is D_{ii} for $\mathbf{a}_i[ii]$.

Alluding to the format of the data at D, it comes in the form of input – output pairs $(\mathbf{x}_k, y_k)$, k=1, 2,..., N which are used to carry out learning in a supervised mode. The previously collected data sets denoted by $D_1, D_2, ..., D_P$ consists of N_1, N_2, and N_P data points. We assume that due to some technical and non-technical reasons, the data available at D_j cannot be shared with D however the communication between the data sites can be realized at the higher conceptual level such as those involved the parameters of the fuzzy models.

9   The Experience-Consistent Development of the Rule-Based Model

Alluding to the formulation of the problem, we consider a rule-based model constructed on a basis of data D where in the construction of the model we are influenced by the models formed with the use of $D_1, D_2, ...,$ and D_P. To realize a mechanism of experience consistency, we introduce several pertinent performance indexes which are crucial in the quantification of this mechanism.

Given the architecture of the rule-based system, it is well known that we encounter here two fundamental design phases, that is (a) a formation of the fuzzy sets standing in the conditions of the rules and (b) the estimation of the corresponding conclusion parts. There are numerous ways of carrying out this construction. Typically, when it comes to the condition parts of the rules, the essence of the design is to granulate data by forming a collection of fuzzy sets. The common technique relates to fuzzy clustering when the condition part of the rule involves a fuzzy set defined in $\mathbf{R}^n$ or a Cartesian product of fuzzy sets defined in $\mathbf{R}$. The conclusion part where we encounter local regression models is formed by estimating the parameters $\mathbf{a}_i$. Such an estimation process is standard to a high extent as it is nothing but a global minimization of the well-known squared error criterion.

The organization of the consistency–driven optimization relies on the reconciliation of the conclusion parts of the rules. We assume that the condition parts, viz. fuzzy sets are developed independently from each other. In other words, we cluster data in the input space of D, D_1, ... , D_P assuming the same number of clusters (c) which results in the same collection of rules. Then the mechanism of experience consistency is realized for the conclusions of the rules. Given the independence of the construction process of the clusters at the individual sites, before moving on with the quantification of the obtained consistency of the conclusion parts of the rules, it becomes necessary to align the information granules obtained at D and the individual data sites D_i.

9.1   The Construction of Information Granules of Conditions of the Rules

Information granules in the input space can be developed in many different ways, cf. [2][15][21]. We are of opinion that they need to be directly reflective of the nature of data being available which brings fuzzy clustering as an intuitively appealing alternative. More specifically, the FCM algorithm comes as a suitable algorithmic vehicle. For the given number of clusters (c), we minimize a standard objective function and as a result obtain a collection of prototypes and a partition matrix. In the ensuing communication schemes of consistency development we will be relying on the exchange of the prototypes.

9.2   The Consistency-Based Optimization of Local Regression Models

To make the ensuing formulas concise, we use a shorthand notation FM, FM[1], FM[2], ..., FM[P] to denote rule-based models pertaining to data D, D[1],... etc.

As usual the optimal parameters of the local models occurring in the conclusions of the rules are chosen in such a way so that they minimize the sum of squared errors

$$Q = \frac{1}{N} \sum_{\substack{x_k \in D \\ y_k \in D}} (FM(\mathbf{x}_k) - y_k)^2 \qquad (20)$$

For given fuzzy sets of conditions, the determination of the parameters of the linear models is standard and well documented in the literature. Considering the form of the rule-based system, the output of the fuzzy model is determined as a weighted combination of the local models with the weights being the levels of activation of the individual rules. More specifically we have

$$\hat{y}_k = \sum_{i=1}^{c} u_i(\mathbf{x}_k)\mathbf{a}_i^T\mathbf{x}_k \qquad (21)$$

where $u_{ik} = u_i(\mathbf{x}_k)$ is a membership degree of the k-th data $\mathbf{x}_k$ to the i-th cluster being computed on a basis of the already determined prototypes in the input space. In a nutshell (21) comes as a convex combination of the local models which aggregates the local models by taking advantage of the weight factors expressing a contribution of each model based upon the activation reported in the input space.

The essence of the consistency-driven modeling is to form local regression models occurring in the conclusions of the rules on a basis of data D while at the same time making the model perform in a consistent manner (viz. close enough) to the rule-based model formed for the respective D_i's. The following performance index strikes a sound balance between the model formed exclusively on a basis of data D and the consistency of the model with the results produced by the models formed on a basis of some other data sites D_i's, that $FM[j](\mathbf{x}_k)$

$$V = \sum_{\substack{x_k \in D \\ y_k \in D}} (FM(\mathbf{x}_k) - y_k)^2 + \alpha \sum_{j=1}^{P} \sum_{\substack{x_k \in D \\ y_k \in D}} (FM(\mathbf{x}_k) - FM[j](\mathbf{x}_k))^2 \tag{22}$$

The calculations of $FM[j](\mathbf{x}_k)$ for some $\mathbf{x}_k$ in D require some words of explanation. The model is communicated to D by transferring the prototypes of the clusters (fuzzy sets) and the coefficients of the linear models standing in the conclusions of the rules refer to Figure 6.

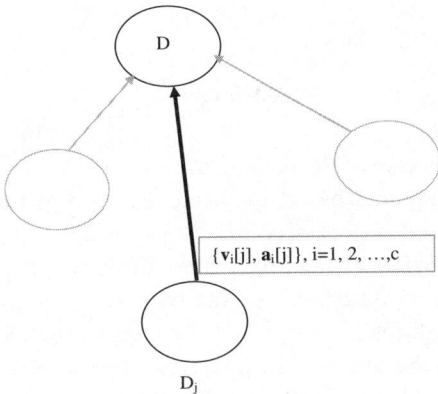

$\{v_i[j], a_i[j]\}$, i=1, 2, ...,c

D_j

Fig. 6. Communication between D and D_j realized by transferring parameters of the rule-based model available at individual data sites D_j

When used at D, the prototypes $v_i[j]$, i=1, 2,...,c give rise to an induced partition matrix in which the k-th column (for data $\mathbf{x}_k$) assumes the following membership values $w_i(\mathbf{x}_k)$ computed in the standard manner as being encountered when running the FCM algorithm, that is

$$w_i(\mathbf{x}_k)[j] = \frac{1}{\sum_{l=1}^{c} \left(\dfrac{\mathbf{x}_k - v_i[j]}{\mathbf{x}_k - v_l[j]} \right)^{1/m-2}} \tag{23}$$

The transferred parameters of the local models obtained at the j-th data site produce the output of the model $FM[j](\mathbf{x}_k)$ obtained at D as a weighted sum of the form

$$FM[j](\mathbf{x}_k) = \sum_{i=1}^{c} w_i(\mathbf{x}_k)[j]\mathbf{a}_i^T(j)\mathbf{x}_k \qquad (24)$$

where $\mathbf{x}_k \in D$.

The minimization of the performance index V for some predefined value of α leads to the optimal vectors of the parameters of the linear models $\mathbf{a}_i(opt)$, i=1, 2,..., c which is reflective of the process of satisfying the consistency constraints. The detailed derivations are a quite standard algebraic exercise. The final result comes in the form

$$\mathbf{a}_{opt} = \frac{1}{\alpha P + 1}\hat{X}^{\#}(\mathbf{y} + \alpha\mathbf{y}_1 + \alpha\mathbf{y}_2 + ... + \alpha\mathbf{y}_P) \qquad (25)$$

where $\mathbf{y}_i$ is a vector of the outputs of the i-th fuzzy model (formed on a basis of D_i) where the corresponding coordinate of this vector the output obtained for the corresponding input, that is

$$\mathbf{y}_i = \begin{bmatrix} FM[i](\mathbf{x}_1) \\ FM[i](\mathbf{x}_2) \\ \\ FM[i](\mathbf{x}_N) \end{bmatrix} \qquad (26)$$

where $\hat{X}^{\#}$ is a pseudoinverse of the data matrix.

An overall balance captured by (22) is achieved for a certain value of α. An evident tendency of increased impact becomes clearly visible: higher values of α stress higher relevance of other models and their more profound impact on the constructed model. First, the model is constructed on the basis of D. Second, the consistency is expressed on a basis of differences between the constructed model and those models coming from D_is where the differences are assessed with the use of data D. There is another interesting view at the format of this performance index under minimization. The second component in V plays a role that is similar to a *regularization* term being typically used in estimation problems however its origin here has a substantially different format from the one encountered in the literature. Here, we consider other data (and models) rather than focusing on the complexity of the model expressed in terms of its parameters to evaluate the performance of the model.

While the semantics of the above performance index (22) is straightforward, a choice of the value of α requires some attention. To optimize the level of contribution coming from the data sets, we may adhere to the following evaluation process which invokes two fundamental components. As usual, the quality of the optimal model is evaluated with respect to data D. The same optimized model (viz. its prototypes and the parameters of the local regression models) are made available at D_i and the quality of the model is evaluated there with the use of the local data present there. We combine the results (viz. the corresponding squared errors) by adding their normalized values. Given these motivating notes, an index quantifying a global behavior of the optimal model arises in the following form

$$VV = \frac{1}{N} \sum_{\substack{x_K \in D \\ y_k \in D}} (FM(\mathbf{x}_k) - y_k)^2 + \sum_{j=1}^{P} \frac{1}{N_j} \sum_{\substack{x_k \in D_j \\ y_k \in D_j}} (FM(\mathbf{x}_k) - y_k)^2 \qquad (26)$$

A schematic view of computing and communication of findings being realized with the aid of (26) is illustrated in Figure 7.

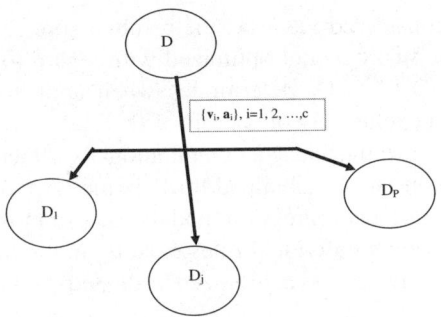

Fig. 7. A quantification of the global behavior of the consistency – based fuzzy model

Note that when the fuzzy model FM(.) is transferred to D_j, as before we communicate the prototypes obtained at D and the coefficients of the local linear models of the conclusion part of the rules. Likewise as shown in (26), the output of the fuzzy model obtained for $\mathbf{x}_k \in D_j$ involves the induced value of membership degree $w_j(\mathbf{x}_k)$ and an aggregation of the local regression models.

Apparently the expression of VV is a function of α and the optimized level of consistency is such for which VV attains its minimal value, namely

$$\alpha_{opt} = \arg \text{Min } VV(\alpha) \qquad (27)$$

The optimization scheme (27) along with its evaluation mechanisms governed by (26) can be generalized by admitting various levels of impact each data D_i might have in the process of achieving consistency. To do so, we introduce some positive weights w_i, i=1, 3, ...p which are afterwards used in the performance index

$$V = \sum_{\substack{x_k \in D \\ y_k \in D}} (FM(\mathbf{x}_k) - y_k)^2 + \alpha \sum_{j=1}^{P} w_j \sum_{\substack{x_k \in D \\ y_k \in D}} (FM(\mathbf{x}_k) - y_k)^2 \qquad (28)$$

Lower values of w_i indicate lower influence of the model formed on a basis of data D_i when constructing the model for data D. The role of such weights is particularly apparent when dealing with data D_i which are in some temporal or spatial relationships with respect to D. In these circumstances, the values of the weights are reflective of how far (in terms of time or distance) the sources of the individual data are from D. For instance, if D_j denotes a collection of data gathered some time ago in comparison to the currently collected data D_i, then it is intuitively clear that the weight w_j is lower than w_i.

As an auxiliary performance index that expresses a quality of the model for which (26) has been minimized with α being selected with regard to (28), we consider the following expression

$$Q^{\sim} = \frac{1}{N} \sum_{\substack{x_k \in D \\ y_k \in D}} (FM(\mathbf{x}_k) - y_k)^2 \tag{29}$$

The values of $Q^{\sim}$ considered vis-à-vis the results expressed by (26) are helpful in assessing an extent the fuzzy model optimized with regard to data D while achieving consistency with D_1, D_2, ..., D_p deteriorates when applied to D over the optimal model being optimized exclusively on a basis of D.

In what follows, we also introduce a computationally effective measure articulating a level of experience consistency obtained for D in the form of *granular* characterization of the parameters of local regression models. Before moving with the details, we elaborate on a way in which individual rules existing in the models formed for D and the data sites D_1, D_2, ..., D_p are "synchronized" (aligned).

9.3 The Alignment of Information Granules

The rules forming each fuzzy model have been formed independently at each data site. If we intend to evaluate a level of consistency of the rules at D vis-à-vis the modeling evidence available at D_j, some alignment of the rules become essential. Such an alignment concerns a way of lining up the prototypes forming the condition part of the rules. We consider the models obtained at D and D_j, j=1, 2, ..., P with their prototypes $\mathbf{v}_1$, $\mathbf{v}_2$, ..., $\mathbf{v}_c$ and $\mathbf{v}_1[j]$, $\mathbf{v}_2[j]$,..., $\mathbf{v}_c[j]$. We say that the rule "i" at D and the rule "*l*" at D_j are aligned if the prototypes $\mathbf{v}_k$ and $\mathbf{v}_l[j]$ are the closest within the collections of the prototypes produced for D and D_j. The alignment process is realized by successively finding the pairs of the prototypes being characterized by the lowest mutual distance. Overall, the alignment process can be described in the following manner:

Form two sets of integers (indexes) **I** and **J**, where $\mathbf{I} = \mathbf{J} = \{1, 2, ...,c\}$. Start with an empty list of alignments, L= $\varnothing$.

Repeat
      Find a pair of indexes i_0 and j_0 for which the distance attains minimum
                    $(i_0, j_0) = \arg\min_{i,l} \|\mathbf{v}_i - \mathbf{v}_l(j)\|$
      The pair (i_0, j_0) is added to the list of alignments, L= L $\cup$ (i_0, j_0)
           Reduce the set of indexes I and J by removing the elements that were placed on the list of alignments, $\mathbf{I} = \mathbf{I} \setminus \{i_0\}$ and $\mathbf{J} = \mathbf{J} \setminus \{j_0\}$
until $\mathbf{I} = \varnothing$

Once the above loop has been completed, we end up with the list of alignment of the prototypes in the form of pairs (i_1, j_1), (i_2, j_2),..., (i_c, j_c)

10 Characterization of Experience-Consistent Models through Its Granular Parameters

Once the mechanism of experience consistency has been completed and the local models have been aligned (following the scheme provided in the previous section), we can now look at the characterization of the set of the related parameters of the local regression models. In essence, through the alignment of the prototypes ad D and D_j, we obtain the corresponding vectors of the parameters of the regression models of the conclusion parts. Denote these vectors corresponding to a certain rule by $\mathbf{a}$, $\mathbf{a}_i$, $\mathbf{a}_k$, ..., and $\mathbf{a}_l$ altogether arriving at P+1 of them. If we now consider the j-th coordinate of all of them, we obtain the numeric values a_j, a_{ij}, ..., a_{lj}. The essence of their aggregation concerns their global representation completed in the form of a single fuzzy set. The employ the same aggregation scheme as presented in Section 6. We intend to span a unimodal fuzzy set A over the set of numeric parameters a_j in such a way that A "represents" these data to the highest possible extent following the principle of justifiable granularity. Consider triangular type of membership functions of the granular parameters of the fuzzy model, the result of the aggregation becomes a triangular fuzzy number of the j-th parameter of the local regression model. Denote it by $A_j =(a_{j-}, a_j, a_{j+})$ with the three parameters denoting the lower, modal, and upper bound of the fuzzy number. Applying the same procedure to all remaining parameters of the vector $\mathbf{a}$, we produce the corresponding fuzzy numbers A_1, A_2, ..., A_{j-1}, A_{j+1}, ..., A_n, and A_0. Given them the rule in D reflects the nature of the incorporated evidence offered by the remaining models D_1, D_2, etc. If there is a fairly high level of consistency, this effect is manifested through a fairly "concentrated" fuzzy number. Increasing inconsistency results in a broader, less specific fuzzy number of the parameters. In summary, a certain fuzzy rule assumes the following format

$$\text{If x is B then } Y = A_0 \oplus A_1 \otimes x_1 \oplus A_2 \otimes x_2 \oplus ... \oplus A_n \otimes x_n \qquad (30)$$

The symbols $\oplus$ and $\otimes$ being used above underline the nonnumeric nature of the arguments standing in the model over which the multiplication and addition are carried out. For given numeric inputs $\mathbf{x} =[x_1, x_2, ..., x_n]^T$ the resulting output Y of this local regression model is again a triangular fuzzy number $Y = <w, y, z>$ where their parameters are computed as follows

Modal value $y = a_0+ a_1x_1+a_2x_2+...+a_nx_n$
Lower bound $w = a_0 +\min(a_{1-}x_1, a_{1+}x_1) + \min(a_{2-}x_2, a_{2+}x_2)+...+ \min(a_{n-}x_n, a_{n+}x_n)$
Upper bound $z = a_0 +\max(a_{1-}x_1, a_{1+}x_1) + \max(a_{2-}x_2, a_{2+}x_2)+...+ \max(a_{n-}x_n, a_{n+}x_n)$

The above process is of the formation of the fuzzy numbers of the local regression model of the rule is repeated for all rules. Finally, we arrive at the rules of the following form

$$\text{If x is } B_1 \text{ then } Y = A_{10} \oplus A_{11} \otimes x_1 \oplus A_{12} \otimes x_2 \oplus ... \oplus A_{1n} \otimes x_n$$
$$\text{If x is } B_2 \text{ then } Y = A_{20} \oplus A_{21} \otimes x_1 \oplus A_{22} \otimes x_2 \oplus ... \oplus A_{2n} \otimes x_n \qquad (31)$$
$$....$$
$$\text{If x is } B_c \text{ then } Y = A_{c0} \oplus A_{c1} \otimes x_1 \oplus A_{c2} \otimes x_2 \oplus ... \oplus A_{cn} \otimes x_n$$

Given this structure, the input vector $\mathbf{x}$ implies the output fuzzy set with the following membership function

$$Y = \sum_{i=1}^{c} w_i(\mathbf{x}) \otimes [A_{i0} \oplus (A_{i1} \otimes x_1) \oplus (A_{i2} \otimes x_2) \oplus \ldots \oplus (A_{in} \otimes x_n)] \tag{32}$$

Owing to the fact of having fuzzy sets of the parameters of the regression model in the conclusion part of the rules, Y becomes a fuzzy number rather than a single numeric value.

11 Conclusions

Distributed architectures of fuzzy models bring yet another dimension to constructs of Computational Intelligence by addressing the rapidly growing needs of system modeling. In this scenario, fuzzy models are formed on a basis of their locally available data sets and findings available from other data. We have stressed the importance of information granules as an efficient vehicle to exchange findings and initiate collaboration while making this communication process viable given a number of technical and non-technical constraints. Fuzzy clustering being regarded as a vehicle to granulate information serves as an important design vehicle in the framework of distributed fuzzy models. We have discussed different schemes of knowledge communication by elaborating on the nature of the knowledge being shared and ways in which specific facets of such knowledge become communicated. While there is a genuine spectrum of collaborative schemes, the study offered a detailed discussion on the two extremes in which all agents are engaged in collaborative pursuits and when one agent in building its fuzzy model benefits from experience available at other data sites which is made available in the form of the parameters of the models formed there. We have demonstrated that all collaboration pursuits contribute to the enhancements of the resulting fuzzy models in a way their parameters are expressed by type-2 fuzzy sets where these fuzzy set constructs are helpful in the quantification of the diversity of findings available throughout the network of agents.

References

1. Acampora, G., Loia, V.: A Proposal of Ubiquitous Fuzzy Computing for Ambient Intelligence. Information Sciences 178, 631–646 (2008)
2. Bezdek, J.C.: Pattern Recognition with Fuzzy Objective Function Algorithms. Plenum Press, New York (1981)
3. Cheng, C.B., Chan, C.C.H., Lin, K.C.: Intelligent Agents for e-Marketplace: Negotiation with Issue Trade-Offs by Fuzzy Inference Systems. Decision Support Systems 42, 626–638 (2006)
4. Costa da Silva, J., Klusch, M.: Inference in Distributed Data Clustering. Engineering Applications of Artificial Intelligence 19, 363–369 (2006)
5. Dudoit, S., Fridlyand, J.: Bagging to Improve the Accuracy of a Clustering Procedure. Bioinformatics 19, 1090–1099 (2003)

6. Dimitriadou, A., Weingessed, K., Hornik, K.: Voting-Merging: An Ensemble Method for Clustering. In: Proc. Int. Conf. on Artificial Neural Networks, Vienna, pp. 217–222 (2001)
7. Genesereth, M.R., Ketchpel, S.P.: Software Agents. Communications of the ACM 37, 48–53 (1994)
8. Hubert, L., Arabie, P.: Comparing Partitions. Journal of Classification 2, 193–218 (1985)
9. Hoppner, F., et al.: Fuzzy Cluster Analysis. J. Wiley, Chichester (1999)
10. Jain, A., Murt, M., Flynn, P.: Data Clustering: a Review. ACM Computing Surveys 31, 264–323 (1999)
11. Jain, A.K., Dubes, R.C.: Algorithms for Clustering Data. Prentice-Hall, Englewood Cliffs (1988)
12. Johnson, E., Kargupta, H.: Collective, Hierarchical Clustering from Distributed, Heterogeneous Data. In: Zaki, M.J., Ho, C.-T. (eds.) KDD 1999. LNCS (LNAI), vol. 1759, pp. 221–244. Springer, Heidelberg (2000)
13. Krogh, A., Vedelsby, J.: Neural Networks Ensembles, Cross Validation, and Active Learning. In: Advances in Neural Information Processing Systems, pp. 231–238. MIT Press, Cambridge (1995)
14. Leung, Y., Zhang, J., Xu, Z.: Clustering by Space-Space Filtering. IEEE Transactions on Pattern Analysis and Machine Intelligence 22, 1396–1410 (2000)
15. Loia, V., Pedrycz, W., Senatore, S.: P-FCM: a Proximity-Based Fuzzy Clustering for User-Centered Web Applications. Int. J. of Approximate Reasoning 34, 121–144 (2003)
16. Merugu, S., Ghosh, J.: A Privacy-Sensitive Approach to Distributed Clustering. Pattern Recognition Letters 26, 399–410 (2005)
17. Nowak, R.: Distributed EM Algorithms for Density Estimation and Clustering in Sensor Networks. IEEE Trans. on Signal Processing 51, 2245–2253 (2003)
18. Oehler, K.L., Gray, R.M.: Combining Image Compression and Classification Using Vector Quantization. IEEE Transactions on Pattern Analysis and Machine Intelligence 17, 461–473 (1995)
19. Pedrycz, W., Vukovich, G.: Clustering in the Framework of Collaborative Agents. In: Proc. 2002 IEEE Int. Conference on Fuzzy Systems, vol. 1, pp. 134–138 (2002)
20. Pedrycz, W.: Collaborative fuzzy clustering. Pattern Recognition Letters 23, 675–686 (2002)
21. Pedrycz, W.: Knowledge-Based Clustering: From Data to Information Granules. J. Wiley, Chichester (2005)
22. Pedrycz, W., Rai, P.: Collaborative Clustering with the Use of Fuzzy C-Means and Its Quantification. Fuzzy Sets & Systems (to appear)
23. Strehl, A., Ghosh, J.: Cluster ensembles: a Knowledge Reuse Framework for Combining Multiple Partitions. Journal of Machine Learning Research 3, 583–617 (2002)
24. Topchy, A., Jain, K., Punch, W.: Clustering ensembles: models of consensus and weak partitions. IEEE Transactions on Pattern Analysis and Machine Intelligence 27, 1866–1881 (2005)
25. Wiswedel, B., Berthold, M.R.: Fuzzy Clustering in Parallel Universes. J. of Approximate Reasoning 45, 439–454 (2007)
26. Zadeh, L.A.: Towards a Theory of Fuzzy Information Granulation and its Centrality in Human Reasoning and Fuzzy Logic. Fuzzy Sets and Systems 90, 111–117 (1997)
27. Zadeh, L.A.: Toward a generalized theory of Uncertainty (GTU)-—an Outline. Information Sciences 172, 1–40 (2005)

Information Fusion for Man-Machine Cooperation

Ronald R. Yager

Machine Intelligence Institute, Iona College
New Rochelle, NY 10801
yager@panix.com

Abstract. We first note that since humans communicate using linguistic terms central to man–machine cooperation is the availability of a common vocabulary understandable by both parties. Here we draw upon structures from granular computing, particularly fuzzy sets, to provide this capability. Having this capability allows the machine to use the types of information humans commonly provide. We then focus on some tools useful for the fusion of information and question answering in the context of man-machine cooperation. We describe methods for fusing information from multiple sources and we provide the capability to have multiple fused values. We also investigate the fusion of probabilistic and possibilistic information.

Keywords: Fuzzy sets, cooperation, uncertainty, fusion.

1 Introduction

Man-machine cooperation is a significant part of our reality. The use of intelligent agents on the Internet is a rapidly growing example of man-machine cooperation. However, in reality almost all our interactions with the Internet involve aspects of man-machine cooperation. Another place of significant man-machine cooperation is in the field of robotics and autonomous vehicles, areas of particular military and commercial interest. Our focus here is on some issues related to the extension of the capabilities for man–machine cooperation.

In trying to get the most from this inter species cooperation we must of course appreciate the capabilities of each of the participants. However, even more importantly, we must never lose sight of the fact that this is not an equal partnership. The sole purpose of man-machine cooperation is to benefit the human. This cooperation is focused on satisfying the human.

While computers are best at searching and processing large amounts of appropriately represented information it is from the perspective of the human that the objectives of their manipulations are derived. An important aspect of human cognition is the role played by language. Human beings understand, reason and communicate in terms of language. Unfortunately human language is not native to digital cognition. Since the priority is the human the machine must obtain the capability to communicate at our level.

We must train the computer to understand our perspective. This involves representing the linguistic concepts that we use in terms of objects from the machines language.

J.M. Zurada et al. (Eds.): WCCI 2008 Plenary/Invited Lectures, LNCS 5050, pp. 140–158, 2008.
© Springer-Verlag Berlin Heidelberg 2008

We must also provide the machine with tools to enable it to manipulate these objects in ways compatible with human reasoning.

At a very fundamental level human language involves a granularization of the world. The field of granular computing [1, 2] has emerged as discipline focused on addressing these problems. Granular computing involves a number components including fuzzy set theory, Dempster-Shafer theory of evidence [3] and rough sets.

Human Communications and comprehension is based a commonly understood vocabulary of words and concepts. As part of the process of making the machine more useful to the human we must teach the machine the meaning of these terms that are used in the human society. One established technology for bridging this gap is fuzzy sets and particularly Zadeh's paradigm of computing with words [4, 5]. Here we can represent linguistic terms as fuzzy subsets, a type of formal objective amenable to computer manipulation and understanding.

With the preeminence of the human in mind in future applications involving man-machine coordination we will be able to assume that the human has given the computer the meaning of the terms in the vocabulary they will use for their communications and other interactions. To put it bluntly the meaning of the terms in the vocabulary has been shoved into the computer.

While each of the disciplines that constitute granular computing has features useful for this task and even more importantly they can easily interact we shall mainly use the fuzzy set representation of our knowledge

Our focus here will be on the task of question answering. This task arises in many situations and different guises, some simple and some complex. A bartender must answer the question of whether a person is at least eighteen in order to determine whether to serve him a drink. The question of whether an adversary is capable of performing a particular action arises in both the military and commercial environment. An Internet shopping agent must answer the question of whether an object meets the criteria his owner has stipulated. A robot working in disaster site must answer questions about the future course of the environment to do its job. These task involve the gathering of information, some of which can be human supplied observations, generally expressed in linguistic terms, as well information based on formal measurements. It must then combine these pieces of information to get a comprehensive picture of the situation. Our objective here is to provide the machinery necessary to enable this question-answering task to be performed by the digital partner.

2   Variables and Question Answering

By a variable we shall mean an attribute associated with some specific object. Thus, if V is a variable then $V \equiv$ attribute (object). John's age is an example of a variable. In this case the attribute is age and the object is John. Typically with a variable, we assume it has a domain, X, consisting of the set of possible values. A common task is the answering of some question about a variable. For example, is John over 65? Another closely related task is that of making a decision in which knowledge about a variable is central to the decision. For example, a bartender deciding whether to serve John a drink must ascertain that his age is at least 21.

We emphasize the distinction between the task of finding the value of a variable and that of answering a question about a variable. There can to some uncertainty and still we can answer a question about a variable with certainty. Knowing the exact value of a variable can help in answering a question but it is not always necessary.

In order to be able to answer a question about the value of a variable, we must draw upon all our sources of information about the variable. The information provided by these sources may be related to the variable of interest in a number of different ways. It may be information directly about the value of the variable of interest, an observation on the age of John. For example, a birth certificate. It may be about the attribute without specific reference to John. Human beings typically live no more than about 85 years. It may be information about the value of another attribute associated with John, "the color of John's hair is gray." It may be information relating the variable of interest to other attributes or variables, "John is five years younger than Mary." These pieces of information may have different degrees of credibility. The information may be obtained from precise measurement or may be based upon perceptions and observations. It may be expressed formally or in linguistic terms. The process of answering the question about the attribute involves a fusing of all this information.

3 Basic Knowledge Representation Using Fuzzy Sets

An important aspect of the question answering process is the representation of the relevant information in a manner that allows their fusion and formal manipulation in ways analogous to human reasoning. The representational language should be rich enough allow the modeling of many types of available information. As noted earlier human beings communicate and reason using linguistic terms that are from a commonly understood vocabulary. A fundamental feature of these linguistic terms is their granular nature. The emerging discipline of granular computing [2] is being developed to provide tools to manage this type of information. Among the disciplines that make up the field of granular computing are fuzzy set theory, Dempster-Shafer theory of evidence and rough sets. Here we shall mainly focus on the role that fuzzy sets can play.

Fuzzy subsets provide the basis for a very expressive framework for the representation of a wide body of knowledge. This knowledge can be either precise or granular. We shall briefly discuss this representational capability, however, we note the extensive literature on this especially the work of Zadeh under his paradigm of computing with words [4] and the related generalized theory of approximate reasoning [6-8].

Within the framework provided by fuzzy sets knowledge about the value of a variable V is expressed using a statement V *is* A where A is a fuzzy subset of the domain X. The use of this type of representation can be seen as a generalization of the idea of imposing a constraint on the value of V, such as saying that V lies in the subset B, when B is a crisp subset of X. An example of this is saying John's age is between 25 and 35. The use of fuzzy subsets in addition to allowing for the granularization allows for a grading of this concept of V lying in the set B. As indicated by Zadeh [6-8] the statement V *is* A provides a constraint on the value of the variable V. This constraint on the variable V induces a possibility distribution on X such that $A(x)$ indicates the possibility that x is the value of V.

An example of this would be the observation that John *is* middle aged. In this case the fuzzy subset A corresponding to the term middle age is such that for $x \in X$ the membership grade A(x) is the compatibility of the age x with the concept being represented, middle-age.

In the types of man-machine coordination applications of interest here we shall assume the machine has been appropriately informed, by his human partner, the meaning of linguistic terms used as fuzzy sets. The man and machine have a commonly understood vocabulary for their interactions.

Some notable examples of fuzzy subsets are worth pointing out. In the case where $A = \{x\}$, then the statement V *is* A is equivalent to saying that V = x. Another special case is when A = X. Here, the statement V *is* X is equivalent to saying that we don't know. If B is some crisp subset of X, then the statement V *is* B is equivalent to saying the value of V lies in B. The situation when $A = \varnothing$, the null set, corresponds to the case where we are saying our knowledge is that V is not in X. This situation indicates a complete conflict with our assumption that V must take its value in X. More generally, if A is such that $Max_X A(x) < 1$ then we have some degree of conflict with the assumption that V has X as its domain. We shall say a fuzzy subset is normal if there exists at least one $x \in X$ so that A(x) = 1. If $Max_X A(x) < 1$ we say A is subnormal. While in most cases observed information is normal subnormality can arise when combining information.

An important tool in human is deduction. Assume we have the knowledge that V lies in B, V *is* B, where B is the crisp subset X. Using this we can infer that V lies in E where $B \subseteq$, here E is any set containing B. Knowing that John's age is between 25 and 35 allows us to infer that John's age is between 10 and 50. In the fuzzy framework that generalizes to what is called the **entailment principle**. This principle states that, from the knowledge that V *is* A, we can infer V *is* F where $A \subseteq F$. We recall that for fuzzy subsets $A \subseteq F$ if $A(x) \leq F(x)$ for all x.

Clearly, the knowledge that V is contained in [25, 35] is more informative less uncertain, then the knowledge that V is contained in [10, 50]. Furthermore the statement that V is 25 is even more informative than either of the preceding as it contains no uncertainty. In [9, 10], we introduced the concept of specificity to measure the amount of information contained in a statement V *is* A. Specificity is inversely related to the idea of uncertainty, the more specific the more certain our knowledge.

Definition. Assume A is a fuzzy subset over X. Let x^* be such that $A(x^*) = Max_X[A(x)]$, it is an element having the maximal membership grade in A. Let $\underline{A}$ be the average membership grade of A over the space X - $\{x^*\}$, the average over all elements except x^*. The specificity of A, Sp(A), is defined as $Sp(A) = A(x^*) - \underline{A}$, it is the difference between the highest membership grade and the average of all the other elements.

Note: The specificity of the statement V *is* A is equal to Sp(A). Thus we use the terms Sp(V *is* A) interchangeably with Sp(A).

We can observe some properties of Sp(A):

      1. It lies in unit interval. $0 \leq Sp(A) \leq 1$.

      2. Sp(A) = 1 iff there exists one element x^* such that $A(x^*) = 1$ and all other elements have A(x) = 0.

3. If $A(x) = c$ for all x, then $Sp(A) = 0$.

4. If A and B are two normal fuzzy subsets, they have one element with membership grade and $B \subseteq A$ then $Sp(B) \geq Sp(A)$. Thus containment in the case of normality means an increase of specificity.

Note: Essentially specificity measures the degree to which V *is* A points to one and only one element as the value of V.

As we shall subsequently see, the measure of specificity can play an important role in the processing of information. Consider the statement V *is* A where A is a normal fuzzy subset, that is there exists at least one element that has full possibility of having the value of V. We earlier noted that if B_1 is such that $B_1 \subset A$ as well as remaining normal then V *is* B_1 provides more information about the value of V than the original statement V *is* A. Essentially in this case with B_1 we reduced the uncertainty by reducing the possibility of some elements while still leaving the possibility of finding a solution. On the other hand, if $B_2 \supset A$ then V *is* B_2 provides less information than V *is* A. In this case, we have reduced our certainty because we have added more possibilities. A third situation is where we have V *is* B_3 but with $B_3 \subset A$ but with B_3 subnormal $Max_x[B_3(x)] < 1$. We don't have a solution completely compatible with the assumption that V lies in X. In this case, we can possibly have less information than the original statement V *is* A, $Sp(B_3) \leq Sp(A)$. We observe that a reduction of specificity (certainty) in our knowledge can come about from two sources, one being increased possibility and the other being an increase in conflict with the assumption that its value lies in the given domain?

4   On the Measures of Possibility and Certainty

We turn to the task of answering a question about some variable such as given that John lives in the pacific northwest does he live near the Cascade Mountains. The formal machine representation of this is given the knowledge that V *is* A our task is the determination of the validity of the statement V *is* B. We now must provide the mechanism for the machine to process this question.

In order to build our intuition, we shall initially consider the case in which the sets A and B are crisp sets. There are two situations regarding our knowledge of A. In the first, we have no uncertainty regarding our knowledge of V, $V = x_1$, here $A = \{x_1\}$. In this situation, we can very clearly answer our question about the truth of the statement V *is* B. If $x_1 \in B$ then the answer is yes, if $x_1 \notin B$ then the answer is no. This exact information with respect to the value of V leads to a crisp answer.

The second case is where A is not a singleton, there exists some uncertainty about the value of V. The uncertainty associated with the knowledge that V *is* A makes the clear determination of whether another statement V *is* B is true or false not always attainable. Using figure #1 can help us understand the situation when A is uncertain.

We see in case 1 knowing that V *is* A assures us that V *is* B is valid. In case 2, knowing that V *is* A assures us that V *is* B is not true. Finally, in case 3, we can't tell.

Thus we observe from this crisp environment that we have the following rules regarding the determination of truth of the statement V *is* B given V *is* A:

If A $\subseteq$ B then the answer is <u>yes</u>
If A $\cap$ B $= \varnothing$ then the answer is <u>no</u>
If A $\cap$ B $\neq \varnothing$ and A $\not\subset$ B then the answer is <u>I don't know</u>

Case 1

Case 2

Case 3

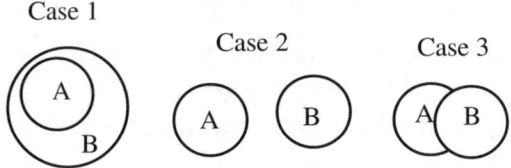

Fig. 1. Different relations between knowledge and question

Thus the attainment of a clear answer to questions in the face of uncertainty in our knowledge is not always attainable. We note this situation holds even in the special case when B is a singleton. We see that asking if V = 30 if we only know that V $\in$ [25, 40] can't be answered yes or no, the appropriate answer is *I don't know*.

In the fuzzy set environment more sophisticated tools are needed to address this problem. Two measures have been introduced by Zadeh [11] to help. These are the measures of possibility and certainty. We note that Dubois and Prade [12] refer to the measure of certainty as the measure of necessity . In the following, we shall, unless otherwise stated, assume A and B are normal

The possibility that V *is* B given V *is* A is defined as

$$\text{Poss}[V \text{ } is \text{ } B \text{ } / \text{ } V \text{ } is \text{ } A] = \text{Max}_x[A(x) \wedge B(x)]$$

We observe that since D = A $\cap$ B where D(x) = A(x) $\wedge$ B(x) we see that Poss[V *is* B / V *is* A] is the maximum degree of intersection between A and B.

The second measure introduced by Zadeh is the measure of certainty. We define this as

$$\text{Cert}[V \text{ } is \text{ } B \text{ } / \text{ } V \text{ } is \text{ } A] = 1 - \text{Poss}[V \text{ } is \text{ } not \text{ } B \text{ } / \text{ } V \text{ } is \text{ } A]$$
$$\text{Cert}[V \text{ } is \text{ } B \text{ } / \text{ } V \text{ } is \text{ } A] = 1 - \text{Max}_x[A(x) \wedge \overline{B} \text{ } (x)]$$

With some manipulation we attain Cert[V *is* B / V *is* A] = $\text{Min}_x[\overline{A}(x) \vee B(x)]$. We observe that Cert[V *is* B / V *is* A] is indicating the degree to which A is contained in B. That is if A is contained in B the knowledge that V is in A assures us that it is in B. If A is a normal fuzzy subset it can be shown that Cert[V *is* B/V *is* A] $\leq$ Poss[V *is* B / V *is* A].

The measures of possibility and certainty provide upper and lower (optimistic and pessimistic) bounds on the answer to the question of whether V *is* B is true given we know that V *is* A. Thus in general the truth lies in an interval.

Let us look at some special cases of A and B. We first consider the case when A and B are crisp. In this case, Cert[B/A] and Poss[B/A] must be either one of zero. If Cert[B/A] = 1 then we must have Poss[B/A] = 1 and this corresponds to the case where V *is* B is true. If Poss[B/A] = 0 then Cert[B/A] = 0 and we know that V *is* B is false. If Cert[B/A] = 0 while Poss[B/A] = 1 then we are in the situation in which the answer is unknown.

Consider the situation where B is a crisp subset and A can be fuzzy. Here we have that

$$\text{Poss}[V \text{ } is \text{ } B \text{ } / \text{ } V \text{ } is \text{ } A] = \text{Max}_{x \in B}[A(x)]$$

$$\text{Cert}[V \text{ } is \text{ } B \text{ } / \text{ } V \text{ } is \text{ } A] = \text{Min}_{x \notin B}[\overline{A}(x)] = 1 - \text{Max}_{x \notin B}[A(x)]$$

An important special case of this is where B = $\{x^*\}$, here we are interested in determining whether V is equal to some particular value. In this case Poss[V *is* B / V *is* A] = $A(x^*)$ and Cert[V *is* B / V *is* A] = $1 - \text{Max}_{x \neq x^*}[A(x)]$. We also observe that if $A(x^*) \neq 1$ then we must have Cert[V *is* x^* / V *is* A] = 0. This follows since with normality there exists some element $x_1 \neq x^*$ with $A(x_1) = 1$ and hence $1 - \text{Max}_{x \neq x^*}[A(x)]. = 0$.

We also observe in the case where X = $\{x_1, x_2\}$ if we ask is V = x_1, we see that Cert[V *is* x_1 / V *is* A] = $1 - A(x_2)$. It is simply the negation of the possibility of the other element.

Consider now the special case where A is a crisp set. Here

$$\text{Poss}[V \text{ } is \text{ } B \text{ } / \text{ } V \text{ } is \text{ } A] = \text{Max}_{x \in A}[B(x)]$$

$$\text{Cert}[V \text{ } is \text{ } B \text{ } / \text{ } V \text{ } is \text{ } A] = \text{Min}_x[\overline{A}(x) \vee B(x)] = \text{Min}_{x \in A}[B(x)].$$

If additionally we assume that A = $\{x_1\}$, the value of V is exactly known, then

$$\text{Poss}[V \text{ } is \text{ } B \text{ } / \text{ } V \text{ } is \text{ } A] = B(x_1) \text{ and Cert}[V \text{ } is \text{ } B \text{ } / \text{ } V \text{ } is \text{ } A] = B(x_1)$$

Here then $B(x_1)$ is the validity of the statement that V *is* B.

Some summarization may be useful here. Generally when our information about the value of the variable V has some uncertainty, V *is* A, the answer to any question about the truth of the statement V *is* B lies in some interval. Thus if A is not a singleton the truth of V *is* B lies in the interval [C, P] where C is the certainty of V *is* B and P is the possibility that V *is* B. On the other hand if A is a singleton then the truth of V *is* B is a precise value C = P. If additionally when A is a singleton B is a crisp subset then this value is one or zero. The important point here is that there are two manifestations lack of precision. One being as a result of our lack of certainty regarding the knowledge of A, it is granular it is not a singleton, this generally results in an interval for our truth-value, granular truth-value. The second issue is related to a lack crispness, the sets involved are fuzzy, this generally introduces aspects of multi-valued logic, the values C and P can be anywhere in unit interval.

In the preceding, we assumed normality with respect to all the sets involved, all sets were assumed to have at least one element with membership grade 1. We now make some comments about the situation with respect to sub-normality, $\text{Max}_x[A(x)] < 1$. While the primary information supplied by the individual sources is generally normal sub-normality can arise from the combination of information from different sources when there is some conflict between the information supplied by the sources. We see then that sub–normality is generally a reflection of some conflict.

In reasoning systems based on classical logic the appearance of conflicting statements results in a situation in which we can infer anything, we conclude that everything is true. Our system has a similar property. Assume V *is* A and A = Ø then for any statement V *is* B we have Cert[V *is* B / V *is* A] = $Min_x[\tilde{A}(x)] \vee B(x)] = 1$. Thus, in the face of complete conflict everything is certain. On the other hand, with A = Ø, Poss[V *is* B / V *is* A] = $Max[A(x) \wedge B(x)] = 0$. Thus nothing is possible but everything is certain. In order to avoid this difficulty of having the certainty greater then the possibility we shall use as our definition certainty

$$Cert[V \ is \ B \ / \ V \ is \ A] = (Min_x[\tilde{A}(x)] \vee B(x)]) \wedge (Max_x[A(x)])$$

If A is normal this just is the definition for certainty we previously used Cert[V *is* B/V *is* A] = $Min_x[\tilde{A}(x)] \vee B(x)$. While if A = Ø we get Cert[V *is* B / V *is* A] = 0. More generally using this definition we always get Cert[V *is* B / V *is* A] ≤ Poss[V *is* B / V *is* A].

One further comment is in order with respect to normality. Previously, we defined the entailment principle as saying from V *is* A we can infer V *is* B where A ⊆ B. This must be modified to say that B must satisfy the $Max_x[B(x)] \leq Max_x[A(x)]$. Thus if A is normal no additional restriction exists on B on the other hand if $Max_x[A(x)]$ = a then any statement V *is* B inferred from this must satisfy both A ⊆ B and $Max_x[B(x)] \leq a$. The inferred set B can't be more possible then the original set A.

5 Hedging on Our Data

In the preceding we introduced V *is* A as a structure for representing knowledge. We indicated that this generalized the idea of knowing that V lies in some subset. More generally this formulation imposes some constraint on the value that V can assume. We now consider the situation where we want to hedge on the knowledge that V *is* A. We consider propositions of the form V *is* A is α– certain where α ∈ [0, 1] indicates the degree of confidence we attribute to the proposition V *is* A.

In [13] we suggested that we can interpret statements of the form V *is* A is α –certain as an qualified proposition V *is* F where F(x) = Max[A(x), $\overline{\alpha}$] = A(x) ∨ $\overline{\alpha}$. Essentially this hedging loosens the constraint on the variable V. We see if α = 1 then F(x) = A(x) and we get our original unqualified proposition. On the other hand if α = 0 then $\overline{\alpha}$ = 1 and F(x) = 1 for all x. Here the statement V *is* A is 0– certain effectively carries no information.

In the following we shall let A^* denote the fuzzy set such that $A^*(x) = 1$ if x = x^* and $A^*(x) = 0$ if x ≠ x^*. Let us see what happens to our measures of possibility and certainty in this hedged situation:

$$Poss[V \ is \ x^* \ / \ V \ is \ A \ is \ \alpha{-}cert] = Max_x[(A(x) \vee \overline{\alpha}]) \wedge A^*(x)] = A(x^*) \vee \overline{\alpha}$$

$$Cert[V \ is \ x^* \ / \ V \ is \ A \ is \ \alpha{-}cert] = Min[\alpha, Min_{x \neq x}{}^*[\overline{A}(x)]].$$

More intuitively we see that

$$\text{Cert}[V \ is \ x^* \ / \ V \ is \ A \ is \ \alpha\text{--cert}] = \text{Min}[\alpha, \text{Cert}[V \ is \ x^* \ / \ V \ is \ A]]$$

In anticipation of what we shall do in the following, we shall refer to these as optimistic and pessimistic measures

$$\textbf{Opt}(V \ is \ x^* \ / \ V \ is \ A \ is \ \alpha\text{--cert}) = \text{Poss}[V \ is \ x^* \ / \ V \ is \ A \ is \ \alpha\text{--cert}] = A(x^*) \vee \overline{\alpha}$$

$$\textbf{Pess}(V \ is \ x^* \ / \ V \ is \ A \ is \ \alpha\text{--cert}) = \text{Cert}[V \ is \ x^* \ / \ V \ is \ A \ is \ \alpha\text{--cert}]$$

$$= \alpha \wedge (1 - \text{Poss}[V \ is \ not \ x^* \ / \ V \ is \ A])$$

We now consider an alternative method for representing a certainty quantified statements using the Dempster-Shafer belief structure [3]. Here we represent the statement V *is* A is α-cert by the proposition V *is* m where m is a D-S belief structure with two focal elements, $B_1 = A$ and $B_2 = X$ having $m(B_1) = \alpha$ and $m(B_2) = 1 - \alpha$. We recall the plausibility and belief measures are respectively the expected possibility and expected certainty and hence we use these measures to obtain our optimistic and pessimistic bounds on the validity of the statement V *is* x*.

$$\text{Pl}[V \ is \ x^* \ / \ V \ is \ m] = \sum_{i=1}^{2} m(B_i) \ \text{Poss}[V \ is \ x^* \ / \ V \ is \ B_i] = = 1 - \alpha \ \overline{A}(x^*)$$

$$\text{Bel}[V \ is \ x^* \ / \ V \ is \ m] = \sum_{i=1}^{2} m(B_i) \text{Cert}[V \ is \ x^* \ / \ V \ is \ B_i] = \alpha \ \text{Min}_{x \neq x}{}^*[\overline{A} \ (x)]$$

We observe that the pessimistic measure can be generalized using a t-norm [14]. Thus if T is any t-norm then

$$\textbf{Pess}[V \ is \ x^* \ / \ V \ is \ A \ is \ \alpha\text{--cert}] = T[\alpha, \text{Min}_{x \neq x}{}^*[\overline{A}(x)].$$

The optimistic measure can be generalized using a t-conorm [14]. Thus if S is any t-conorm then

$$\textbf{Opt}[V \ is \ x^* \ / \ V \ is \ A \ is \ \alpha\text{--cert}] = S(\overline{\alpha}, A(x^*))]$$

6 Multi-source Information Fusion

We now turn to the issue of aggregation of information from multiple sources.

If V *is* A and V *is* B are two pieces of information then their conjunction (fusion) is V *is* D where D = A ∩ B, that is D(x) = Min[A(x), B(x)]. More generally, if V *is* A_i, for i = 1 to q, are a collection of normal propositions then their conjunction is V *is* D where $D = \bigcap_{i=1}^{q} A_i$ here D(x) = Min$_i$[A_i(x)]. We observe a fundamental feature of this conjunction process, D ⊆ A_i that is or all x, D(x) ≤ A_i(x). More generally, if $D = \bigcap_{i=1}^{q} A_i$ and E = D ∩ A_{q+1} then E ⊆ D, E(x) ≤ D(x) for all x. Thus we see the more information we get the smaller the fuzzy subsets.

In using multiple sources of information our objective is to increase the amount of information we have about the variable of interest. We desire to increase the specific-ity. We observe that if $D = \bigcap_{i=1}^{q} A_i$, is normal then $Sp(D) \geq Sp(A_i)$ for any i and we have we have gained information. Thus if the information is not conflicting, D is normal, then fusing the information supplied by the multiple sources is a process which <u>can't decrease</u> the information we have from any of the individual sources. Normally D is more informative than any of the individual sources.

If some of the pieces of information are conflicting this may result in a situation in which D is subnormal, $Max_x D(x) \leq 1$. In this case, the fusion of the pieces of infor-mation may result in a situation in which our informativeness, specificity, has de-creased. Generally it is difficult dealing with situations in which we have conflicting source information. One approach in this situation is not to use all the information. That is, we can selectively choose which information to use to find our fused value. This requires adjudicating between the information supplied by the different sources. The choice of the appropriate manner of adjudication often requires the use of subjec-tive considerations on the part of the agent responsible for fusing the information. In the following, we shall suggest one approach to addressing this problem. We should note that other approaches are possible.

The process suggested involves a tradeoff between selecting a subset of the avail-able information that is not conflicting and yet large enough to provide a credible fu-sion of the available information. The technique we shall suggest will make use of the concept a credibility measure to help in this process.

Let P_i denote V *is* A_i, a be piece of data about the variable V. We refer to the col-lection of these as $P = \{P_1, ..., P_q\}$. We associate with P a measure $\mu: 2^P \rightarrow [0, 1]$ such that for each subset B of P, $\mu(B)$ indicates the credibility of using as our fused knowledge the conjunction of the data in B. We shall call μ the credibility measure. We can associate with μ some basic properties: $\mu(\varnothing) = 0$ and $\mu(P) = 1$. Additionally μ must be monotonic, if $B_1 \subset B_2$ then $\mu(B_2) \geq \mu(B_1)$.

Assume B is a subset of P. Let $D_B = \bigcap_{P_i \in B} A_i$, it is the fusion of the knowledge in B. We observe using the subset B leads to the statement V *is* D_B. However, any statement obtained by using only the information in B only has a credibility of $\mu(B)$.

In order to determine the quality of the knowledge obtained by using the subset B we must consider two criteria. One criterion is that the knowledge provided fusing the data in B is informative and the other criteria is that B is credible. The degree of satis-faction to the criteria of informativeness, Inf(B), can be obtained using the measure of specificity, thus $Inf(B) = Sp(D_B)$. We recall $Sp(D_B) = D_B(x^*) - \underset{X-\{x^*\}}{Ave} (D_B)$ where x^* is any element having maximal membership grade in D_B. The credibility of using the subset B, Cred(B), can be measured by $\mu(B)$. Since our measure of quality is an *anding* of these criteria we can define the measure of the quality of the result obtained using the subset B as $Qual(B) = Inf(B) Cred(B)$, thus $Qual(B) = Sp(D_B) \mu(B)$.

We make some observations about this process of multi source fusion. First, observe that if the whole collection of data $\mathcal{P}$ is such that $D_{\mathcal{P}}$ is normal then for all B since $D_{\mathcal{P}} \subseteq D_B$ then D_B is also normal. Hence, in this case, Sp($D_{\mathcal{P}}$) $\geq$ Sp(D_B). Furthermore, since $\mu(\mathcal{P}) = 1 \geq \mu(D_B)$ then

$$\text{Qual}(\mathcal{P}) = \text{Sp}(\ D_{\mathcal{P}}) \geq \mu(D_B)\ \text{Sp}(D_B) \geq \text{Qual}(B).$$

Thus, in the case where the fusion of the data from all the sources doesn't induce any conflict the most informative thing to do is to use fusion of all the data in $\mathcal{P}$.

More generally ,we make the following observation.

Observation: If B_1 is a subset of $\mathcal{P}$ such that D_{B_1} is normal then for all subsets B_2 of $\mathcal{P}$ such that $B_2 \subset B_1$ then $\text{Qual}(B_1) \geq \text{Qual}(B_2)$.

Definition: We shall call a subset B where D_B is a normal a non-conflicting subset. Furthermore, we call a subset B **maximally non-conflicting** if B is non-conflicting and the addition of any other piece of data to B results in sub-normal fusion.

Observation: Any subset B of data containing a maximally non-conflicting subset can't provide the best fusion.

We shall now consider some examples of the credibility measure One special class of credibility measure are those we call cardinality based measures. For these measures no distinction is made between credibility the different pieces of data, $\mu(B)$ just depends on how many pieces of data are in B, the cardinality of B. We can define a cardinality-based measure using a function $h:[0,\ 1] \rightarrow [0,\ 1]$ that satisfies $h(0) = 0$, $h(1) = 1$ and is monotonic, $h(r_1) \geq h(r_2)$ if $r_1 > r_2$. Using h we can define $\mu(B) = h(\frac{\overline{|B|}}{|\mathcal{P}|})$. These types of functions are often obtained as a representation of some linguistic quantifier such as *most*, "*at least about half*".

Another class of credibility measures are those that are completely additive. Here we associate with each piece of data P_i a value $\alpha_i \in [0,\ 1]$ and assume $\sum\limits_{i=1}^{q} \alpha_i = T$.

In this case $\mu(B) = \dfrac{1}{T} \sum\limits_{j \in B} \alpha_j$.

Let G_k, $k = 1$ to g, be a collection of subsets of $\mathcal{P}$ that provides a partition. Using this we can obtain credibility measure where $\mu(B) = 1$ if B contains at least piece of data from each of the G_k and $\mu(B) = 0$ otherwise. Closely related to this is a measure in which we associate with each G_k a nonnegative value g_k and define $\mu(B) = \sum\limits_{k=1}^{g} g_k \dfrac{|B \cap G_k|}{|B|}$. Here we also assume the g_k sum to one.

Another of type credibility measure is one that contains a crucial piece of data. We say that P_j is crucial if $\mu(B) = 0$ if $P_j \notin B$.

Another interesting example of credibility measure is the following. Let B_1 be a subset of $\mathcal{P}$. Consider a measure such that $\mu(B) = 0$ if $B_1 \cap B \neq \varnothing$ and $B_1 \subseteq B$. This measure, which we can call a balanced measure, requires that if we include any data from B_1 in our fusion we must use all the data in B_1.

Let us summarize the procedure we suggested for providing a user with quality fusion of the data in the collection $\mathcal{P}$. The first step is to calculate the subset B^* of $\mathcal{P}$ with the highest quality conjunction of its component data. That is we find B^* such that $\mathrm{Qual}(B^*) = \underset{B \subseteq \mathcal{P}}{\mathrm{Max}} [\mathrm{Qual}(B)$ where $\mathrm{Qual}(B) = \mathrm{Sp}(D_B)\, \mu(B)$. Having found this subset B^* we indicate to the client that V *is* D_B* is the result of our multi-source data fusion and that the credibility of this information is $\mu(B^*)$.

7   Multiple Fused Values from Multi-source Data

In some situations, the presentation of a single fused value may not be sufficient or appropriate. Here we shall suggest a process that will allow us to provide multiple fused values over the data set $\mathcal{P}$.

Our point of departure is again a collection of multi-source data $\mathcal{P} = \{P_1, ..., P_q\}$. Each piece of data P_j being of the form V *is* A_j where A_j is a fuzzy subset of the domain of V, X. In addition, we have a credibility measure $\mu: 2^{\mathcal{P}} \to [0, 1]$ where $\mu(B)$ is the degree credibility assigned to a fusion using the data in the subset B of $\mathcal{P}$.

In the preceding, we defined a process for obtaining an optimal subset B_1 and which provided a fused value V *is* D_{B_1} with credibility $\mu(B_1)$. Here $D_{B_1} = \underset{j,\, P_j \in B_1}{\bigcap} A_j$. This approach finds the subset of data B_1 such that that $\mathrm{Qual}(B_1) = \mu(B_1)\, \mathrm{Sp}(D_{B_1}) = \underset{B_1 \subseteq \mathcal{P}}{\mathrm{Max}} [\mu(B)\, \mathrm{Sp}(D_B)]$. We shall refer to this process as **Qual-Fuse**$(\mathcal{P}, \mu)$. Thus Qual–Fuse$(\mathcal{P}, \mu)$ returns B_1, which enables the determination of D_{B_1} and $\mu(B_1)$.

In the following, we shall suggest a procedure which allows the for generation of multiple fusions from the pair $(\mathcal{P}, \mu)$. For notational convenience in the following we shall find it convenient to denote the fuzzy subsets A_j as A_j^1, thus our data is still $\mathcal{P} = \{P_1, ..., P_q\}$ where P_j corresponds to the observation V *is* A_j^1. μ is still a credibility measure over $\mathcal{P}$.

The basic algorithm of our procedure is as follows.

1. Initialize our system with $\mathcal{P}$, μ and set $i = 1$.
2. Apply Qual-Fuse($\mathcal{P}$, μ) this returns B_1 and D_{B_1} and V *is* D_{B_1} with credibility $\mu(B_1)$.
3. Revise each of the P_j to V *is* A_j^2 where $A_j^2 = A_j^1 - D_{B_1}$. That is we remove the subset D_{B_1} from the subset A_j^1. We recall $A_j^1 - D_{B_1} = A_j^1 \cap \overline{D}_{B_1}$ and therefore $A_j^2(x) = Min[A_j^1(x), 1 - D_{B_1}(x)]$
4. Set $i = 2$
5. Let $\mathcal{P} = [P_1, ..., P_1]$ with P_j such that V_j *is* A_j^i
6. Apply Qual-Fuse($\mathcal{P}$, μ). This returns B_i and the statement V *is* D_{B_i} with credibility $\mu(B_i)$. Here $D_{B_i} = \underset{j, P_j \in B_i}{\cap} A_j^i$
7. Additional fusion desired ? No - stop, Yes - continue
8. Set $i = i + 1$
9. Calculate $A_j^i = A_j^{i-1} - D_{B_{i-1}}$
10. Go to step 5.

The final result of this process is a collection of fused values of the form

$$V \text{ } is \text{ } D_{B_1} \text{ with credibility } \mu(B_1)$$

$$V \text{ } is \text{ } D_{B_2} \text{ with credibility } \mu(B_2)$$

$$V \text{ } is \text{ } D_{B_k} \text{ with credibility } \mu(B_k)$$

The key idea we suggested here is the removal of the already presented fused value from the data remaining to be used to fuse. This is very much in the spirit of the Mountain Clustering method [15]. This removal process tends to result in situation where the D_{B_j} are disjoint .

8   Fusing Probabilistic and Possibilistic Data

An important issue in the field of data fusion concerns itself with the combination of two pieces of information where one is expressed in terms of a fuzzy subset (possibility distribution) and the other is expressed in terms of a probability distribution. Here we shall introduce some ideas related to this problem.

Let G be an attribute which is associated with some class of objects Z. Let X be the domain in which this attribute takes its value. Our interest here is on the determination of the value of the attribute G for some specific entity, z^*, from this class. Thus we are interested in the determination of the value of variable $G(z^*)$. We shall denote this variable as G_*.

Consider a piece of data about G_* such a G_* *is* A where is a fuzzy subset of X. Let's look at this data more carefully. First, we see it is directly about the variable of interest. That is it is a statement about the attribute for the object of interest.   It of course reflects some uncertainty with respect to the sources observation. In the situation in which A is assumed normal this uncertainty can be measured by the cardinality subset A, $\Sigma_X A(x)$. In the case where we must deal with subnormality more sophisticated measures such as $Un(A) = 1 - Sp(A) = 1 - (Max(A) - Ave(A))$ should be used. We see if $Max(A) = 1$ then $Un(A) = Ave(A)$ which is essentially $\Sigma_X A(x)$.

Consider now the situation in which we have additional probabilistic information consisting of a probability distribution P over the space X where $P(x_i)$ is the probability associated with the attribute value x_i. In order to find a basis for fusing these two pieces of information we shall take advantage of a view proposed by Coletti and Scozzafava [16] who suggested that an elements membership grade in a fuzzy, $A(x_i)$, can be viewed as the conditional probability of A given x_i, $P(A/x_i) = A(x_i)$. Having this allows us to use Bayes' rule to generate the fused information. Let P(x/A, P) indicate the probability of x given are two pieces of knowledge. In particular, P(x/A, P) $= \dfrac{P(A/x)}{P(A)}$ P(x). Using P(A/x) = A(x) we have P(x/A, P) $= \dfrac{P(A/x)}{P(A)}$ P(x). Furthermore, since $P(A) = \sum_{i=1}^{n} P(\dfrac{A}{x_i}) P(x_i)$ then P(A) can be expressed as $\sum_{i=1}^{n} A(x_i) P(x_i)$. Using this, we get P(x/A, P) $= \dfrac{A(x) P(x)}{\Sigma_i A(x_i) \cdot P(x_i)}$. At times we shall find it convenient to express this as P(x/A, P) $= \dfrac{A(x)}{\Sigma_i A(x_i) \cdot \dfrac{P(x_i)}{P(x)}}$

Thus the result of fusing these two pieces of data is a probability distribution with respect to the value of G_*. Using the notation suggested by Zadeh in [7] we can express this as G_* *isp* **R** where **R** indicates a probability distribution on X such that P(x/A, P) is the probability that G_* assumes the value x. The fact that this is the case is not surprising since the knowledge in the possibility distribution is actually saying that the value of the variable G_* lies in a set, A. So we are actually finding the probability of x conditioned on the knowledge that G_* lies in a set.

We look at this for some special cases. Consider the case where $P(x_i) = \dfrac{1}{n}$. Here, the probability distribution is essentially providing no information. Here $\dfrac{P(x_i)}{P(x)} = 1$ for all x_i and hence P(x/A, P) $= \dfrac{A(x)}{\sum_{i=1}^{n} A(x_i)}$. Thus here we obtain P(x/A, P) as simply a normalization of the possible distribution.

Consider now the case in which $A(x_i) = 1$ for all x_i. Here the possible distribution is providing no information. In the case $P(x/A, P) = \dfrac{P(x)}{\Sigma_i P(x_i)} = P(x)$. We get back the original probability distribution.

Consider the case where A corresponds to some crisp subset B of X. That is $A(x_i) = 1$ for $x_i \in B$. In this case $P(x/A, P) = \dfrac{P(x)}{\sum\limits_{x_i \in B} P(x_i)}$. This is the classic case of conditional probability.

One issue that must be addressed is conflicting information. Consider the case where we have $A(x_1) = 1$ and $A(x_j) = 0$ for all other x_j and where $P(x_1) = 0$. In this case, we see that $P(x_i)A(x_i) = 0$ for all x_i and our aggregation leads to a kind of indeterminism. Here, we essentially must decide, do we believe the possibility distribution which says the answer is definitely x_1 or do we believe the probability distribution which says the answer is definitely <u>not</u> x_i.

Another form of conflict can be seen in the following case. Let $A = \{\dfrac{1}{x_1}, \dfrac{0.1}{x_2}, \dfrac{0}{x_3}\}$ and let the probabilistic information be such that $P(x_1) = 0$, $P(x_2) = 0.1$ and $P(x_3) = 0.9$. In this case we obtain $P(x_1/A, P) = 0, P(x_2/A, P) = 1$ and $P(x_3/A, P) = 0$. This may be somewhat disturbing. Here while both pieces of information lend little support to x_2 their combination leads to its strong support

In order to address this issue of conflict we must first consider the context in which we obtain probabilistic information. We can envision two situations when we obtain probabilistic information. One of these is in the frequentist spirit and the other is of a subjective kind.

One situation where we have probabilistic information is where the probability distribution is a reflection of some observation about the attribute G over the objects in the class Z. Thus here $P(x_j)$ is the probability that an object in Z has value for attribute G equal to x_j. The point we want to emphasize here is that this information is <u>not</u> directly about the entity of interest z^*. It is not information about our variable of interest G_{x^*}. Although it is useful and valuable information, it is not directly about the object of interest. The important observation here is that the information contained in this type of probabilistic information is of a lower priority than the direct information contained in a statement G_* is A. Thus, here there is a priority ordering with respect to our information and in the face of conflict we want to give preference to the direct information, G_* is A.

The use of a probabilistic representation can also occur in the case in which the source is providing information directly about the attribute value for the object of interest. Consider the situation where the source has some uncertainty with regard to the actual value of the variable G_*. Here, he uses the probability framework to express his perception of the uncertainty. He is saying that my feeling about the uncertainty associated with the value of G_* is similar to that of a random experiment in which $P(x_i)$ is the probability that $G_* = x_i$. Again, in this situation, the information provided by the source is also less direct that that provided by the observation that G_* *is* A.

The overall point we want to make here is that often the information provided using a probabilistic representation has a lesser priority than that provided using the fuzzy representation. This is not to say that fuzzy sets are better then probability but only that the type of information represented by a probability distribution is less directly relevant.

This distinction in the priority of the two different kinds of information allows us to provide a reformulation of the aggregation of these two kinds of information to allow for an intelligent adjudication of conflicts. As a first step in this process, we shall turn to the issue of measuring the conflict or conversely the consistency between a probability distribution and a possibility distribution.

Let $\Pi : X \rightarrow [0, 1]$ be a possibility distribution over the X. We shall assume this is normal, there exists some x^* such that $\Pi(x^*) = 1$. Let $P: X \rightarrow [0, 1]$ be a probability distribution over X. The probability distribution has the added requirement that $\Sigma_i P(x_i) = 1$. Let $p^* = \text{Max}_i[P(x_i)]$ it is the maximal probability associated with P. We can observe that $\frac{1}{n} \leq p^* \leq 1$, where n is the cardinality of X. It is well known that the negation of the Shannon entropy, $\Sigma_i P(x_i) \ln[P(x_i)]$, provides a measure of information content of a probability distribution. What is worth pointing out is the $\text{Max}_i(P(x_i))$ provides an alternative measure of this information content. While Shannon measure has some properties that make it preferred, especially when we consider multiple distributions, in the case when we are focusing on one probability distribution, $\text{Max}_i(P(x_i))$ provides a simple and acceptable measure of the information content of a probability distribution.

We now introduce a measure called the consistency of Π and P

$$\text{Consist}(\Pi, P) = \text{Max}_i[\Pi(x_i) \wedge \tilde{P}(x_i)] \text{ where } \tilde{P}(x_i) = \frac{P(x_i)}{p^*}$$

We observe that if P is such that if $P(x_i) = \frac{1}{n}$ for all x_i then $p^* = \frac{1}{n}$ and $\tilde{P}(x_i) = 1$ for all x_i. In this case both $\Pi(x^*) = 1$ and $\tilde{P}(x^*) = 1$ and hence $\text{Consist}(\Pi, P) = 1$. Thus, the situation when P has maximal uncertainty it is consistent with any possibility distribution. On the other hand we see that if $P(x_1) = 1$ and $\Pi(x_1) = 0$ then $\text{Consist}(\Pi, P) = 0$ they are in complete conflict. In the case where $X = \{x_1, x_2, x_3\}$ and $\Pi(x_1) = 1$, $\Pi(x_2) = 0.1$ and $\Pi(x_3) = 0$ while $P(x_1) = 0$, $P(x_2) = 0.1$ and $P(x_3) = 0.9$ we get $\tilde{P}(x_1) = 0$, $\tilde{P}(x_2) = 0.11$ and $\tilde{P}(x_3) = 1$ and hence $\text{Consist}(\Pi, P) = 0.1$

We now consider the modification of the procedure for aggregating possibility and probability distributions that uses this measure of consistency to aid in the adjudication of conflicting information.

In the preceding, we defined the aggregation of V *is* A and the probability distribution P as inducing a probability distribution where $P(x/A, P) = \dfrac{A(x)P(x)}{\sum\limits_{i=1}^{n} A(x_i)P(x_i)}$.

We now provide a modification of this to account for conflicts between the input distributions. As we shall see, this is will give a priority to the information V *is* A.

Letting $\alpha = \text{Consist}(P, A)$ we define

$$P(x/A, P) = \frac{A(x)[\alpha P(x) + \bar{\alpha}\,\dfrac{1}{n}]}{\sum\limits_{j=1}^{n} A(x_j)[\alpha P(x_j) + \bar{\alpha}\,\dfrac{1}{n}]}$$

Let us see how this works. If the two sources are consistent., $\alpha = 1$, then $P(x/A, P) = \dfrac{A(x)P(x)}{\sum\limits_{j=1}^{n} A(x_j)P(x_j)}$ and we get our original formulation. If the two pieces of

information are completely conflicting, $\alpha = 0$ we get $P(x/A, P) =$

$\dfrac{A(x)\dfrac{1}{n}}{\sum\limits_{j=1}^{n} A(x_j)\dfrac{1}{n}} = \dfrac{A(x)}{\sum\limits_{j=1}^{n} A(x_j)}$. Here we completely discount the information contained in

the probability distribution P and simply obtain P(x/A, P) as a normalization of A.

Here, we shall refer to $F(\alpha, P_j) = \alpha P(x_j) + \bar{\alpha}\,\dfrac{1}{n}$ as the probability transform and refer to $\lambda(x_j) = F(\alpha, P(x_j))$ as the transformed probabilities. We see that in the face of conflict the transformed probabilities move toward $\dfrac{1}{n}$.

We further observe that if $A(x_j) = 0$, then $P(x/A, P) = 0$.

Example: Assume $X = \{x_1, x_2, x_3\}$, $A = \{\dfrac{1}{x_1}, \dfrac{0.1}{x_2}, \dfrac{0}{x_3}\}$ and $P(x_1) = 0$, $P(x_2) = 0.1$ and $P(x_3) = 0.9$, Here we get $\tilde{P}(x_1) = 0$, $\tilde{P}(x_2) = 0.11$ and $\tilde{P}(x_3) = 1$ and hence Consist$(\Pi, P) = 0.1$. In this case the transformed probabilities are:

$$\lambda(x_1) = (0.9)\frac{1}{3} = 0.3$$
$$\lambda(x_2) = (0.1)(0.1) + (0.9)\frac{1}{3} = 0.31$$
$$\lambda(x_3) = (0.1)(0.9) + (0.9)\frac{1}{3} = 0.39$$

In this case $\sum\limits_{i=1}^{3} A(x_i)\lambda(x_i) = 0.3 + 0.031 = 0.331$ and hence $P(x_1/A, P) = \dfrac{0.3}{0.331} = 0.906$, $P(x_2/A, P) = \dfrac{0.031}{0.331} = 0.094$ and $P(x_3/A, P) = \dfrac{0}{0.331} = 0$.

We must consider one other issue here. We have implicitly assumed that the possibility distribution is normal, $\text{Max}_j(x_j) = 1$. If this is not the case some problems can arise. Since $\text{Consist}(A, P) = \text{Max}_j[A(x_j) \wedge \dfrac{P(x_j)}{p^*}] \leq \text{Max}_j[A(x_j)]$ our maximal possible consistency goes down. Here the problems of reduced consistency may be an issue related to the internal conflict of the possibility distribution rather then its incompatibility with probability distribution.

It may be interesting to consider a slight modification in the case where we have $\text{Max}_j[A(x_j)] = a^* < 1$. Here, instead of the end result being a probabilistic distribution we end up with a Dempster-Shafer belief structure m. This belief structure has n + 1 focal elements $B_j = \{x_j\}$ for j = 1 to n and $B_{n+1} = X$. Furthermore for j = 1 to n we have $m(B_j) = a^* P(x_j/A, P)$ where the $P(x_j/A, P)$ are calculated as in the preceding, For $B_{n+1} = X$ we have $m(X) = 1 - a^*$. We shall not pursue this but leave it as a suggestion.

9 Conclusion

We first observed that since humans communicate using linguistic terms central to man–machine cooperation is the availability of a common vocabulary understandable by both parties. We have drawn upon structures from granular computing, particularly fuzzy sets, to provide this capability. We then focused on some tools useful for the fusion of information and question answering in the context of man-machine cooperation. We described methods for fusing information from multiple sources and we provided the capability to have multiple fused values. We also investigated the fusion of probabilistic and possibilistic information.

References

[1] Lin, T.S., Yao, Y.Y., Zadeh, L.A.: Data Mining, Rough Sets and Granular Computing. Physica-Verlag, Heidelberg (2002)
[2] Bargiela, A., Pedrycz, W.: Granular Computing: An Introduction. Kluwer Academic Publishers, Amsterdam (2003)
[3] Yager, R.R., Liu, L.: Classic Works of the Dempster-Shafer Theory of Belief Functions (A. P. Dempster and G.Shafer, Advisory Editors). Springer, Heidelberg (to appear)
[4] Zadeh, L.A.: Fuzzy logic = computing with words. IEEE Transactions on Fuzzy Systems 4, 103–111 (1996)
[5] Zadeh, L.A., Kacprzyk, J.: Computing with Words in Information/Intelligent Systems 1. Physica-Verlag, Heidelberg (1999)
[6] Zadeh, L.A.: A theory of approximate reasoning. in Machine Intelligence. In: Hayes, J., Michie, D., Mikulich, L.I. (eds.), vol. 9, pp. 149–194. Halstead Press, New York (1979)
[7] Zadeh, L.A.: Toward a generalized theory of uncertainty (GTU)-An outline. Information Sciences 172, 1–40 (2005)
[8] Zadeh, L.A.: Generalized theory of uncertainty (GTU)-principal concepts and ideas. Computational Statistics and Data Analysis 51, 15–46 (2006)

[9] Yager, R.R.: Entropy and specificity in a mathematical theory of evidence. Int. J. of General Systems 9, 249–260 (1983)

[10] Yager, R.R.: On measures of specificity. In: Kaynak, O., Zadeh, L.A., Turksen, B., Rudas, I.J. (eds.) Computational Intelligence: Soft Computing and Fuzzy-Neuro Integration with Applications, pp. 94–113. Springer, Berlin (1998)

[11] Zadeh, L.A.: Fuzzy sets and information granularity. In: Gupta, M.M., Ragade, R.K., Yager, R.R. (eds.) Advances in Fuzzy Set Theory and Applications, pp. 3–18. North-Holland, Amsterdam (1979)

[12] Dubois, D., Prade, H.: Possibility Theory: An Approach to Computerized Processing of Uncertainty. Plenum Press, New York (1988)

[13] Yager, R.R.: Approximate reasoning as a basis for rule based expert systems. IEEE Trans. on Systems, Man and Cybernetics 14, 636–643 (1984)

[14] Klement, E.P., Mesiar, R., Pap, E.: Triangular Norms. Kluwer Academic Publishers, Dordrecht (2000)

[15] Yager, R.R., Filev, D.P.: Approximate clustering via the mountain method. IEEE Transactions on Systems, Man and Cybernetics 24, 1279–1284 (1994)

[16] Coletti, G., Scozzafava, R.: Probabilistic Logic in a Coherent Setting. Kluwer Academic Publishers, Dordrecht (2002)

Bio-inspired Self-Organizing Relationship Network as Knowledge Acquisition Tool and Fuzzy Inference Engine

Takeshi Yamakawa[1,2] and Takanori Koga[1]

[1] Department of Brain Science and Engineering,
Faculty of Life Science and Systems Engineering,
Kyushu Institute of Technology,
2-4 Hibikino, Wakamatsu, Kitakyushu 808-0196, Japan
[2] Fuzzy Logic Systems Institute (FLSI),
680-41 Kawazu, Iizuka, Fukuoka 820-0067, Japan
{yamakawa, koga}@brain.kyutech.ac.jp

Abstract. Since the SOM visualizes the similarity of raw information on the competitive layer, it can be utilized in the field of pattern classification, data analysis, and so on. However, it cannot model the input-output characteristics of the system of interest. In order to squeeze out the input-output relationship from the data set with evaluation obtained by trial and error, the novel modeling tool was developed by the author (1999), which is the extension of SOM and in which the input-output relationship of the system is mapped onto the competitive layer. The system is named as self-organizing relationship network (SOR network). A set of units on the competitive layer of the SOR network after learning exhibits a set of typical input-output characteristics of the system of interest and thus the network achieves the knowledge acquisition (IF-THEN rules) from the raw data with evaluation and the effective fuzzy inference with defuzzification. The plenary talk presents the tutorial aspects of the SOR network and an application to an intelligent control.

Keywords: self-organizing relationship (SOR) network, self-organizing maps (SOM), attractive/repulsive learning, knowledge acquisition, fuzzy IF-THEN rules, fuzzy inference, defuzzification, intelligent control.

1 Introduction

Synaptic junctions are tightly connected by passing through of the signals and the similar external signals activate the synaptic junctions located nearby. Thus the external complicated information can be mapped on the surface of the brain where the signals similar to each other activate the neurons or neuron populations located near by. For instance, somatosensory stimulus is mapped on the central groove of a brain and the map is widely known as "somatotopic map"[1]. This feature of a brain was modeled by Teuvo Kohonen as Self-Organizing Maps in 1982 [2], [3], the number of papers related to this topics run to a huge amount and the world congress on SOM is held every other year.

The SOM has the architecture of one or two dimensional alignment of "units" on the competitive layer which are characterized by the so-called "reference vector"

J.M. Zurada et al. (Eds.): WCCI 2008 Plenary/Invited Lectures, LNCS 5050, pp. 159–180, 2008.

constructed with the same number of elements to those of the input vectors. After the competitive learning with numerous input vectors, the reference vectors of units represent the quantized vectors of the input data. The SOM exhibits the following distinctive features;

(1) Vector quantization
    Reference vector distribution after learning approximates the distribution of input vectors.
(2) Topological mapping
    Reference vectors of neighboring units in the competitive layer are similar to each other.
(3) Visualization of similarities
    Multi-dimensional complicated vectors are mapped onto one-or two-dimensional space.

Since the Self-Organizing Maps (SOM) can statistically squeeze out the feature from an input data set, it can be applied to pattern classification, data analysis, and so on. Therefore it may be extended to summarize a large number of cause and results or input-output data set to obtain the know-how. If the know-how is obtained in the form of IF-THEN rules, inference can be preferably achieved. In order to achieve the knowledge acquisition and inference in the same system, the self-organizing relationship (SOR) network was proposed [4]. The SOR network preserves three features described above and facilitates the fuzzy inference with interpolation between the neighboring IF-THEN rules. The learning of SOR network is achieved with input-output pairs and their evaluations. The evaluation can be given by intuitively [5] or objectively [6].

This tutorial plenary talk presents the learning and execution modes of the SOR network and its application to back-up control of a trailer-truck.

2 Self-Organizing Relationship (SOR) Network

When we create the model of some real system or ideal system, we collect the input-output pairs by the trial-and-error method. Various input signals or actions are applied to the real system at random or by curiosity-driven trial, and the output signals are carefully watched and compared with the real system of interest and evaluated (Fig.1). In the ordinary modeling, the difference between the outputs of the real system and the model is fed back to change the characteristic parameters of the model. However in some cases, we often meet a case when the real system exhibits bad or dangerous reaction for the given input and the preferable model should be created. In order to meet this requirement the evaluation value can be negative as well as positive.

In the supervised learning system, a set of input-output pairs (teaching data) which exactly describes the system under consideration are necessary for the learning. However the teaching data can not be necessarily obtained. Even in this case, the input-output pairs by trial-and-error method and their evaluations are easily obtained, and thus employed as learning data for the SOR network.

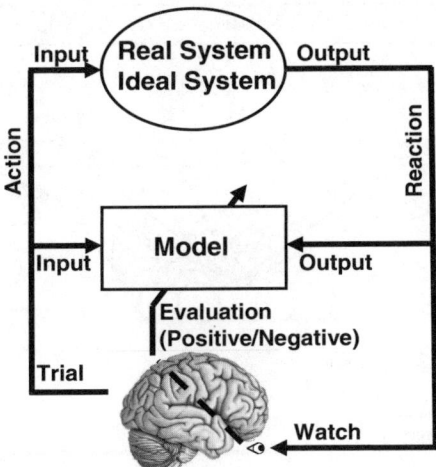

Fig. 1. Input (Action) is applied to the real/ideal system as a trial, and then its Output (Reaction) is watched to collect the input-output pair with evaluation for the learning of the SOR network as the model

The SOR network consists of a set of reference vectors v_i which are associated with the units on the competitive layer (Fig. 2 (a)). Usually, the competitive layer is 1D or 2D for visualization of state of the reference vectors. The v_i is the Cartesian product of reference vectors in the input space w_i and in the output space u_i. The w_i and the u_i have N_x and N_y elements, respectively. The operation of the SOR network is divided into two modes, the learning mode (Fig. 2 (b)) and the execution mode (Fig.2 (c)). The network can be established, by learning, in order to approximate a desired function $y^* = f(x^*)$.

After learning, an input is applied to the SOR network and then it produces the output. Thus the behavior of the SOR network is divided to two modes, *learning mode* and *execution mode*.

2.1 Learning Mode of the SOR Network

In the learning mode, a set of learning vectors obtained by trial and error is applied to the SOR network together with evaluation values (Fig.2 (b)). The learning algorithm of the SOR network is summarized as follows.

[Step 0]
All reference vectors $v_i = (w_i, u_i)(i = 1, ..., N_v)$ are initialized by random numbers.

[Step 1]
A best matching unit c_l for each learning vector $I_l = (x_l, y_l)(l = 1, ..., L)$ is selected with the smallest Euclidean Distance as follows.

$$c_l = arg\,\min_i \left\| I_l - v_i \right\|. \tag{1}$$

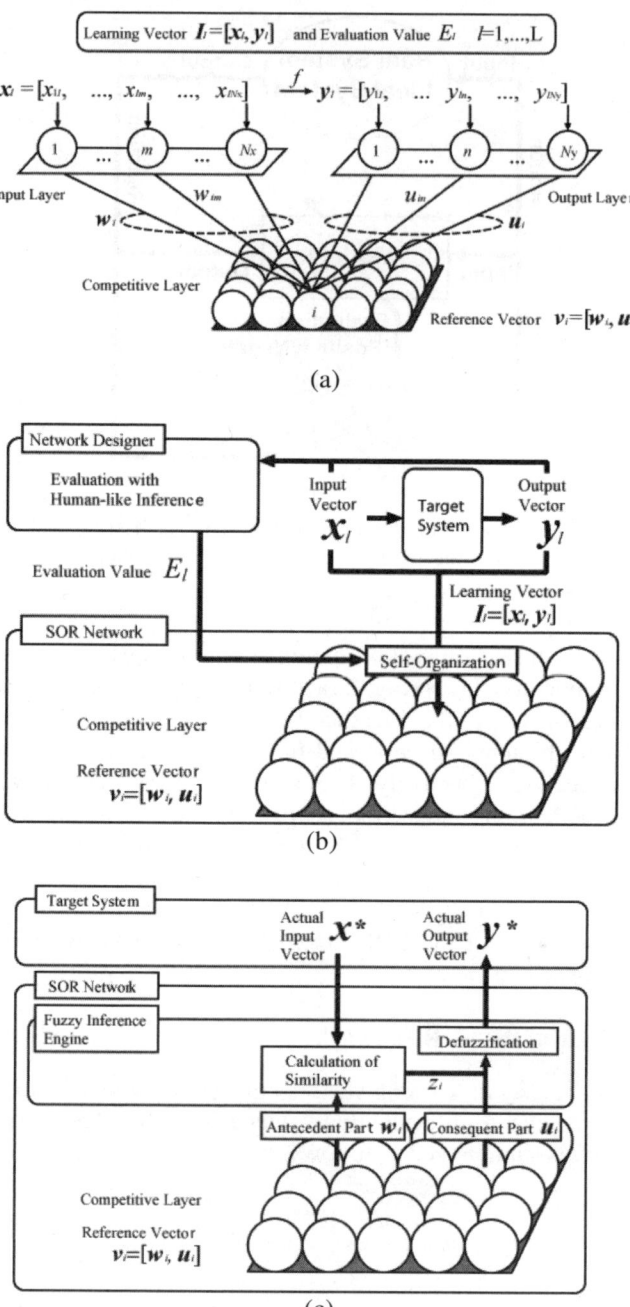

Fig. 2. (a) Architecture of SOR network. (b) Learning mode. (c) Execution mode.

[Step 2]
A Gaussian neighborhood function is calculated as follows.

$$h_{cl,i}(t) = exp\left\{-\frac{\|r_i - r_{cl}\|^2}{2\sigma^2(t)}\right\}. \tag{2}$$

where r_i and r_{c_l} are the positions of a unit i and the best matching unit c_l on the competitive layer. $\sigma(t)$ is the width of the neighborhood function, the value of which decay with time.

[Step 3]
Updating values of each reference vector are calculated as follows.
    For attractive learning $(E_i > 0)$:

$$\Delta v_{i,l}^{(posi)} = \alpha(t)E_l^{(posi)}\left\{I_l^{(posi)} - v_i(t)\right\}. \tag{3}$$

    For repulsive learning $(E_l < 0)$:

$$\Delta v_{i,l}^{(nega)} = \beta(t)h_{cl,i}(t)E_l^{(nega)} exp\left\{-\frac{\left\|I_l^{(nega)} - v_i(t)\right\|}{\sigma_r}\right\}\frac{I_l^{(nega)} - v_i(t)}{\left\|I_l^{(nega)} - v_i(t)\right\|}. \tag{4}$$

Here, $\Delta v_{i,l}^{(posi)}$ and $\Delta v_{i,l}^{(nega)}$ denote the updating values for the l-th learning vector with positive and negative evaluation values, respectively. $I_l^{(posi)}$ and $I_l^{(nega)}$ are learning vectors with a positive evaluation value $E_l^{(posi)}$ and a negative evaluation value $E_l^{(nega)}$, respectively. And σ_r denotes a coefficient deciding the extent of the repulsive effect, $\alpha(t)(> 0)$ a coefficient for the attractive learning and $\beta(t)(> 0)$ a coefficient for the repulsive learning. Eqs. (3) and (4) are basically same to the on-line updating laws of the original SOR network [4]. Then, each reference vector is finally updated as follows.

$$v_i(t+1) = v_i(t)$$
$$+\frac{\sum_{l=1}^{L(posi)} \Delta v_{i,l}^{(posi)} + \sum_{l=1}^{L(nega)} \Delta v_{i,l}^{(nega)}}{\sum_{l=1}^{L(posi)} \alpha(t)h_{cl,i}(t)E_l^{(posi)} + \sum_{l=1}^{L(nega)} \beta(t)h_{cl,i}(t)\left|E_l^{(nega)}\right|} \tag{5}$$

where $L^{(posi)}$ and $L^{(nega)}$ are the numbers of the learning vectors with positive and negative evaluation values, respectively. All updating values given by Eqs. (3) and (4) are accumulated and normalized in this process.

[Step 4]
Steps 1 to 3 are repeated with decreasing $\sigma(t)$, $\alpha(t)$ and $\beta(t)$ monotonically. Usually, these values are calculated at each learning step t by the following equations.

$$\sigma(t) = \{\sigma(0) - \sigma(\textit{final})\} \times exp\left(-\frac{t}{\tau_{\sigma}}\right) + \sigma(\textit{final}).$$ (6)

$$\alpha(t) = \{\alpha(0) - \alpha(\textit{final})\} \times exp\left(-\frac{t}{\tau_{\alpha}}\right) + \alpha(\textit{final}).$$ (7)

$$\beta(t) = \{\beta(0) - \beta(\textit{final})\} \times exp\left(-\frac{t}{\tau_{\beta}}\right) + \beta(\textit{final}).$$ (8)

where τ_{σ}, τ_{α} and τ_{β} denote each decay rate.

2.2 Execution Mode of the SOR Network

The execution mode of the SOR network corresponds to a simplified fuzzy inference with Nv if-then rules (Fig. 1 (c)). The output of the network $y^* = [y_1^*, ..., y_n^*, ..., y_{Ny}^*]$ is the weighted average of u_i by the similarity measure z_i, that is, n-th element of y^* is calculated as follows.

$$y_n^* = \frac{\sum_{i=1}^{Nv} z_i u_{in}}{\sum_{i=1}^{Nv} z_i}.$$ (9)

Similarity measures z_i between an actual input vector $x^* = [x_1^*, ..., x_m^*, ..., x_{Nx}^*]$ and all reference vectors in input space w_i of the units on the competitive layer are calculated by:

$$z_i = exp\left\{-\frac{\|x^* - w_i\|^2}{2\gamma_i^2}\right\}.$$ (10)

where, γ_i is a parameter representing fuzziness of similarity. The parameter γ_i is decided in consideration of the distribution of the reference vectors by:

$$\gamma_i = \frac{1}{S}\sum_{s=1}^{S}\|w_i - w_i^{(s)}\|.$$ (11)

where, $w_i^{(n)}$ is the s-th nearest reference vector according to the distance to the reference vector w_i. The value of S can be assigned empirically, which is explained at section 5.

3 Trailer-Truck Back-Up Control

In the learning mode of the SOR network, the network is established based on the learning vectors and their evaluation values. In the case that the SOR network is

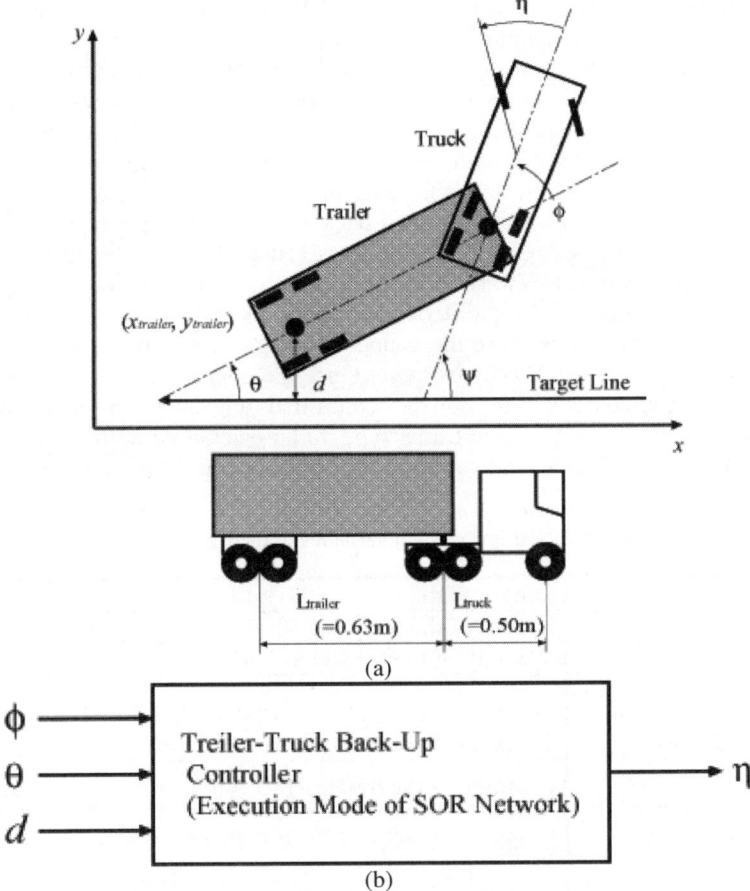

Fig. 3. (a) Model of the semi-trailer type trailer-truck. (b) Trailer-truck back-up controller constructed in this paper.

applied to a control problem, the evaluation values should be quantitatively calculated. The evaluation values, in the power system stabilization [4] and adaptive control of a DC motor [6], are simply defined by the decrease of error. However, in more complicated control systems, e.g. trailer-truck back-up control, it is difficult to describe the designer's knowledge with mathematical expressions. A mobile robot such as trailer-truck cannot be directly regulated from any initial positions to target position. This is because car-like mobile robots are nonholonomic systems (e.g. [7]). Nonholonomic systems are characterized by constraint equations involving the time derivatives of the system configuration variables. And, these equations are non-integrable. Nonholonomic systems cannot be stabilized by continuous and time-invariant state feedback system [8]. Generally, this kind of system should be regulated to target state via appropriate subgoals. In the field of control theory, there are many papers in which the controllers for nonholonomic systems are designed by using nonlinear control theories (e.g. [9], [10], [11], [12], [13]). Generally, these methods need the advanced mathematical knowledge. On the contrary, there are papers in which soft-computing techniques are employed for nonholonomic systems, in particular trailer-truck back-up control, (e.g [14], [15], [16], [17], [18]).

In order to apply the SOR network to this nonholonomic problem, we propose a new method where the learning vectors are evaluated by fuzzy inference with fuzzy control strategies. The description in detail appears in the next section.

A semi-trailer type trailer-truck is used in this study. The trailer is connected to the truck with a rotatable shaft, and is pushed back at one point when he truck backs up. The control objective is to make the trailer-truck follow the target line from any initial state by only backward movement at constant velocity. The trailer-truck control is a severe nonlinear control. The trailer-truck easily meets unstable states, so-called "jackknife state". This is because the connection angle (angle from alignment of the trailer and the truck) easily becomes large. Once the connection angle is excessively enlarged, the trailer-truck can not be controlled any more by only backward movement. The model of the semi-trailer type trailer-truck used in this study is shown in Fig. 3 (a). The parameters and variables of the trailer-truck are shown in Table. 1.

Table 1. Parameters and variables of the trailer-truck

$L_{trailer}$	Length of the trailer(=0.63m)
L_{truck}	Length of the truck(=0.50m)
v	Velocity of the truck(=-0.1m/s)
η	Front wheel angle of the truck $-30^\circ \leq \sigma \leq 30^\circ$
ψ	Angle of the truck $-180^\circ \leq \psi \leq 180^\circ$
θ	Angle of the trailer $-180^\circ \leq \theta \leq 180^\circ$
ϕ	Connection angle of the trailer-truck $-45^\circ \leq \phi \leq 45^\circ$
$x_{trailer}$	X-coordinate of the trailer
$y_{trailer}$	Y-coordinate of the trailer
d	Position error against the target line

The kinematic model of the trailer-truck [16], [17] is described by:

$$\psi[k+1] = \psi[k] + \frac{v \cdot \Delta t \cdot tan\, \eta[k]}{L_{truck}}. \tag{12}$$

$$\theta[k+1] = \theta[k] + \frac{v \cdot \Delta t \cdot tan\, \phi[k]}{L_{trailer}}. \tag{13}$$

$$\phi[k] = \psi[k] - \theta[k]. \tag{14}$$

$$x_{trailer}[k+1] = x_{trailer}[k] + v \cdot \Delta t \cdot cos\, \phi[k] \cdot cos\, \frac{\theta[k+1] + \theta[k]}{2}. \tag{15}$$

$$y_{trailer}[k+1] = y_{trailer}[k] + v \cdot \Delta t \cdot cos\,\phi[k] \cdot sin\,\frac{\theta[k+1] + \theta[k]}{2}. \qquad (16)$$

where k is a time step, and Δt a sampling period ($\Delta t = 1.0$[sec]). In the proposed system, the connection angle ϕ, the angle of the trailer θ, and the distance between the trailer-truck and the target line d are the inputs to the SOR network, and the SOR network generates the front wheel angle η (Fig. 3(b)).

4   Acquisition of Learning Vectors and Their Evaluation by Fuzzy Inference

In order to employ the SOR network to design a controller for a complicated system, i.e. trailer-truck back-up control, we propose the new evaluation method by employing fuzzy inference (e.g. [19], [20], [21], [22], [23], [24]). The fuzzy inference has many features, for example, (1) experiential knowledge which involves uncertainty is easily described by simple form of fuzzy if-then rules, (2) the practical nonlinear system can be modeled by a set of fuzzy if-then rules rather than by mathematical expressions, (3) each of fuzzy if-then rules can be revised individually if it is incorrect.

Fig. 4 shows the procedure of the learning vector acquisition. First, a state $(\varphi(k), \theta(k), d(k))$ of the trailer-truck at time step k is randomly given. Then, the front wheel angle $\eta(k)$ is randomly given as an operation at time k. These values are elements of the learning vectors. As a result of operation, the states at $k+1$ and $k+2$ are calculated using the kinematic model given by Eqs. (12)-(16). The designer evaluates the I/O relationship of the controller with fuzzy inference by watching the change of the state of the trailer-truck for two sampling intervals. In order to make the fuzzy IF-THEN rules for evaluation, four fundamental control strategies are heuristically contrived as follows.

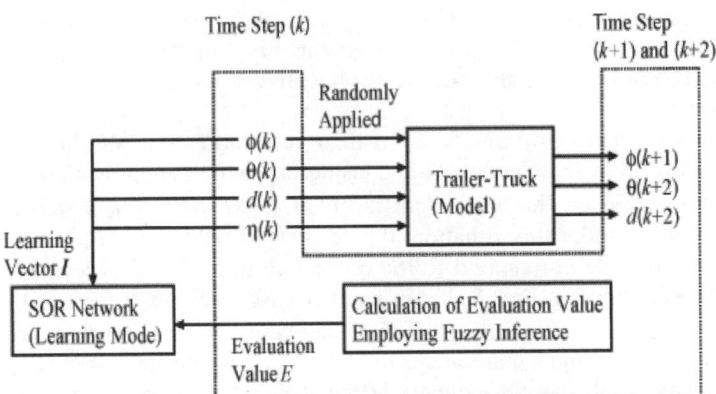

Fig. 4. Procedure of learning vector acquisition and evaluation

Control Strategy 1
If the connection angle between the trailer and the truck ϕ is large, it should be decreased to avoid falling into jackknife state.

Control Strategy 2
If the trailer is directed away from the target line, the direction should be corrected to make the trailer-truck approach the target straight line.

Control Strategy 3
If the trailer-truck moves in a direction opposite to the target direction, the direction of movement should be corrected to make the trailer-truck move in the target direction.

Control Strategy 4
After control strategies 1-3 are accomplished, the trailer-truck should be controlled to follow the target line considering the distance d and the angle θ. Once the trailer-truck follows the target line, the connection angle ϕ should be kept zero to avoid divergence.

We should emphasize that it is not so difficult for designers to consider these control strategies even if they do not have enough knowledge about dynamics of the trailer-truck. In addition, in these control strategies, it is not referred at all how to operate the front wheel angle of the truck η.

These control strategies are represented by fuzzy if-then rules (Eqn. (17)) and membership functions as shown in Fig. 5.

$$\text{IF } \phi(k) \text{ is } A_{1j}, \ \theta(k) \text{ is } A_{qj} \text{ and } d(k) \text{ is } A_{Ninj},$$
$$\text{THEN Ej=E}_\phi \text{ or E}_\theta \text{ or E}_d \ \ j=1, \ldots, N_r . \tag{17}$$

where j and q are suffixes of rule and input variable, respectively. N_r and N_{in} are the numbers of rule and input variable, respectively. A_{qj} is the label of a fuzzy set. And please note that the suffix of learning vector l is omitted to clarify the explanation. The antecedent variables of the rules are the state of the trailer-truck. Five fuzzy sets labeled as "PL (Positively Large)", "PS (Positively Small)", "ZR (Approximately Zero)", "NS (Negatively Small)" and "NL (Negatively Large)" are arranged for each input variable.

In actual construction of the fuzzy if-then rule table, one of three consequent variable in Eqn. (17) is selected with considering which input variable should be decreased. For example, the antecedent is given as $\phi(k)=PL$, $\theta(k)=$"don't care" and $d(k)=$"don't care". Under this situation, the designer should select E_ϕ as a consequent. This fuzzy if-then rule correspond to the control strategy 1 and is located in the top square of the rule table in Fig. 5 (b). In another case, the consequent of the rule "IF $\phi(k)$ is PS, $\theta(k)$ is PS and $d(k)$ is PL, THEN ?" can be assigned to be E_d, because $d(k)$ is PL, which is shown in the square under the top square in the rule table (Fig. 5 (b)). In the similar manner, all the consequents of the if-then rules are assigned as shown in Fig. 5(b).

The consequent part is given as a constant value given by Eqs. (18)-(20).

$$E_\phi = tanh\left\{\frac{|\phi(k)| - |\phi(k+1)|}{a_\phi}\right\}. \tag{18}$$

$$E_\theta = tanh\left\{\frac{|\theta(k)| - |\theta(k+2)|}{a_\theta}\right\}. \tag{19}$$

$$E_d = tanh\left\{\frac{|d(k)| - |d(k+2)|}{a_d}\right\}. \tag{20}$$

These evaluation values are defined as the decrease of the error normalized with the sigmoid function ranging from -1.0 to +1.0. The reason why the sigmoid function is used is to emphasize the evaluation near 0. The coefficients a_ϕ, a_θ and a_d decide the slope of the function, and the values are decided as 0.6 times of maximum decrease of

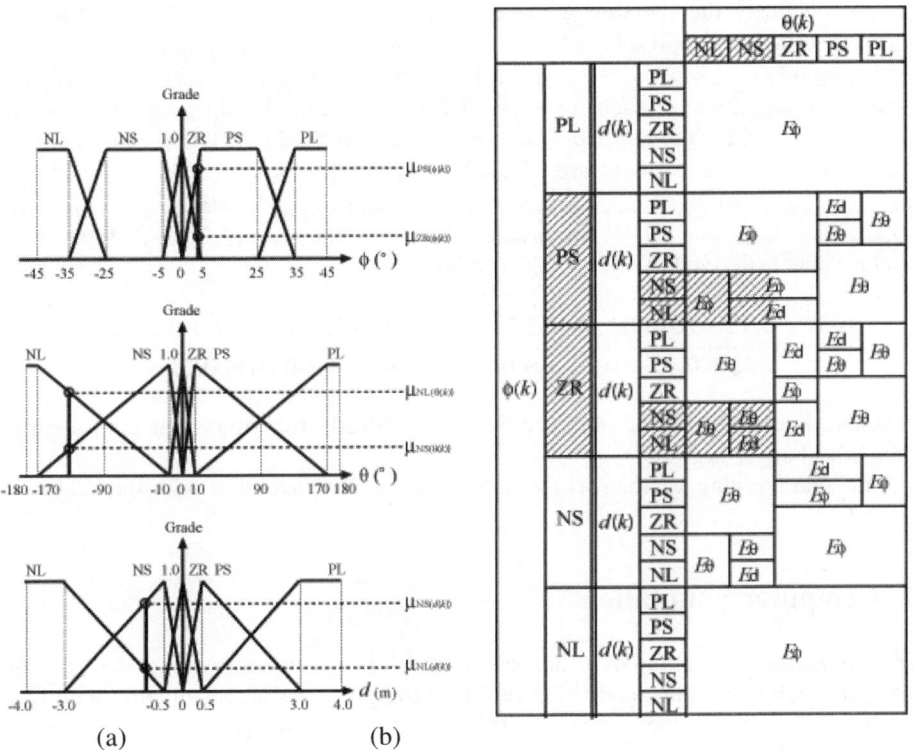

(a) (b)

Fig. 5. Fuzzy IF-THEN rules for evaluation. (a) Membership function of antecedent part. (b) Fuzzy IF-THEN rule table.

the error (a_ϕ =3.0°, a_θ =3.8°and a_d=0.06m). In the fuzzy inference, the product-sum-gravity method (e.g. Mizumoto, 1990) is employed. First, the matching grade of each rule λ_j is calculated as follows.

$$\lambda_j = \prod_{q=1}^{N_{in}} \mu_{Aqj}(x_q).$$ (21)

Here, $\mu_{Aqj}(x_q)$ is membership grade of the fuzzy set A_{qj}, where x_q is the input variable, i.e. $x_1 = \phi(k)$, $x_2 = \theta(k)$ and $x_3 = d(k)$. Finally, the evaluation value E is calculated by the following equation.

$$E = \frac{\sum_{j=1}^{N_r} \lambda_j E_j}{\sum_{j=1}^{N_r} \lambda_j}.$$ (22)

Below is a demonstration of concrete example of the fuzzy inference process. Suppose that a state at time step k is given by $\phi(k)$=4.0°, $\theta(k)$= −130.0°and d = −1.0m. First, each fuzzy membership grade $\mu_{Aqj}(x_q)$ is derived as shown in Fig.5(a). The concrete values are $\mu_{ZR}(\phi(k)$=4.0) = 0.2, $\mu_{PS}(\phi(k)$=4.0) = 0.8, $\mu_{NS}(\theta(k)$= −130.0) = 0.25, $\mu_{NL}(\theta(k)$= −130.0) = 0.75, $\mu_{NS}(d(k)$= −1.0) = 0.8 and $\mu_{NL}(d(k)$= −1.0) = 0.2. Next, matching grade of each fuzzy IF-THEN rule is calculated by Eqn. (21). In this case, eight rules, i.e. the meshed-square in the rule table (Fig.5(b)), have non-zero values. The concrete values are $\lambda_{(\phi=PS,\theta=NL,d=NS)}$ =0.8 × 0.75 × 0.8=0.48, i.q. $\lambda_{(\varphi=PS,\theta=NL,d=NL)}$ =0.12, $\lambda_{(\varphi=PS,\theta=NS,d=NS)}$ =0.16, $\lambda_{(\varphi=PS,\theta=NS,d=NL)}$ =0.04, $\lambda_{(\varphi=ZR,\theta=NL,d=NS)}$ =0.12, $\lambda_{(\varphi=ZR,\theta=NL,d=NL)}$ =0.03, $\lambda_{(\varphi=ZR,\theta=NS,d=NS)}$ =0.04 and $\lambda_{(\varphi=ZR,\theta=NS,d=NL)}$ =0.01. Thus, final E value is derived by Eqn. (22) as follows.

$$E = \frac{(0.48 + 0.12 + 0.16)E_\phi + (0.12 + 0.03 + 0.04)E_\theta + (0.04 + 0.01)E_d}{0.48 + 0.12 + 0.16 + 0.04 + 0.12 + 0.03 + 0.04 + 0.01}.$$ (23)

Here, E_φ, E_θ and E_d can be obtained by Eqs. (18)-(20), two time steps after a certain front wheel angle is given.

Thus one learning vector $(\varphi(k),\theta(k),d(k),\eta(k))$ is now acquired with the evaluation value E.

5 Computer Simulation

The computer simulation of trailer-truck back-up control employing the proposed method is achieved. The acquisition of the learning vectors and evaluation with fuzzy inference is done by the method of the description in the previous section. In the learning of the SOR network, 6,561 learning vectors are regularly sampled. In this regard, 100 learning vectors, which is assigned to be (ϕ, θ, d, η, E_l)= (0.0,0.0,0.0,0.0,1.0), are deliberately added in order to ensure the stability around the target state. The number of learning iterations is 200, the number of units on the 2D

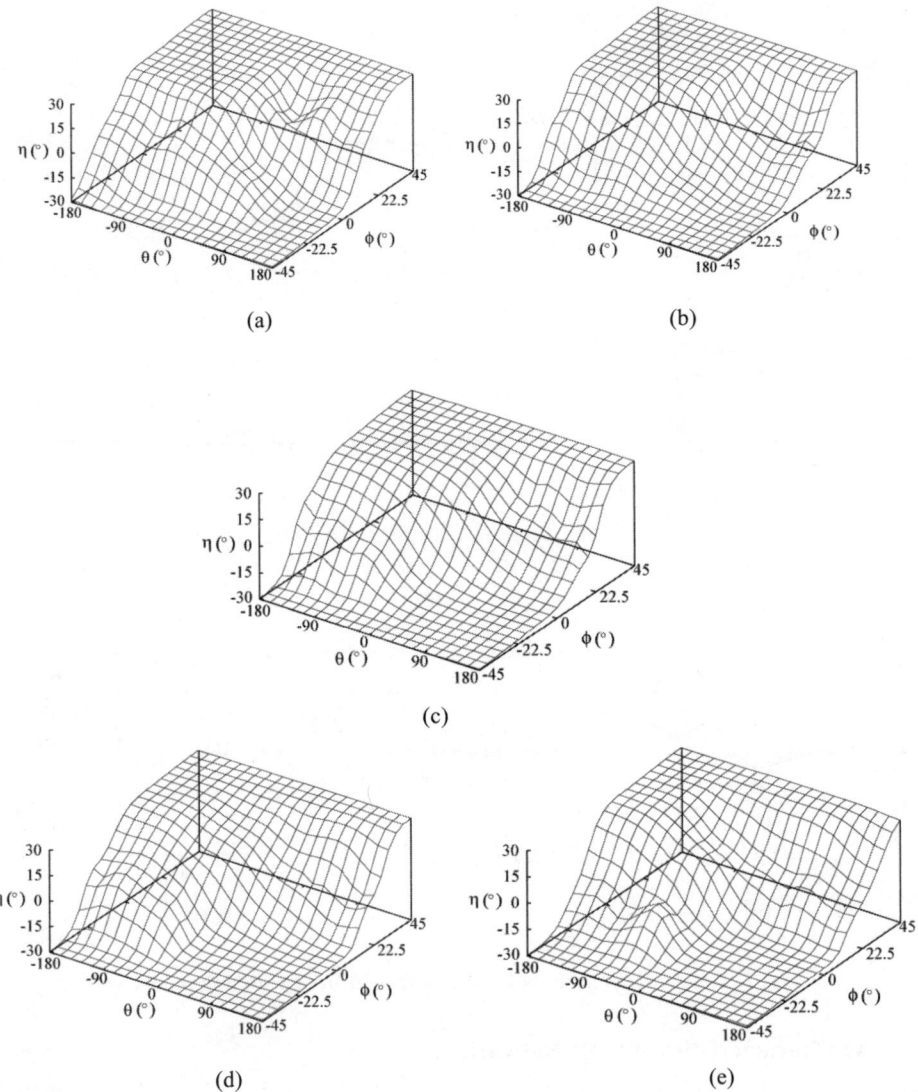

Fig. 6. I/O characteristics of the designed controller. (a) d=4.0m. (b) d=2.0m. (c) d=0.0m. (d) d=-2.0m. (e) d=-4m.

competitive layer is 625 (25×25), $\alpha(0) = 1.0$, α (final)=0.01, $\beta(0) = 0.1$, $\beta(final)=0.005$, $\sigma(0) = 25.0$, $\sigma(final)=0.25$, $\sigma_r =0.1$, $\tau_\alpha=30.0$, $\tau_\beta =30.0$ and $\tau_\sigma =30.0$. In the execution mode, the parameter S in Eqn. (11) is empirically assigned to be $S =3$, by considering the distribution of the reference vectors.

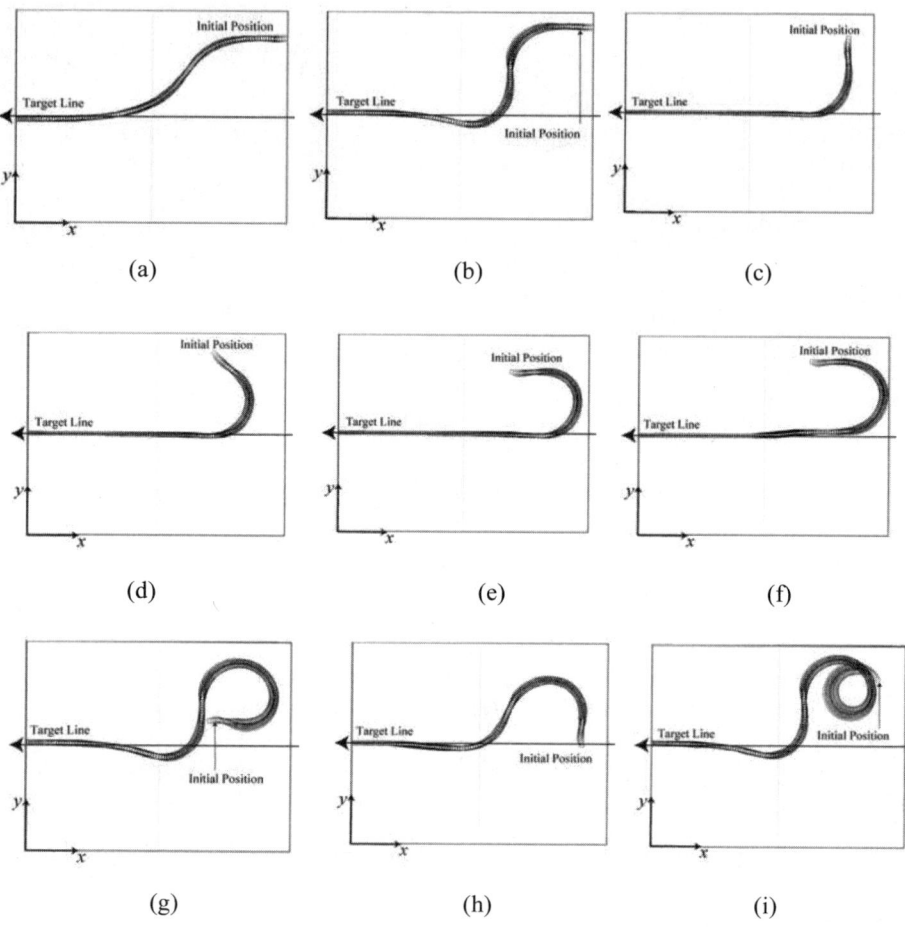

Fig. 7. Trajectories of the trailer –truck from different starting points

5.1 I/O Characteristics of SOR Network

I/O characteristic of the designed controller is represented in 4-dimensional space (3-input 1-output), hence, the I/O characteristic is shown for five cases of d in Fig. 6 (a)-(e). All figures show the saturation in the front wheels angle η, which limits the larger angle to avoid the jackknife state. This avoidance action shows the first priority over any other operations (Control Strategy 1). Once the trailer-truck falls into the jackknife state, it becomes uncontrollable, even if any other variables are assigned well. The curvature of each figure in Fig. 6 represents the operation to correct the direction of trailer-truck movement (Control Strategies 2 and 3), and the operation to regulate the position d and angle θ (Control Strategy 4). It is emphasized that the nonlinear I/O relationship necessary for the adequate control can be self-organizingly established by the learning vectors with evaluation, which are obtained by trial and error.

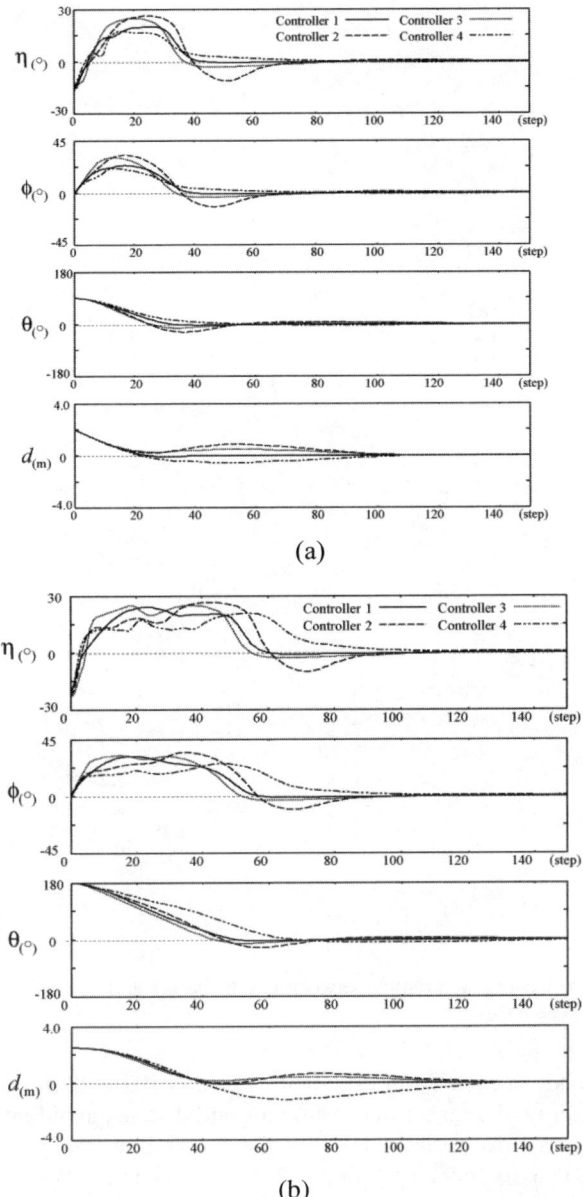

Fig. 8. Comparison of control results using Controllers 1-4

5.2 Trajectories of Trailer-Truck

The trajectories of the trailer-truck controlled by the designed controller are shown in Fig. 7 (a)-(i), where sampled positions and directions of the trailer-truck are illustrated

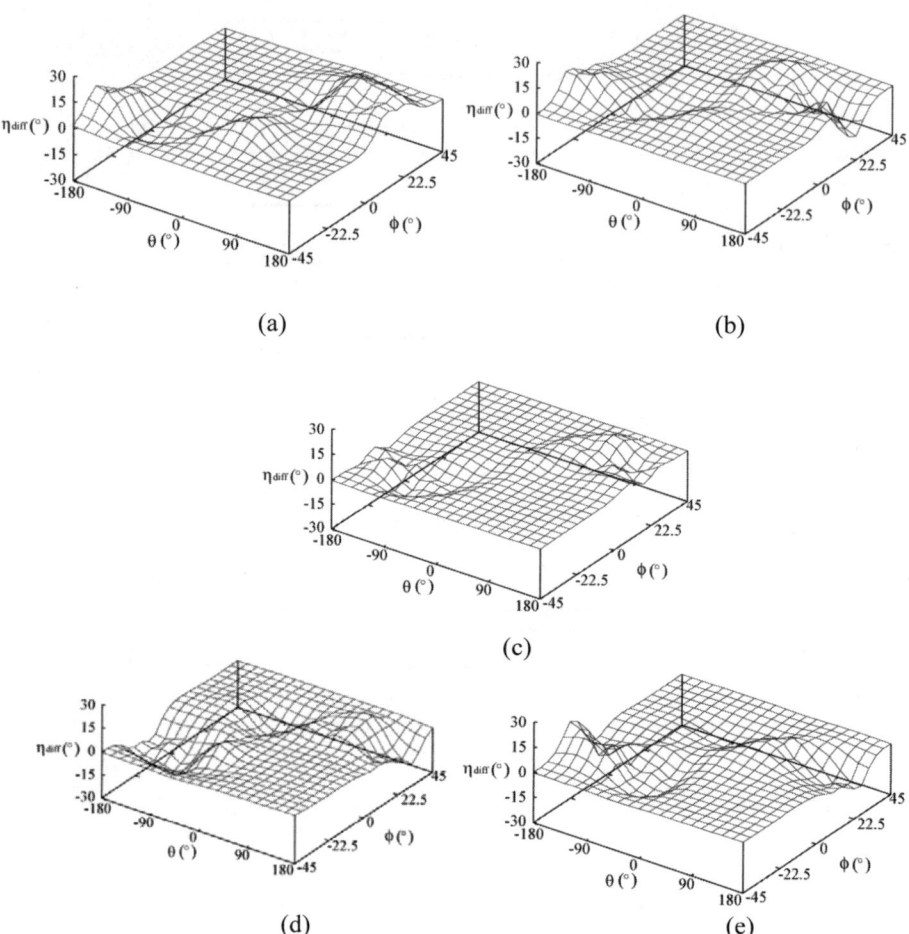

Fig. 9. Difference of I/O characteristics between Controllers 1 and 2. (a) d=4.0m, (b)d=2.0, (c) d=0.0, (d)d=-2.0, (e)d=-4.0m.

at every one second. In each figure, the work space is 10.0m×8.0m. The trailer-truck is controlled to follow the target line from any initial states avoiding jackknife state. From the line-symmetric initial states for the target line, the trailer-truck can be completely regulated to the target line. These results show that the sophisticated operation is successfully earned by the learning of the SOR network without any expert's knowledge, even though the designer does not give the detail of operation expressly.

5.3 Effectiveness of Repulsive Learning

Figs. 8-10 show experimental results which verify the effectiveness of repulsive learning. In the repulsive learning, the ratio of the final values of α and β is the dominant parameter. In these experiments, four controllers are established. Controller 1 is the above-mentioned one, i.e. finely-tuned controller ($\beta(0)$=0.1and $\beta(final)$=0.005). Controllers 2-4 are established under the basically same conditions to Controller 1. The Controller 2 is established by only attractive learning ($\beta(0)$=0.0 and $\beta(final)$=0.0). Controller 3 is established under somewhat excessive repulsive effect, $\beta(0)$=0.1 and $\beta(final)$ =0.01. Controller 4 is established under extremely excessive repulsive effect, $\beta(0)$=0.1 and $\beta(final)$ =0.05.

The comparison of control results among four controllers is shown in Fig. 8. Two typical initial states are set, Fig. 8 (a) φ=0°, θ=90°, d=2.0m, Fig. 8 (b) φ=0°, θ=180°, d=2.5m. All the controllers can finally regulate the trailer-truck to the target line. In particular, the SOR network can regulate the trailer-truck to the target line even if the repulsive learning is omitted (i.e. using Controller 2). The lack of repulsive learning cannot be fatal.

Fig. 9 shows the difference of I/O characteristics between Controller 1 and Controller 2. In the figure, η_{diff} represents the difference of the front wheel angle η between two controllers, i.e. η of Controller 1 minus η of Controller 2. This result

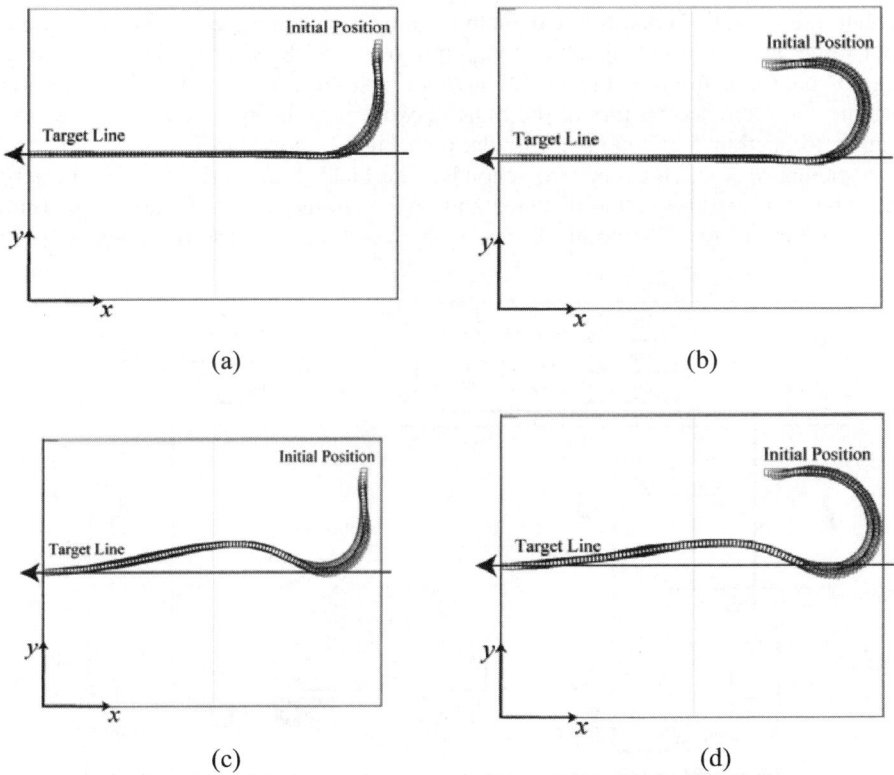

(a)   (b)

(c)   (d)

Fig. 10. Comparison of trajectories of the trailer-truck (a), (b) and (c), (d) are results controlled by Controller 1 and Controller 2,respectively

shows that the obtained I/O relationships of two controllers involve notable differences. The trajectories of the trailer-truck which are controlled by Controller 1 and Controller 2 are shown in Fig. 10. Fig. 10 (a), (b) are the results by using Controller 1 and Fig. 10 (c), (d) are ones by using Controller 2. Under the situation without employing repulsive learning, the SOR network cannot control the trailer-truck to avoid actively 'undesirable' state. Therefore, some undesirable phenomena such as overshoot, low-speed convergence are observed. On the contrary, extreme excessive repulsive learning also breaks down the successful control (i.e. using Controller 4 in Fig. 8). The result using Controller 3 is similar to the result using Controller 1.

In this experiment, as the parameter $\beta(final)$ is close to 0.005, i.e. the finely-tuned parameter, the performance of the controller tends to become well. In this regard, a quantity of difference around this parameter is permissible.

6 Practical Experiment

The experimental system is shown in Fig. 11 and the flow of processing is as follows. The motion capture system with two CCD cameras captures the coordinates of the three markers attached to the trailer-truck, and the coordinates are input to the PC through the digital I/O board to calculate the angles ϕ, θ and the distance d. Then the front wheel angle η is calculated as the output of the SOR network. The value of η is sent to the truck through the D/A converter and the remote control proportional system. The reference vectors of the SOR network have been established in the same manner to section 5 (Computer Simulation). Fig.12 shows an experimental result. Each picture was taken every two seconds. The black line on the floor is the target line. The initial values of the distance and angles are $\varphi = 0°$, $\theta = 90°$, and d = 2.0m as shown in Fig. 12 (a). The point of view is switched after a lapse of 16 seconds. The

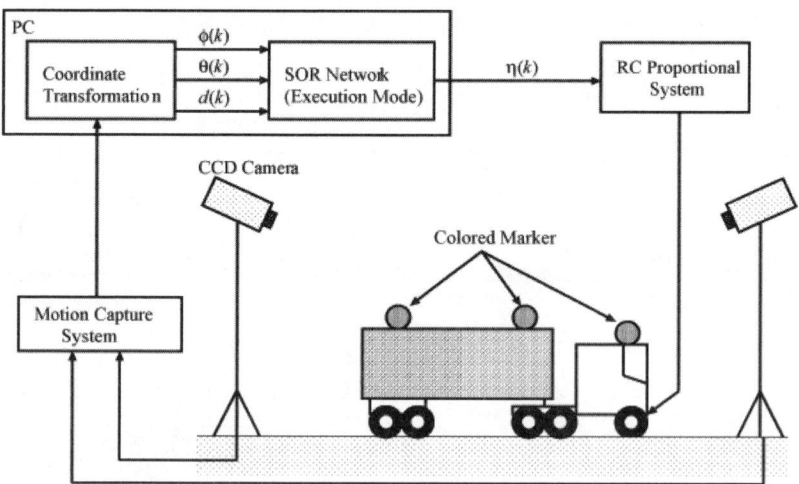

Fig. 11. Trailer-truck control system for the practical experiment

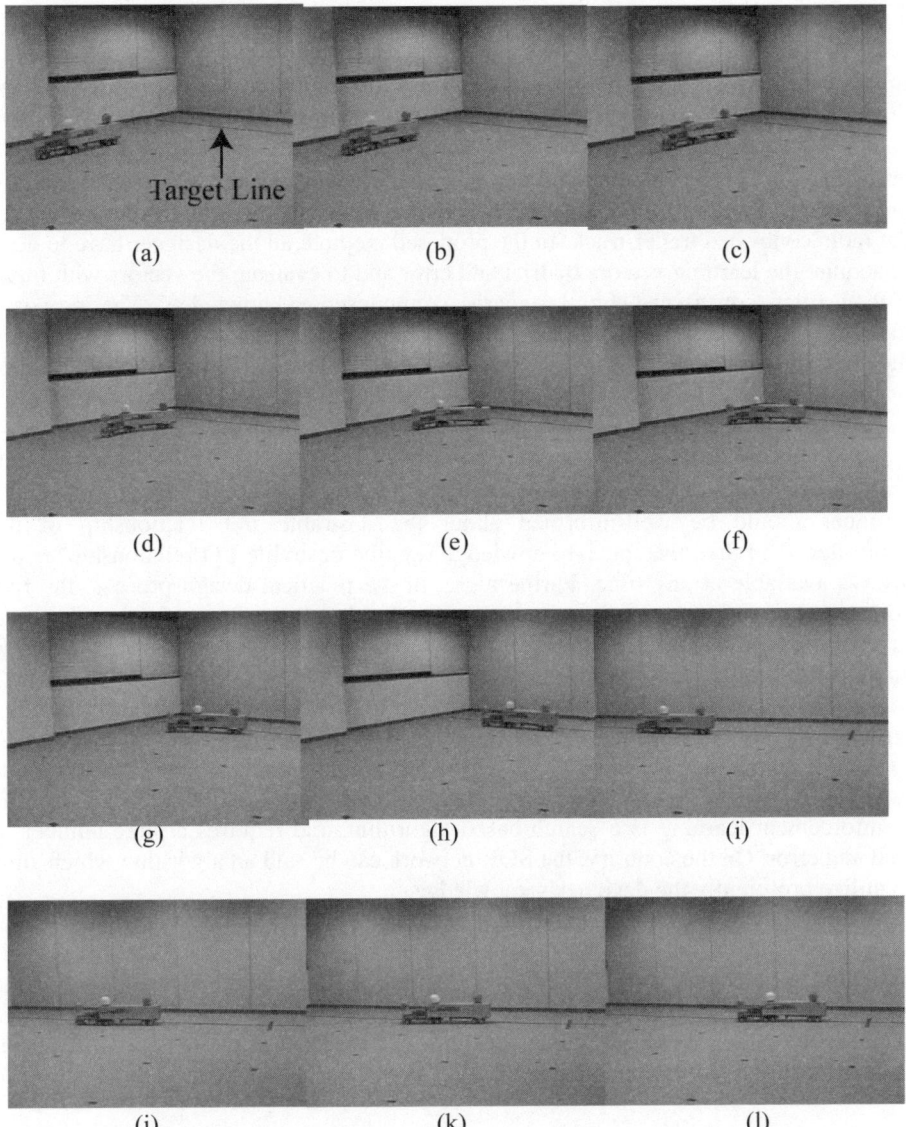

Fig. 12. Practical experimental result. The black line on the floor is the straight target line. (a) Initial position. (b)-(k) Intermediate steps. (l) Final Position. Each photograph was taken at intervals of two seconds.

right side of Fig. 12 (h) indicates almost the same position as the one in the left side of Fig. 12 (i). The angle of the trailer becomes smaller and the trailer-truck follows the target line finally. It is confirmed that the trailer-truck can follow the target line even in the case of other initial values.

7 Conclusions

In this paper, we proposed the new evaluation algorithm of the learning vectors for the SOR network, in which the fuzzy inference was employed. And, the proposed method is applied to the trailer-truck back-up control which is a well-known benchmark problem as a control of nonholonomic system. We verified the effectiveness of the proposed method both in the computer simulation and the practical experiment with the radio controlled trailer-truck. In the proposed method, all the designer have to do is to acquire the learning vectors by trial and error and to evaluate the vectors with fuzzy if-then rules constructed by designer's commonsense knowledge. The proposed method is much easier than the controller design method employing nonlinear control theory. This is because the proposed method does not require any advanced mathematical knowledge and expressions. Furthermore, the proposed method is a model-less approach.

Seemingly, to employ the rule-based fuzzy logic controller which should control trailer-truck is effective because the design is linguistically achieved. However, the designer should be well-informed about the desirable I/O relationship of the controller. And that, the special knowledge (i.e. the desirable I/O relationship) is not always available at any time. Furthermore, in the practical design process, the fine adjustment of parameters is required. On the other hand, the fuzzy if-then rules used in the proposed algorithm can be constructed without expert's special knowledge about the desirable I/O relationship of the controller.

In the learning of the SOR network, undesirable I/O relationship obtained by trial and error are actively utilized, i.e. repulsive learning. This approach is somewhat similar to the Reinforcement learning [25]. In the Reinforcement learning, a controller is established by the interaction with the environments. Generally, however, the Reinforcement learning is a search-based algorithm, and requires a huge number of trial and error. On the contrary, the SOR network can be said an algorithm which aims to utilize profoundly the designer's knowledge.

In our future work, we aim to develop the SOR network not only as a controller but as a knowledge acquisition tool. The learning algorithm of the SOR network is basically same to the SOM. Therefore, the information stored in the units on the competitive layer can be analyzed as mentioned in [26], [27]. Finally, we aim at a construction of a system in which the designer and the SOR network act on each other in order to improve their knowledge.

References

1. Penfield, W., Rasmussen, T.: The Cerebral Cortex of Man. Macmillan, New York (1952)
2. Kohonen, T.: Self-organizing formation of topologically correct feature maps. Biological Cybernetics 43(1), 59–69 (1982)
3. Kohonen, T.: Analysis of a simple self-organizing process. Biological Cybernetics 44(2), 135–140 (1982)
4. Yamakwa, T., Horio, K.: Self-organizing relationship (SOR) network. IEICE Transactions on Fundamentals E82-A, 1674–1678 (1999)

5. Horio, K., Haraguchi, T., Yamakawa, T.: An Intuitive Contrast Enhancement of an Image Data Employing the Self-Organizing Relationship (SOR) Network. In: Proceedings of International Joint Conference on Neural Networks (IJCNN 1999), pp. 10–16 (1999)
6. Horio, K., Yamakawa, T.: Adaptive Self-Organizing Relationship Network and Its Application to Adaptive Control. In: Proceedings of the 6th International Conference on Softcomputing and Information/Intelligent Systems (IIZUKA 2000), pp. 299–304 (2000)
7. Laumond, J.P., Sekhavat, S., Lamiraux, F.: Guidelines in Nonholonomic Motion Planning for Mobile Robots. Robot Motion Planning and Control, 229 (1998)
8. Brockett, R.W.: Asymptotic stability and feedback stabilization. In: Brockket, R., Millman, R., Sussman, H. (eds.) Differential Geometric Control Theory, pp. 181–191. Birkhauser, Boston (1983)
9. Pomet, J.B.: Explicit design of time-varying stabilization control laws for a class of controllable systems without drift. Systems & Control Letters 18(2), 147–158 (1992)
10. Tilbury, D., Murray, R.M., Sastry, S.S.: Trajectory Generation for the N-Trailer Problem Using Goursat Normal Form. IEEE Transactions on Automatic Control 40(5), 802–819 (1995)
11. Sampei, M., Tamura, T., Kobayashi, T., Shibui, N.: Arbitrary path tracking control of articulated vehicles using nonlinear control theory. IEEE Transactions on Control Systems Technology 1(4), 587–592 (1995)
12. Kolmanovsky, M., Reyhanoglu, M., McClamroch, N.H.: Switched mode feedback control laws for nonholonomic systems in extended power form. Systems & Control Letters 27(1), 29–36 (1996)
13. Prieur, C., Astolfi, A.: Robust stabilization of chained systems via hybrid control. In: Proceedings of the 41st IEEE Conference on Decision and Control, pp. 522–527 (2002)
14. Nguyen, D., Widrow, B.: The truck backer-upper: an example of self-learning in neural network. In: Proceedings of International Joint Conference on Neural Networks (IJCNN 1989), pp. 357–363 (1989)
15. Kong, S.G., Kosko, B.: Adaptive fuzzy-systems for backing-up a truck-and-trailer. IEEE Transactions on Neural Networks 3(2), 211–223 (1992)
16. Tanaka, K., Sano, M.: A robust stabilization problem of fuzzy control systems and its application to backing up control of a truck-trailer. IEEE Transactions on Fuzzy Systems 2(2), 119–133 (1994)
17. Ichihashi, H., Miyoshi, T., Nagasaka, K., Tokunaga, M., Wakamatsu, T.: A Neurofuzzy Approach to Variational Problems by Using Gaussian Membership Functions. International Journal of Approximate Reasoning 13(4), 287–302 (1995)
18. Hong, C.S., Won, J.M., Lee, J.S.: Multi-Thread Evolutionary Programming and Its Application to Truck-and-Trailer Backer-Upper Control. IEICE Transactions on Fundamentals E84-A(2), 597–603 (2001)
19. Zadeh, L.A.: Fuzzy Sets. Information and Control 8, 338–353 (1965)
20. Zadeh, L.A.: The concept of a new approach to the analysis of complex systems and decision processes. IEEE Transactions on Systems, Man and Cybernetics 3(1), 28–44 (1973)
21. Zadeh, L.A.: Knowledge representation in fuzzy logic. IEEE Transactions on Knowledge and Data Engineering 1, 89–100 (1989)
22. Mamdani, E.H., Assilian, S.: An experiment in linguistic synthesis with a fuzzy logic controller. International Journal of Man-Machine Studies 7(1), 1–13 (1975)
23. Mamdani, E.H.: Applications of fuzzy logic to approximate reasoning using linguistic synthesis. IEEE Transactions on Computers 26(12), 1182–1191 (1977)

24. Takagi, T., Sugeno, M.: Fuzzy Identification of Systems and Its Applications to Modeling and Control. IEEE Transactions on Systems, Man and Cybernetics 15(1), 116–132 (1985)
25. Sutton, R.S., Barto, A.G.: Reinforcement Learning: An Introduction. MIT Press, Cambridge (1998)
26. Koga, T., Horio, K., Yamakawa, T.: Applications of brain-inspired SOR network to controller design and knowledge acquisition. In: Proceedings of 5th Workshop on Self-Organizing Maps (WSOM 2005), CD-ROM (2005)
27. Koga, T., Horio, K., Yamakawa, T.: The Self-Organizing Relationship (SOR) Network Employing Fuzzy Inference Based Heuristic Evaluation. Neural Networks 19, 799–811 (2006)

Type-2 Fuzzy Logic Controllers: A Way Forward for Fuzzy Systems in Real World Environments

Hani Hagras

The German University in Cairo, New Cairo City, Egypt and the Department of Computing
and Electronic Systems, University of Essex, Colchester, CO4 3SQ, United Kingdom
hani@essex.ac.uk

Abstract. Type-1 Fuzzy Logic Controllers (FLCs) have been applied to date
with great success to different applications. However, for many real-world
applications, there is a need to cope with large amounts of uncertainties. The
traditional type-1 FLCs that use crisp type-1 fuzzy sets cannot directly handle
such uncertainties. Type-2 FLCs that use type-2 fuzzy sets can handle such uncertainties to produce a better performance. Hence, type-2 FLCs will have the
potential to overcome the limitations of type-1 FLCs and produce a new generation of fuzzy controllers with improved performance for many applications
which require handling high levels of uncertainty. This chapter will provide an
overview of the interval type-2 FLCs and their advantages over type-1 FLCs.
We will also present different techniques to avoid the computational overheads
and thus enabling the type-2 FLCs to produce a good real time response.
Furthermore, we will present various successful real world applications of
type-2 FLCs.

Keywords: Fuzzy Logic, Interval Type-2 Fuzzy Logic Controllers, Uncertainty
Handling.

1 Introduction

Fuzzy control is regarded as the most widely used application of fuzzy logic [1]. A
Fuzzy Logic Controller (FLC) is credited with being an adequate methodology for
designing robust controllers that are able to deliver a satisfactory performance in the
face of uncertainty and imprecision.

FLCs have successfully outperformed the traditional control systems (like PID
controllers) and have given a satisfactory performance similar (or even better) to the
human operators. According to Mamdani [2]: " *When tuned, the parameters of a PID
controller affect the shape of the entire control surface. Because fuzzy logic control is
a rule-based controller, the shape of the control surface can be individually manipu-
lated for the different regions of the state space, thus limiting possible effects to
neighbouring regions only* ".

FLCs have been applied with great success to many applications where the first
FLC was developed in 1974 by Mamdani and Assilian for controlling a steam generator [3]. In 1976, Blue Circle Cement and SIRA in Denmark developed a cement kiln

J.M. Zurada et al. (Eds.): WCCI 2008 Plenary/Invited Lectures, LNCS 5050, pp. 181–200, 2008.

controller— which is the first industrial application of fuzzy logic. The system went to operation in 1982 [4]. Following this, the 1980s and 1990's have witnessed a wide scale deployment of FLCs to several successful applications.

However, there are many sources of uncertainty facing the FLC in dynamic real-world unstructured environments and many real-world applications; some of the uncertainty sources are as follows:

- Uncertainties in inputs to the FLC, which translate into uncertainties in the antecedents' membership functions as the sensors measurements are affected by high noise levels from various sources. In addition, the input sensors can be affected by the conditions of observation (i.e. their characteristics can be changed by the environmental conditions such as wind, sunshine, humidity, rain, etc.).
- Uncertainties in control outputs, which translate into uncertainties in the consequents' membership functions of the FLC. Such uncertainties can result from the change of the actuators' characteristics, which can be due to wear, tear, environmental changes, etc.
- Linguistic uncertainties as the meaning of words that are used in the antecedents' and consequents' linguistic labels can be uncertain, as words mean different things to different people [1]. In addition, experts do not always agree and they often provide different consequents for the same antecedents. A survey of experts will usually lead to a histogram of possibilities for the consequent of a rule; this histogram represents the uncertainty about the consequent of a rule [1].
- Uncertainties associated with the change in the operation conditions of the controller. Such uncertainties can translate into uncertainties in the antecedents' and/or consequents' membership functions.
- Uncertainties associated with the use of noisy training data that could be used to learn, tune or optimise the FLC.

All of these uncertainties translate into uncertainties about fuzzy set membership functions [1]. The vast majority of the FLCs that have been used to date were based on the traditional type-1 FLCs. However, type-1 FLCs cannot fully handle or accommodate the linguistic and numerical uncertainties associated with dynamic unstructured environments as they use type-1 fuzzy sets. Type-1 fuzzy sets handle the uncertainties associated with the FLC inputs and outputs by using *precise and crisp* membership functions that the user believes capture the uncertainties. Once the type-1 membership functions have been chosen, all the uncertainty disappears because type-1 membership functions are precise [1]. The linguistic and numerical uncertainties associated with dynamic unstructured environments cause problems in determining the exact and precise antecedents' and consequents' membership functions during the FLC design. Moreover, the designed type-1 fuzzy sets can be sub-optimal under specific environment and operation conditions; however, because of the environment changes and the associated uncertainties, the chosen type-1 fuzzy sets might not be appropriate anymore. This can cause degradation in the FLC performance, which can result in poor control and inefficiency and we might end up wasting time in frequently redesigning or tuning the type-1 FLC so that it can deal with the various uncertainties.

A type-2 fuzzy set is characterised by a fuzzy membership function, i.e. the membership value (or membership grade) for each element of this set is a fuzzy set in [0,1], unlike a type-1 fuzzy set where the membership grade is a crisp number in [0,1] [1]. The membership functions of type-2 fuzzy sets are three dimensional and include a footprint of uncertainty. It is the new third dimension of type-2 fuzzy sets and the footprint of uncertainty that provide additional degrees of freedom that make it possible to directly model and handle uncertainties [1]. The type-2 fuzzy sets are useful where it is difficult to determine the exact and precise membership functions. Type-2 Fuzzy Logic Systems (FLSs) that use type-2 fuzzy sets have been used to date with great success where the type-2 FLSs have outperformed their type-1 counterparts as was shown in the fields of equalization of nonlinear time-varying channels [5], overcoming time varying co-channel interference [6], connection admission control in ATM Networks [7] and several medical applications as in [8], [9].

The successful type-2 FLSs applications highlighted above have encouraged the application of type-2 FLSs in control in what is known as type-2 FLCs. In the next section, we will introduce the interval type-2 FLC and highlight its benefits.

2 Type-2 Fuzzy Sets

Type-1 FLCs employ the crisp and precise type-1 fuzzy sets. For example consider a type-1 fuzzy set representing the linguistic label of "*Low*" temperature in Fig 1a. Hence if the input temperature x is $15°$ C, then the membership of this input to the "*Low*" type-1 set will be the certain and crisp membership value of 0.4. However, the centre and endpoints of this type-1 fuzzy set will vary with the season, the human preferences, the country, the context as well as other factors and uncertainties. Hence, if this linguistic label was employed with a FLC to control temperature, then the type-1 FLC would need to be continuously tuned to handle all the faced uncertainties. Alternatively, we would need to have a group of separate type-1 sets and type-1 FLC where each will handle a certain situation.

From the above example, it is clear that the type-1 FLCs have the common problem that they cannot fully handle or accommodate for the linguistic and numerical uncertainties associated with changing and dynamic unstructured environments.

On the other hand, a type-2 fuzzy set is characterised by a fuzzy membership function, i.e. the membership value (or membership grade) for each element of this set is a fuzzy set in [0,1]. For example if the linguistic label of "*Low*" temperature is represented by a type-2 fuzzy set as shown in Fig. 1b, then the input x of $15°$ C will no longer have a single value for the membership function. Instead, the membership function takes on values wherever the vertical line intersects the blurred area shaded in grey. Hence $15°$ C will have primary membership values that lie in the interval [0.2, 0.6]. Each point of this interval will have also a weight associated with it. Hence this will create an amplitude distribution in the third dimension to form what is called a secondary membership function which can be a triangle as shown in Fig. 1c. In case the secondary membership function is 1 for all the points in the primary membership and if this is true for $\forall x \in X$, we have the case of an interval type-2 fuzzy sets (drawn in dotted lines in Fig.1c). Hence the input x of $15°$ C will have primary membership interval and an associated secondary membership function. Doing this for all $x \in X$, we create a three-dimensional membership function (as shown in Fig. 1d) —a type-2 membership function—that characterizes a type-2 fuzzy set. The membership functions of type-2 fuzzy sets are three dimensional and include a Footprint of Uncertainty

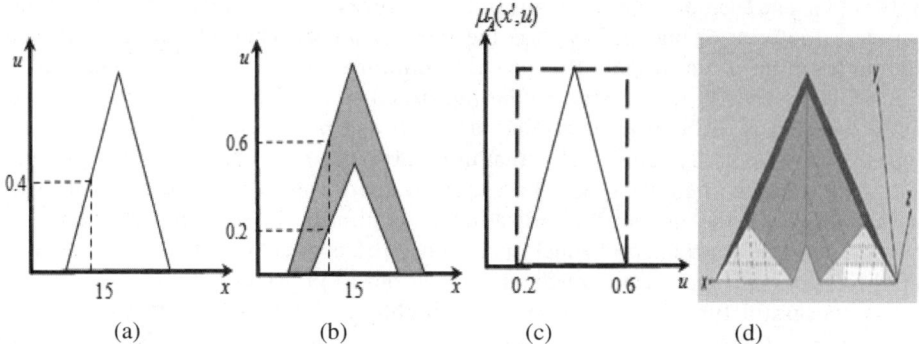

Fig. 1. a) A type-1 fuzzy Set. b) A type-2 fuzzy set- primary membership function. c) A type-2 fuzzy set- secondary membership function. d) 3-d view of a type-2 fuzzy set.

(FOU) (shaded in grey in Fig.1b). It is the new third-dimension of type-2 fuzzy sets and the FOU that provide additional degrees of freedom that make it possible to directly model and handle uncertainties [1], [10].

3   Interval Type-2 FLC

The interval type-2 FLC uses interval type-2 fuzzy sets to represent the inputs and/or outputs of the FLC. The interval type-2 FLC is a special case of the general type-2 FLC. The vast majority of type-2 FLC applications to date employ the interval type-2 FLC. This is because the general type-2 FLC is computationally intensive and the computation simplifies a lot when using the interval type-2 FLC which will enable us to design a FLC that operates in real time.

The interval type-2 FLC is depicted in Fig. 2 and it consists of a Fuzzifier, Inference Engine, Rule Base, Type-reducer and Defuzzifier. The interval type-2 FLC operate as follows: the crisp inputs from the input sensors are first fuzzified into input type-2 fuzzy sets; singleton fuzzification is usually used in interval type-2 FLC applications due to its simplicity and suitability for embedded processors and real-time applications. The input type-2 fuzzy sets then activate the inference engine and the rule base to produce output type-2 fuzzy sets. The type-2 FLC rules will remain the same as in the type-1 FLC, but the antecedents and/or the consequents will be represented by interval type-2 fuzzy sets. The inference engine combines the fired rules and gives a mapping from input type-2 fuzzy sets to output type-2 fuzzy sets. The type-2 fuzzy outputs of the inference engine are then processed by the type-reducer, which combines the output sets and performs a centroid calculation that leads to type-1 fuzzy sets called the type-reduced sets. The type-reduction process use the iterative Karnik-Mendel (KM) procedure to calculate the type-reduced fuzzy sets [1]. The KM procedure convergence is proportional to the number of fired rules and hence this can cause a computational bottleneck for the type-2 FLC. After the type-reduction process, the type-reduced sets are then defuzzified (by taking the average of the type-reduced set) to obtain crisp outputs that are sent to the actuators. More information about the interval type-2 FLC can be found in [11].

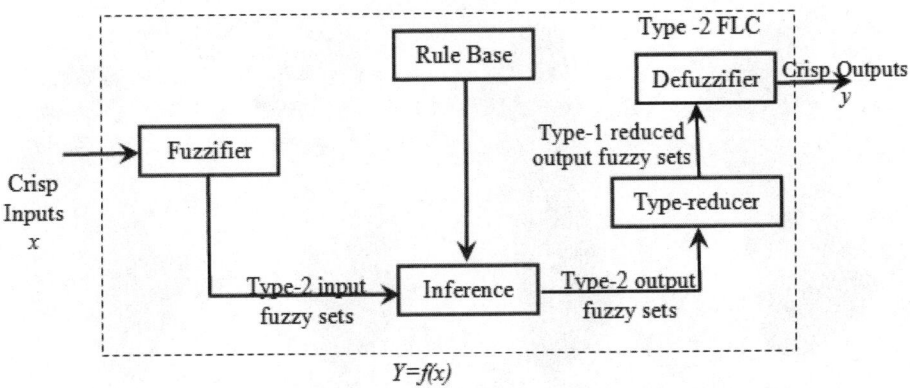

Fig. 2. The type-2 FLC

It has been argued that using interval type-2 fuzzy sets to represent the inputs and/or outputs of FLC has many advantages when compared to the type-1 fuzzy sets; we summarize some of these advantages as follows:

- As the type-2 fuzzy sets membership functions are fuzzy and contain a FOU, hence they can model and handle the linguistic and numerical uncertainties associated with the inputs and outputs of the FLC. Therefore, FLCs that are based on type-2 fuzzy sets will have the potential to produce a better performance than the type-1 FLCs when dealing with uncertainties [11].

- Using type-2 fuzzy sets to represent the FLC inputs and outputs will result in the reduction of the FLC rule base when compared to using type-1 fuzzy sets, as the uncertainty represented in the FOU of the type-2 fuzzy sets lets us cover the same range as type-1 fuzzy sets with a smaller number of labels and the rule reduction will be greater when the number of the FLC inputs increases [1].

- Each input and output will be represented by a large number of type-1 fuzzy sets, which are embedded in the type-2 fuzzy sets [1], [10]. The use of such a large number of type-1 fuzzy sets to describe the input and output variables allows for a detailed description of the analytical control surface as the addition of the extra levels of classification give a much smoother control surface and response. In addition, according to Karnik and Mendel [12], the type-2 FLC can be thought of as a collection of many different embedded type-1 FLCs.

- It has been shown in [13] that the extra degrees of freedom provided by the FOU enables a type-2 FLC to produce outputs that cannot be achieved by type-1 FLSs with the same number of membership functions. It has been shown that a type-2 fuzzy set may give rise to an equivalent type-1 membership grade that is negative or larger than unity. Thus, a type-2 FLC is able to model more complex input-output relationships than its type-1 counterpart and, thus, can give better control response.

The above points could be shown in Fig. 3 which shows for an outdoor mobile robot how a type-2 FLC with a rule base of only 4 rules could produce a smoother control surface as shown in Fig. 3a and hence better result than its type-1 counterpart that

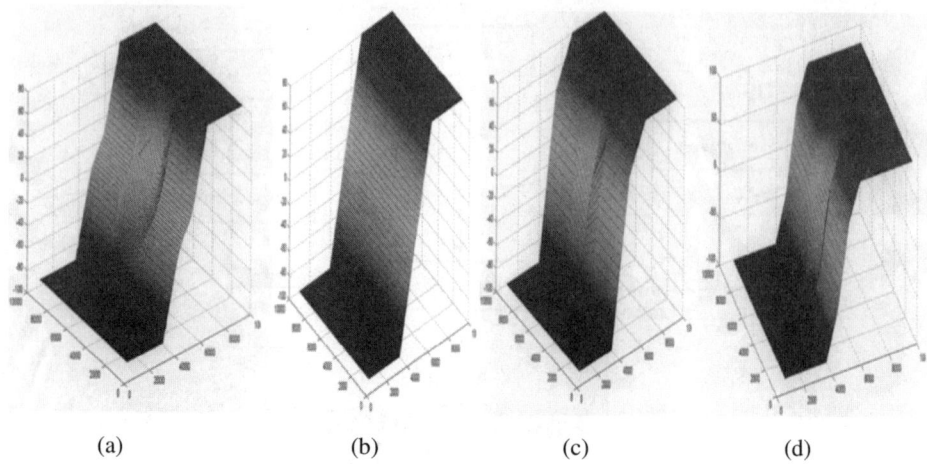

(a)	(b)	(c)	(d)

Fig. 3. (a) Control surface of a robot type-2 FLC with 4 rules. (b) Control surface of a robot type-1 FLC with 4 rules. (c) Control surface of a robot type-1 FLC with 9 rules. (d) Control surface of a robot type-1 FLC with 25 rules.

used a rule base of 4, 9 and 25 rules as shown in Fig. 3b, Fig.3c and Fig. 3d respectively [11]. It is also shown that as the type-1 FLC rule base increases, its response approaches that of the type-2 FLC which encompasses a huge number of embedded type-1 FLCs.

4   Avoiding the Computational Overheads of Type-2 FLCs

Type-2 FLCs are more computionally demanding when compared to type-1 FLCs. This is because type-2 FLCs involve dealing with interval type-2 fuzzy sets which are more complex than the corresponding type-1 fuzzy sets. Hence, the fuzzification and inference mechanisms are more computationally demanding when compared to their type-1 counterparts. In addition, the type-2 FLC involve one more process which is type-reduction. Type-reduction is regarded as the major computational bottleneck for type-2 FLCs as type-reduction employ the iterative KM procedure whose convergence is directly proportional to the number of fired rules. This computational overhead translates to a protracted controller response which can diminish the robustness and the real-time performance of the type-2 FLC especially when operating on industrial embedded platforms (this is more evident for FLCs with large rule bases). Hence, although the type-2 FLC introduced a very promising system for uncertainty handling for many real world problems, it was perceived as a complex computationally intensive system. This perception has hindered the widespread deployment of type-2 FLCs to many real world applications. As a result, several approaches have been presented to speed the type-2 FLC performance and alleviate the various computational overheads to allow the FLC to produce fast real time response. The following subsections will introduce some of the approaches introduced to speed the type-2 FLC response.

4.1 Type-Reduction Approximation

Wu and Mendel [14] introduced a method to approximate the type-reduced set by the inner and outer bound sets, thus avoiding the use of the iterative KM procedure. This method provides equations that compute the bound sets with no need to have any iterative procedures. It has been shown that the type-2 FLCs that employed the Wu-Mendel (WM) uncertainty bounds method had the potential to provide a valid approximation from the control point of view to the type-2 FLC using the iterative KM type-reduction procedure [15], [16].

Fig. 4 shows the computational savings as a result of employing the WM method as opposed to the KM iterative type-reduction procedure in a marine diesel engine control application. Fig.4 shows the number of fired rules plotted against the time required to generate crisp outputs from given crisp inputs. Fig. 4 also illustrates the mean (solid lines) and standard deviations (dashed upper and lower error bars) of these calculated times (expressed in micro-seconds) for a varying number of fired rules. It was shown that for the WM based type-2 FLCs the mean and the standard deviation intervals for the computational time were always less than the type-2 FLC employing the KM procedure. For example in the instance of 25 rules firing the WM type-2 FLC achieved a comparative reduction of 28.5% relative to the KM type-2 FLC. It is shown that as number of fired rules increase, the KM type-2 FLC real time performance will degrade. On the other hand, the Wu-Mendel type-2 FLC will produce faster computation times where for example the WM type-2 FLC is 48.8% and 54% faster than the KM type-2 FLC in the instance of 100 and 200 rules firing respectively and these computational savings will increase as the number of firing rules increase. This WM type-2 FLC faster response will map to a better real-time response that will impact not only our ability to meet hard real-time deadlines but also to free the processor for executing other high priority components such as alarms and monitoring whilst still having the superior type-2 FLC control response.

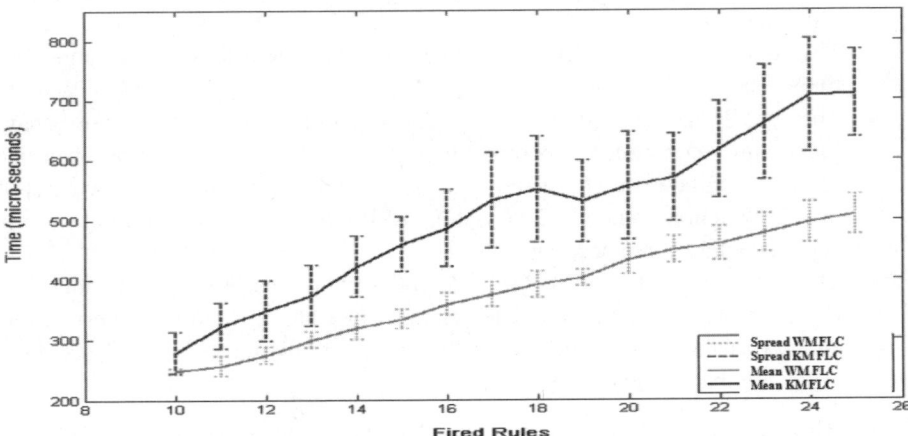

Fig. 4. Mean and Standard Deviation for the timings of the WM type-2 FLC and the KM type-2 FLC

4.2 Type-2 Hierarchical Fuzzy Logic Controllers

Single rule base FLC (type-1 or type-2) suffer from the serious limitation that the number of rules increases *exponentially* with the number of variables involved. In addition, for interval type-2 FLCs computing the type-reduced fuzzy set is directly proportional to the number of fired rules.

Type-2 Hierarchical Fuzzy Logic Controllers (HFLCs) have been introduced to cope with *the rule explosion problem* and its effects on the design and real-time operation of the interval type-2 FLCs. In type-2 HFLCs (depicted in Fig. 5), the control problem is hierarchically decomposed by breaking down the input space for analysis by sharing it amongst multiple low level type-2 FLCs. Each FLC responds to specific types of situations and then a high level type-2 coordination layer integrates the recommendations of these low level type-2 FLCs. Each low level type-2 FLC has a small number of inputs and outputs and a small rule base and it serves a single purpose. The low level type-2 FLCs will typically (but not necessarily) map different inputs sensors to common actuators outputs. Such low level type-2 FLCs are building blocks for more intelligent composite behaviours, i.e. their capabilities can be combined through synergistic coordination by the high level type-2 fuzzy coordination layer to obtain an overall coherent behaviour that achieves the intended task(s). Type-2 fuzzy coordination allows the ability to express partial and concurrent activations of behaviours; and the smooth transition between behaviours.

The hierarchical fuzzy systems have a nice property that the total number of rules increases linearly rather than exponentially as in the single rule base FLC. Hence type-2 HFLCs offer the following advantages:

- It simplifies the design of the controller and reduces the number of rules to be determined so we can have a real time operation of the interval type-2 FLC.
- It uses the benefits of type-2 fuzzy logic to deal with the large amounts of imprecision and uncertainty present in changing and dynamic unstructured environments.
- The controllers can achieve multiple goals, whose priorities may change with time.
- The type-2 fuzzy coordination provides a smooth transition between behaviours with a consequent smooth output response which allows more than one behaviour to be active to differing degrees thereby avoiding the drawbacks of on-off switching schema (i.e. dealing with situations where several criteria need to be taken into account).
- This hierarchical structure offers a flexible structure where new behaviours can be added or modified easily. The system is capable of performing very different tasks using identical behaviours by changing only the context rules and coordination parameters.

Type-2 HFLCs have been applied with great success to autonomous robot control as in [11].

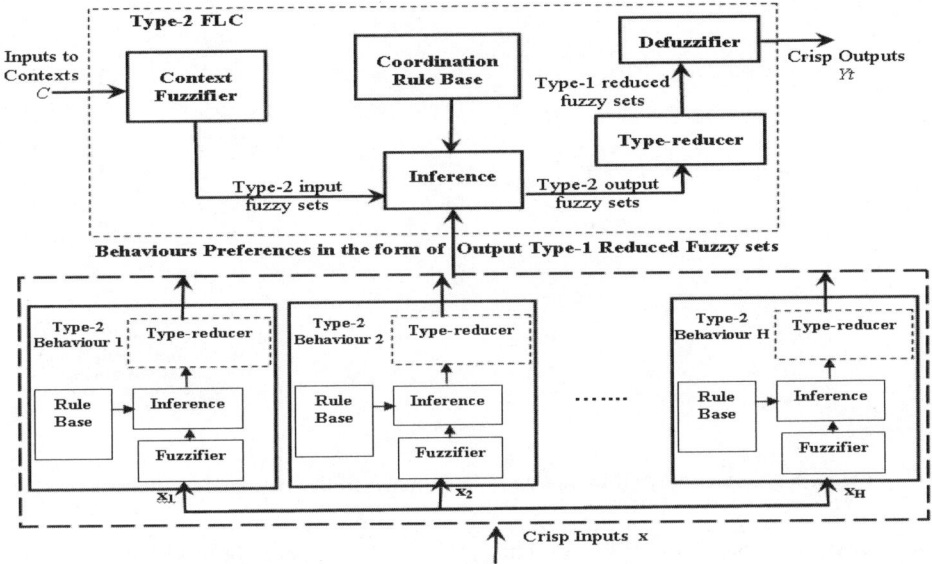

Fig. 5. The type-2 HFLC architecture

4.3 Hardware Implementations and Type-2 Co-processors

Most of the current commercial controllers, attempt to address the faced uncertainties through averaging the sensor inputs and using gain scheduled control algorithms such as the gain scheduled PID controller with numerous non-linear gain functions integrated in the embedded controller. Despite the additional complexity of applying these supplementary functionalities, the commercial controllers are still computationally efficient. Where for example in marine diesel engines, the Viking 25 commercial controller require about ten thousand clock cycles to perform all of its speed control functions, and using the remaining clock cycles to perform other engine management features such as signal conditioning, communications, alarm and monitoring etc. On the other hand, a type-2 FLC requires the equivalent number of clock cycles to perform type-reduction alone [16]. Thus despite any control performance improvements type-2 FLC may offer, these computational bottlenecks remain as a barrier to the type-2 FLC deployment in commercial embedded control systems. As a result, alternative hardware solutions were presented to exploit the high level of parallelism offered by the type-2 FLC.

Currently there are only some hardware implementations of the interval type-2 FLC available. The first implementation was presented in [17] and they produced a VLSI implementation where the type-2 FLC was designed at the transistor level on a single chip for a dual input single output controller supporting up to 64 rules. This approach whilst offering a tailored solution does not present the flexibility nor reprogrammability of a micro-processor based solution. Alternatively Melgarejo et al [18] designed a type-2 FLC for an adaptive filter with a rule base of nine rules using the Wu-Mendel approximation. This implementation was embedded on a Field

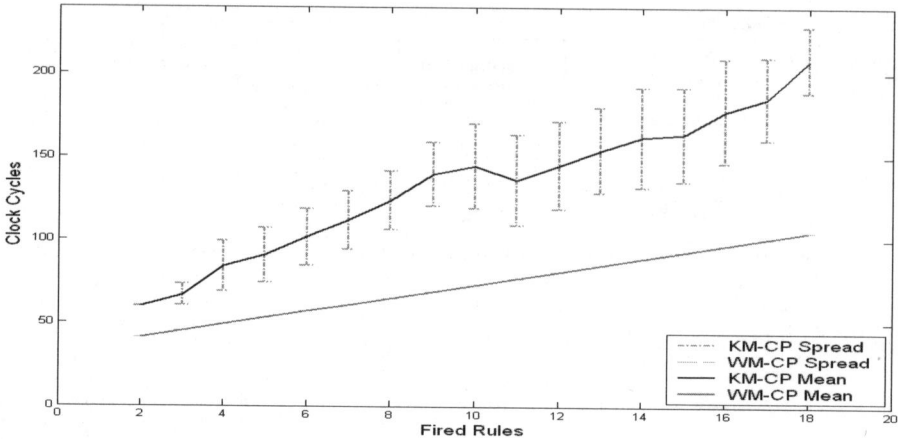

Fig. 6. Mean and Standard Deviation for the timings of the WM CP and the KM CP

Programmable Gate Array (FPGA). This approach is a highly optimised and pipelined solution offering a type reduced set in 9 clock cycles at the expense of being highly memory intensive; making use of memory base fuzzification, reciprocal division and distributed arithmetic each of which require large amount of on chip memory. In [19], Lynch et al. presented parallel hardware implementations of the interval type-2 FLC for the purpose of industrial control which can accommodate much larger rule bases (supporting up to 64 fired rules and hence supporting much larger rule bases) and use cheap hardware solutions. The developed solution was based on bespoke co-processors that can perform functions such as fuzzification and type reduction. Fig. 6 illustrates the results obtained by the bespoke Co-Processors (CPs) where the x-axis represents the number of fired rules and the y-axis represents the time the type-2FLC takes to produce crisp outputs given the crisp inputs. Fig. 6 shows the linear relationship between the Wu-Mendel CP (WM-CP) to the number of fired rules. This is a very advantageous trait as the WM-CP consistency and predictability allow the system to transfer all data to the WM-CP and then continue executing other code, rather than waiting for the WM-CP to return the type reduced sets. Conversely the KM-CP is dominated by the number of fired rules and the number of iterations required to complete type reduction. Thus it is difficult to predict in advance the total clock cycles the KM-CP will need, as the number of required iterations is unknown. It is also clearly shown that the WM-CP requires much less clock cycles relative to the KM-CP. The type-2 CPs achieved an approximate computational reduction of approximately 100% relative to their sequential counterparts. This means that the developed CP hardware solution can achieve 100 % faster response compared to the normal interval type-2 FLC implementation. In the marine diesel engine domain, the CP based type-2 FLC was ten times quicker than the employed Viking 25 commercial controller. Thus the hardware acceleration offered by the co-processors removes any significant bottlenecks from type-2 FLC and make it even faster than the sequential type-1 FLC and the commercial controllers whilst the type-2 FLC offers a superior control performance. Hence, the proposed co-processors enable to fully explore the potential of interval and possibly general type-2 FLCs in applied commercial embedded applications.

Most recently, Coupland et al. developed a hardware implementation of general type-2 FLC.

5 Successful Applications of Type-2 FLCs

5.1 Applications of Type-2 FLCs to Industrial Control

Due to the high uncertainty levels facing industrial control systems, type-2 FLCs had much attention in industrial control applications. In [15], [16], [19], interval type-2 FLC was applied to the speed control of marine diesel engines which was regarded as the first heavy industrial application of type-2 FLC. Marine diesel engines are huge engines that are classified according to their speeds into three main categories: slow speed engines, medium speed engines and high speed engines. Slow speed engines are massive and can produce up to 80 MW of power such as the 2300 ton Wartsila-Sulzer engine shown in Fig. 7, which operates at a few hundred rpm and is designed mainly for large container ships [20]. Due to their vast sizes and large power outputs, marine diesel engines require accurate and robust speed control/governing. Accurate speed control of marine diesel engines is of critical importance as significant deviations from the speed set point could be detrimental and damaging to the engine and the respective loads. Moreover, for applications such as power generation sets, the engine speed in rpm must be stable multiples of the generated base frequency, i.e. 50Hz frequency would require the engine to operate at 1000 rpm, 1500 rpm, etc. Hence, significant speed deviation can cause the generation of incorrect frequencies resulting in loss of synchronisation between the generator and associated power grid, which is obviously very problematic for any power generation system and coupled loads. Robustness in speed control is required to overcome and recover quickly from the inherent instabilities and disturbances associated with marine diesel engines which operate in highly dynamic and uncertain environments, experiencing vast changes in ambient temperature, fuel, humidity and load.

Hence, choosing an appropriate speed control mechanism that is able to model and handle these uncertainties to produce an accurate and robust control is of vital importance. Moreover, this control mechanism has to be computationally undemanding to be able to operate on the industrial embedded electronic controllers, which have limited computational and memory capabilities that must typically be shared among speed control, alarm and monitoring, speed measurements, communications and signal conditioning. Due to their simplicity and their suitability for the industrial embedded controllers, various forms of the PID controller have been used for the speed control in marine diesel engines. It has also been shown that Fuzzy Logic and Fuzzy PID controllers can provide improved control and robustness over traditional PID controllers. However, as mentioned above, a type-1 FLC cannot fully handle the uncertainties associated with the marine diesel engines. Therefore, in practice, these uncertainties can be handled by continuously tuning the type-1 FLC or by using a group of type-1 FLCs to handle the faced uncertainties. Hence, the type-2 FLC

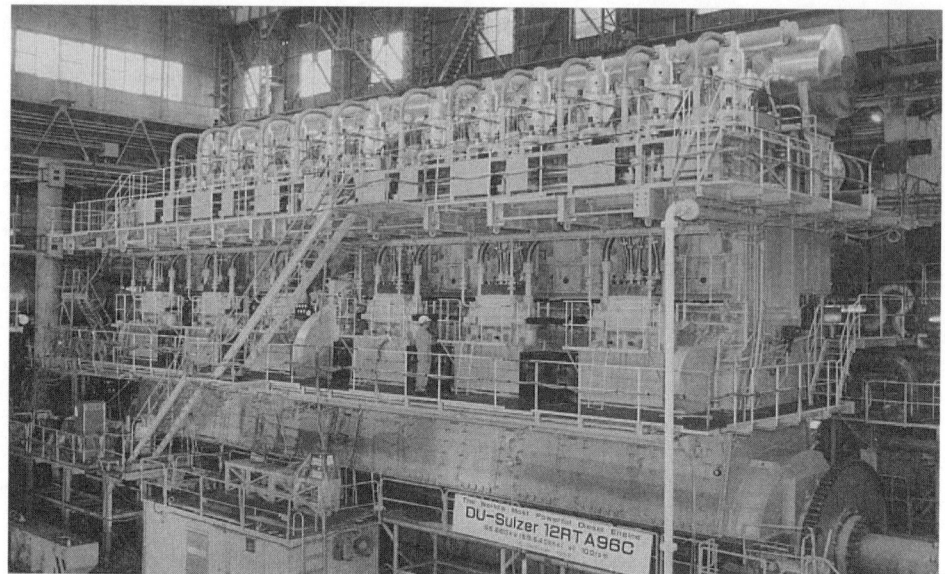

Fig. 7. The low speed Wartsila-Sulzer RTA96-C turbocharged two-stroke diesel engine [20]

appeared to be a very attractive control mechanism to handle the uncertainties faced by the marine diesel engines. In [15], a type-2 FLC has been presented that is suited for the embedded controllers operating in marine diesel engines. This type-2 FLC was based on using the Wu-Mendel uncertainty bounds method.

Fig. 8a confirms the same results shown in Fig. 3 where the control surface of the type-1 FLC is steep and non-smooth, especially near the set point where the error

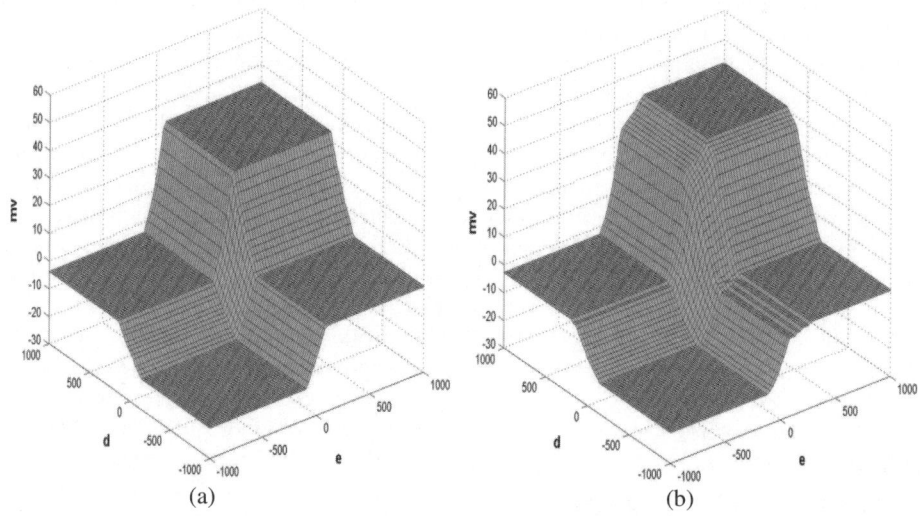

Fig. 8. (a) Control surface of a type-1 FLC. (b) Control surface of a type-2 FLC using Wu-Mendel uncertainty bounds method.

(e) between the speed set point and the actual value as well as the change of error (d) should be equal zero. Consequently, any small variation of e and d can cause considerable changes in the manipulating variable (mv), which means that the type-1 FLC is vulnerable to noise and uncertainties. Moreover, the larger the variations in e and d due to the uncertainties, the larger will be the disturbances to mv, which can cause instability and destruction of the engine. The control surface of the type-2 FLC shown in Fig. 8b shows a very smooth response as it is (in theory) aggregating the outputs of a large number of embedded type-1 FLCs. This smooth response will consequently give a very good control performance that can handle the uncertainties and disturbances as near the set point where $e= 0$ and $d=0$; small variations in e and d will not cause significant changes to mv and the response goes on gradually and smoothly with no steep changes.

Recently, a Real-Time Type-2 Neuro-Fuzzy Controller (RT2NFC) was developed, which uses the Wu-Mendel uncertainty bounds method [16]. The performance of the RT2NFC was compared with a T2NFC (Type-2 Neuro Fuzzy Controller that uses the KM iterative type-reduction procedure). Both the type-2 neuro fuzzy controllers were compared with type-1 FLCs and the Viking 25 commercial controller. In the representative experiments shown in Fig. 9a and Fig. 9b, both the Viking 25 and the type-1 FLC were tuned so that they can handle disturbances that were equivalent to 20 % of the full load (which is common disturbance that can face the engines at normal sea condition) [16]. The data used to train the RT2NFC and the T2NFC were obtained from the tuned Viking 25. It was noticed that the performances of the RT2NFC and the T2NFC are similar to the Viking 25 and type-1 FLC when introducing the disturbance of 20 % load that they were tuned to handle. However, as the uncertainty associated with the change of load increases to 100 % load as shown in Fig. 9a and Fig. 9b, the performance of both the Viking 25 and type-1 FLC degrades significantly, producing large overshoots/undershoots as well as long settling times. Hence, the performance of Viking 25 and the type-1 FLC will be unacceptable under these levels of uncertainties and this will not satisfy the desired standards. On the other hand, both the RT2NFC and the T2NFC produced type-2 FLCs that handled effectively the uncertainties associated with the change of the load and operation condition to give a very good performance that has small overshoots/undershoots as well as short settling times. The performance of both the RT2NFC and the T2NFC satisfy the required standards and, thus, they will require no further tuning. Therefore, the RT2NFC and T2NFC could be used effectively to produce accurate and robust speed controllers for marine diesel engines. This shows the power of type-2 FLCs as though the RT2NFC and T2NFC were trained to mimic the commercially used Viking 25 controller; however, as the level of uncertainties increase, the type-2 FLCs were able to handle the uncertainties and outperform the Viking25 controller. As noted in Fig. 9b, the similarity in control response between the RT2NFC and the T2NFC is a valid indication that they will both produce the same outputs. However, it was shown in [16] that the RT2NFC will produce a much faster response (especially as the number of used rules increases), which will enable real-time operations.

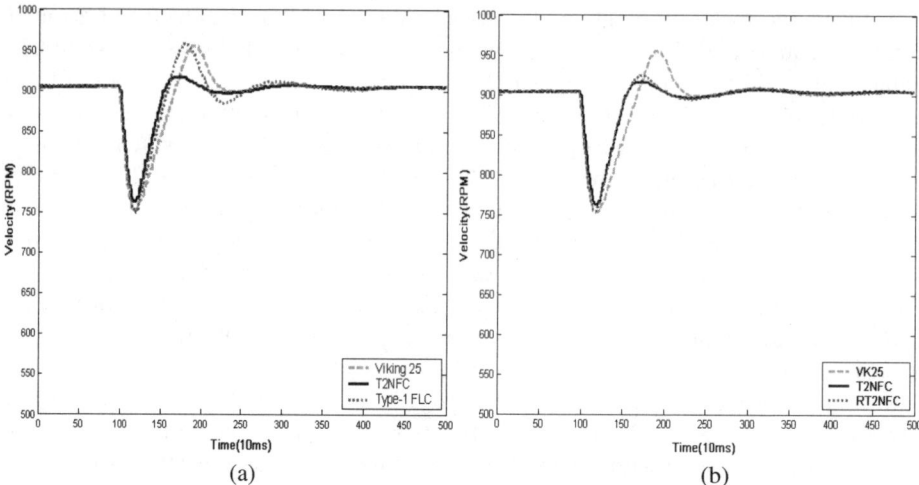

Fig. 9. (a) Comparison of the response of 20 % type-1 FLC and Viking 25 against a T2NFC with 100 % load addition. (b) Control response of the RT2NFC, T2NFC and Viking 25 to the uncertainties associated with load changes of 100 %.

In [21], a type-2 FLC was applied to the control of a buck DC-DC converter. DC-DC converters are power electronic systems that convert one level of electrical voltage into another level by switching action. The DC-DC converters are used extensively in personal computers, computer peripherals, and adapters of consumer electronic devices. The DC-DC converters are an intriguing subject from the control point of view due to their intrinsic nonlinearity. The control technique for DC-DC converters must cope with their wide input voltage and load variations to ensure stability in any operating condition while providing fast transient response. The control problem is to control the duty cycle so that the output voltage can supply a fixed voltage in the presence of the input voltage uncertainty and load variations. It has been shown that the performance of the type-2 FLC is better than its type-1 counterpart where the rise time response of type-2 FLC is faster than that of type-1 FLC with no overshoot in the type-2 FLC controlled system [21].

In [22], a Genetic Algorithm was used to evolve a type-2 FLC to control a liquid-level process. It was observed that both the type-1 and the type-2 FLCs are able to attenuate the oscillations when the modelling uncertainties are small. The liquid level in the tank will eventually reach the desired set-point, though the settling time is shorter when the type-2 FLC is employed. When the modelling uncertainties are larger, the type-1 FLC will give rise to persistent oscillations while the type-2 FLC has the ability to eliminate these oscillations and the liquid level reaches its desired height at steady state. It was concluded that the type-2 FLC is more robust than the type-1 FLC as the type-2 FLC outperforms its type-1 counterpart, especially when the uncertainty is large [22].

As in the above industrial applications, it was also noticed, for the feedback control system reported in [23], that without uncertainties, the type-1 FLC and the type-2 FLC responses are very similar. However, as the level of uncertainties increases, the type-2 FLC produces a much improved performance with lower overshoot errors and better settling times than the type-1 counterparts. Thus, it was concluded that using a type-2 FLC in real-world applications can be a better choice since the amount of uncertainty in real systems most of time is difficult to estimate [23].

In [24] a type-2 FLC was presented for the desulphurization process of a real steel industry in Canada. It is shown that the proposed type-2 fuzzy logic system is superior in comparison to multiple regression and type-1 fuzzy logic systems in terms of robustness, and error reduction.

5.2 Applications of Type-2 FLCs to Robot Control

Autonomous mobile robots navigating in real-world unstructured environments must be able to operate under the conditions of imprecision and uncertainty present in such environments. Hence, type-2 FLCs have found good grounds for application in mobile robot control. In [11], the interval type-2 FLC was presented to robot control involving indoor and outdoor robots and it was noticed that the type-2 FLC always outperformed its type-1 counterpart while using less number of rules. This was shown through the robot paths and the control surfaces, which graphically represent the unknown function articulated by the FLC.

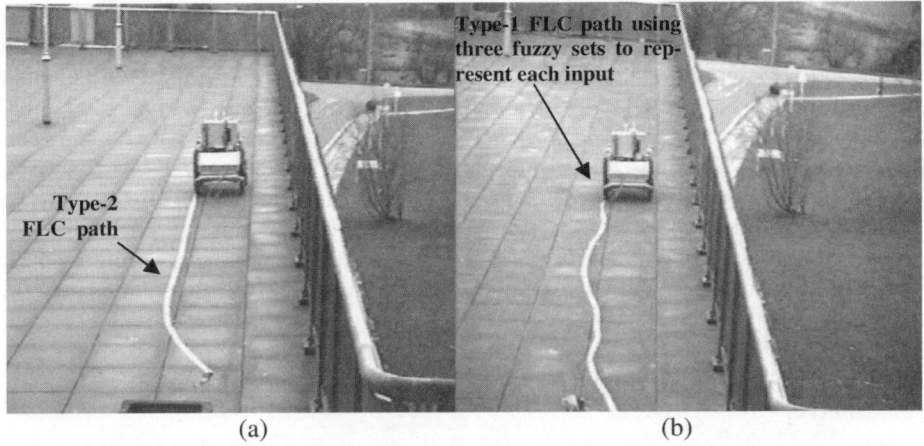

(a)   (b)

Fig. 10. (a) The outdoor robot path using type-2 FLC to implement the right edge following behaviour to follow an irregular edge. (b) The robot path using a type-1 FLC, which gave a bad response when the environment changed (windy weather).

In [11], experiments with robots in outdoor unstructured environments were presented to evaluate the real-time performance of the robot type-2 FLC and how it handles the large amounts of uncertainty and imprecision facing the mobile robots in such changing and dynamic environments. The robots were tested under different environmental conditions (like rain, wind, sunshine, etc.) and different ground conditions (such as slippery and dry grounds) as well as different times of the day. These experiments also involved the use of different challenging environmental features like metallic and plant edges, which offer bad sonar response. It has been shown that the type-2 FLC shown in Fig. 10a can handle uncertainties to give a better response while using a smaller rule base. It was also noticed that the type-1 FLC can give a good response under specific weather, ground and robot conditions, but if any of these conditions change like operating in a windy weather condition then the type-1 FLC with nine rules shown in Fig. 10b will fail and give a bad response as it cannot handle the uncertainties associated with the outdoor environments.

The type-2 HFLCs were applied also to outdoor robots in which they gave a very good performance that outperformed its type-1 counterparts.

In [25], an interval type-2 FLC was presented for a robotic agent intended to track a mobile object in the context of robot soccer games. In this domain, there are many sources of uncertainty, which include the image processing algorithms (which could be classified as uncertainty about the FLC inputs) as well as uncertainties in the actuators and networking resources. It was shown that the type-2 FLC is able to cope with the involved uncertainty in a better way than the type-1 counterparts and it was also noted that with the type-2 FLC, it is not necessary to include more rules or fuzzy sets like the type-1 FLC and, hence, the type-2 FLC uses a smaller rule base.

In [26], a generalised type-2 FLC has been introduced and applied to the control of mobile robots. The robots were given the task of navigating around a curved obstacle. The task of the FLC is essentially to minimise the deviation from the ideal path between the start and finish lines. Three controllers were compared, which are the type-1 FLC, the interval type-2 FLC and a hybrid generalised type-2 FLC that used geometric type-2 fuzzy logic in order to achieve the execution speeds required by the robot control system. It was noticed from the visual inspection of the robot paths that the generalised type-2 FLC performed most consistently. The interval type-2 FLC had a wider but consistent spread while the type-1 FLC had spread of paths somewhere between the two with a few paths quite far outside the main spread. It was concluded in [26] that the generalised type-2 FLC had the best performance among the interval type-2 FLC and the type-1 FLC. It was also noticed that the generalised type-2 FLC performed consistently well while the interval type-2 FLC performed quite well but was a little inconsistent, and the type-1 FLC performed relatively badly but was consistent in the level of error.

5.3   The Application of Type-2 FLCs to Ambient Intelligent Environments Control

This section present the application of interval type-2 FLC to the control of Ambient Intelligent Environments (AIEs) as an example of domestic environments control. One of the main underlying challenges facing intelligent environments lies in the ability to manage the short-term and long-term uncertainties that arise due to changes in the environmental conditions and the user behaviour and activities during time. In [27], an agent architecture was presented for AIEs that used type-2 FLC for the control of AIEs. The presented agent architecture used a one-pass (non iterative) method to learn online the user's particular behaviours and preferences for controlling the AIE in a non intrusive and seamless manner. The system learns the user behaviour by learning his particular rules and type-2 membership functions required by the type-2 fuzzy agent. These can then be adapted incrementally in a life-long learning mode to suit the changing environmental conditions and user preferences. The presented

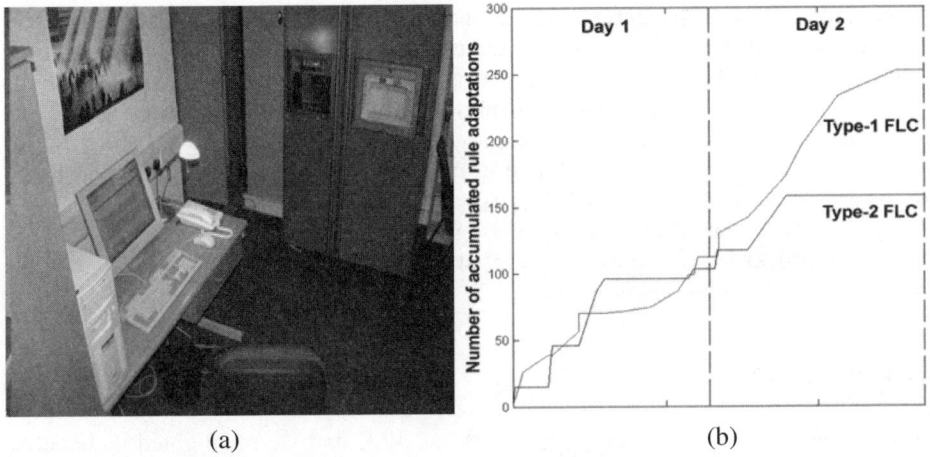

Fig. 11. (a) The iDorm. (b) Number of accumulated online user adaptations

type-2 agent architecture is suited for the embedded platforms used in AIEs, which have limited computational and memory capabilities. The type-2 FLC based agent was evaluated in the Essex intelligent Dormitory (iDorm), which is shown in Fig. 11a. The iDorm is a multi-user inhabited space that is fitted with a plethora of embedded sensors, actuators, processors and heterogeneous networks that are cleverly concealed (buried in the walls and underneath furniture) so that the user is completely unaware of the hidden intelligent infrastructure of the room. Unique experiments were conducted with various users during an extended period (spanning the course of the year) where it was possible to evaluate and demonstrate how the agent can adapt in a life-long learning mode and handle the faced short and long-term uncertainties. The type-2 FLC based agents were compared with type-1 FLC based agents in their ability to model the user's behaviour while handling the long-term uncertainties. The results showed that the type-2 FLC was better able to model the user behaviour and handle the short and long-term uncertainties while using fewer rules. In addition, online experiments were conducted in the iDorm where the user satisfaction was measured by monitoring how well the agents adjusted the iDorm environment to the user's preferences such that the user intervention was reduced during time, which can be used as a measure of the user's satisfaction. Fig. 11b shows, for a two-day experiment, the number of rules that were adapted online every time the user had to override the agent's decision. From Fig. 11b, it can be seen that the type-2 FLC based agent required significantly less user interaction than the type-1 agent. The plot for the type-2 agent shows that the user intervention was initially high but then stabilised by the second day. Therefore, the type-2 agent only required a very short online tuning period of approximately one day. This is because the type-2 agent better modeled the user behavior and handled the short and long-term uncertainties. The plot for the type-2 agent also shows it to be more stable than the type-1 agent in controlling the environment between the points when the user had to intervene in the agent's decisions and adapt the rules. In comparison, the plot for the type-1 agent shows that the user intervention continued to increase and did not properly stabilise by the end of the second day. As a result, the frequency of the user interaction with the type-1 agent was quite high compared to the type-2 agent as the user had to continually adjust the agent's decisions. It was also shown that type-2 agents can adapt to user behaviours and that they always generated fewer rules compared with the type-1 agents. It was shown that this rule reduction will lead to faster processing and more efficient memory usage when compared with the type-1 agents where it was found that the type-2 agent was able to outperform the type-1 agent performance while achieving a 60 % increase in processing speed as a result of attaining a 50 % reduction in the size of the rule base, thus reducing memory usage [27].

6 Conclusions

In this chapter, we presented the interval type-2 FLC and we highlighted its benefits, especially in highly uncertain environments. We have also highlighted the various techniques presented inorder to avoid the computational overheads of the interval type-2 FLC and thus speeding its response to achieve satisfactory real time performance.

Through the review of the various type-2 FLC applications, it has been shown that as the level of imprecision and uncertainty increases, the type-2 FLC will provide a powerful paradigm to handle the high level of uncertainties present in real-world environments. It has been also shown in various applications that the type-2 FLCs have given very good and smooth responses that have always outperformed their type-1 counterparts. Thus, using a type-2 FLC in real-world applications can be a better choice than type-1 FLCs since the amount of uncertainty in real systems most of the time is difficult to estimate [23].

Current research has started to explore the general type-2 FLC. Recent research is looking at generating general type-2 FLCs that embed a group of interval type-2 FLCs. This will enable to build on the existing theory of interval type-2 FLC while exploring the power of general type-2 FLCs.

Thus with the latest developments in interval type-2 FLCs, we can see that type-2 FLC overcomes the limitations of type-1 FLCs and will present a way forward to fuzzy control and especially in highly uncertain environments, which includes most of the real-world applications. Hence, it is envisaged to see a wide spread of type-2 FLCs in many real-world application in the next decade.

References

1. Mendel, J.: Uncertain Rule-Based Fuzzy Logic Systems: Introduction and New directions. Prentice-Hall, Englewood Cliffs (2001)
2. Mamdani, E.: Fuzzy control - a misconception of theory and application. IEEE Expert 9(4), 27–28 (1994)
3. Mamdani, E., Assilian, S.: An Experiment in Linguistic Synthesis with a Fuzzy Logic Controller. International Journal of Machine Studies 7(1), 1–13 (1975)
4. Holmblad, L., Ostergaard, I.: Control of a Cement Kiln by Fuzzy Logic. In: Gupta, M., Sanchez, E. (eds.) Fuzzy Information and Decision-Processes, pp. 389–399. North-Holland, Amsterdam (1982)

5. Liang, Q., Mendel, J.: Equalization of Nonlinear Time-Varying Channels using Type-2 Fuzzy Adaptive Filters. IEEE Transaction on Fuzzy Systems 8, 551–563 (2000)

6. Liang, Q., Mendel, J.: Overcoming Time-Varying Co-Channel Interference Using Type-2 Fuzzy Adaptive Filter. IEEE Transactions on Circuits and Systems 47, 1419–1428 (2000)

7. Liang, Q., Karnik, N., Mendel, J.: Connection Admission Control in ATM Networks using Survey-Based Type-2 Fuzzy Logic Systems. IEEE Transactions on Systems, Man and Cybernetics Part C: Applications and Reviews 30, 329–339 (2000)

8. Innocent, P., John, R.: Type-2 Fuzzy Representations of Lung Scans to Predict Pulmonary Emboli. In: Proceedings of the Joint 9th IFSA World Congress and 20th NAFIPS International Conference, Vancouver, Canada, pp. 1902–1907 (2001)

9. John, R., Lake, S.: Type-2 Fuzzy Sets for Modelling Nursing Intuition. In: Proceedings of the Joint 9th IFSA World Congress and 20th NAFIPS International Conference, Vancouver, Canada, pp. 1920–1925 (2001)

10. Mendel, J., John, R.: Type-2 Fuzzy Sets Made Simple. IEEE Transactions on Fuzzy Systems 10, 117–127 (2002)

11. Hagras, H.: A Hierarchical Type-2 Fuzzy Logic Control Architecture for Autonomous Mobile Robots. IEEE Transactions on Fuzzy Systems 12, 524–539 (2004)

12. Karnik, N., Mendel, J.: An Introduction to Type-2 Fuzzy Logic Systems. USC report (1998), http://sipi.usc.edu/~mendel/report

13. Wu, D., Tan, W.: Type-2 FLS Modeling Capability Analysis. In: Proceeding of the 2005 IEEE International Conference on Fuzzy Systems, Reno, USA, pp. 242–247 (2005)

14. Wu, H., Mendel, J.: Uncertainty Bounds and Their Use in the Design of Interval Type-2 Fuzzy Logic Systems. IEEE Transactions on Fuzzy Systems 10, 622–639 (2002)

15. Lynch, C., Hagras, H., Callaghan, V.: Embedded Type-2 FLC for Real-time Speed Control of Marine and Traction Diesel Engines. In: Proceeding of the 2005 IEEE International Conference on Fuzzy Systems, Reno, USA, pp. 347–352 (2005)

16. Lynch, C., Hagras, H., Callaghan, V.: Using Uncertainty Bounds in the Design of an Embedded Real-Time Type-2 Neuro-Fuzzy Speed Controller for Marine Diesel Engines. In: Proceeding of the 2006 IEEE International Conference on Fuzzy Systems, Vancouver, Canada, pp. 7217–7224 (2006)

17. Huang, S., Chen, Y.: VLSI Implementation of Type-2 Fuzzy Inference Processor. In: Proceedings of the IEEE International Symposium on Circuits and Systems, pp. 3307–3310 (2005)

18. Melgarejo, M., Garcia, A., Pena-Reyes, C.: Pro-Two: A Hardware Based Platform for Real Time Type-2 Fuzzy Inference. In: Proceedings of the IEEE International Conference on Fuzzy Systems, Budapest, Hungary, pp. 977–982 (2004)

19. Lynch, C., Hagras, H., Callaghan, V.: Parallel Type-2 Fuzzy Logic Co-Processors for Engine Management. In: Proceedings of the 2007 IEEE International Conference on Fuzzy Systems, London, UK, pp. 907–912 (2007)

20. Walke, T.: Wartsila-Sulzer RTA96-C, The Most Powerful Diesel Engine in the World (2005), http://www.bath.ac.uk/~ccsshb/12cyl/

21. Lin, P., Hsu, C., Lee, T.: Type-2 Fuzzy Logic Controller Design for Buck DC-DC Converters. In: Proceedings of the 2005 IEEE International Conference on Fuzzy Systems, Reno, USA, pp. 365–2370 (2005)

22. Wu, D., Tan, W.: Type-2 Fuzzy Logic Controller for the Liquid-Level Process. In: Proceeding of the 2004 IEEE International Conference on Fuzzy Systems, Budapest, Hungary, pp. 248–253 (2004)

23. Sepúlveda, R., Castillo, O., Melin, P., Diaz, A., Montiel, O.: Handling Uncertainty in Controllers Using Type-2 Fuzzy Logic. In: Proceeding of the 2005 IEEE International Conference on Fuzzy Systems, Reno, USA, pp. 248–253 (2005)
24. Zarandi, F., Turksen, B., Kasbi, T.: Type-2 Fuzzy Modeling for Desulphurization of Steel Process. Expert Systems with Applications 32(1), 157–171 (2007)
25. Figueroa, J., Posada, J., Soriano, J., Melgarejo, M., Roj, S.: A Type-2 Fuzzy Logic Controller for Tracking Mobile Objects in the Context of Robotic Soccer Games. In: Proceeding of the 2005 IEEE International Conference on Fuzzy Systems, Reno, USA, pp. 359–364 (2005)
26. Coupland, S., Gongora, M., John, R., Wills, K.: A Comparative Study of Fuzzy Logic Controllers for Autonomous Robots. In: Proceedings of the 2006 Information Processing and Management of Uncertainty in Knowledge-based Systems conference, Paris, France, pp. 1332–1339 (2006)
27. Hagras, H., Doctor, F., Lopez, A., Callaghan, V.: An Incremental Adaptive Life Long Learning Approach for Type-2 Fuzzy Embedded Agents in Ambient Intelligent Environments. IEEE Transactions on Fuzzy Systems 15(1), 41–55 (2007)

The Burden of Proof: Part II

David B. Fogel

Natural Selection, Inc.
9330 Scranton Rd., Suite 150
San Diego, CA 92130 USA
dfogel@natural-selection.com

Abstract. Standards of evidence in scientific work, by the very term "standards," should be consistent, but they are not. Often, well-known "facts" or claims turn out to be wrong, disagreements over the interpretation of data and methods yield to political motivations. Even people who would have us strive for the highest aspirations of scientific quality defend arguments from *vox populi*, or at least majority rule. This chapter will discuss the standards of evidence in scientific work, with particular emphasis on evolutionary computation and modeling complex adaptive systems. Evidence shows that some models of seemingly simple systems are really quite complicated. In other cases, adjusting assumptions about a model leads to results that are at significant variance from what is commonly accepted. The implications of accepting well-known models of these systems are explored. Two common concepts are identified as being associated with potential problematic models: expectation and equilibrium.

1 Introduction

For those who are not professional scientists, science may seem a lofty profession, a profession of standards, where evidence rules the outcome and objective decisions lead to an advancement of knowledge. There is no doubt that the scientific method is the best method devised for adding to the collective knowledge of humanity, at least so far. Yet science is a human endeavor, requiring a purpose-driven observer who must invent hypotheses, collect data, interpret the findings, and come to conclusions. Science is driven in part by free market dynamics, which implies a selective process akin to natural selection in the living world. There is a finite set of available resources (e.g., funding, time on a telescope, page limits on journals, computer equipment) and stringent competition for those resources. In nature, those creatures that do not find adequate resources while avoiding predation die. In science, an analogous condition occurs, and the axiom of "publish or perish" in academia is real. As a consequence, the human endeavor of science is often political, reflecting the aims of the scientist and perhaps not the primary principle of science itself: To increase knowledge.

The burden of proof is always on the scientist who proposes an explanation for some observation. By convention, the harshest critic of such an explanation should be the one who proposes it, for he or she is the one who presumptively knows it the best. Yet too often we see scientists who are advocates for hypotheses and critical of only those hypotheses proposed by other scientists. Too often we see scientific movements

J.M. Zurada et al. (Eds.): WCCI 2008 Plenary/Invited Lectures, LNCS 5050, pp. 201–231, 2008.

begin and persist even in the face of contradictory data (or no real data at all), and persist for a very long time.

Many scientists, particularly upon receiving a doctoral degree, will admit that even though they are experts in their fields, they really only know a small fraction of what there is to know about those fields, and an even smaller fraction of everything else. When reading a publication in their area of expertise, they may think "no, that's not right at all." But upon reading a publication that is not in their area of expertise, and having no basis for criticizing the content, they may be far more trusting: At least the material has survived peer review, unless the publication is a newspaper, commercial magazine, or perhaps merely a website. This trust is easily misplaced. Thus it becomes necessary to ask constantly "How do we know what we think we know?"

For example, how do we know that giraffes have long necks to reach the leaves at the top of trees? That giraffes have long necks is obvious. But why do they have long necks? The answer most of us are taught early on is so that giraffes can outcompete other herbivores by reaching to eat leaves at the top of trees that are out of reach of other animals (an idea attributed to Darwin [1, 2]). More precisely, the argument is that natural selection has favored the longer neck of the giraffe (selecting against shorter necks), reinforcing the genetic foundation that creates this longer neck; presumably then, giraffes with shorter necks face greater competition for resources and have a lower survival rate. But how do we know this to be true? Perhaps it is not true. Simmons and Scheepers [3] observed that even in times when food is scarce, giraffes rarely eat with their necks fully stretched and instead usually eat from low shrubs; however, male giraffes fight in contests that involve swinging their heads and necks at each other, with those possessing longer necks having an advantage. Thus the long neck of the giraffe may be a product of sexual selection. The necks of male giraffes are also longer than those of female giraffes, and this dimorphism adds credibility to this alternative explanation.

Perhaps the most controversial current political and social topic is anthropogenic global warming, but it is also a topic of scientific disagreement. For the past decade, at least back to [4], there has been a vocal group of scientists arguing in favor of models that support the contention that the Earth is warming and also that this warming is a function of human activity. (There is a much longer history of study of global climate change generally.) Most recently, the 2007 Nobel Peace Prize was awarded to the United Nations Intergovernmental Panel on Climate Change and former U.S. Vice President Al Gore, "for their efforts to build up and disseminate greater knowledge about man-made climate change," bringing increasing political attention to the issue. (It may be interesting to note that more recent discussions have been framed in the context of "global climate change" rather than "global warming." Since the Earth has always undergone global climate change, this revised terminology is considerably weaker.) Many mathematical models have shown a correlation between human activity and recent global warming. Yet it is interesting to note that one of the most common arguments heard in the media in favor of this conclusion has been of the form: The scientific community has reached a consensus. Anyone who doubts this is probably on the payroll of an oil company. Maybe there are still some of these people who also believe the Earth is flat. Gore asserted this on NBC's Today show on November 5, 2007. He also asserted on June 6, 2006: "The debate's over. The people who dispute the international consensus on global warming are in the same category now with the people who think the moon landing was staged on a movie lot in Arizona." These are not the tactics of a scientist who is his own harshest critic. They more like the tactics used by elementary school children who settle arguments by getting the most other children to agree that they are right.

Today, there are many scientists who are skeptical of the conclusions that have been reached by prior analysis of global climate change data suggesting that the Earth is warming as a function of human activity. (There is little doubt that the Earth has been in a period of recent warming, and it has gone through thousands of warming and cooling periods in its 4.5 billion years of history. What this recent warming portends for the future of the planet is the more pertinent question.) Some of these scientists are even members or former members of the United Nations panel that received the Nobel Prize. But it is not the size of the constituency on one side or the other of a scientific argument that is important. What is important is which side is right (if either), a matter that can only be decided by vigorous analysis of the data, models, and conclusions. There appears now to be more willingness to have such a strong debate, and plainly baseless assertions are met with more direct criticism rather than reluctant deference. For example, after a wave of tornadoes in the United States in early 2008, U.S. Senator John Kerry suggested that this spate of violent weather was a function of global warming [5]. Roger Edwards, a climatologist from the National Weather Center, was quickly quoted [5] from earlier statements in 2007 indicating "…no scientific studies solidly relate climatic global temperature trends to tornadoes… ." It is regrettable that debate on this crucial issue of global importance, particularly in the public arena, has been stifled by arguments from authority, arguments from *vox populi*, and simple ridicule.

Unfortunately, *vox populi* is not constrained to matters here on Earth. Consider even something as simple as how many planets orbit the Sun. The matter pertains to the definition of a planet, with significant implications on how science is conducted to study our solar system. In August, 2006, the International Astronomical Union (IAU) met in Prague and voted to adopt a definition that reclassified Pluto as a "dwarf planet" and not the "ninth planet" in our solar system. The rules that were adopted to define a planet were: (1) it must be in orbit around the Sun, (2) it must be large enough so that it takes on a nearly round shape, and (3) it has cleared its orbit of other objects. Because Pluto's orbit overlaps Neptune's it does not clear other objects and is therefore not a planet, according to the definition. There are at least two problems with this. First, Neptune does not clear its orbit either (because it overlaps with Pluto!) and yet no one is suggesting that Neptune is not a planet. In fact, in the IAU statement on the matter, Neptune is explicitly included as a planet without mention of this obvious exception to the rule. Second, although 2,411 people attended the meeting, only 424 voted on this issue (out of nearly 10,000 members in total). Approximately 800 to 1,000 people attended the session that discussed the issue (which was held on the final day of the meeting). The vote was 237-157 with 30 abstentions. It might be puzzling that only 237 votes in favor of a motion in a society of close to 10,000 is sufficient to determine something as important as whether or not all humankind should consider Pluto to be a planet. Yet, it seems that the IAU does not require a quorum of its members at meetings of the general assembly, while at the same time its by-laws do provide for voting on "issues of a primarily scientific nature." An impertinent question can hardly be avoided: Since when are issues of a primarily scientific nature decided by voting?

Deciding whether or not Pluto is a planet is more controversial than many might anticipate. So perhaps returning to something less controversial would be appropriate. For example, it is well known that increasing salt intake is associated with hypertension (high blood pressure). The U.S. Department of Health and Human Services publishes a public circular (revised in 2006) through the National Institutes of Health (NIH) that specifically addresses this in dietary approaches to stop hypertension (DASH) and encourages eating less salt because of its sodium content [6]. There is an explicit claim that studies show lowering sodium consumption has a direct effect on lowering blood pressure: "The lower your salt intake is, the lower your blood pressure." That could not be more clear. The circular does not identify any reference to examine, but there is a description of a study on 412 individuals who were assigned to eat three different sodium levels per day (3,300 mg, 2,300 mg, and 1,500 mg), with the result being that lower sodium levels resulted in lower blood pressure readings. This is certainly the research published in Sacks et al. [7], which identified a statistically significant 2.1 mm Hg drop in systolic pressure from the high sodium level to the intermediate level, and another statistically significant 4.6 mm Hg drop in systolic pressure from the intermediate level to the low level when combined with a typical American diet.

It would be easy to believe that there never has been any controversy about salt intake and hypertension, but such a belief would be wrong. Controversy continues today. In 1998, a review titled "The (Political) Science of Salt" published in *Science* examined the conclusions of meta-analyses of various salt studies [8]. The results of over 50 such studies combined were inconclusive. Data could be interpreted to support a link between salt intake and increased blood pressure, or to support a conclusion that no such link exists. Gary Taubes, who authored the *Science* article, wrote "This situation is exacerbated by a remarkable inability of researchers in this polarized field to agree on whether any particular study is believable. Instead, it is common for studies to be considered reliable because they get the desired result." Recent research [9] has suggested that some people are "salt sensitive" and that for these individuals it is important to reduce sodium intake, but not generally in the global population. Other research has addressed an apparent decrease in insulin sensitivity associated with a restricted sodium diet [10, 11, 12]. The question of whether or not this implies an increased likelihood of diabetes remains open. If the answers are to be found in personalized prescriptions for action, the answers are not likely to be found conclusively when studying large populations. This highlights a significant flaw in the way medical trials are conducted currently, and one that will likely be of much greater significance as personalized genetic information becomes more widely available so that medical actions can be tailored to individuals.

Against the backdrop of this conundrum of controversy and confusion on matters that might at first appear simple or "decided," it is easy to imagine that the study of complex adaptive systems and evolutionary games would pose its own set of challenges. Again, it is appropriate to ask "How do we know what we think we know?" Has the burden of proof been met? For example, one explanation of adaptation in natural systems views organisms as receiving payoffs from different behaviors and asserts that organisms seek to minimize expected losses from these behaviors. But how do we know this is a reasonable assertion? Another example views organisms in competition for payoffs and asserts that the only viable explanations for the behaviors that the organisms adopt must be solutions to an equation that places the expected payoffs of alternative behaviors at an equilibrium point. But how do we know this is a reasonable assertion? In fact, there is evidence against these and other assertions, which this chapter will offer for consideration in the hope of encouraging more critical thinking regarding these and other important matters of science.

2 *K*-Armed Bandits and Minimizing Expected Losses

Evolution is viewed commonly as an optimizing process [13, 14]. The question of what is being optimized has remained open to interpretation. One idea is that selection favors those behaviors that minimize expected losses over a series of decisions, where each decision receives a stochastic payoff [15]. Such a criterion is convenient mathematically because it is amenable to an analysis of optimal strategies. But such mathematics cannot address the principal question of the suitability of the criterion itself; it merely assumes the criterion as given. However, the aptness of this criterion can be assessed by computer simulations in which individuals compete for survival based on the reward they receive while employing different strategies for sampling from a *k*-armed bandit.

The *k*-armed bandit serves as a familiar analogy that can provide insight into animal behavior in diverse environments [14]. Variations of the analogy are typical in optimality models that presume individuals can adopt alternative behavioral strategies, each having a random distribution of payoffs with certain likelihoods, much like pulling a one-armed bandit (a slot machine). These payoffs can be measured in terms of food obtained, shelter, reproductive success, or other suitable standards [16, 17, 18]. In traditional evolutionary game theory, expected payoffs translate linearly into reproductive success or "fitness" [19]. When interest is focused on the specific behaviors that provide the foundation for that reproductive success, however, the relationship between behaviors and fitness may be nonlinear. Alternative behavioral strategies can be compared in light of a chosen mathematical criterion, with those that are found to be optimal in turn being compared to those observed in nature. It is hoped that with appropriate abstraction, this mathematical device can provide insight into the fundamental dynamics that underlie the observed behaviors.

One widespread abstraction of the *k*-armed bandit problem to adaptation in natural settings is found in the canonical genetic algorithm [15], in which the problem of adaptation is framed as a series of decisions on how to best allocate trials to alternative bandits in light of payoffs received from previous trials. The principal assumption underlying the mathematical framework of genetic algorithms is that natural selection minimizes expected losses while sampling from alternative bandits (described as "schemata" [15], which are subsections of complete solutions). Under this assumption, Holland [15] offered what was presumed to be an optimal sampling strategy of devoting an exponentially increasing number of trials to the observed best bandit(s) (the one(s) with the greatest observed average payoff). Independent analyses [20, 21], however, proved that the development in [15] is mathematically flawed: The preferred strategy of the genetic algorithm is not, in truth, optimal for the criterion of minimizing expected losses.

Rather than seek to fix this circumstance by discovering alternative strategies that are indeed optimal for this criterion (also see [21] for the mathematical challenge that this poses), a more fundamental question can be raised: Does selection in fact favor those behavioral strategies that seek to minimize expected losses, and if not, what are the conditions that determine which tactics will be favored by selection? To begin to address this question, a two-armed bandit can be studied using a simulated population of competing strategies that are subject to random variation and selection. The hypothesis that selection favors minimizing expected losses can then be tested statistically by examining the strategies that survive over a large number of generations. As will be seen, the results of these simulations do not offer general support for this hypothesis.

2.1   Methods on a 2-Armed Bandit Problem

A population of N individuals was constructed where each faced the choice of sampling from either of two slot machines. The first machine offered a payoff that was Gaussian distributed with a mean of 1.0 and standard deviation of $\sigma_1 = 1.0$. The second machine was also Gaussian distributed, but with zero mean and standard deviation that was parameterized by the symbol σ_2. After each individual sampled from either of the bandits, all individuals were ranked in order of decreasing payoff and the subset K of these with the greatest payoffs were selected to generate subsequent progeny. In this way, K represented the carrying capacity of the environment. This process was then iterated over several thousands of "generations."

The behavior of each individual was defined by a single parameter, p_i, $i = 1, ..., N$, which corresponded to the probability that it would sample from the first bandit (i.e., the one with the greater mean). This protocol is typical of phenotypic optimality models in which the underlying genetics of a particular behavior is abstracted out of consideration [19]. A surplus of offspring was generated from surviving individuals through a slight random variation of each parent's parameter. Each surviving individual was given an equal probability of being selected to generate each next offspring ($i = 1, ..., N$). Specifically, each offspring's parameter p'_i was set equal to its parent's parameter p_i, with the addition of zero mean Gaussian noise with standard deviation of 0.01. This choice was deemed reasonable for representing a small amount of persistent random variation. If any offspring's p'_i became greater than 1.0 or less than 0.0 it was set to the limit it exceeded, thereby maintaining its interpretation as a probability. Note that each parent could generate more than one offspring.

Consideration was given to case of stringent selection pressure, in which the maximum population size, N, was significantly larger than the carrying capacity, K. That is, a great percentage of offspring do not survive to reproduce and the reproductive strategy of generating a tremendous surplus of offspring is adopted, rather than having parental investment be constrained to only a small number of offspring. Experiments were conducted for the cases where (1) $K = 1$ and $N = 100$, and (2) $K = 100$ and $N = 10,000$ (i.e., on average, each parent generated 100 offspring) at various settings of σ_2 ranging from 0.01 to 5.0 at selected intervals. Each individual was defined to have a maximum lifespan of one generation, that is, all parents were removed from the population at each iteration, and only the best K out of the N offspring were selected for further reproductive attention. In both the (1, 100) and (100, 10000) cases, the initial population of N individuals was selected with each individual's probability parameter being chosen uniformly at random over the interval [0, 1].

Attention was focused on the mean of all surviving parents' probability parameters at each generation. To avoid initial transient effects, data were recorded from generations 100,000 to 10 million for the case of (1, 100) and from generations 20,000 to 120,000 for the case of (100, 10000). The mean probability parameters over these generations were then averaged to generate a single datum representing the mean probability of choosing the first bandit for each investigated setting of σ_2. *Under the hypothesis that selection favors strategies that minimize expected losses, the anticipated results would indicate a strong tendency to sample from the first bandit regardless of σ_2 because it has the higher average payoff* [15].

2.2 Results on the 2-Armed Bandit Simulation

Figures 1(a) and (b) show the results for both cases. The figures indicate a clear shift of optimal (i.e., selected) behavior away from sampling the bandit with the higher mean as the standard deviation of the second bandit was increased beyond a particular threshold in the range of $1 < \sigma_2 < 2$. This threshold point can be analyzed mathematically (discussed below). The natural outcome of these simple evolutionary systems was a selection for risky behavior even when the average payoff for that risk was lower than that offered by the less variable option. Note also that even at very low values of σ_2, there was a saturation of the mean probability of selecting the first bandit at values significantly lower than 1.0. That is, even when the second bandit had very low variance, there was insufficient selection pressure to drive the mean probability of choosing the bandit with a higher average payoff to complete certainty. Even under these conditions, selection did not favor strategies that minimized expected losses exclusively. The mathematical analysis of this result is provided in [22].

2.3 Discussion on Minimizing Expected Losses

Several design choices were made for the simulation, but chief among these is the stringency of selection pressure coupled with a large surplus of offspring. This situation necessitates risky behavior because as the possible reward for taking that risk becomes greater, the conservative strategy of opting for the less variable but greater average payoff becomes untenable. Given a sufficient number of risk takers, a sufficient subset of those will get lucky and gain a payoff that is larger than is likely to be gained when choosing conservatively. Most risk takers will be losers, but at the same time most of the winners in this lottery will also be risk takers.

It is of interest to identify conditions in natural settings that are similar to those incorporated in the models studied here. The situation of a high selection pressure and larger surplus of offspring is not uncommon in nature (r-selection). Individual mortality rates are often difficult to estimate, but there have been some assessments. The daily rate of larval mortality in northern anchovies has been estimated between 16 and 20% [23, 24]. It has also been estimated that 70% of the eggs of Atlantic herring in a

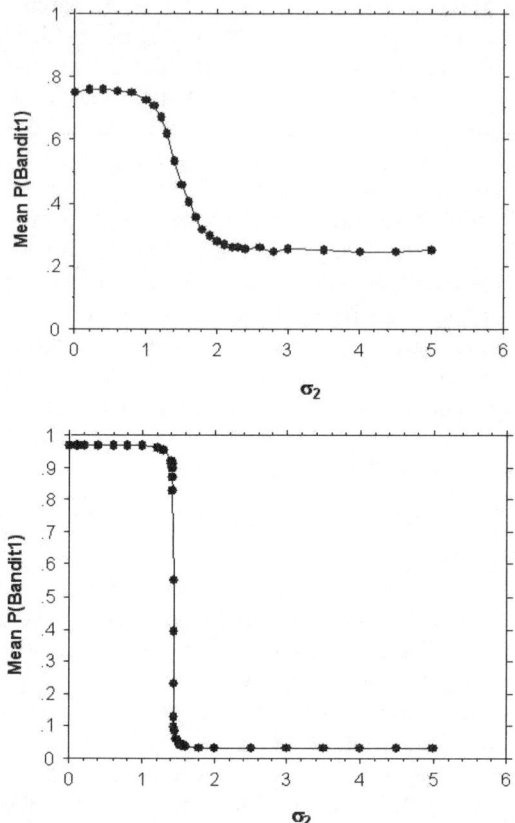

Fig. 1. (a-top)(b-bottom). The mean probability of choosing the bandit with greater average payoff as a function of the standard deviation of the bandit with lower average payoff. (a) $K = 1$ and $N = 100$, and (b) $K = 100$ and $N = 10000$. Sampling was conducted from generation 100,000 to 10 million for case (a), and 20,000 to 120,000 for case (b), so as to avoid transient effects that might result from the uniform initialization of strategies. Under the hypothesis that selection would favor strategies that minimize expected losses, the anticipated result would be a flat line indicating complete certainty of choosing the first bandit with fixed mean and standard deviation regardless of the value of σ_2. In contrast, the results indicate not only a lack of convergence to strategies that always choose the first bandit when σ_2 is small but also a consistent shift that favors strategies that choose the second bandit with lower mean payoff as σ_2 is increased. Each datum is the average of on the order of 10^5 or 10^7 trials and thus the 95% confidence limits around each point are not visible at the scale shown. The results indicate statistically significant evidence rejecting the hypothesis that selection favors strategies that minimize expected losses ($P \ll 10^{-6}$).

patch off the Canadian coast were eaten by predatory flounders (one examined flounder contained 16,000 eggs) [25]. These rates concern only immature fishes and therefore the percentage of individuals that survive to maturity and further go on to reproduce must be even lower. Similar observations can be made for a great number of insect species and flora (e.g., ferns generate billions of spores annually [26]) in contrast with, say, mammals which are often K-selected; however, even in this latter case, predation is often a serious threat [27, 28, 29], and sexual selection may be very stringent, with few of the males actually taking part in reproduction (e.g., with the intense competition between male elephant seals less than one-third of the beach-resident males copulate during a breeding season and most matings are accomplished by only a few males [30, 31]; moreover, only about 10% of the male pups are still alive when they are mature enough to compete seriously for rookeries with other males [26].

The idea that the variability of payoffs plays a role in choosing optimal behaviors has been examined in depth by animal behaviorists in different settings [32, 33]. For example, Caraco [34] offered juncos choices between being fed with a fixed or variable number of seeds, with both choices constrained to have the same mean number of seeds (i.e., 2 vs. 0 or 4). The juncos were observed to prefer the more variable choice after being deprived of food for a prolonged period of time. Similar tendencies toward more variable payoffs when under stress, even though the mean payoff for each choice was identical, have been observed in other studies [35, 36]. This is prima facie evidence contradicting the explanation that the behaviors have evolved to minimize expected losses, for if this criterion were true there would be no expected bias for making either choice given identical mean payoffs.

The results presented here go further by indicating that risky behavior can be preferred even when the mean payoff is sacrificed if the potential rewards are sufficiently high and sufficiently likely. Moreover, selection may not completely eliminate behaviors that yield a lower expected reward even when the variance of this subpar payoff is small. When payoffs do not translate linearly into reproductive success, minimizing expected losses does not imply maximizing reproductive success. Thus, assuming this criterion is not an appropriate starting point for a general theory that seeks to explain adaptation in natural (and artificial) settings.

3 Evolutionary Unstable Strategies

Assumptions are integral to all models. Often, assumptions are made merely for mathematical convenience (e.g., Gaussian distributed noise in a linear regression analysis). When mathematical convenience supersedes the main objective of modeling – to gain insight about a real-world phenomenon – the result is often the right answer but to the wrong problem. When the assumptions are adjusted even slightly, quite different answers may appear. Such is the case with models based on the concept of evolutionary stable strategies.

Evolutionary stable strategies (ESSs) have become routine in the explanations of the long-term dynamics of complex adaptive systems. Thousands of papers have been contributed to the archive literature on this concept. An evolutionary stable strategy is defined on an evolutionary game with various possible strategies for each player and prescribed payoffs that depend on the simultaneous play of each participant. The equilibrium conditions of the game are determined, and it is assumed that once the players' strategies arrive at such an equilibrium they will tend to remain in that condition, barring external influences. Thus the equilibrium states are regarded as the likely resulting behaviors of the complex coevolutionary system with the caveat that such states are accessible [37, 38] and that the game provides an adequate description of real-world conditions.

A most basic game is a mathematical construction involving pairwise contests among an infinite collection of players over resources of specified values. The requirement for an infinite collection is of critical importance, as will be observed shortly. Each competing player can adopt one of two alternative strategies, A or B. A payoff is defined for each alternative combination of behaviors. The respective worth for a player adopting a certain behavior is determined by the expected payoff for the behavior, given the distribution of behaviors in the population. The payoff for a single encounter is denoted as $E(A, B)$, where the payoff is to the individual adopting strategy A against an opponent adopting strategy B, with a similar notation for other possible pairs of strategies.

For a strategy I to be an ESS, it must satisfy either of the following conditions:

1. $E(I, I) > E(I, J)$
2. $E(I, I) = E(J, I)$ and $E(I, J) > E(J, J)$

where J is any other strategy, $J \neq I$ [19, 39]. Essentially, an ESS is a strategy, or set of strategies, that cannot be invaded by any other strategy. When members of a population adopt an ESS, their expected payoffs are always greater than the payoff awarded by any new member adopting an alternative policy.

ESS analysis has been applied for over three decades, including early reports to predict the behavior and characteristics of naturally evolved organisms (e.g., expected sex ratio [40, 144-145], courtship strategies [40, 150-151], searching for suitable feeding areas [41], superparasitism [42], mating tactics [43]), and more recently [44, 45, 46] and others.

The principal assumption for analyzing systems in terms of their ESSs is an infinite population [19, p. 20] (cf. [47, 48] for addressing finite populations, which is of particular importance to the discussion that follows). Under this assumption, if individual payoffs reflect random effects (e.g., when two equivalent strategies meet, one wins and the other loses with equal probability), these effects can be collapsed to their expectation (i.e., the variability of the distribution of the sample mean goes to zero in the limit); however, under a finite population, regardless of population size, these random effects are instead described by a probability mass function (cf. [47], which continued to treat the payoff only on the average). Sampling from such a probability distribution can have a marked effect on the trajectory of a population over the course of many iterations.

In an adaptation of the simple hawk-dove game [40], simulations that incorporated a finite population and random payoffs demonstrated limit cycle behavior (Figure 2) and population trajectories that were not associated with the ESS found in the infinite game [49]. Fogel et al. [50] extended these results to include various levels of selection pressure (i.e., the fraction of the population culled by selection at each generation) for populations of size 600 (Figure 3). The results indicated that the mean fraction of hawks was qualitatively different from the ESS for selection pressures of 16% and above (the percentage of the population replaced in a generation), and was statistically significantly different from the ESS for pressures $\geq 7\%$. Fogel et al. [51] extended the results to small populations (e.g, fewer than 100 individuals), which are most relevant for field work in evolutionary biology.

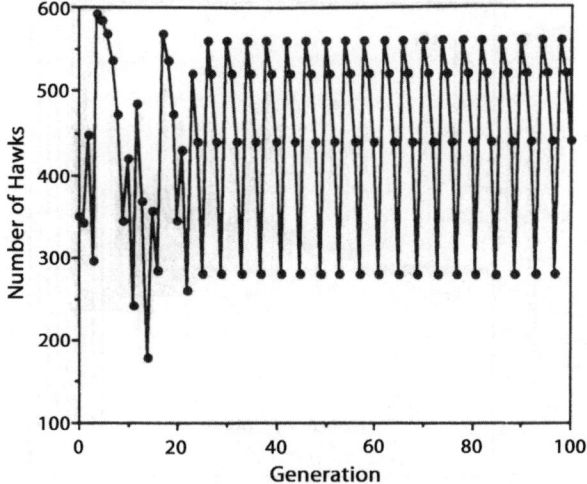

Fig. 2. The results of a typical simulation [49] of the hawk-dove game with a finite population and random payoffs. A population of 600 individuals was initialized at the ESS (350 hawks). At each generation, all individuals met in a round-robin competition (all possible pairs) and the 300 individuals with the lowest scores were replaced by copies of the 300 individuals with the highest scores (50% selection). The population did not remain at the ESS, but instead diverged away from it and fell into an apparent limit cycle for an indeterminate length of time. The particular cycle of (560, 520, 440, 280) shown above, describing the number of hawks in the population over successive generations, occurred in each of 10 trials. The mean of this cycle is 450 hawks, which does not correspond well with the ESS. Another frequently generated cycle was (400, 200), which again does not correspond well with the ESS. A variety of other results were generated for populations of 60 and 6000 individuals, but in no case did the ESS provide a useful description of the population's trajectory.

3.1 Background on the Hawk-Dove Game

The hawk-dove game involves two players who may choose between strategies of hawk or dove. A hawk is always aggressive and only retreats when injured. A dove, in contrast, merely adopts a threatening posture but never causes physical harm to an opponent. If a hawk fights a dove, the dove flees. If a hawk fights another hawk, they continue to fight until one of them is injured. If a dove meets a dove, neither is harmed; both adopt threatening positions for a long time until one retires. It is assumed that there are no recognition mechanisms that would enable either player to discern the opponent's strategy before an encounter (which is another significant simplification).

Points are awarded for encounters as follows. A win is worth 50 points, a loss is worth 0 points, being injured is worth −100 points, and wasting time in a long contest is worth −10 points. These values are taken from [40, p. 70] and are somewhat arbitrary but are meant to reflect the reproductive potential in light of the above descriptions. Encounters between a hawk and dove always yield 50 points for the hawk and 0 for the dove. Encounters between doves yield 40 points to one dove (50 points for the

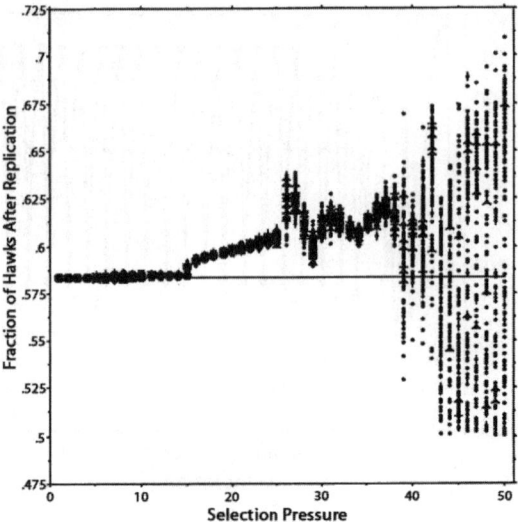

Fig. 3. A scatterplot of the mean fraction of hawks after replication for each of 100 trials at each level of selection pressure (i.e., the fraction of the population that is eliminated at each generation) using a population of 600 players (from [50]). Coincident points are represented by larger star symbols. As the selection pressure was increased, the mean behavior of the population after replication tended to diverge away from the ESS (which is depicted by the horizontal line). For selection pressure up to ≈15%, hawks comprised the majority of individuals both eliminated and replicated (i.e., they possessed scores in both tails of the distribution). After 15% selection, the distribution of the fraction of hawks after replication diversified, and the mean behavior of the population tended to drift sharply away from the ESS until it was ultimately dominated by limit cycles. At very high selection pressure, the data suggest that the probability distribution of the mean fraction of hawks may be bimodal.

win and −10 for wasting time) and −10 to the other, with the winner chosen with equal probability. Encounters between hawks yield 50 points to the victor and −100 to the vanquished, again with the winner chosen with equal probability. Thus the expected payoffs for encounters are:

$$E(H, H) = -25$$
$$E(H, D) = 50$$
$$E(D, H) = 0$$
$$E(D, D) = 15$$

where H and D are the respective hawk and dove strategies, and the payoff is the mean payoff for the strategy listed first in parentheses. Operating on these expected values, the ESS for the game is a population consisting of 5/12 doves and 7/12 hawks. In the infinite population game, if the fraction of hawks started to increase above 7/12, doves would begin to gain an extra advantage, and the stable 7:5 ratio of hawks to doves would reappear. But under more realistic conditions of a finite population, with high selection pressure (i.e., the fraction of the population that is eliminated), where the payoffs for individual encounters are not taken as the expected payoffs but rather are sampled as random variables, the ratio of hawks to doves can diverge from the ESS and may, as illustrated in Figure 2, exhibit limit cycles.

3.2 Experimental Results Showing ESSs Are Not Stable

Fogel et al. [50, 51] offered two sets of experimental results showing that populations in the simple hawk-dove game did not converge to the supposed evolutionary stable strategy, and diverged from that strategy even when the population was initialized at that solution to the game.

In one experimental design, 100 trials were conducted for a variety of selection pressures (Figure 4) within a finite population of 60 individuals playing a hawk-dove game. In each trial, the population was initialized at the ESS: 35 hawks and 25 doves. Each individual in the population competed in a round-robin tournament (i.e., each individual paired once with each other individual), and payoffs were awarded in each encounter by sampling from the appropriate random variable, rather than using its expected value. For example, if two hawks met, one would receive a payoff of 50 for the win while the other would receive −100 for fighting and losing, instead of assigning both hawks the statistical expectation of −25. The results of any encounter did not affect the probabilities of winning or losing in subsequent encounters (i.e., encounters between like individuals were always decided with equal chances for each to win). After all contests were completed and point totals accumulated, a selected percentage of the population with the lowest scores was removed from the population and replaced by copies of the corresponding percentage of highest scoring individuals. For example, if the selection pressure were 10%, then the six lowest scoring individuals would be replaced by replicas of the six highest scoring individuals. Selection levels covered the range from 1%-50% by single percentage points. Each trial was iterated for 200 iterations of replication and selection, this being chosen to minimize any initial transient effects on the mean population trajectory.

In another experimental design, round-robin pairing was replaced by a randomized mixing procedure. Rather than compute all pairwise encounters, a random mixing level was described as a percentage of the population size. The expected number of encounters for each individual per iteration was determined by multiplying the mixing percentage by the population size. For example, if the mixing level were 5% and the population size consisted of 60 individuals, this would indicate that each individual should be expected to engage in three encounters. The expected number of encounters per individual was multiplied by the total population size to determine a total number of encounters for each iteration. Individuals were then selected completely at random for each encounter up to the prescribed maximum number. Thus it was possible for an individual to have more than the expected number of engagements, or even no engagements. Each individual was initialized with zero points at the start of each round of competition, and selection was imposed based on the number of points per individual after all pairwise competitions had been completed. The mixing level was increased from 5% to 100% in 5% increments and the selection percentage was simultaneously stepped from 5% to 50% (i.e., 100 trials for 200 iterations for each pair of settings for mixing level and selection pressure). In addition, a selection pressure of 1% (which was set equal to one individual) was also executed.

Conceptually, the simulation operated in discrete phases of selection and replication, thus consideration should be given to the mean fraction or number of hawks in the population at the completion of both phases. Experiments with 600 individuals in [50] indicated that the ESS was not qualitatively relevant to the fraction of hawks after selection, and attention was therefore focused on the fraction or number of hawks in the population after replication. For the first experiment, this is displayed in Figure 4. The mean fraction of hawks after replication was statistically significantly different from the ESS ($P < 0.01$) at all selection levels $\geq 5\%$. The mean fraction of hawks after replication moved increasingly away from the ESS at successively higher levels of selection pressure. None of the sets of 100 trials for any selection pressure above 10% generated a distribution of the mean fraction of hawks that bounded the ESS, nor did the ESS appear to be a useful point estimate of the distribution of trials for selection pressure at $\geq 5\%$.

For the second experiment, with a population size of 60, Figures 5 and 6 show the combined effects of varying the selection percentage and the mixing level (noted as encounter percentage), respectively. When the mixing level was held constant and selection percentage was varied, systematic deviations away from the ESS were observed. These deviations were exaggerated for a mixing level of only 5%, where the mean number of hawks after replication in 100 trials of 200 iterations appeared to take on a decidedly nonlinear pattern as selection percentage is varied from 1-30 individuals (i.e., 1%-50%). For any particular mixing level, the ESS did not serve as a useful point estimate of the population's mean behavior at any selection percentage greater than 5%. There was often considerable variability even at only 5% selection. For constant selection pressure of 5%, 25%, and 50% (as shown in Figure 5), as the encounter percentage was increased, the distribution of the mean number of hawks after replication (across all iterations) appeared to stabilize. But the ESS served as a potentially useful point estimate of the mean population only for the relatively slight selection pressure of 5%. Note that as the mixing level was reduced the results typically became more variable. Moreover, the directionality of the variability was not consistent (i.e., low mixing at low selection pressure generated a low mean number of hawks, while low mixing and medium and high selection pressures generated a high mean number of hawks).

3.3 Discussion of Evolutionary Unstable Strategies

These results, as well as those in [50], provide evidence that the equilibrium conditions associated with ESSs may not be stable in simple evolutionary games involving finite populations and random payoffs based on individual encounters. (The essential aspects of the research have also been replicated in [52, 53], with a thorough analysis of the mathematics that generates the observations.). The simulation demonstrated the potential for generating qualitatively different results than would be expected under the assumption of an infinite population simply by varying fundamental characteristics such as the selection pressure and mixing rate. Cavalieri and Kocak [54] showed that in a model simulation study with parameter values based on field data, a population undergoing regular periodic cycles can become chaotic in the absence of changes

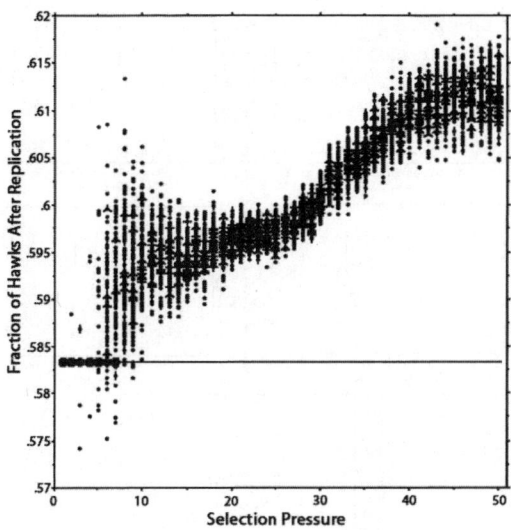

Fig. 4. A scatterplot of the mean fraction of hawks after replication for each of 100 trials at each level of selection pressure (i.e., the percentage of the population that is eliminated at each generation) using a population of 60 players. Coincident points are represented by larger star symbols. As the selection pressure was increased, the mean behavior of the population after replication tended to diverge away from the ESS (the horizontal line). At and above 5% selection, the mean behavior of the population tended to drift sharply away from the ESS.

in environmental factors. Analysis by Dieckmann et al. [55] also indicated that evolutionary limit cycles may be a natural outcome to coevolutionary dynamics, as opposed to points of stability.

If attention is given to the long-term behavior in equilibrium of the hawk-dove game in the framework of the finite population simulations indicated above, there is an obvious congruence between the equilibrium conditions and the ESS for the classic game with an infinite population. Indeed, the definition of an ESS follows the concept of putting strategies into equilibrium conditions [56, p. 162]. The ergodic equilibrium conditions for the simulations conducted here can be determined in a manner similar to [56].

Given H hawks in a population of size N, and therefore $N - H$ doves, the expected payoff to each hawk and dove under panmictic conditions is:

$$E(H) = -25(H - 1) + 50(N - H)$$
$$E(D) = 0(H) + 15(N - H - 1)$$

Note that each individual cannot play against itself. At equilibrium:

$$-25H + 25 + 50N - 50H = 15N - 15H - 15$$

thus

$$-60H = -35N - 40$$

and

$$H/N = 7/12 + 2/(3N).$$

216   D.B. Fogel

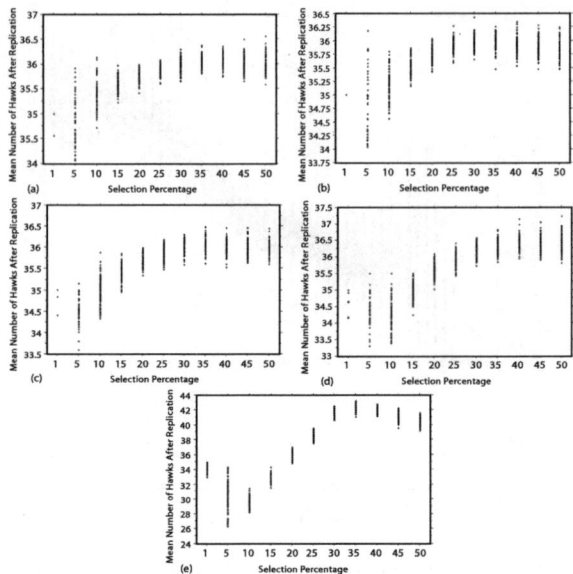

Fig. 5. A series of scatterplots of the mean number of hawks after replication for each of 100 trials with the mixing level held constant and selection pressure varying in a population of size 60. Each data point is the mean of 200 iterations in independent trials. Plots (a)-(e) are for mixing levels (encounter percentage) of 100%, 75%, 50%, 25%, and 5%, respectively, with selection percentage ranging from [1, 50] in increments of five units. At mixing levels of 75% (b) and 100% (a), for selection pressure greater than 5%, the distribution of results drifts away from the ESS. At mixing levels of 25% (d) and 50% (c), the ESS is not a useful point estimate of the distribution of results even at selection pressures of only 5%. When the mixing level is lowered to 5% (e), the distribution takes on a nonmonotonic characteristic as selection pressure is increased. Note that in plots (a)-(d), a vast majority of trials at the 1% selection level never deviated from the ESS (e.g., for the case of mixing level of 100% (a), 99 of 100 trials remained at the ESS).

H/N defines the proportion of hawks in the population, and as N tends to infinity, this ratio tends to 7/12, which is the ESS for the infinite population case.

Space does not allow a complete review of the additional experiments described in [50, 51, 52]. These experiments used larger population sizes (100 and 600) and studied various levels of selection pressure and mixing within the population, as well as forms of selection, including both truncation and proportional selection. For truncation selection, there was no tendency for a population to stay at an ESS. For proportional selection, the mean fraction of hawks could vary by as much as 3% from the ESS even at populations as large as 500. Most field studies involve populations on the order of 100 or fewer individuals, thus calling into question the utility of ESSs as an explanation of observed behaviors. If an ESS cannot be relied upon as an explanation of the dynamics of something as simple as the hawk-dove game, there would seem to be little reason to expect it to be reliable in predicting the dynamics of more complicated real-world settings. Indeed, chaos and limit cycles have been observed in these settings [54, 55, 57].

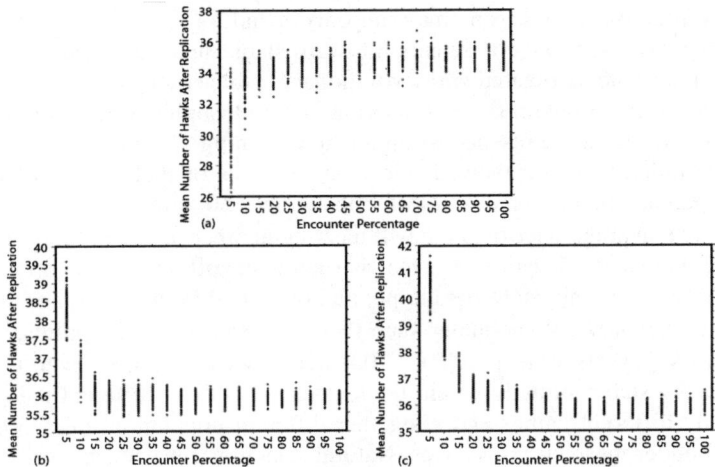

Fig. 6. Scatterplots of the mean number of hawks (see Figure 5) for the cases where selection pressure is held constant. As the mixing level is increased, (a) 5%, (b) 25%, (c) 50%, the distribution of trials appears to stabilize, but the ESS is only useful as a point estimate in the case of 5% selection pressure (a). For low mixing levels, none of the results shows reasonable agreement with the ESS.

Much of the work in ESSs relates only to mathematical problems posed under a variety of factors (in other words, "in theory"), such as mate desertion [58], individual condition and tactic frequency [59], renewing resources [60], learning rules [61], and others. Many of these efforts make analogies to real organisms but unfortunately no real data are offered in support of the derived models.

In many cases where field data are offered, the population sizes have been relatively small. For example, Brockmann et al. [62], Davies and Halliday [63], Gross and Charnov [64], Gross [65, 66], and Sinervo and Lively [67], each relied on field data from an effective population size numbering less than 500, and in most cases less than 100. The current results suggest that such population sizes are insufficient to verify or falsify the existence of an ESS (cf. [19, 68]). Further, the simulations reported in this chapter were conducted over 200 generations. In contrast, in most field studies, only a very few generations are ever observed. Certainly, this smaller sample size must also lead to greater variability in the mean behavior of a population and leads to even greater skepticism concerning the general utility of ESSs.

There have been at least two longstanding misconceptions within the literature of evolutionary stable strategies. First, Brockmann et al. [62] suggested that two alternative strategies cannot coexist in a population unless they are equally successful on the average. The experiments reported here demonstrate that this assumption is not correct: Two alternative strategies can co-exist under conditions in which, on average, one strategy is more successful than the other. Second, Brockmann and Dawkins [69] offered that populations should be expected to be found in evolutionary stable states because little time is required in the dynamic process of evolving toward stable states. The experiments do not support this claim. They instead actually provide evidence rejecting the premise: populations may not only spend a great deal of time in transition, they may not even make a transition toward stable states at all; they may in fact diverge from the state associated with evolutionary stable strategies.

At the core of the problem lies that fact that the philosophical framework of evolutionary stable strategies represents essentialistic reasoning[1]. The outcomes of encounters between individuals are viewed solely on the basis of their expectations. In the hawk-dove game, when two hawks meet, one receives 50 points while the other loses 100 points. The average payoff to a hawk meeting another hawk is −25 points. However, no individual hawk can ever receive such a payoff in any encounter. Worse perhaps, the expected end state for the population would be the same for any pair of payoffs in the hawk-hawk encounter such that the mean were −25 points. That is, if the hawk-hawk payoffs were not (50, −100) but rather (0, −50) or even (−25, −25), the evolutionary stable strategy would be the same. Yet these different payoffs represent distinctly different games and altogether different population trajectories should be expected under these circumstances. Natural selection always acts against the specific individual, not against the average individual, or even further removed, not against the mere concept of an "average individual."

4 Expecting the Unexpected at the El Farol

By their very nature, complex adaptive systems are difficult to analyze and their behavior is difficult to predict. These systems, which include ecologies and economies, involve a population of purpose-driven agents, each acting to obtain required resources in an environment. The conditions these agents face vary in time both as a consequence of external disturbances (e.g., weather) and internal cooperative and competitive dynamics. Moreover, such systems are often extinctive, where those agents that consistently fail to acquire necessary goods (e.g., food, shelter, monetary capital) are eliminated from the population. The essential mechanisms that govern the dynamics of complex adaptive systems are evolutionary: random variation of agents' behavior (this in contrast to the hawk-dove game described earlier) coupled with selection in light of a nonlinear, possibly chaotic, environment. By consequence, reductionist, linear piecemeal dissection of complex adaptive systems rarely provides significant insight. The behavior of each agent is almost always more than can be assembled from the "sum of its parts" and interactions with its predators and prey, its enemies and allies. The fabric of these complex systems is tightly woven, and no examination of single threads of the fabric in isolation, no matter how exacting, can provide a sufficient understanding of the integrated tapestry.

One such complex system that has received considerable attention is the market economy [70, 71]. The traditional view of human behavior as being completely rational has given way to an alternative perspective of bounded rationality [72]. It is

---

[1] Essentialism was the view that variations in individuals were nothing more than "errors" around mean values. In contrast "populational thinking" stresses the uniqueness of everything in the living world. Differences between individuals are real, whereas mean values are human constructs. See Mayr [84] for a historical discussion of these views. As Mayr [84, p. 47] offered "Darwin could not have arrived at a theory of natural selection if he had not adopted populational thinking."

recognized that economic decision making, like most human judgment, is made in the face of incomplete knowledge both of the extrinsic market conditions and the expected actions of other entities. The cascade of suppositions about how other actors in the environment will react to current and projected circumstances might be best described as an "arms race of uncertainty." Rather than expect a stable outcome in the face of perfect information that is available globally to all agents, where each reasons correctly that there is only one best allocation of resources and each allocation is obvious to all involved, the more likely outcome for such market dynamics would seem to be characterized by chaotic transgressions and instability.

Surprisingly then, one simulation of inductive reasoning and bounded rationality that gained attention evidenced no such chaotic behavior [73]. Instead, the "economy" varied consistently around a stable point, and it was conjectured that aggregate behaviors in complex adaptive systems serve to bring about "mutual attractors" where these systems will tend to return to a stable point when disturbed from equilibrium. If true, this would be a remarkable insight because it would imply that that it may be possible to leverage traditional analytic tools of evolutionary stable systems [19] and game theory [74] to determine the equilibrium conditions of complex adaptive systems. Being aware of the preceding results regarding the instability of so-called "evolutionary stable strategies," this insight would be all the more remarkable. Unfortunately, experiments in [75] indicate that this is unlikely to be the case.

4.1 The El Farol Problem

The economic system under investigation is an idealized model of agents who must decide whether or not to commit a resource in light of the likely commitment of other agents in the environment. Commitment is time dependent, iterated over a series of interactions between the agents in which previous behavior can affect future decisions. To distill the essential aspects of the model, Arthur [73] suggested the following setting based on a bar, the El Farol in Santa Fe, NM, which offers Irish music on Thursday nights.

Let each of N Irish music aficionados choose independently whether or not to go to the El Farol on a certain Thursday night. Further, suppose that each attendee will enjoy the evening if no more than a certain percentage of the population N is present, otherwise the bar is overcrowded. To make the considerations specific, let $N = 100$ and let the maximum number of people in the bar before becoming overcrowded be 60. Each agent interested in attending cannot collude with others to determine or estimate the density of the bar *a priori*; instead, they must predict how busy the bar will be based on previous attendance. Presume that data on prior weeks' attendance are available to all N individuals. Based on these data, each person makes a prediction about the likely attendance at the bar on the coming Thursday night. If the prediction indicates fewer than 60 bargoers then the person will choose to attend; otherwise the person will stay home. The potential for paradoxical outcomes is clear: If everyone believes that the bar will be relatively vacant then they will attend, and instead it will be crowded; conversely if everyone believes that the bar will be crowded, it will be empty. Of interest are the dynamics of attendance over successive weeks.

Arthur [73] offered the following procedure for determining this attendance. Each individual has k predictive models and chooses whether or not to attend the bar based on the prediction offered by its current best (or active) model measured in terms of how well it fit the available weekly attendance. The active model is dependent on the historical attendance, and in turn the attendance is dependent on each individual's active model. It is evident that the class of models used for predicting the likely attendance can have an important effect on the resulting dynamics. The specifics in [73] are not clear on which models were used, but some were suggested, including (1) use the last week's attendance, (2) use an average of the last four weeks, (3) use the value from two weeks ago (a period two cycle detector), and so forth. Starting from a specified set of models assigned to each of the N individuals, the dynamics were completely deterministic. The results indicated a consistent tendency for the mean attendance over time to converge to 60 (see Figure 7). Curiously, a mixed strategy of forecasting above 60 with probability 0.4 and below 60 with probability 0.6, which would engender a mean attendance of 60 individuals, is a Nash equilibrium when the situation is viewed in terms of game theory [74]. This result implies that traditional game theory may be useful in explicating the expected outcomes of such complex systems.

But people do not reason with a fixed set of models, deterministically iterated over time. Indeed, inductive reasoning requires the introduction of potentially novel models that generalize over observed data; restricting attention to a fixed set of rules appears inadequate. A more appropriate model of the El Farol problem would therefore include both a stochastic element, in which new models were created by randomly varying existing ones, and a selective process that served to eliminate models that were relatively ineffectual. Individuals would thereby improve their predictive models in a manner akin to the scientific method and evolution [76]. The results of this variant on the method of [73], published in [75], were quantitatively different and did not reflect any tendency toward stability in the limit or on the average.

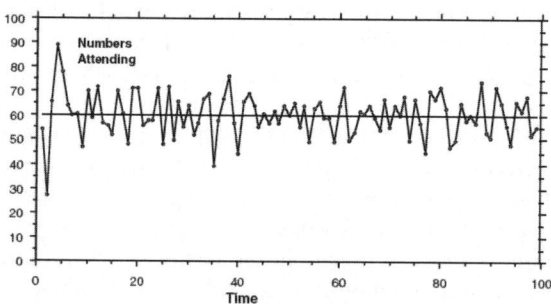

Fig. 7. Bar attendance in the first 100 weeks of simulated time in [73] using unspecified deterministic models for predicting the attendance

4.2 Experimental Methods

Following [73], N was set to 100 and the bar was considered overcrowded if attendance exceeded 60. Each individual was given $k = 10$ predictive models. For simplicity, these models were autoregressive (AR) with their output made unsigned and rounded. For the ith individual, its jth predictor's output was given by:

$$x^i_j(n) = \text{round}\,(|\,a^i_j(0) + a^i_j(t)x(n-t) + \ldots + a^i_j(t^i_j)x(n-t^i_j)\,|)$$

where $x(n - t)$ was the attendance on week $(n - t)$, l^i_j was the number of lag terms in the jth predictor of individual i, $a^i_j(t)$ was the coefficient for the lag t steps in the past, and $a^i_j(0)$ represented a constant bias term. Taking the absolute value and rounding the model's output ensured nonnegative integer values. Any predictions greater than 100 were set equal to 100 (predictions greater than the total population size N were not allowed). For each individual, the number of lag terms for each of its ten models was chosen uniformly at random from the integers $\{1, ..., 10\}$. The corresponding lag terms (including the bias) were uniformly distributed over the continuous range $[-1, 1]$.

Prior to predicting the current week's attendance, each individual evolved its set of models for 10 generations. This was arbitrary, but was chosen to allow a minimal number of iterations for improving the existing models. Details of the evolutionary process are offered in [75]. In general, each parent model created an offspring model with potential variation of both the number of lag terms and the associated coefficients of the lags. Each of the 20 models was evaluated based on the sum of its squared errors made in predicting the attendance at the bar over the past 12 weeks. The 10 models with the best fit were selected to be parents of the next generation. When all generations were exhausted, the best model (lowest error) was used to make a prediction about the attendance and an action to go or stay home was taken. The first 12 weeks of attendance were generated by sampling from a Gaussian random variable with mean 60 and standard deviation of 5. This was meant to start the system with a sufficient sample for each individual's predictors while not biasing the mean away from the previously observed average [73]. The simulation was run 300 times independently, each trial being executed over 982 weeks (18.83 years) in order to observe the long-term dynamics of the evolutionary system.

4.3 Results of Evolving Predictors for the El Farol Problem

Figure 8 shows the mean weekly attendance at the bar averaged over all 300 trials. The first 12 weeks exhibited a mean close to 59.5 resulting from the random initialization. For roughly the next 50 weeks, the mean attendance exhibited large oscillations. This "transient" state transpired completely by about the 100th week, with weeks 101-982 displaying more consistent statistical behavior. For notational convenience, consider this period to be described as the "steady state." The mean attendance for the steady state was 56.3155, with a standard deviation of 1.0456. This is statistically significantly different ($P < 0.01$) from the previously observed mean attendance of 60 offered in [73]. Further, as a mean over 300 trials, the variability depicted in Figure 8 is more than an order of magnitude lower than that of each single trial and the individual dynamics of each trial have been averaged out. Figure 9 depicts the results of a typical trial having a mean steady-state attendance of 56.3931 and a standard deviation of 17.6274. None of the 300 trials showed convergence to an equilibrium behavior around the crowding limit of 60 as observed in [73], nor were any obvious cycles or trends apparent in the weekly attendance. The introduction of evolutionary learning to the system of agents had a marked impact on the observed behavior: The overall result was one of chaos and large oscillations rather than stability and equilibria. Indeed, describing the dynamics of a system with behavior as shown in Figure 9 in terms of its mean does not appear useful.

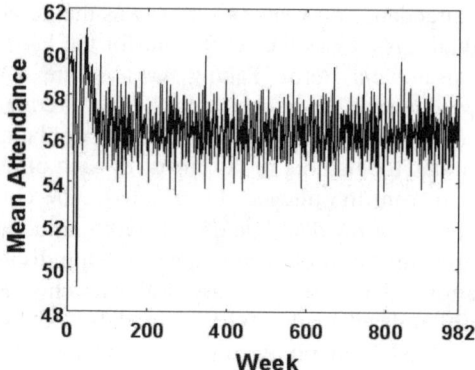

Fig. 8. The mean weekly attendance in the evolutionary El Farol simulation averaged across 300 trials

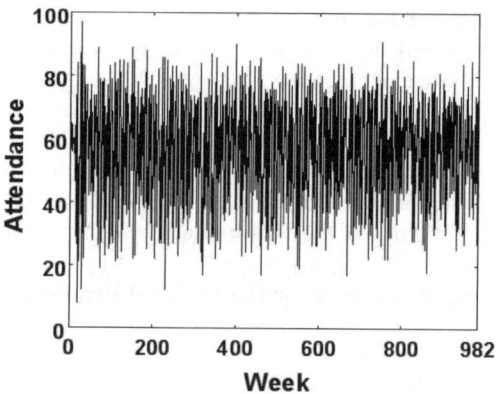

Fig. 9. The attendance observed in a typical trial of the evolutionary simulation of the El Farol problem

4.4 Discussion on El Farol Results

The system studied here is only slightly more complex than that offered in [73]. It is certainly a highly idealized simulation of a market economy. Each agent in a constant-size population was only allowed linear predictive models with an AR form and a window into the past that was restricted to no more than three months. Moreover, the process for generating new models was a relatively simple mutation of existing coefficients and model structure. One could easily imagine variations that allowed agents to migrate to and from the city, employ generalized nonlinear predictive models, collaborate or collude with other agents, and so forth. Yet none of these more sophisticated procedures were required to generate statistically significantly different behavior from that obtained in [73].

Arthur [73] recognized the potential deficiency of mandating strictly deterministic models:

"It might be objected that I lumbered the agents in these experiments with fixed sets of clunky predictive models. If they could form more open-ended, intelligent predictions, different behavior might emerge. We could certainly test this using a more sophisticated procedure, say genetic programming ... . This continually generates new hypotheses – new predictive expressions – that adapt 'intelligently' and often become more complicated as time progresses. But I would be surprised if this changes the above results in any qualitative way."

It is said that the most information is gained when a scientist finds a surprising result. In retrospect, however, there really should be no surprise here. In every case of simulating complex adaptive systems, the emergent properties are strictly dependent on the "rules" preprogrammed by the investigator. Unfortunately, the results of the interactions of agents in light of even mildly complicated rules can lead to behaviors that are "surprising." This merely reflects our own ignorance, our own inability to foresee what was predestined. This inability is heightened when faced with stochastic as opposed to deterministic models. Consequently, the traditional approach in such circumstances is to either assume away the noise or average it out of consideration (as is done in the evolutionary stable strategies approach, both by assuming away mutation and averaging out random payoffs). When random effects are known to exist in the physical system being modeled, there must be compelling reasons for abstracting out that randomness in simulation; otherwise, the results should be viewed with caution, if not skepticism.

5 Conclusions

All human activities are subject to human bias and human error. Science is not immune from this law. Our reservoir of knowledge advances due to the endeavor of science, and it also at times suffers from that endeavor. With a goal of maximizing advances and easing suffering, it is necessary always to take a skeptical view of hypotheses and ask if the burden of proof has been met. This skepticism is especially important for those who are offering the hypotheses, and not just to those who may be viewed as "critics," for it is most often those who offer the hypotheses who are in the best position to be critical of them. Hubris is one of the greatest of human errors.

Several recent scientific activities have not done well in measuring up to this goal. In part that is because of what may best be described as the "politics of science," but in part it is because things that seem simple are often deceptively complex. Why should it be so difficult to discover the true effects of salt on human blood pressure? Salt has been around a long time, and so have observations of blood pressure. It would be easy and reasonable to believe that science should have this figured out by now; however, the system that is the human body is remarkably complex, with remarkable individual variation.

To date, the vast majority of Phase III medical assays have been based on showing a mean deviation from a control result (e.g., based on a placebo) across a large sample of individuals. To the extent that variations in individuals are truly present but acting in opposite directions, such variations will likely never be detected by this approach, because they are averaged out. Fortunately, personalized medicine based on the genetic composition of individuals may offer a significant advance in this regard. Yet, it will take innovative actions to move away from the traditional large-sample approaches to testing for efficacy to smaller and carefully selected samples of people. In the meantime, because of a perceived need to prove things "on the average," many people will suffer who might have otherwise been helped by the more rapid advancement of knowledge.

The concept of averaging out noise is ubiquitous in engineering. Sometimes it is done actively [77] to reduce background interference. Sometimes it is done merely for convenience to be able to use statistical tools that rely on known distributions of sums of random variables. When studying a complex adaptive system, however, averaging out randomness offers at best a restricted view of the system's dynamics. At worst it leads to a highly misleading view. The level of illusion can be amplified by iterating the system's average dynamics, expecting the system to reach an equilibrium, thereby compounding deception as a function of time. The level of illusion can be stepped up even further by incorrectly assessing the nature of the agents' goals, taken on the average, and iterated over time.

In complex adaptive systems, agents act to achieve goals while interacting with other agents and the environment. Individual variation is a key component to the behaviors of these systems as a whole. The agents never stop searching for better ways to achieve their goals. Behaviors reflect randomized exploration. Rewards reflect random happenstance. Mathematicians may compute "rational" outcomes from the interactions of agents' behaviors, "solutions" for which any deviation would yield a lower payoff. Yet we often observe real behaviors (human and otherwise) that deviate considerably from this supposed rational end-point. There is a certain beauty in a mathematically rational solution. *The beauty of a mathematical solution is irrelevant to whether or not that solution is right. Nature is right. The models are wrong.* All models are wrong because they leave out details of the true system (else they would be just as complicated as the system itself), but some models are more useful than others [78].

As identified here, two modeling concepts that deserve a great deal more scrutiny are expectations and equilibrium: more directly, the use of iterated expectations to arrive at a future predicted end state of a system and the notion that once in such an end state, a condition of equilibrium will prevail. These two concepts are central to much of the thought in mathematical biology, particularly in animal behavior, but they are now also central to concepts in ecology, economics, sociology, and other fields as well. As observed in the case studies offered here, the notion that an expected payoff can be used iteratively to determine the probabilities or percentages of adopting alternative behaviors in natural settings is suspect. So too is a notion that any of these natural settings involving adaptive purpose-driven organisms will be found in equilibrium. As John Holland is said to have offered so insightfully: "If you find a complex adaptive system in equilibrium, it is probably dead."

Thousands of papers have been published on evolutionary stable strategies (ESSs) in the past 40 years. Yet, it is truly difficult to find articles that use these concepts to provide verifiable in-the-blind predictions about what real organisms will do. Within the field of computational intelligence, there are common notions of training, testing, and validation. By way of analogy, comparably few examples of testing are available in use of ESSs, and even fewer of validation. Evolutionary stable strategies have been accepted too easily as explanations of observed data.

For example, Davies and Halliday [63] studied competitive mate searching in male toads (*Bufo bufo*). During migration to a spawning pond, 363 males and 77 females were observed. Males obtained females by (1) encountering an available female and pairing up (riding her back), or (2) dislodging a paired male. Consideration was given to modeling the likelihoods for successful pairing by searching at a spawn site or away from a spawn site, with the belief that the individuals act to equalize these probabilities. Predictions from the equilibrium model suggest the percentage of males to utilize either location on a daily basis over 12 days. Davies and Halliday [63] recognized the ability of the model to predict a seasonal trend for a increase in the percentage of males searching at the spawn site, yet also described the overall fit to the observed data as "reasonably good," despite 3 or 4 of the 12 observations undershooting and overshooting the actual percentages in the range of 10-25%. This "reasonably good" variability leads naturally to the question of what level of variability would have been required to falsify the predictive model. Many other examples could be indicated.

The central problem of evolutionary stable strategies is the use of expected outcomes in place of outcomes taken from an appropriate temporal probability density function. Recall in the hawk-dove game that when two hawks met, both were considered to receive a penalty of −25 points, instead of one hawk gaining a resource (+50) and the other getting injured (−100). In the real world, when you get injured, not only do you generally not get some part of the reward you would have received had you not gotten injured, you also do not compete as effectively in future contests. For example, in some spiders, male competition leaves the combatants with fewer legs, and by consequence handicapped for future jousts with other males [79] (also see [80]). It cannot be surprising that "averaging out" these considerations would lead to disparities between a given model and the reality it seeks to explain. What would be surprising is if a model actually explained that reality in spite of averaging out such important details.

Ficici et al. [53] noted that the use of truncation selection in the models examined here (from [49, 50, 51]) mirrors truncation processes in nature (i.e., natural selection) but, importantly, the specific choice of how truncation is implemented is relevant to how plausible a model may be in explaining a particular observed set of behaviors in nature. Simply replacing the proportional selection that is found in evolutionary game theory with an arbitrary choice of truncation selection is not sufficient to give confidence that such a modified evolutionary model will do better in explaining natural phenomena. Coevolutionary models can be sensitive to choices of how selection is implemented [53]. It is important to recognize that it would be just as misleading to model a natural setting with a hawk-dove game using expected payoffs and some idealized level of truncation selection pressure as it would be to model that setting with expected payoffs and proportional selection. The experiments provided here do not seek to explain any real-world condition; instead, they seek only to provide a counterexample to claims that evolutionary stable strategies, as offered consistently in the scientific literature, are likely to be sufficient models for complex animal behaviors, particularly in small populations.

Models that rely on successive iterations of expected outcomes, particularly under stochastic effects (but even under deterministic effects) appear in various forms. Skepticism regarding models of evolutionary stable strategies also leads to skepticism about models of parental investment and kin selection. In each case, a random variable is collapsed into an expectation and iterated. The random variable in question pertains to the degree of relatedness that a parent has with an offspring, or that any individual has with another. For example, in kin selection, there is a concept that human siblings are related at 50 percent, denoted by $r = 0.5$. This is the same degree of relatedness that human children have with their parents, and also with their own offspring, owing to sexual recombination of each parents' chromosomes. But this is just an expectation. The real degree of relatedness between any two siblings, or a parent and child depends on the random variation that occurs during the assortment of chromosomes, as well as the degree to which a father and mother are themselves related (sharing genes in common). That is, relatedness is a random variable. Unraveling this thread further, skepticism extends immediately to the entire philosophy of selfish genetics [40], for which parental investment and kin selection are merely special cases.

If all this modeling activity were used only to estimate the number of people who might hear Irish music on a Thursday night in Santa Fe, New Mexico, it would not be very important. We could get everything wrong and it would not matter. It would not matter if we assumed that these music aficionados tried to minimize their expected losses, or if they tried to be "happy" on average, or if they even decided not to listen to Irish music anymore and started listening to Gershwin. But we face matters of considerable consequence in diverse areas.

Species are going extinct at an alarming rate. The economies of some nations are booming while others are waning, or in shambles. The Earth's climate is changing. Problems sometimes literally cry out for solutions. It is critical that before we implement "solutions" to address these matters that we understand not only the systems in question but also the effects that those solutions will have. All models are wrong, but we cannot afford to get these models wrong by too much. The unintended consequences of some solutions may be worse than no solution at all (see [81, 82, 83]). If we are to do our best to avoid these undesirable outcomes, the scientific method must be allowed to be practiced freely, without trepidation.

The fields of complex adaptive systems and evolutionary games are not immune to human bias and human error. On the contrary, there are many examples. The examples include cases in which a hypothesis that could be falsified is instead taken as a given. This chapter presents a very few of these cases, indicating that the assumptions that have guided previous study of each particular problem in question are in fact not assumptions but beliefs that can be shown to be false. The cases are not exhaustive of all such examples and are intended only to provide inspiration for more critical analysis of all scientific endeavors within these fields and others, and a constant need to ask "How do we know what we think we know?" In asking this question – asking it often, without hesitation or fear – we may arrive at better answers, and with that knowledge help create a better future.

Acknowledgments

Readers who are curious about the subtitle to the paper may appreciate knowing that Part I of this discussion was offered as the keynote lecture at the 1997 Genetic Programming Conference in Stanford, California. The author would like to thank the conference organizers for the invitation to present this plenary lecture. The paper is based on and reprints sections from several earlier publications [22, 51, 75], for which the author gratefully acknowledges copyright permissions from Elsevier (for sections on k-armed bandits and evolutionary unstable strategies) and IEEE (for sections on the El Farol problem). The author also thanks his co-authors on these works, including Gary Fogel, Peter Andrews, Pete Angeline, Kumar Chellapilla, and Hans-Georg Beyer. The author also thanks the many people who helped review and comment on the author's research that is offered in this chapter. The people include Wirt Atmar, Larry Fogel, Charles Marshall, Ernst Mayr, Charles Taylor, Bill Porto, Jim Haefner, Michael Conrad, John Maynard Smith, Sevan Ficici, Lee Altenberg, Russell Anderson, Bill Macready, and Tom English. Special thanks are owed to Lee Altenberg, Tom English, Sevan Ficici, Gary Fogel, Joanne Fogel, Garry Greenwood, David Wolpert, and Don Wunsch for commenting on this draft directly, as well as brief remarks from three anonymous reviewers of sections of the draft, and to Jacquelyn Fogel for her encouragement and support.

References

1. Darwin, C.: The Origin of Species, 6th edn (1872)
2. Gould, S.J.: The Tallest Tale. Natural History 105(5), 18–23, 54–57 (1996)
3. Simmons, R.E., Scheepers, L.: Winning By a Neck: Sexual Selection in the Evolution of Giraffe. Amer. Nat. 148, 771–786 (1996)
4. Hansen, J.E., Sato, M., Ruedy, R., Lacis, A., Glascoe, J.: Global Climate Data and Models: A Reconciliation. Science 281, 930–932 (1998)
5. http://www.businessandmedia.org/articles/2008/20080206170159.aspx
6. http://www.nhlbi.nih.gov/health/public/heart/hbp/dash/new_dash.pdf
7. Sacks, F.M., Svetkey, L.P., Vollmer, W.M., Appel, L.J., Bray, G.A., Harsha, D., Obarzanek, E., Conlin, R.P., Miller III, E.R., Simons-Morton, D.G., Karanja, N., Lin, P.-H.: (DASH-Sodium Collaborative Research Group): Effects on Blood Pressure of Reduced Dietary Sodium and the Dietary Approaches to Stop Hypertension (DASH). Diet. N. Engl. J. Med. 344, 3–10 (2001)
8. Taubes, G.: The (Political) Science of Salt. Science 281, 898–905 (1998)
9. Weinberger, M.H., Fineberg, N.S., Fineberg, S.E., Weinberger, M.: Salt Sensitivity, Pulse Pressure, and Death in Normal and Hypertensive Humans. Hypertension 37, 429–432 (2001)
10. Petrie, J.R., Morris, A.D., Minamisawa, K., Hilditch, T.E., Elliott, H.L., Small, M., McConnell, J.: Dietary Sodium Restriction Impairs Insulin Sensitivity in Noninsulin-Dependent Diabetes Mellitus. J. Clin. Endocrin. Metabol. 83(5), 1552–1557 (1998)
11. Vecchione, C., Morisco, C., Fratta, L., Argenziano, L., Trimarco, B., Lembo, G.: Dietary Sodium Restriction Impairs Endothelial Effect of Insulin. Hypertension 31, 1261–1265 (1998)

12. Perry, C.G., Palmer, T., Cleland, S.J., Morton, I.J., Salt, I.P., Petrie, J.R., Gould, G.W., McConnell, J.: Decreased Insulin Sensitivity During Dietary Sodium Restriction is Not Mediated by Effects of Angiotensin II on Insulin Action. Clin. Sci. London 105, 187–194 (2003)
13. Mayr, E.: Toward a New Philosophy of Biology: Observation of an Evolutionist. Belknap, Harvard (1988)
14. Alexander, R.M.: Optimal For Animals (revised). Princeton Univ. Press, Princeton (1996)
15. Holland, J.H.: Adaptation in Natural and Artificial Systems. University of Michigan Press, Ann Arbor (1975)
16. Krebs, J.R., Kacelnik, A., Taylor, J.P.: Test of Optimal Sampling by Foraging Great Tits. Nature 275, 27–31 (1978)
17. Green, R.F.: Bayesian Birds: A Simple Example of Oaten's Stochastic Model of Optimal Foraging. Theor. Popul. Biol. 18, 244–256 (1980)
18. Bateson, M., Kacelnik, A.: Starlings' Preferences for Predictable and Unpredictable Delays to Food. Anim. Behav. 53, 1129–1142 (1997)
19. Maynard Smith, J.: Evolution and the Theory of Games. Cambridge University Press, Cambridge (1982)
20. Rudolph, G.: Reflections on Bandit Problems and Selection Methods in Uncertain Environments. In: Baeck, T. (ed.) Proc. Seventh Intern. Conf. Genetic Algorithms, pp. 166–173. Morgan Kaufmann, San Francisco (1997)
21. Macready, W.G., Wolpert, D.H.: Bandit Problems and the Exploration/Exploitation Tradeoff. IEEE Transactions on Evolutionary Computation 2, 2–22 (1998)
22. Fogel, D.B., Beyer, H.-G.: Do Evolutionary Processes Minimize Expected Losses? J. Theoret. Biol. 206, 117–123 (2000)
23. Hewett, R.: The Value of Pattern in the Distribution of Young Fish. In: Lasker, R., Sherman, K. (eds.) The Early Life History of Fish, vol. 178, pp. 229–236 (1981)
24. Blaxter, J.H.S., Hunter, J.R.: The Biology of Clupeoid Fishes. In: Blaxter, J.H.S., Russell, J., Young, C. (eds.) Advances in Marine Biology, vol. 20, pp. 1–203. Academic Press, New York (1982)
25. Tibbo, S.N., Scarratt, D.J., McMullion, P.W.G.: An Investigation of Herring. In: Clupea harengus, L. (ed.) Spawning Using Free-Diving Techniques, J. Fishieries Res. Board Canada, vol. 20, pp. 1067–1079 (1963)
26. Gould, J.L., Gould, C.G.: Sexual Selection. Scientific American Library, New York (1989)
27. Cheney, D.L., Seyfarth, R.M., Andelman, S.J., Andelman, L.P.: Reproductive Success in Vervet Monkeys. In: Clutton-Brock, T.H. (ed.) Reproductive Success, pp. 382–402. Chicago University Press, Chicago (1988)
28. Boesch, C.: The Effects of Leopard Predation on Grouping Patterns in Chimpanzees. Behaviour 117, 220–242 (1991)
29. Stanford, C.B., Wallis, J., Matama, H., Goodall, J.: Patterns of Predation by Chimpanzees on Colobus Monkeys in Gombe National Park, 1982-1991. Am J. Phsy. Antrhopol. 94, 213–228 (1994)
30. McCann, T.S.: Aggression and Sexual Activity of Male Southern Elephant Seals, Mirounga Leonina. J. Zool. 195, 295–310 (1981)
31. Andersson, M.: Sexual Selection. Princeton University Press, Princeton (1994)
32. Stephens, D.W., Krebs, J.R.: Foraging Theory. Princeton University Press, Princeton (1986)
33. McNamara, J.M., Houston, A.I.: Risk-Sensitive Foraging: A Review of the Theory. Bull. Math. Biol. 54, 355–378 (1992)

34. Caraco, T., Martindale, S., Whittham, T.S.: An Empirical Demonstration of Risk-Sensitive Foraging Preferences. Anim. Behav. 28, 820–830 (1980)
35. Caraco, T., Blanckenhorn, W.U., Gregory, G.M., Newman, J.A., Recer, G.M., Zwicker, S.M.: Risk-Sensitivity: Ambient Temperature Affects Foraging Choice. Anim. Behav. 39, 338–345 (1990)
36. Cartar, R.V., Dill, L.M.: Why Are Bumble Bees Risk-Sensitive Foragers? Behav. Ecol. Socio-biol. 26, 121–127 (1990)
37. Nowak, M.: An Evolutionary Stable Strategy Be Inaccessible. J. Theoret. Biol. 142, 237–241 (1990)
38. Takada, T., Kigami, J.: The Dynamic Attainability of ESS in Evolutionary Games. J. Math. Biol. 29, 513–529 (1991)
39. Maynard Smith, J., Price, G.R.: The Logic of Animal Conflict. Nature 246, 15–18 (1973)
40. Dawkins, R.: The Selfish Gene, 2nd edn. Oxford University Press, Oxford (1989)
41. Motro, U.: Co-operation and Defection: Playing the Field and the ESS. J. Theoret. Biol. 151, 145–154 (1991)
42. Visser, M.E., Van Alphen, J.J.M., Hemerik, L.: Adaptive Superparasitism and Patch Time Allocation in Solitary Parasitoids: An ESS Model. J. Anim. Ecol. 61, 93–101 (1992)
43. Wolf, L.L., Waltz, E.C.: Alternative Mating Tactics in Male White-Faced Dragonflies: Experimental Evidence for a Behavioural Assessment ESS. Anim. Behav. 46, 325–334 (1993)
44. Ensminger, A.L., Crowley, P.H.: Strangers and Brothers: A Paternity Game Between House Mice. Animal Behaviour 74, 23–32 (2007)
45. Hovestadt, T., Mitesser, O., Elmes, G.W., Thomas, J.A., Hochberg, M.E.: An Evolutionary Stable Strategy Model for the Evolution of Dimorphic Development in the Butterfly Maculinea rebeli, a Social Parasite of Myrmica Ant Colonies. Amer. Naturalist 169, 466–480 (2007)
46. Fenton, A., Paterson, S., Viney, M.E., Gardner, M.P.: Determining the Optimal Developmental Route of Strongyloides ratti: An Evolutionarily Stable Strategy Approach. Evolution 58, 989–1000 (2004)
47. Neill, D.B.: Evolutionary Stability for Large Populations. J. Theoret. Biol. 227, 397–401 (2004)
48. Schaffer, M.E.: Evolutionary Stable Strategies for a Finite Population and Variable Contest Size. J. Theoret. Biol. 132, 469–478 (1988)
49. Fogel, D.B., Fogel, G.B.: Evolutionary Stable Strategies Are Not Always Stable Under Evolutionary Dynamics. In: McDonnell, J., Reynolds, R.G., Fogel, D.B. (eds.) Evolutionary Programming IV, pp. 565–577. MIT Press, Cambridge (1995)
50. Fogel, D.B., Fogel, G.B., Andrews, P.C.: On the Instability of Evolutionary Stable Strategies. BioSystems 44, 135–152 (1997)
51. Fogel, G.B., Andrews, P.C., Fogel, D.B.: On the Instability of Evolutionary Stable Strategies in Small Populations. Ecol. Modelling 109, 283–294 (1998)
52. Ficici, S.G., Pollack, J.B.: Evolutionary Dynamics of Finite Populations in Games with Polymorphic Fitness Equilibria. J. Theoret. Biol. 247, 426–441 (2007)
53. Ficici, S.G., Melnik, O., Pollack, J.B.: A Game-Theoretic and Dynamical-Systems Analysis of Selection Methods in Coevolution. IEEE Transactions on Evolutionary Computation 9(6), 580–602 (2005)
54. Cavalieri, L.F., Kocak, H.: Intermittent Transition Between Order and Chaos in an Insect Pest Population. J. Theoret. Biol. 175, 231–234 (1995)
55. Dieckmann, U., Marrow, P., Laws, R.: Evolutionary Cycling in Predator-Prey Interactions: Population Dynamics and the Red Queen. J. Theoret. Biol. 176, 91–102 (1995)

56. Maynard Smith, J., Parker, G.A.: The Logic of Asymmetric Contests. Anim. Behav. 24, 159–175 (1976)
57. Hori, M.: Frequency-Dependent Natural Selection in the Handedness of Scale-Eating Ciclid Fish. Science 260, 216–219 (1993)
58. Grafen, A., Sibly, R.: A Model of Mate Desertion. Anim. Behav. 26, 645–652 (1978)
59. Repka, J., Gross, M.R.: The Evolutionary Stable Strategy Under Individual Condition and Tactic Frequency. J. Theoret. Biol. 176, 27–31 (1995)
60. Houston, A.I., McNamara, J.M., Webb, J.N.: The Evolutionary Stable Exploitation of a Renewing Resource. J. Theoret. Biol. 177, 151–158 (1995)
61. Tracy, N.D., Seaman, J.W.: Properties of Evolutionarily Stable Learning Rules. J. Theoret. Biol. 177, 193–198 (1995)
62. Brockmann, H.J., Grafen, A., Dawkins, R.: Evolutionarily Stable Nesting Strategy in a Digger Wasp. J. Theoret. Biol. 77, 473–496 (1979)
63. Davies, N.B., Halliday, T.R.: Competitive Mate Searching in Male Common Toads. Bufo Bufo. Anim. Behav. 27, 1253–1267 (1979)
64. Gross, M.R., Charnov, E.L.: Alternative Mate Life History in Bluegill Sunfish. PNAS 77, 6937–6940 (1980)
65. Gross, M.R.: Evolution of Alternative Reproductive Strategies: Frequency-Dependent Sexual Selection in Male Bluegill Sunfish. Philos. Trans. R. Soc. London B 332, 59–66 (1991)
66. Gross, M.R.: Sneakers, Satellites, and Parentals: Polymorphic Mating Strategies in North American Sunfishes. Z. Tierpsychol. 60, 1–20 (1982)
67. Sinervo, B., Lively, C.M.: The Rock-Paper-Scissors Game and the Evolution of Alternative Male Strategies. Nature 380, 240–243 (1996)
68. Maynard Smith, J.: The Games Lizards Play. Nature 380, 198–199 (1996)
69. Brockmann, H.J., Dawkins, R.: Joint Nesting in a Digger Wasp as an Evolutionarily Stable Preadaptation to Social Life. Behaviour 71, 230–245 (1979)
70. Tesfatsion, L.: An Evolutionary Trade Network Game with Preferential Partner Selection. In: Fogel, L.J., Angeline, P.J., Baeck, T. (eds.) Evolutionary Programming V, pp. 45–54. MIT Press, Cambridge (1996)
71. Vriend, N.: Self-Organization of Markets: An Example of a Computational Approach. Comput. Econ. 8, 205–231 (1995)
72. Simon, H.: Models of Bounded Rationality. MIT Press, Cambridge (1982)
73. Arthur, W.B.: Inductive Reasoning and Bounded Rationality: The El Farol Problem. Amer. Econ. Assn. Papers Proc. 84, 406–411 (1994)
74. Nash, J.: Equilibrium Points in N-Person Games. PNAS 36, 48–49 (1950)
75. Fogel, D.B., Chellapilla, K., Angeline, P.J.: Inductive Reasoning and Bounded Rationality Reconsidered. IEEE Transactions on Evolutionary Computation 3(2), 142–146 (1999)
76. Fogel, L.J., Owens, A.J., Walsh, M.J.: Artificial Intelligence through Simulated Evolution. John Wiley, New York (1966)
77. Fogel, L.J.: Method of Improving Intelligence Under Random Noise Interference, U.S. Patent #2,866,848 (1958)
78. Box, G.E.P., Draper, N.R.: Empirical Model-Building and Response Surfaces, p. 424. Wiley, New York (1987)
79. Austad, S.: A Game Theoretical Interpretation of Male Combat in the Bowl and Dolly Spider (Frontinella pyramitela). Anim. Behav. 31, 59–73 (1983)
80. Brown, W.D., Smith, A.T., Moskalik, B., Gabriel, J.: Aggressive Contests in House Crickets: Size, Motivation and the Information Content in Aggressive Songs. Anim. Behav. 72, 225–233 (2006)

81. Norton, R.: Unintended Consequences,
    http://www.econlib.org/library/Enc/UnintendedConsequences.html
82. Perry, G.: Disabling America: The Unintended Consequences of the Government's Protection of the Handicapped. WND Books, TN (2003)
83. Tenner, E.: Why Things Bite Back: Technology and the Revenge of Unintended Consequences. Vintage Books, New York (1997)
84. Mayr, E.: The Growth of Biological Thought: Diversity, Evolution, and Inheritance. Belknap Press, Harvard (1982)

Evolution of Altruistic Robots

Dario Floreano[1], Sara Mitri[1], Andres Perez-Uribe[2], and Laurent Keller[3]

[1] Laboratory of Intelligent Systems, EPFL, Lausanne, Switzerland
[2] University of Applied Sciences, Yverdon, Switzerland
[3] Department of Ecology and Evolution, University of Lausanne, Switzerland

Abstract. In this document we examine the evolutionary methods that may lead to the emergence of altruistic cooperation in robot collectives. We present four evolutionary algorithms that derive from biological theories on the evolution of altruism in nature and compare them systematically in two experimental scenarios where altruistic cooperation can lead to a performance increment. We discuss the relative merits and drawbacks of the four methods and provide recommendations for the choice of the most suitable method for evolving altruistic robots.

1 Altruistic Cooperation in Nature

The competition for survival and reproduction postulated by Darwin seems at odds with the observation that some organisms display cooperative behaviors. In order to understand the evolutionary conditions when cooperation can emerge, Lehmann and Keller [14] suggested to distinguish between two types of cooperation (figure 1), namely the situations where a cooperator does not pay a fitness cost from helping other individuals and the situations where a cooperator must pay a fitness cost for helping other individuals. Let us remember that in biology fitness benefits and costs translate into the number of genetic copies that an individual can produce or loose with respect to its baseline reproduction rate.

The situation where cooperation generates a fitness benefit without any cost to the cooperator is relatively common in nature. This situation can be further divided in two cases, when the benefit is immediate or direct and when the benefit is indirect. Examples of cooperation with direct benefits include nest building and group hunting. Whenever a cooperator obtains an immediate and direct benefit from helping another individual, cooperation will always evolve and remain stable, no matter whether the receiving individuals belong to another species or have never been seen before.

If the benefit is indirect, i.e., the act of helping is not immediately reciprocated or the benefit appears only in the long term, cooperation evolves only if individuals have an initial tendency to cooperate, interact together several times, and can both recognize the partner and remember the outcome of previous interactions. If these conditions are satisfied, cooperation will always evolve and remain stable even if cooperating individual belong to different species.

It has also been shown that recognition of other individuals and memorization of the outcomes of the interactions is not necessary if there is a reputation system

J.M. Zurada et al. (Eds.): WCCI 2008 Plenary/Invited Lectures, LNCS 5050, pp. 232–248, 2008.

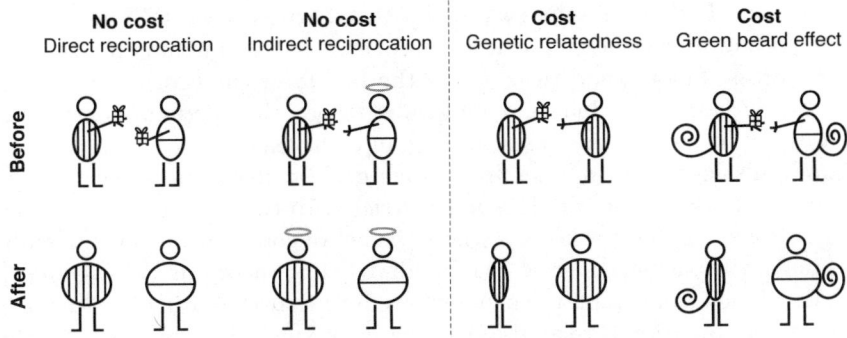

Fig. 1. Conditions for the evolution of cooperation according to the classification suggested by Lehmann and Keller [14]. When there is no cost for the cooperator, cooperation can evolve if there is direct reciprocation or indirect reciprocation (in the latter case, a reputation system may help). When there is a cost for the cooperator, cooperation can evolve if individuals have a high level of genetic relatedness or if they both have greenbeard genes. The pattern indicates the genetic similarity between individuals. The size change after cooperation indicates the cost or benefit of cooperation. Figure loosely inspired by figure 5.9 in [29].

that informs how cooperative an individual is [21]. The way in which animals and people decide to cooperate has been studied extensively in game theory, notably within the framework of the Prisoner's Dilemma game.

On the other hand, the situation where cooperation implies a fitness cost for the cooperator is less common. Cooperation with a cost is also known as *altruism* because the cooperator helps other individuals at its own expense. Parental care is an instance of altruism directed towards offspring of the individual because it implies an energetic cost for the parent. The specialization of ant colonies into large numbers of sterile workers (for food collection, nest defense, rearing of the pupae of the queen, etc.) is yet another instance of altruistic cooperation where the helping workers incur the highest fitness cost because they cannot reproduce.

Building on earlier intuitions by Haldane [10], Hamilton [11] suggested that altruism can evolve if the cooperator is genetically related to the recipient of help. In this case, even if the cooperator cannot propagate its own genes to the next generation, its altruistic act will increase the probability that a large portion of those genes will be propagated through the reproduction of the recipient of the altruistic act. Hamilton [11] proposed the notion of *inclusive fitness*, which is the sum of the individual fitness and of the fitness effects caused by its own act on the portion of genes shared with other individuals. The portion of shared genes between two individuals is known as *genetic relatedness*. He [11] predicted that altruistic cooperation will evolve if the inclusive fitness of the helper is larger than zero

$$rb - c > 0 \tag{1}$$

where r is the coefficient of genetic relatedness, b is the fitness benefit of the recipient(s) of help, and c is the fitness cost of the helper. To use an example suggested by Haldane, in the case of brothers, where $r = 1/2$, an individual may be willing to sacrifice its own life and thus pay the maximum cost $c = 1$ if its act increases more than twice $b > 2$ the fitness of the brother. For cousins, where $r = 1/8$, an individual may be willing to pay the maximum cost if its act increases the fitness of the cousin more than eight times.

Hamilton's inequality applies to average genetic relatedness over the entire genotype and population, i.e. it is not restricted to the sharing of a specific set of genes. It also applies to the case where the act of cooperation benefits multiple individuals with various degrees of relatedness. The theory of *kin selection* [16], which developed from Hamilton's model, predicts that the ratio of altruistic individuals in a population is related to the degree of kinship, or genetic relatedness, among individuals. Although the theory is widely accepted, its quantitative validation in nature has not yet been done because it is difficult to precisely measure the values of the three variables in equation 1.

For evolution of altruism to occur, helping should be directed towards related individuals. This is more likely to happen when individuals share the same geographical space, such as a nest, for social activities. Indeed, most cases of altruistic cooperation are found in families of social insects [12]. Kin selection does not require that individuals recognize kin individuals or know their degree of genetic relatedness. As long as the act of altruism preferentially benefits genetically-related individuals, altruism will spread throughout the population and remain stable.

A particular case of altruism occurs when individuals share few specific genes that favor cooperating behaviors only between individuals having a specific phenotypic character, such as a green beard [7], and that express the same phenotypic character. However, altruism due to *greenbeard effects* can be disrupted if the linkage between the genes responsible for the green beard and the genes responsible for altruistic behavior is disrupted. For example, a mutant individual with a green beard but without the altruistic behavior will have larger inclusive fitness than individuals who have both types of genes; consequently, it will spread in the population and destroy altruistic cooperation [14].

The four conditions for the evolution of cooperation, direct or indirect reciprocity, genetic relatedness and greenbeard genes, which can all be included within a single model [14], hold only if cooperation brings a net fitness advantage to the individuals. In some societies, the actual values of benefits and costs are distorted by means of coercion and punishment to ensure maintenance of cooperative behavior.

Yet another explanation for the evolution of altruistic cooperation is provided by the theory of *levels of selection*, which argues that altruistic cooperation may also evolve in colonies of genetically unrelated individuals that are selected and reproduced all together at a higher rate than the single individuals composing the colony [31]. This could happen in situations where the synergetic effect of cooperation by different individuals provides a higher fitness to the group with respect to other competing groups.

However, the colony-level selection has been criticized because genetic mutations at the level of the individual are more likely and frequent than mutations at the level of the colony, thus creating stronger competition among individuals than among colonies. It has also been argued that the transition from unicellular to multi-cellular organisms can be explained by kin selection because all cells share the same genotype [30]. Although proponents of colony-level selection respond to these criticisms by pointing to evidence for the evolution of colony-level features that decrease individual conflict (such as a reduced mutation rate of individual organisms or cells that compose the colony), the theory of colony-level selection is still widely debated. Furthermore, colony-level selection may eventually lead to high genetic relatedness, thus making the disambiguation between the original driving forces that led to altruistic cooperation even more difficult.

2   Artificial Evolution of Cooperation

In robotics, the evolution of collective behaviors has been studied in several experiments, but often without attention to whether it involves only behavior coordination or also cooperation and whether cooperation involves a cost for the individuals. In those situations where cooperation is explicitly mentioned, it is described as a situation where robots obtain an advantage by working together rather than working in isolation.

When it comes to evolving teams of robots, the experimenter is presented with two design choices: 1) whether robots should be genetically identical or different; and 2) whether the fitness used for selection should take into account the performance of the entire group or only that of single individuals. These two choices are analogous to the issues of genetic relatedness and of level of selection that were discussed above in the context of the biological literature. If we consider only the extreme cases of each design choice, robots in a team can be genetically homogeneous (clones) or heterogeneous (they differ from each other); and the fitness can be computed at the level of the team (in which case, the entire team of individuals is reproduced) or at the level of the individual (in which case, only individuals of the team are selected for reproduction).

Biological theory tells us that the evolution of genetically related robots should lead to cooperative behaviors, but the question of the appropriate level of selection, or fitness computation, is still open for discussion. Furthermore, biological theory does not make any prediction on the comparative performances that we may expect from robots evolved under different conditions.

The majority of current approaches to the evolution of multi-agent systems use genetically homogeneous teams evolved with team-level selection (a comparative survey can be found in [27]). Where the reasons for the choice of genetically homogeneous teams are made explicit, it is argued that homogeneous teams are easy to use [3,26], require fewer evaluations [15,25], scale more easily [6], and are more robust against the failure of team members [6,24] than heterogeneous teams.

Fig. 2. A swarm-bot composed of four interconnected s-bots in chain formation

The choice of level of selection is rarely discussed explicitly despite the fact that fitness distribution leads to credit assignment problems [9,19] in many cooperative multi-agent tasks because individual contributions to team performance are often difficult to estimate or difficult to monitor [23].

Let us consider the case of evolving control systems for a population of identical robots, the s-bots shown in figure 2, which can self-connect to form a swarm-bot [20]. In a simple case, a swarm-bot of four s-bots assembled in chain formation were evolved for the ability to move coordinately on a flat terrain. Each s-bot was provided with a neural controller where sensory neurons were directly connected to the motors neurons that controlled the desired speed of the tracks. The sensory neurons received information from distance sensors around the body of the robot and from a torque sensor that measured the amount of torsional force exerted by other robots. In this case, all s-bots in the swarm-bot were genetically identical and the fitness measured the progress of the entire swarm-bot on the ground. Evolved controllers were also capable of producing coordinated movement also when the swarm-bot was augmented by additional s-bots and re-organized in different shapes. Swarm-bots also dynamically rearranged their shape so as to effectively negotiate narrow passages and were capable of moving on rough terrains over holes or slopes that could not be passed by a single robot. Such robots also collectively avoided obstacles and coordinated to transport heavy objects [1,2,26].

The choice of team-level selection in this case was imposed by the difficulty to assign fitness values to individual s-bots that composed the swarm-bot. However, the choice of genetically related teams was not duly justified because it may have prevented the emergence of specialized individuals.

The question therefore remains of what is the best performing set of choices for tasks that benefit from cooperative behaviors when there is both a choice between genetic relatedness and level of selection. In the remainder of this chapter, we will describe the systematic comparison of these design choices for two sets of experiments that can benefit from the evolution of altruistic cooperation.

Level of Selection

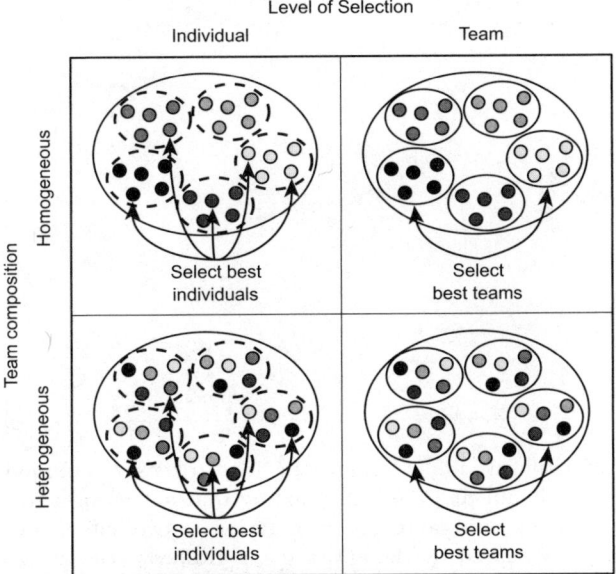

Fig. 3. Four conditions for the evolution of robot collectives. A population (large oval) is composed of several teams (medium ovals), each of which is composed of several robots (small circles). Genetic team composition is varied by either composing teams of robots with identical genomes (homogeneous, identical shading), or different genomes (heterogeneous, different shading). The level of selection is varied by either measuring team performance and selecting teams (team-level selection) or measuring individual performance and selecting individuals independently of their team affiliation (individual-level selection).

2.1   Evolutionary Conditions

We compared four evolutionary conditions (figure 3): genetically homogeneous teams evolved with team-level selection; genetically homogeneous teams evolved with individual-level selection; genetically heterogeneous teams evolved with team-level selection; and genetically heterogeneous teams evolved with individual-level selection. Team-level selection (akin to colony-level selection) consisted of computing the fitness of the team and reproducing the robots in the best teams to create a new population of robot teams. Individual-level selection instead consisted of computing the fitness of individual robots (notice that even robots with identical genomes can obtain different fitness because they are exposed to different situations) and reproducing the best ones independently of their team affiliation to recreate new teams.

The comparisons were carried out in situations where both selfish and altruistic behaviors could produce fitness increments over generations, but altruistic behavior corresponded to larger fitness increments, that is to a larger quantity of work accomplished by the team of robots. In a first set of experiments, we resorted to simplified behaviors and simulated environments in order to disentangle

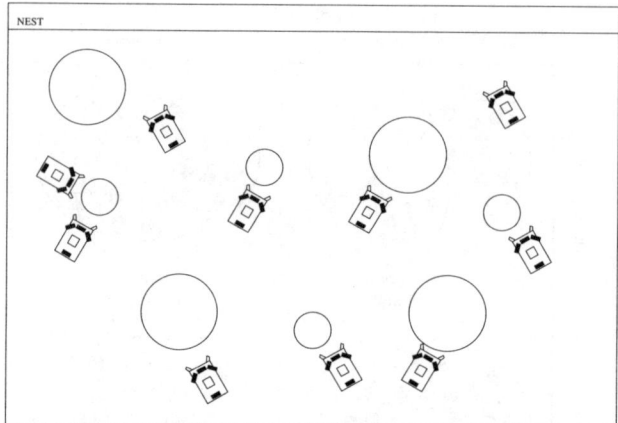

Fig. 4. A team of artificial ants is foraging for food tokens. Small food tokens can be transported by a single ant and are consumed by that ant when it manages to get to the nest. Large food tokens require the cooperation of two ants to be transported to the nest, but they are shared by the entire team. However, the share of a large food token provides less food intake to each individual than a small token. For the sake of simplicity, in this figure we are only showing 10 artificial ants.

fitness differences due to the effects of the evolvability of control systems in situated environments from the effects of the four evolutionary conditions. In a second set of experiments, we resorted to neural controllers in real and simulated robots.

2.2 Altruistic Foraging

In the first set of experiments, we used an agent-based model of a team of artificial ants performing a foraging task (figure 4). The agents or artificial ants (e.g., robots) are supposed to look for food items randomly scattered in a foraging area. There are two kinds of food items, small food items which can be transported by single agents to the nest, and large food items, which can only be transported if two ants cooperate. When a cooperative foraging ant happens to find a large food item, it sends a local message asking for help. Given the local nature of the help message, another cooperative individual will only be able to help the first one if it happens to be close to it and hear its message. For sake of simplicity large food items can only be transported by a pair of ants and we have not included a pheromone-like communication among ants.

Each ant is endowed with a set of three genes encoding three threshold values that are used to determine if one or more predefined behaviors (b_0, b_1 or b_2) are activated at each step of a foraging trial, as shown in the table.

b_0	b_1	b_2	Behavioral strategies
0	0	0	do nothing
1	0	0	if a small food item is found, bring it to the nest, ignore large food items, and do not help other ants
0	1	0	if a large food item is found, stay and ask for help, ignore small food items, and do not help other ants
0	0	1	if a help message is perceived, go and help, ignore small and large food items
1	1	0	if a small food item is found, bring it to the nest, if a large food item is found ask for help, but do not help other ants
1	0	1	if a small food item is found, bring it to the nest, help other ants, but ignore large food items
0	1	1	if a large food item is found, stay and ask for help, ignore small food items, and help other ants
1	1	1	if a small food item is found, bring it to the nest, if a large food item is found, stay and ask for help, and help other ants

The expression of a given behavior b_i depends on the number of foragers already engaged in that behavior and is mediated by the thresholds values that are genetically encoded, as suggested by the response threshold value of [4]. For example, if the proportion of members of the team having activated a given behavior j is smaller than the corresponding threshold of ant k, behavior b_j^k is set to '1' (i.e., it is activated).

The agents were not physically simulated; the model assumed a random walk and took into account the probability of finding a food token at each time step, which decreased in proportion to the number of token collected by the agents. The model also included a probabilistic function of perception and action.

We used 20 agents foraging for 4 large food tokens and 4 small food tokens. The performance of the robot teams was measured using the average score obtained during 20 foraging trials. The small food items provided a score of 1.0 to the single ant who transported it to the nest, while the large food items provided a total score of 16.0. However, since the large food items were shared with the whole team, each individual obtained a score of 0.8 for any large food item taken to the nest. According to these payoffs, all individuals, including those that do not cooperate, can get 0.8 points for every large food item transported by other individuals of the team, whereas the individuals that cooperate in foraging for large food items, pay a cost of 0.2 points compared to the score 1.0 that they would made if they foraged on small food items. The total performance of the team, or total energy brought to the nest, was highest when individuals were altruist rather than selfish.

Performance differences appeared to be cause mainly by genetic relatedness (figure 5). Homogeneous colonies displayed significantly higher mean fitness than heterogeneous colonies. The difference between homogeneous and heterogeneous fitness depends on the relative cost and benefit ratios, as postulated by Hamilton's inequality. However, there was no significant difference between the mean performance of homogeneous colonies evolved using team-level selection

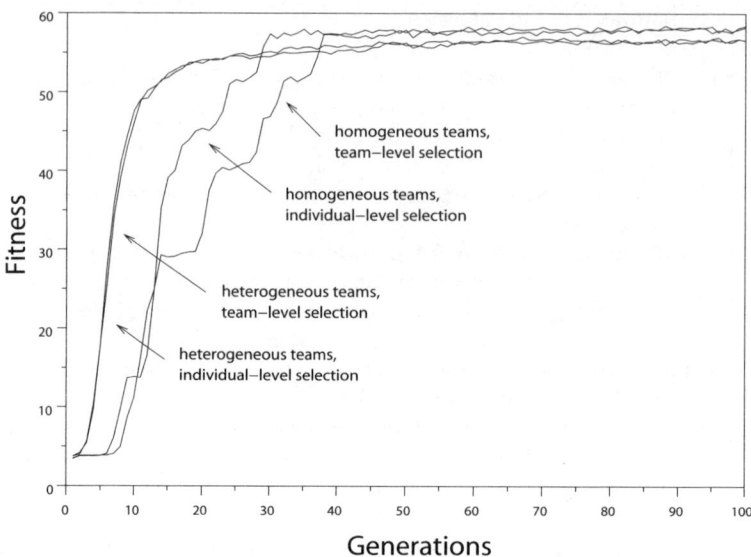

Fig. 5. Evolution of the mean performance of homogeneous and heterogeneous colonies under individual and team-level selection (each curve is the average over 10 different evolutionary runs of mean population fitness)

and mean performance of homogeneous colonies evolved using individual-level selection.

The use of pre-defined behaviors allowed us to precisely measure the amount of altruistic individuals in the evolving teams in each of the four evolutionary conditions (figure 6). We considered an individual to be "altruistic" when it expressed behaviors that did not "pay attention" to small food items and concentrated only on large food items, either by searching for large food items or by helping other individuals to transport large food items (see table above).

As expected, the frequency of altruistic individuals within populations of heterogeneous teams evolved using individual-level selection remained below 10%. However, in all other three conditions we observed a gradual dominance of altruistic individuals in the population. In particular, the resulting number of altruistic individuals is higher when using team-level selection (Figure 6b and Figure 6d). This is understandable because team-level selection favors the individuals that work for the team and not the ones that specialize in the foraging of small food items for their own benefit.

This set of experiments indicated that homogeneous teams were conducive to higher performances in a scenario that could benefit from altruistic behavior and that team-level selection tended to produce more altruistic individuals than individual-level selection. Therefore, it came with no surprise that teams of heterogeneous individuals evolved with individual-level selection produced very few altruistic individuals and obtained lower fitness. The question however remained of why heterogeneous teams evolved with team-level selection produced a

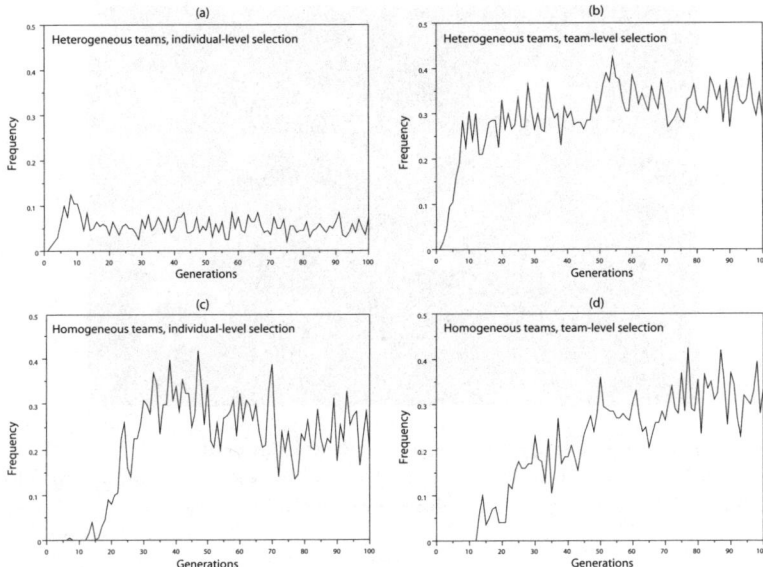

Fig. 6. Evolution of the frequency of altruistic individuals in the simulated ant populations (average of 10 runs) given the following experimental setups: (a) Heterogeneous teams, individual-level selection, (b) Heterogeneous teams, team-level selection, (c) Homogeneous teams, individual-level selection, and (d) Homogeneous teams, team-level selection

majority of altruistic agents, but did not result in better fitness than heterogeneous teams evolved with individual-level selection.

We will get back to this issue in the next set of experiments where we repeated our comparison of the four evolutionary conditions in a more realistic scenario both with physics-based robot simulations and with real robots.

2.3 Altruistic Communication

The evolution of communication is a particularly challenging problem both in biological and in robotic systems because efficient communication requires tight co-evolution between the signal emitted and the response elicited [17]. Furthermore, most communication systems are also costly because of the energy required for signal production [32] and/or increased competition for resources resulting from the transmitted information. For example, if organisms decide to communicate the location of a limited food source, individuals may pay a cost due to decreased food intake. In these situations, communication is another example of altruism and its evolvability and efficiency may depend on the four evolutionary conditions mentioned above.

We therefore set up an experimental scenario for comparing the four evolutionary conditions where communication provides both benefits and costs [8]. We used teams of 10 s-bots that could forage in an environment containing a

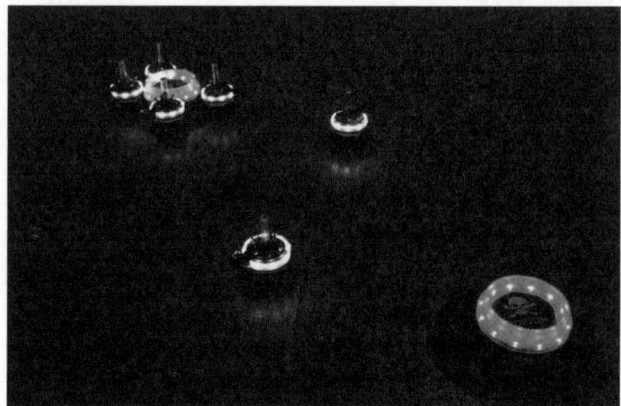

Fig. 7. A team of s-bots engaged in cooperative communication. A team of four s-bots feed on the food objects while they are lit up in blue color. Two s-bots in white color are attracted by the blue signal and move away from the poison object.

food and a poison source that both emitted red light (figure 7). Under such circumstances, foraging efficiency could potentially be increased if robots transmitted information on food and poison location. However, such communication also incurred direct costs to the signaler because it resulted in higher robot density and increased competition and interference nearby the food (i.e., spatial constraints around the food source allowed a maximum of 8 robots out of 10 to feed simultaneously and resulted in robots sometimes pushing each other away from the food). Thus, while beneficial to other team members, signaling of a food location effectively constituted a costly act because it decreased the food intake of signaling robots. This setting thus mimics the natural situation where communicating almost invariably incurs costs in terms of signal production or increased competition for resources.

The experiments were conducted multiple times using a physics-based simulator which accurately models the dynamical properties of the s-bots. The results were then verified by running a single evolutionary experiment for each of the four conditions with the physical robots. The robots had a translucent ring around the body that could emit blue light and a 360° vision system that could detect the amount and intensity of red and blue light. A circular piece of gray paper was placed under the food source and a similar black paper under the poison source. These paper circles could be detected by infrared ground sensors located between the tracks underneath the robot and thus allowed discrimination of food and poison.

The robots were equipped with a neural network to process the visual information and ground sensor input in order to set the direction and speed of the two tracks and control the emission of blue light accordingly every $50ms$ cycle. During each cycle, a robot gained one performance unit if it detected food with its ground sensors and lost one performance unit if it detected poison. The performance of each robot at the end of a trial was computed as the

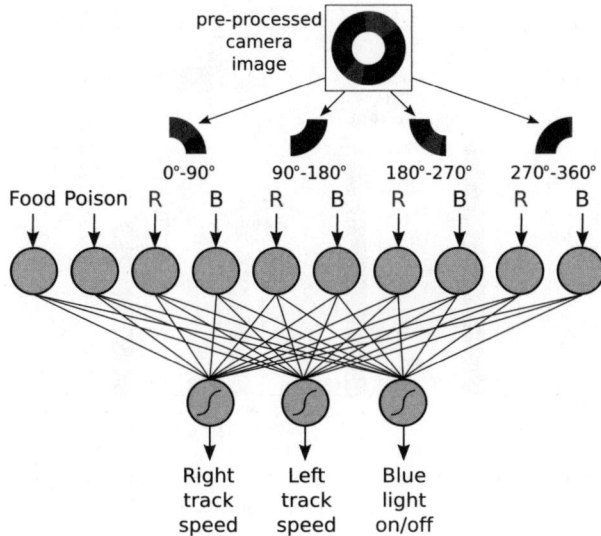

Fig. 8. The neural network architecture used in the experiments on communication

sum of performance units obtained during that trial (1200 sensory motor cycles of $50ms$) and the robot performance was quantified as the sum of performance units over all 10 trials. Team performance was equal to the average performance of all robots in the team.

The feed-forward neural controller had 10 input and 3 output neurons (figure 8). Once a robot had detected the food or poison source, the corresponding neuron was set to 1. This value decayed to 0 by a factor of 0.95 every $50ms$, thereby providing a short-term memory even after the robot's sensors were no longer in contact with the gray and black paper circles placed below the food and poison. The remaining 8 neurons were used to encode the 360° visual input image, which was divided into four sections of 90° each. For each section, the average of the blue and red channels was calculated and normalized within the range of 0 and 1, such that one neural input was used for the blue and one for the red value. The activation of each of the output neurons was computed as the sum of all inputs multiplied by the weight of the connection and passed through the continuous $\tanh(x)$ function (i.e., their output was between -1 and 1). Two of the three output neurons were used to control the two tracks, where the output value of each neuron gave the direction of rotation (forward if > 0 and backward if < 0) and velocity (the absolute value) of one of the two tracks. The third output neuron determined whether to emit blue light, which was the case if the output was greater than 0. The genotype of an individual encoded the synaptic weights of the neural network in a bit string. Each synaptic weight was encoded in 8 bits, giving 256 values that were mapped onto the interval $[-1, 1]$.

For each of the four conditions, we ran 20 independent evolutionary experiments with 100 colonies of 10 robots. Furthermore, as a control situation, we repeated all experiments (4 times 20 runs) by disabling the light ring of the

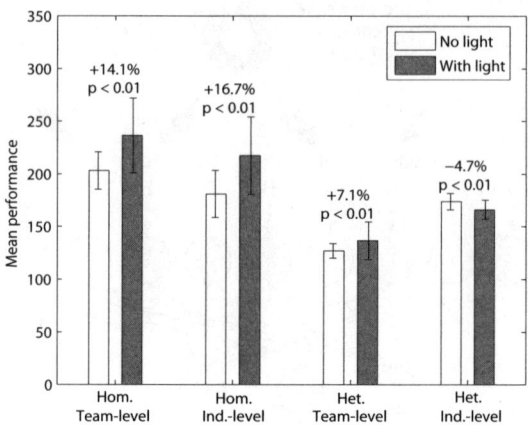

Fig. 9. Mean (+ S.D.) performance of robots during the last 50 generations for each condition when robots could versus could not emit blue light (20 experiments per condition)

robots, but the neural architecture and genotype were the same as in the normal condition.

To compare team performance between treatments, we calculated the average performance of the 100 colonies over the last 50 generations for each of the 20 experiments per condition (figure 9). In evolving teams where robots could produce blue light, foraging efficiency greatly increased over generations and was significantly greater compared to control experiments for all evolutionary conditions, except for the condition of heterogeneous teams under individual-level selection. An analysis of the robot behavior revealed that this performance increment in the three conditions of genetic relatedness or team-level selection was associated with the evolution of effective systems of communication [8].

In teams of genetically related robots with team-level selection, two distinct communication strategies evolved. In 12 of the 20 evolutionary experiments, robots preferentially produced light in the vicinity of the food and were attracted by blue light (figure 10, left). Instead, in the other 8 evolutionary experiments, robots tended to emit light near the poison and were repulsed by blue light (figure 10, right). Teams of robots that signaled food resulted in higher team performance. Interestingly, once one type of communication was well established, there was no transition to the alternate and more efficient strategy. This was because a change in either the signaling or response strategy would completely destroy the communication system and result in a performance decrease. Thus, each communication strategy effectively constituted an adaptive peak separated by a valley with lower performance values.

Heterogeneous teams evolved with team-level selection reliably established communication protocols and displayed increased performance with respect to the control situation. However, their performance was similar to that of heterogeneous teams evolved with individual-level selection, who did not communicate.

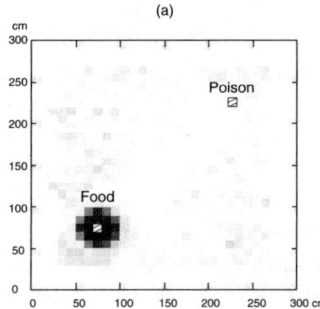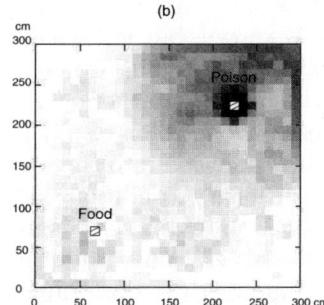

Fig. 10. Signaling frequency measured in each area of the arena for robots from two different evolved teams. a) The team was one where robots signal the presence of food. b) In this team robots signal the presence of poison. The darkness of each square is proportional to the amount of signaling in that area of the arena. From [8].

This result was analogous to the previous example where heterogeneous team evolved with team-level selection displayed a high number of altruistic foragers, but their performance was similar to that of heterogeneous teams evolved with team-level selection, who had very few altruistic foragers.

3   Conclusion

We have presented four algorithms for evolving robot collectives that are presented with situations where altruistic cooperation can lead to a performance increment. Only three of the four algorithms lead to altruistic cooperation, as predicted by kin selection and levels of selection. Heterogeneous teams of robots evolved with individual-level selection do not display altruistic cooperation and consequently result in lower fitness values in tasks that require altruistic cooperation.

Heterogeneous teams evolved with team-level selection represent a special case because in both examples they did evolve stable altruistic cooperators, but their fitness was lower than that of homogeneous teams. We think that this was due to the fact that after making copies of the individuals belonging to the best teams, those individuals were mated with individuals from other teams and randomly re-grouped in new teams. Although this was biologically plausible and necessary to prevent the genetic convergence of inbreeding teams, which would have rapidly led to homogeneous teams and thus confused the experimental design, it resulted in sub-optimal performance because combinations of well-integrated diverse individuals were disrupted at every generation.

From a practical perspective, homogeneous teams evolved with team-level selection are recommended for tasks that can benefit from altruistic cooperation. Not only do they bring together both conditions for the emergence of reliable altruism and thus result in higher performance, but they also do not require the need for separately computing the individual performance of each individual in a team. This is particularly useful in robotic tasks where only the resulting work of the team is known, but not what each robot in the team did and how.

We would like to emphasize that the results described in this chapter are specific to the case where there is an opportunity for altruistic cooperation and where altruistic cooperation results in higher fitness. We are currently expanding this line of investigation into three directions. First, we systematically compare the four evolutionary conditions described in this paper across experimental scenarios that require different degrees of cooperation, ranging from simple co-ordination to cooperation without a cost all the way to altruistic cooperation. Second, we compare these evolutionary conditions with other evolutionary methods in tasks that can benefit from non-trivial division of labor. Third, we compare the four evolutionary conditions in situations where the individuals in the team have a specific identity and can recognize each other, which was not the case in these experiments.

The study of the evolution of robotic collectives is not only promising for developing efficient control systems and testing biological hypotheses, but may also have an impact in a larger number of areas that require an optimal trade-off between the good of the individual and that of the society, such as internet agents, plant optimization, logistics, and economics.

Acknowledgments

The authors gratefully acknowledge support from the Swiss National Science Foundation and from the European projects IST-FET ECAgents and IST-FET Swarmanoids.

References

1. Baldassarre, G., Nolfi, S., Parisi, D.: Evolving mobile robots able to display collective behaviour. Artificial Life 9, 255–267 (2003)
2. Baldassarre, G., Parisi, D., Nolfi, S.: Coordination and behavior integration in cooperating simulated robots. In: Schaal, S., Ijspeert, A., Billard, A., Vijayakumar, S., Hallam, J., Meyer, J.A. (eds.) From Animals to Animats 8: Proceedings of the VIII International Conference on Simulation of Adaptive Behavior. MIT Press, Cambridge (2003)
3. Baray, C.: Evolving cooperation via communication in homogeneous multi-agent systems. In: Intelligent Information Systems, IIS 1997. Proceedings, pp. 204–208 (1997)
4. Bonabeau, E., Theraulaz, G., Schatz, B., Deneubourg, J.-L.: Response threshold model of division of labour in a ponerine ant. Proc. Roy. Soc. London B 265, 327–335 (1998)
5. Bourke, A.F.G., Franks, N.R.: Social Evolution in Ants. Princeton University Press, Princeton (1995)
6. Bryant, B.D., Miikkulainen, R.: Neuroevolution for adaptive teams. In Proceedings of the 2003 Congress on Evolutionary Computation, CEC 2003 3, 2194–2201 (2003)

7. Dawkins, R.: The Selfish Gene. Oxford University Press, Oxford (1976)
8. Floreano, D., Mitri, S., Magnenat, S., Keller, L.: Evolutionary conditions for the emergence of communication in robots. Current Biology 17, 514–519 (2007)
9. Grefenstette, J.J.: Credit assignment in rule discovery systems based on genetic algorithms. Machine Learning 3, 225–245 (1988)
10. Haldane, J.B.S.: Population genetics. New Biology 18, 34–51 (1955)
11. Hamilton, W.D.: The genetical evolution of social behavior, I and II. Journal of Theoretical Biology 7, 1–52 (1964)
12. Keller, L., Chapuisat, M.: Eusociality and cooperation. In: Encyclopedia of Life Sciences, pp. 1–8. Macmillan, Basingstoke (2002)
13. Keller, L., Surette, M.G.: Communication in bacteria. Nature Reviews Microbiology 4, 249–258 (2006)
14. Lehmann, L., Keller, L.: The evolution of cooperation and altruism – a general framework and a classification of models. Journal of Evolutionary Biology 19, 1365–1376 (2006)
15. Luke, S., Hohn, C., Farris, J., Jackson, G., Hendler, J.: Co-evolving soccer softbot team coordination with genetic programming. In: Proceedings of The First International Workshop On RoboCup, IJCAI 1997, Nagoya, pp. 214–222. Springer, Heidelberg (1997)
16. Maynard-Smith, J.: Group selection and kin selection. Nature 201, 1145–1147 (1964)
17. Maynard-Smith, J., Harper, D.: Animal Signals. Oxford University Press, Oxford (2003)
18. Michod, R.E.: Darwinian Dynamics: Evolutionary Transitions in Fitness and Individuality. Princeton University Press, Princeton (1999)
19. Minsky, M.: Steps toward artificial intelligence. Proceedings of the Institute of Radio Engineers 49, 8–30 (1961); Reprinted in Computers and Thought, McGraw-Hill (1963)
20. Mondada, F., Pettinaro, G., Guignard, A., Kwee, I., Floreano, D., Deneubourg, J.-L., Nolfi, S., Gambardella, L.M., Dorigo, M.: Swarm-bot: A new distributed robotic concept. Autonomous Robots 17, 193–221 (2004)
21. Nowak, M.A., Sigmund, K.: Evolution of indirect reciprocity by image scoring. Nature 393, 573–577 (1998)
22. Nowak, M.A., Sigmund, K.: Evolution of indirect reciprocity. Nature 437, 1291–1298 (2005)
23. Panait, L., Luke, S.: Cooperative multi-agent learning: The state of the art. Autonomous Agents and Multi-Agent Systems 11, 387–434 (2005)
24. Quinn, M., Smith, L., Mayley, G., Husbands, P.: Evolving teamwork and role-allocation with real robots. In: Artificial Life VIII: Proceedings of the Eighth International Conference, pp. 302–311. MIT Press, Cambridge (2002)
25. Richards, M.D., Whitley, D., Beveridge, J.R., Mytkowicz, T., Nguyen, D., Rome, D.: Evolving cooperative strategies for UAV teams. In: Proceedings of the 2005 Conference on Genetic and Evolutionary Computation, pp. 1721–1728 (2005)
26. Trianni, V., Nolfi, S., Dorigo, M.: Cooperative hole-avoidance in a swarm-bot. Robotics Autonomous Systems 54, 97–103 (2006)
27. Waibel, M., Keller, L., Floreano, D.: Genetic team composition and level of selection in the evolution of cooperation. IEEE Transactions in Evolutionary Computation (under review, 2008)

28. Wilson, E.O.: The Insect Societies. The Belknap Press of Harvard University Press, Cambridge, MA (1971)
29. Wilson, E.O.: Sociobiology: The New Synthesis, Twenty-Fifth Anniversary Edition. The Belknap Press of Harvard University Press, Cambridge (2000)
30. Wolpert, L., Szathmary, E.: Multicellularity: Evolution and the egg. Nature 420, 745–745 (2002)
31. Wynne-Edwards, V.C.: Evolution Through Group Selection. Blackwell, Palo Alto (1986)
32. Zahavi, A., Zahavi, A.: The Handicap Principle. A Missing Piece of Darwin's Puzzle. Oxford University Press, New York (1997)

Simulated Evolution under Multiple Criteria Conditions Revisited

Günter Rudolph and Hans-Paul Schwefel

Dortmund University of Technology
Faculty of Computer Science
44221 Dortmund, Germany

Abstract. Evolutionary Algorithms (EAs) as one important sub-domain of Computational Intelligence (CI) have conquered the field of experimental as well as difficult numerical optimization despite the lack of addresses of welcome half a century ago. Meanwhile, they go without saying into the toolboxes of most practitioners who have to solve real-world problems. And an overwhelming number of theoretical results underpin at least parts of the practice. More recently, even vector optimization problems can be tackled by means of specialized EAs. These multiobjective evolutionary algorithms (MOEAs or EMOAs) help decision makers to reduce the number of design possibilities to the subsets that make the best of the situation in case of conflicting objectives. This article briefly describes the problem setting, the most important solution approaches, and the challenges that still lie ahead in their improvement. Most sophisticated algorithms in this domain have somehow lost their character of mimicking natural mechanisms found in organic evolution. That is why a couple of more bio-inspired aspects are mentioned in the second part of this contribution that may help to diversify further research and practice in multiobjective optimization (MOO) without forgetting to foster the interdisciplinary dialogue with natural scientists.

1 Introduction

For sure, organic evolution is confronted with a couple of difficulties that are nightmares for traditional optimization algorithms. Among those difficulties are situations in which survival demands the ability to affront not only one peril, but several ones at the same time. More often than not some of the survival criteria are in conflict with each other. Evolutionary algorithms, which have entered successfully the market of solving difficult optimization problems, therefore have been extended to the more general class of multiple criteria optimization - not from the very beginning, but al least during the last decade.

The first part of this article tries to present an overview of different approaches that have been proposed, implemented, analyzed, and used in practice.

The question is then raised, whether these approaches, though successful, reflect mechanisms that are found in nature, so that they can be called bio-inspired - or not. If not, the next question is, how organic evolution deals with

J.M. Zurada et al. (Eds.): WCCI 2008 Plenary/Invited Lectures, LNCS 5050, pp. 249–261, 2008.

multiple objectives, represented for example by different predator species or challenges like diseases and environmental stresses and even different opportunities. Though no final answers can be presented, the attempt is made to highlight at least one direction of further efforts to create algorithms that both solve multi-criteria problems effectively and deliver an explanation how organic evolution really works.

2   Characteristics of Multiobjective Optimization

The main difference between single- and multi-objective optimization rests on the fact that two elements are not guaranteed to be comparable in the latter case. To understand the problem to full extent it is important to keep in mind that the values $f_1(x), \ldots, f_d(x)$ of the d objective functions represent *incommensurable* quantities that cannot be minimized simultaneously since they are conflicting. While f_1 may measure production costs, f_2 may measure the capacity of a battery, f_3 the pressure of some boiler, and so forth. As a consequence, the notion of the "optimality" of some solution needs a more general formulation as in the single-criterion case. It seems reasonable to regard those elements as being optimal which cannot be improved with respect to one criterion without getting a worse value in another criterion. Elements with this property are said to be Pareto-optimal in this context. This concept will be formalized next.

Definition 1. *Let F be a set. A reflexive, antisymmetric, and transitive relation $\preceq$ on F is termed a* partial order relation *whereas a* strict partial order relation *$\prec$ must be antireflexive, asymmetric, and transitive. The latter relation may be obtained by the former relation by setting $a \prec b \Leftrightarrow a \preceq b \wedge a \neq b$.*   □

Here, the set $F = \{f(x) : x \in X \subseteq \mathbb{R}^n\}$ is formed via the vector-valued objective function $f : X \to \mathbb{R}^d$ with $d \geq 2$. The partial order relation is given by

$$f(x) \preceq f(y) \quad \Leftrightarrow \quad \forall i = 1, \ldots, d : f_i(x) \leq f_i(y).$$

Evidently, in case of the strict order relation at least one component must fulfill the strict inequality. It is easy to see that there are pairs of distinct elements $a, b \in F$ that are not comparable, i.e., we have neither $a \preceq b$ nor $b \preceq a$. In this case the elements are termed *incomparable*, denoted by $a \parallel b$. After these preparations we can turn to the concept of optimality.

Definition 2. *Let $X \subseteq \mathbb{R}^n$ and $f : X \to \mathbb{R}^d$ with $d \geq 2$. The task*

$$f(x) \to \min! \quad s.t. \ x \in X$$

is called the multiobjective optimization problem *(MOP). A solution $x^* \in X$ is called* Pareto-optimal *if there is no $x \in X$ with $f(x) \prec f(x^*)$. In this case $f(x^*)$ is termed* efficient. *If $f(x) \prec f(y)$ then we say that x dominates y and also that $f(x)$ dominates $f(y)$. The set of all Pareto-optimal solutions is called the* Pareto set *whereas the set of all efficient points is termed the* efficient set *or the* Pareto front.   □

For later purpose we introduce the sets

$$B(z^0) = \{z \in F : z \preceq z^0\} \qquad (\text{"better"})$$
$$I(z^0) = \{z \in F : z \parallel z^0\} \qquad (\text{"incomparable"})$$
$$W(z^0) = \{z \in F : z \succeq z^0\} \qquad (\text{"worse"})$$

where $B(z^0)$ denotes all points that are better than z^0, $I(z^0)$ denotes all points that are incomparable to z^0 and $W(z^0)$ denotes all points that are worse than z^0.

3   Evolutionary Multiobjective Optimization (EMO)

During the last decade the field of multiobjective optimization (MOO) via evolutionary computation has received considerable attention in theory and practice which is witnessed by a rapid increase in the number of publications. Therefore the overview presented in this section cannot be complete. Rather, it is intended to sketch the main line of developments and the current trends as a preparation of the discussion in the subsequent sections. For a more comprehensive overview see for example [1].

Evolutionary algorithms (EAs) for MOO usually pursue the so-called *a posteriori* approach, i.e., the MOEA (multiobjective evolutionary algorithm) approximates the Pareto front before the decision maker selects an efficient solution from the approximated Pareto front. Needless to say, this approximation should fulfill some conditions: The points should be as close as possible to the true Pareto front, they should be a complete cover of the true Pareto front and they should be uniformly distributed. These three goals are not easy to achieve and several mechanisms have been developed for this purpose.

3.1   Individual-Based Approaches

The simplest approach uses a single individual that is changed at random and accepted if it is better than the current solution. It can be rejected if it is worse. But what should be done with solutions that are incomparable? Moreover, since each run yields only one solution at the end, we have to archive all solutions found before we filter out all dominated solutions in the final step.

3.1.1   Multiobjective (1+1)-Evolutionary Algorithm

The (1+1)-EA known from optimization of a single objective function is easily generalized for multiobjective optimization [2]: We can keep the mutation operation unchanged and the selection operation simply uses a different comparison operation. In the pseudo code below, m_k denotes the mutation vector at iteration $k \geq 0$.

```
choose x_0 ∈ X at random; set k = 0
repeat
    y_k = x_k + m_k
    if f(y_k) ∈ B(f(x_k)) then x_{k+1} = y_k
    otherwise                x_{k+1} = x_k
    k = k + 1
until stopping condition fulfilled
```

Evidently, any point that is better than the current point is accepted whereas worse and even incomparable points are rejected. It is clear that this algorithm gets stuck prematurely if we have many objectives since the cone of potential better solutions takes only a fraction of 2^{-d} in the objective space. And since the cone of potential worse solutions covers only a fraction of 2^{-d} in the objective space, almost all objective space is incomparable to a current solution! As a consequence, something useful should be done with incomparable solutions. The version analyzed in [3] also accepted incomparable solutions in a randomized manner:

choose $x_0 \in X$ at random; set $k = 0$
repeat
    $y_k = x_k + m_k$
    choose an index $i \in \{1, \ldots, d\}$ at random
    if $f_i(y_k) < f_i(x_k)$ then $x_{k+1} = y_k$
    otherwise $x_{k+1} = x_k$
    $k = k + 1$
until stopping condition fulfilled

This kind of selection is equivalent to accepting a point if it is better, and if it is incomparable then we accept the point with probability $1/d$. This approach also works for large d but the distribution of solutions found has several accumulation points along the Pareto front even for simple problems.

3.1.2   Multiobjective Threshold Accepting

The threshold accepting technique also accepts point that are slightly worse than the current point. The version for the multiobjective case could work as follows:

choose $x_0 \in X$ at random; set $k = 0$
repeat
    $y_k = x_k + m_k$
    if $f(y_k) \in B(f(x_k) + T_k)$ then $x_{k+1} = y_k$
    otherwise $x_{k+1} = x_k$
    $T_{k+1} = \gamma(T_k), \; k = k + 1$
until stopping condition fulfilled

Here, T_k is a sequence of vectors where each component decreases to zero via some function $\gamma(\cdot)$, for example $\gamma(T) = c \cdot T$ with some constant $c \in (0, 1) \subset \mathbb{R}$. But as soon as the threshold T_k is close to zero this algorithm behaves almost identically to the simple $(1+1)$-EA and inherits its deficiencies. A remedy might be as follows: Let $\delta_i = f_i(y_k) - f_i(x_k)$ and accept y_k if

$$\sum_{i=1}^{d} w_i \cdot \delta_i \leq T_k$$

for some convex combination of positive weights w_i. Apparently, these ideas have not been tested or analyzed until now.

3.1.3   Multiobjective Simulated Annealing

Another way to handle incomparable and worse points is realized in multiobjective versions of simulated annealing [4,5].

choose $x_0 \in X$ at random; set $k = 0$
repeat
$\quad y_k = x_k + m_k$
$\quad$ if $f(y_k) \in B(f(x_k))$ then $x_{k+1} = y_k$
$\quad$ if $f(y_k) \in W(f(x_k)) \cup I(f(x_k))$
$\quad\quad$ if $u < \exp(-\Delta_k / T_k)$ then $x_{k+1} = y_k$
$\quad\quad$ otherwise $\quad\quad\quad\quad\quad x_{k+1} = x_k$
$\quad T_{k+1} = \gamma(T_k); \ k = k + 1$
until stopping condition fulfilled

Here, $\Delta_k = \max\{f_i(y_k) - f_i(x_k) : i = 1, \ldots, d\}$ or some other formula proposed in [4] and $T_k \to 0$ as $k \to \infty$. As soon as T_k gets close to zero, the algorithm behaves practically like the multiobjective $(1 + 1)$-EA inheriting the problems already discussed.

3.1.4   Deployment of an Archive

Since all these individual-based approaches have problems to generate a well-spread approximation of the Pareto-front it is useful to add an archive to the algorithm [6] that records the non-dominated solutions previously found. This archive is used as a reference set against which each mutated individual is being compared. In this manner, a better distribution of efficient individuals can be achieved. The overhead to maintain the archive, however, grows exponentially with the number of objectives.

3.2   Population-Based Approaches

3.2.1   Early Approaches

The earliest population-based approaches date back at least to the mid-1980s, when Schaffer [7] introduced the vector-evaluated genetic algorithm (VEGA): Here, the population is partitioned into d subpopulations and each subpopulation P_i runs a single-objective GA with objective function f_i for $i = 1, \ldots, d$. After some generations the individuals of all subpopulations are collected, shuffled, and assigned to new subpopulations at random.

Kursawe [8] ran a (μ, λ)-ES (evolution strategy) with a randomly weighted sum of objectives and used an archive to record non-dominated individuals. Other early approaches [9,10,11,12] used different concepts like fitness sharing and niching to end up with a population that is a good representation of the Pareto front. These approaches were important steps toward more efficient and effective MOEAs described next.

3.2.2   Established Approaches

Currently, the probably most popular MOEA is NSGA-2 [13]. The population is partitioned in a hierarchy of anti-chains (i.e., sets of mutually incomparable solutions) as suggested by Goldberg [14] already. This technique guarantees that solutions at the boundary of the current approximation of the Pareto front are not deleted if a dominating super-individual is found. In order to achieve a uniform distribution, the so-called crowding distance is used to decide which element should be deleted from the population among individuals of equal quality.

Another popular MOEA is SPEA-2 [15], which combines several ideas from other MOEAs. It uses an external archive and assigns a so-called strength to each individual that is based on the number of individuals dominated and a nearest-neighbor density estimation value yielding a scalar measure to rank and select the best individuals.

These MOEAs have been used extensively in practice. Recently it was recognized that these approaches work well up to three objectives but as soon as more than three objectives are to be optimized the performance decays rapidly [16]. The conjectured reason for this behavior is seen in the ranking of individuals according to the Pareto partial order: As soon as the number d of objectives increases almost all individuals become incomparable and there is not enough selection pressure to drive the population towards the Pareto front. As a consequence, other selection schemes and/or progress accelerators are necessary.

3.2.3   Current Trends

There are several indicators to measure and assess the quality of a population achieved in a run of an MOEA. But if these measures should attain a maximum value at the end of the run then it is probably a good idea to use these indicators to select those individuals from parents and offspring that optimize this indicator value [17]. A well established indicator is the S-metric or dominated hyper-volume. This leads to the SMS-EMOA [18,19], which does not have the performance problems for more than three objectives. Recently, this algorithm was extended by calculating the gradient [20] of the S-metric value that is based on all individuals. Once this gradient is known all individuals can be changed simultaneously by moving them into the direction of the gradient. Numerical tests have shown a rapid increase in performance.

Another idea to speed-up the evolution towards the Pareto front is the use of the covariance matrix adaptation method [21] known from single-objective optimization. Recent numerical tests [22] have shown that this approach accelerates the convergence rate to the Pareto front considerably.

Summing up, the EMOAs (evolutionary multi-criterion optimization algorithms) become more and more sophisticated by integrating mathematical tools to increasing extent. As a consequence, EMOAs experience a steady loss of their biological inspiration and ideas. It might be fruitful to explore tools and principles of nature for new ideas to improve EMOAs.

4   More Bio-inspired Algorithms

Already at their first appearance, EAs had not only been thought of as optimization tools, but also as simulation models of organic evolution - simplified models,
of course. Choosing the proper level of abstraction has always been one of the
most challenging choices for modeling systems and processes of the *real world*,
and Darwin has ingeniously found the right one to enhance the understanding
of evolutionary processes on the level of individuals and species.

John Holland [23] has not become tired to emphasize this side of GAs, and
Hans-J. Bremermann [24] as well as Lawrence Fogel [25] had introduced their
evolutionary adaptation / optimization models sometimes explicitly as *simulated evolution*. Looking onto contemporary incarnations of evolutionary optimization tools, one must admit that this aspect has more or less disappeared.
Many features have been added to the basic operators of variation and selection
that do not have analoga in nature, for example long term archives and tedious
arithmetic operations. The question may be raised, whether more sophisticated
models of organic evolution cannot be both successful optimizers and helpful
in understanding natural phenomena. Some examples of such models adding
features found in nature to basic EAs are:

4.1   Diploid Genomes

In a first attempt to create an evolution strategy for vector optimization with
conflicting objectives, individuals were equipped with a diploid genome. This
helped in keeping up the necessary diversity in the population throughout the
whole search. Intermediate best solutions were gathered in an external archive
that finally presented the sought for answer to the problem of finding nondominated positions in the search space and the Pareto front in the space of
objectives [26].

4.2   Multicellular Individuals

In binary optimization one may attach individual mutation rates to all bits and
represent them in the individuals' genomes as well as the bits themselves. Mul-
ticellular individuals may then be conceived as being composed of clusters of
cells or tissues that develop from one genetically coded stem cell via several cell
duplications. Somatic mutations - according to the also genetically coded mutation rates - lead to tissues containing a mix of cells of both types, e.g. black and
white, with the result of a lighter or darker grey tissue. The consequence of this
setting is, that for those bits which need to be inverted during the optimization
process the mutation rates increase until the corresponding bit flips, and thereafter rapidly decrease again so that the genotype of object bits stabilizes in its
proper state. Object bits that need not be altered receive mutation rates rapidly
falling below a dangerous level if necessary [27].

4.3 Gene Duplication and Gene Deletion

A variable length genome had already been used for an early application of the simplest evolution strategy, a (1+1)-ES, for the experimental optimization of a supersonic one-component two-phase nozzle, the length of which could thus be conceived to be variable besides of the diameters at different positions [28]. More recently, gene deletion and gene duplication has shown up to be important not only at the level of the decision or object variables, but even more so on the level of so-called strategy variables, i.e., the number of active search directions. Not changing all the object variables at a time, but a varying subset only, has turned out to be advantageous in highly multi-modal optimization tasks, where one likes to end up at the global optimum or at least at a very good local one [29].

4.4 Gender Dimorphism

Constrained optimization problems often have solutions with one or more active constraints at the optimum, i.e., in narrow corners of the feasible region, so that accessing the goal with precision turns out to be very difficult if one does not accept infeasible search points in the interim. To overcome this difficulty, one idea has been to split the population into two 'sexes', which underlie different selection procedures, so that one of them can approach the goal from the subspace of infeasible solutions, whereas the other is forced to maintain feasibility. Intermediary recombination of parents in pairs of both sexes are expected to lead to better approximations of the problem than their ancestors. Though some open questions remain, the idea seems to work, in general [30].

4.5 Predators and Prey

There are good reasons for leaving the traditional scheme of consecutive generations with sequential phases for mating selection, recombination, mutation, and environmental selection. A predator-prey approach to evolutionary optimization is not only more natural, but also serves two goals as pinpointed below.

In the following, we concentrate on the latter way of adding realism to $(\mu, \kappa, \lambda, \rho)$ evolution strategies (ESs) with mutative self-adaptation of standard deviations for gaussian mutations of real-valued decision variables [31]. There are μ parents in pairs of $1 \leq \rho \leq \mu$ - most often $\rho = 2$ - that produce $\lambda > \mu$ offspring by means of recombination and subsequent mutation. Originally, μ out of the whole set of μ parent and λ offspring individuals are selected according to their age, measured in reproduction cycles, and their fitness. This has been a generalization of the elitist $(\mu + \lambda)$-ES version, in which parents could survive forever, if no offspring with better fitness were found, and the (μ, λ)-ES, where parents are not allowed to produce grand-children, i.e., they live for just one generation or reproduction cycle only. Now, $1 \leq \kappa \leq \infty$ is the maximal number of cycles an individual can participate in the game if no offspring outperform it earlier already. The extreme cases $\kappa = 1$ and $\kappa = \infty$ resemble the plus- and comma-ES versions, respectively, a $(\mu+1)$-ES usually being called a steady state procedure. Small values of κ are helpful in rapidly changing environments. Even more important for all $(\mu, \kappa, \lambda, \rho)$ versions are genetically represented parameters, perhaps like introns, that do not influence the phenotypes, but control the variation process, mainly the mutation strength, either one common or even individual ones for each object variable.

Such generational scheme has a big disadvantage in case of fitness tests (objectives and constraints) that require vastly different computational effort, which is often the case when simulation tools are used to evaluate the feasibility and the quality of the tested design parameter settings. The generation cannot be completed before the last offspring has finished its simulation phase. Especially parallelized versions of this strategy suffer from idling processors under such conditions.

Trying to escape from wasting computer power leads to look for some kind of asynchronous procedure and thus to think of breaking the traditional generational cycle. Offspring production and selection could take place simultaneously on different processors as soon as these have finished their preceding task. In nature, birth and death processes also take place at the same time, but at different places. Just think of predators hunting prey at some place and (other) prey giving birth to children at some other place.

Immediately, this scheme can be expanded with respect to the number of different types of predators, each type challenging the prey according to its specific criterion. In this sense, diseases and environmental conditions can be thought of as some kind of predators, as well.

4.5.1 First Predator-Prey Approach Toward MOO
A first step into the direction of predator and prey ES was taken already ten years ago [32], with prey individuals inhabiting a toroidal grid and predators performing random walks on that grid. But, since each act of killing (eating) a prey individual by the predator was followed by filling the emptied grid place by means of creating an offspring from neighboring survivors, this strategy resembled a steady-state EA and, consequently, disabled the self-adaptation of mutation strengths from working properly, because reductions of step sizes were always rewarded via inevitably higher survival probabilities (which go up to 50% with vanishing standard deviations of normally distributed mutations on continuously differentiable response surfaces).

4.5.2 Further Improvements
The latter phenomenon can be avoided through alternating phases of partly emptying and subsequently refilling grid places and using several predators of each type as selectors [33]. This does not mean to return to the generational scheme, because different phases can happen at the same time but at different locations. Since the population underlies pressures toward the set of non-dominated regions of the search space, it was expected to finally assemble on the Pareto front in the space of objectives. Without an archive of so far best solutions the population size has to be large enough to present sufficiently dense points on that front, of course. No tests are necessary to predict that panmictic recombination is

prohibitive for the outcome wanted. Besides an uneven spread of the individuals on the Pareto front, another danger is premature stagnation before reaching the front with sufficient accuracy. The latter can be more or less avoided by means of sufficient birth surplus and finite life span of the prey. Observations during this kind of artificial evolution show a rather good spread shortly before reaching the Pareto front with sufficient precision, but thereafter the prey form clusters closer to the front and with gaps in between.

4.5.3   Open Questions

Many kinds of tricks to enhance the approximation quality and the uniform distribution at the same time have been proposed and tested already, see e.g. [34], but at lot of open questions still remain for currently preferred algorithms. The same holds for predator-prey approaches [35].One wants to know, for example:

1. Should the predators all roam on the whole grid, or should they own particular hunting grounds - possibly modeled by means of a gaussian hunting frequency around a fixed home place? In the latter case one could expect that the prey specialize and concentrate on subsets of the Pareto set, which would nicely fit into the well-known allopatric speciation scheme observed in nature. Instead of a simple population-based search, this would resemble a multi-species approach.
2. On what kind of grid should the prey reside, and should they also move around or not? A toroidal grid is not necessarily the best choice, because selfish ends of the Pareto front may need more isolated species. The best way to design the prey's habitat may finally depend on the number of objectives, too.
3. Should recombination be confined to mating in the neighborhood only or at least mainly, but with less frequent exemptions?

It will be fascinating to see forthcoming attempts of inventing bio-inspired *and* effective versions of multi-objective evolutionary algorithms.

5   Conclusions

The ideas presented above are obviously half baked and not yet proven to work as expected, neither theoretically nor by exhaustive tests. They are presented here to enhance the broadness of future efforts, since even the most advanced versions of multi-criteria EAs are still far from being perfect [34]. Many features of these strategies are superficial and not gleaned from nature. That would not matter from the perspective of optimization alone, but from the other perspective of understanding organic evolution by means of properly modeling natural processes. Especially situations with multiple selection criteria might enhance mutual understanding among biologists and algorithm engineers. One old misunderstanding resulted from ignoring that vector optimization with a whole set of solutions instead of just one single outcome is the general case both in technical as well as economical settings - and in nature. Another unnecessary controversy sometimes arose from seeing a fundamental difference between adaptation and optimization. This is unnecessary because a change from well adapted to better adapted is amelioration or a step toward optimization, of course under the given circumstances and perhaps only for a while until environmental conditions change. We hope to have given some stimulation toward a better mutual understanding and further fruitful discussions.

References

1. Coello Coello, C.A.: Evolutionary multiobjective optimization: A historical view of the field. IEEE Computational Intelligence Magazine 1, 28–36 (2006)
2. Rudolph, G.: Evolutionary Search for Minimal Elements in Partially Ordered Finite Sets. In: Porto, V., Saravanan, N., Waagen, D., Eiben, A. (eds.) Evolutionary Programming VII, Proceedings of the 7th Annual Conference on Evolutionary Programming, pp. 345–353. Springer, Berlin (1998)
3. Rudolph, G.: On a Multi-Objective Evolutionary Algorithm and Its Convergence to the Pareto Set. In: Proceedings of the 5th IEEE Conference on Evolutionary Computation, pp. 511–516. IEEE Press, Piscataway, New Jersey (1998)
4. Serafini, P.: Simulated Annealing for Multiple Objective Optimization Problems. In: Tzeng, G., Wang, H., Wen, U., Yu, P. (eds.) Proceedings of the Tenth International Conference on Multiple Criteria Decision Making: Expand and Enrich the Domains of Thinking and Application, vol. 1, pp. 283–294. Springer, Berlin (1994)
5. Engrand, P.: A multi-objective optimization approach based on simulated annealing and its application to nuclear fuel management. In: Proceedings of the Fifth International Conference on Nuclear Engineering, Nice, France, American Society of Mechanical Engineering, pp. 416–423 (1997)
6. Knowles, J.D., Corne, D.W.: Approximating the Nondominated Front Using the Pareto Archived Evolution Strategy. Evolutionary Computation 8, 149–172 (2000)
7. Schaffer, J.D.: Multiple Objective Optimization with Vector Evaluated Genetic Algorithms. PhD thesis, Vanderbilt University (1984)
8. Kursawe, F.: A Variant of Evolution Strategies for Vector Optimization. In: Schwefel, H.-P., Männer, R. (eds.) PPSN 1990. LNCS, vol. 496, pp. 193–197. Springer, Heidelberg (1991)
9. Hajela, P., Lin, C.Y.: Genetic search strategies in multicriterion optimal design. Structural Optimization 4, 99–107 (1992)
10. Fonseca, C.M., Fleming, P.J.: Genetic Algorithms for Multiobjective Optimization: Formulation, Discussion and Generalization. In Forrest, S., ed.: Proceedings of the Fifth International Conference on Genetic Algorithms, San Mateo, California, University of Illinois at Urbana-Champaign, Morgan Kauffman Publishers (1993) 416–423
11. Horn, J., Nafpliotis, N., Goldberg, D.E.: A Niched Pareto Genetic Algorithm for Multiobjective Optimization. In: Proceedings of the First IEEE Conference on Evolutionary Computation, IEEE World Congress on Computational Intelligence, vol. 1, pp. 82–87. IEEE Computer Society Press, Piscataway, New Jersey (1994)
12. Srinivas, N., Deb, K.: Multiobjective Optimization Using Nondominated Sorting in Genetic Algorithms. Evolutionary Computation 2, 221–248 (1994)
13. Deb, K., Agrawal, S., Pratab, A., Meyarivan, T.: A Fast Elitist Non-Dominated Sorting Genetic Algorithm for Multi-Objective Optimization: NSGA-II. In: Deb, K., Rudolph, G., Lutton, E., Merelo, J.J., Schoenauer, M., Schwefel, H.-P., Yao, X. (eds.) PPSN 2000. LNCS, vol. 1917, pp. 849–858. Springer, Heidelberg (2000)

14. Goldberg, D.E.: Genetic Algorithms in Search, Optimization and Machine Learning. Addison-Wesley Publishing Company, Reading (1989)
15. Zitzler, E., Laumanns, M., Thiele, L.: SPEA2: Improving the Strength Pareto Evolutionary Algorithm. In: Giannakoglou, K., Tsahalis, D., Periaux, J., Papailou, P., Fogarty, T. (eds.) EUROGEN 2001. Evolutionary Methods for Design, Optimization and Control with Applications to Industrial Problems, Athens, Greece, pp. 95–100 (2002)
16. Wagner, T., Beume, N., Naujoks, B.: Pareto-, Aggregation-, and Indicator-Based Methods in Many-Objective Optimization. In: Obayashi, S., Deb, K., Poloni, C., Hiroyasu, T., Murata, T. (eds.) EMO 2007. LNCS, vol. 4403, pp. 742–756. Springer, Heidelberg (2007)
17. Zitzler, E., Künzli, S.: Indicator-Based Selection in Multiobjective Search. In: Yao, X., Burke, E.K., Lozano, J.A., Smith, J., Merelo-Guervós, J.J., Bullinaria, J.A., Rowe, J.E., Tiňo, P., Kabán, A., Schwefel, H.-P. (eds.) PPSN 2004. LNCS, vol. 3242, pp. 832–842. Springer, Heidelberg (2004)
18. Emmerich, M., Beume, N., Naujoks, B.: An EMO Algorithm Using the Hypervolume Measure as Selection Criterion. In: Coello Coello, C.A., Hernández Aguirre, A., Zitzler, E. (eds.) EMO 2005. LNCS, vol. 3410, pp. 62–76. Springer, Heidelberg (2005)
19. Beume, N., Naujoks, B., Emmerich, M.: SMS-EMOA: Multiobjective selection based on dominated hypervolume. European Journal of Operational Research 181, 1653–1669 (2007)
20. Emmerich, M., Deutz, A., Beume, N.: Gradient-Based/Evolutionary Relay Hybrid for Computing Pareto Front Approximations Maximizing the S-Metric. In: Bartz-Beielstein, T., Blesa Aguilera, M.J., Blum, C., Naujoks, B., Roli, A., Rudolph, G., Sampels, M. (eds.) HM 2007. LNCS, vol. 4771, pp. 140–156. Springer, Heidelberg (2007)
21. Igel, C., Hansen, N., Roth, S.: Covariance Matrix Adaptation for Multi-objective Optimization. Evolutionary Computation 15, 1–28 (2007)
22. Voß, T., Beume, N., Rudolph, G., Igel, C.: Scalarization versus indicator-based selection in multi-objective CMA evolution strategies. In: Proceedings of the IEEE Conference on Evolutionary Computation (CEC 2008), IEEE Service Center, Piscataway, New Jersey (in press, 2008)
23. Holland, J.: Outline for a logical theory of adaptive systems. Journal of the Association for Computing Machinery NN, 297–314 (1962)
24. Bremermann, H.: Optimization through evolution and recombination. In: Yovits, M., Jacobi, G., Goldstein, G. (eds.) Self-Organizing Systems 1962. Proceedings of the Conference on Self-Organizing Systems, Chicago, Illinois, 22.-24.5 1962, pp. 93–106 (1962)
25. Fogel, L., Owens, A., Walsh, M.: Artificial intelligence through a simulation of evolution. In: Maxfield, M., Callahan, A., Fogel, L. (eds.) Biophysics and Cybernetic Systems, pp. 131–155 (1965)
26. Bäck, T., Hoffmeister, F., Kursawe, F., Rudolph, G., Schwefel, H.P.: Four contributions to the development of evolution strategies. Technical Report "Grüne Reihe" Nr. 368, University of Dortmund, Department of Computer Science (1990)
27. Schwefel, H.P., Kursawe, F.: On natural life's tricks to survive and evolve. In: Fogel, D.B., Schwefel, H.P., Bäck, T., Yao, X. (eds.) Proc. Fifth IEEE Conf. Evolutionary Computation (ICEC 1998), Anchorage AK, pp. 1–8. IEEE Press, Piscataway NJ (1998)

28. Klockgether, J., Schwefel, H.P.: Two-phase nozzle and hollow core jet experiments. In: Elliott, D.G. (ed.) Proc. Eleventh Symp. Engineering Aspects of Magnetohydrodynamics, Pasadena CA, California Institute of Technology, pp. 141–148 (1970)
29. Schmitt, K.: Using gene deletion and gene duplication in evolution strategies. In: Beyer, H.-G., et al. (eds.) Proc. Genetic and Evolutionary Computation Conf (GECCO 2005), Washington D.C, vol. 1(poster), pp. 919–920. ACM Press, New York (2005)
30. Kramer, O., Schwefel, H.P.: On three new approaches to handle constraints within evolution strategies. Natural Computing 5, 363–385 (2006)
31. Schwefel, H.P., Rudolph, G.: Contemporary evolution strategies. In: Morán, F., Merelo, J.J., Moreno, A., Chacon, P. (eds.) ECAL 1995. LNCS, vol. 929, pp. 893–907. Springer, Heidelberg (1995)
32. Laumanns, M., Rudolph, G., Schwefel, H.P.: A spatial predator-prey approach to multi-objective optimization. In: Eiben, A.E., Bäck, T., Schoenauer, M., Schwefel, H.-P. (eds.) PPSN 1998. LNCS, vol. 1498, pp. 241–249. Springer, Heidelberg (1998)
33. Schmitt, K., Mehnen, J., Michelitsch, T.: Using predators and preys in evolution strategies. In: Beyer, H.-G., et al. (eds.) Proc. Genetic and Evolutionary Computation Conf (GECCO 2005), Washington D.C, vol. 1(poster), pp. 827–828. ACM Press, New York (2005)
34. Purshouse, R.C., Fleming, P.J.: On the evolutionary optimization of many conflicting objectives. IEEE Trans. Evolutionary Computation 11, 770–784 (2007)
35. Grimme, C., Lepping, J.: Designing Multi-objective Variation Operators Using a Predator-Prey Approach. In: Obayashi, S., Deb, K., Poloni, C., Hiroyasu, T., Murata, T. (eds.) EMO 2007. LNCS, vol. 4403, pp. 21–35. Springer, Heidelberg (2007)

Handling Uncertainties in Evolutionary Multi-Objective Optimization

Kay Chen Tan[1] and Chi Keong Goh[2]

[1] National University of Singapore, 4 Engineering Drive 3, Singapore 117576
eletankc@nus.edu.sg
[2] Data Storage Institute, Agency for Science
Technology and Research, 5 Engineering Drive 1, Singapore 117608
Goh_chi_keong@dsi.a-star.edu.sg

Abstract. Evolutionary algorithms are stochastic search methods that are efficient and effective for solving sophisticated multi-objective (MO) problems. Advances made in the field of evolutionary multi-objective optimization (EMO) are the results of two decades worth of intense research, studying various topics that are unique to MO optimization. However many of these studies assume that the problem is deterministic and static, and the EMO performance generally deteriorates in the presence of uncertainties. In certain situations, the solutions found may not even be implementable in practice. In this chapter, the challenges faced in handling three different forms of uncertainties in EMO will be discussed, including 1) noisy objective functions, 2) dynamic MO fitness landscape, and 3) robust MO optimization. Specifically, the impact of these uncertainties on MO optimization will be described and the approaches/modifications to basic algorithm design for better and robust EMO performance will be presented.

1 Introduction

Multi-objective optimization is a challenging research topic because it involves the simultaneous consideration of several complex objectives in the Pareto optimal sense and requires researchers to address many issues that are unique to MO problems. Multi-objective evolutionary algorithms (MOEAs) are a class of biologically-inspired optimization techniques that have been shown to be very effective in solving multi-objective problems. The advances made in the field of MOEA are the result of two decades of intense research examining topics such as fitness assignment [23,50], diversity preservation [44], balance between exploration and exploitation [6], and elitism [45] in the context of MO optimization. However, as we start to apply evolutionary multi-objective optimization (EMOO) to real-world problems [16,67], it is necessary to consider practical challenges beyond those encountered in existing test functions.

One of the challenges faced in most optimization problems is the presence of uncertainties. Interestingly, many researchers assume that the optimization problems are deterministic, and studies investigating the issues of uncertainties are

J.M. Zurada et al. (Eds.): WCCI 2008 Plenary/Invited Lectures, LNCS 5050, pp. 262–292, 2008.

only gaining attention from the EMOO community recently. The optimization of solutions in uncertain environments is of particular importance in real-world problems, such as scheduling, vehicle routing and engineering design optimization, where certain characteristics of the environment may not be known with absolute certainty, problem characteristics may be changing or the decision space can be sensitive to parametric variations. In this chapter, we consider three different forms of uncertainties, depending on how they affect the optimization process [40,54]: 1) noise, 2) robustness, and 3) dynamic fitness functions. Specifically, the impact of these uncertainties on MO optimization will be described and the approaches/modifications to basic algorithm design for better and robust EMOO performance will be presented.

The organization of the chapter is as follows: Section 2 provides some background information on the concepts of multi-objective optimization. The rest of this work is divided into three parts, with each part considering a different form of uncertainties. Section 3 includes an overview of existing evolutionary techniques for noisy multi-objective optimization and describes a heuristical approach from our previous work. The design issues of MOEAs in dynamic environments and a number of dynamic MOEAs are presented in Section 4. Simulation results demonstrating the effectiveness of the competitive-cooperative paradigm for tracking the dynamic solution set is also presented. Section 5 begins with a discussion of the ideal properties of robust MO test functions followed by the description of a robust multi-objective test suite. An overview of a number of existing approaches for robust optimization is also provided. Subsequently, a hybrid multi-objective evolutionary algorithm which is capable of finding robust routes for the vehicle routing problem with stochastic demands is presented. Conclusions are drawn in Section 6.

2   Background Information

Many real-world tasks involves the simultaneous optimization of several competing specifications, and instances of such multi-objective problems can be found in diverse fields ranging from engineering to economics. These problems are typically represented by its mathematical model and the specification of multi-objective criteria captures more information about the modeled problem as several problem characteristics are taken into consideration. Without any loss of generality, a minimization problem is considered in this chapter and the multi-objective problem can be formally defined as

$$\min_{\boldsymbol{x} \in \boldsymbol{X}^{n_x}} \boldsymbol{f}(\boldsymbol{x}) = \{f_1(\boldsymbol{x}), f_2(\boldsymbol{x}), ..., f_M(\boldsymbol{x})\} \tag{1}$$

$$\text{s.t. } \boldsymbol{g}(\boldsymbol{x}) \geq 0, \boldsymbol{h}(\boldsymbol{x}) = 0$$

where $\boldsymbol{x}$ is the vector of decision variables bounded by the decision space, $\boldsymbol{X}^{n_x}$ and $\boldsymbol{f}$ is the set of objectives to be minimized. The terms "solution space" and "search space" are often used to denote the decision space and will be used interchangeably throughout this work. The functions $\boldsymbol{g}$ and $\boldsymbol{h}$ represents the set of inequality and equality constraints that defines the feasible region of the n_x-dimensional continuous or discrete feasible solution space.

One of the key differences between single-objective and multi-objective optimization is that multi-objective problems constitute a multi-dimensional objective space, $\boldsymbol{F}^M$. The conventional notion of optimality is no longer applicable and the concepts of Pareto dominance and Pareto optimality are fundamental in multi-objective optimization. There are three possible relationship between the solutions that are defined by Pareto dominance.

Definition 1: Weak Dominance: $\boldsymbol{f_1} \in \boldsymbol{F}^M$ weakly dominates $\boldsymbol{f_2} \in \boldsymbol{F}^M$, denoted by $\boldsymbol{f_1} \preceq \boldsymbol{f_2}$ iff $x_{1,i} \leq x_{2,i}$ $\forall i \in \{1, 2, ..., M\}$

Definition 2: Strong Dominance: $\boldsymbol{f_1} \in \boldsymbol{F}^M$ strongly dominates $\boldsymbol{f_2} \in \boldsymbol{F}^M$, denoted by $\boldsymbol{f_1} \prec \boldsymbol{f_2}$ iff $x_{1,i} \leq x_{2,i}$ $\forall i \in \{1, 2, ..., M\}$ and $x_{1,j} < x_{2,j}$ $\exists j \in \{1, 2, ..., M\}$

Definition 3: Incomparable: $\boldsymbol{f_1} \in \boldsymbol{F}^M$ is incomparable with $\boldsymbol{f_2} \in \boldsymbol{F}^M$, denoted by $\boldsymbol{f_1} \sim \boldsymbol{f_2}$ iff $x_{1,i} > x_{2,i}$ $\exists i \in \{1, 2, ..., M\}$ and $x_{1,j} < x_{2,j}$ $\exists j \in \{1, 2, ..., M\}$

3   Noisy MO Optimization

Noise stems from several sources, including sensor measurement errors, incomplete simulations of computational models and stochastic simulations. Apart from these external sources, noise can also be intrinsic to the problem. A good example is the evolution of neural networks where the same network structure can give rise to different fitness values due to different weight instantiations [40,30]. A distinctive feature of noisy fitness function is that each evaluation of the same solution results in different objective values. The noisy multi-objective optimization can be written as

$$\min_{\boldsymbol{x} \in \boldsymbol{X}^{n_x}} \boldsymbol{F}(\boldsymbol{x}) = \{f_1(\boldsymbol{x}) + \delta_1, f_2(\boldsymbol{x}) + \delta_2, ..., f_M(\boldsymbol{x}) + \delta_M\} \tag{2}$$

where δ_i is a scalar noise parameter added to the original objective function of f_i and $\boldsymbol{F}$ is the resultant objective vector. The optimization of noisy problems is greatly influenced by the noise model adopted and the level of noise intensity. Most studies of evolutionary optimization in single-objective noisy environments [1,3,7,8,12,26,53,58,61,62] are done on the basis of Gaussian noise. The investigation conducted by Arnold and Beyer [2] revealed significant differences between the influence of Gaussian, Cauchy and χ^2 distributed noise on the performance of $(\mu/\mu, \lambda)$-ES. In the context of multi-objective optimization, Teich [68] considers a uniform noise model while Buche et al [14] incorporates the effects of outliers on the optimization process.

Extensive studies have been performed in [27] to examine the impact of noisy environments in evolutionary multi-objective optimization, particularly on the population dynamics of fitness and diversity. It has been observed that the impact of noise on MOEA is different for the various benchmark problems, i.e., MOEA tends to evolve better solutions for some of the problems in the presence

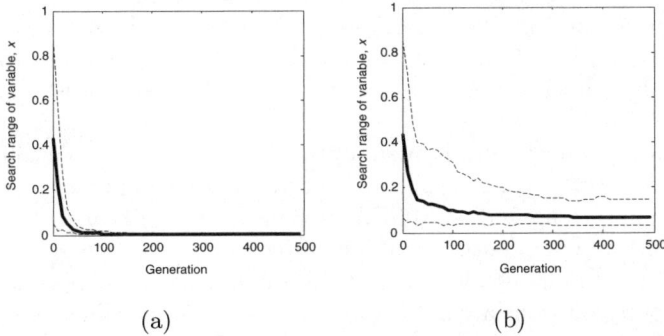

Fig. 1. Search range of an arbitrary decision variable for ZDT1 at (a) 0% and (b) 20% noise. The thick line denotes the trace of the population mean along an arbitrary decision variable space, while the dashed line represents the bounds of the decision variable search range along the evolution.

of low-level noise, while the evolutionary optimization process degenerates into a random search under increasing level of noise. Based on the observation of the decision errors made over time, it seems that the selection process is more reliable in the early stage of evolution. Furthermore, Fig. 1 shows that the evolution defines a population distribution with a mean value that remains relatively invariant in the decision space despite the different environmental conditions.

3.1   Handling Noisy Multi-Objective Optimization

Based on the literature, three basic approaches have been proposed to suppress the detrimental effects of noise [40], 1) explicit averaging, 2) implicit averaging, and 3) selection modification. In explicit averaging, each solution is evaluated a number of times and averaged to compute the expected objective values. Increasing the number of samples (H) reduces the degree uncertainty by a factor of $\sqrt{H}$ at the expense of higher computational cost. Intead of re-evaluating and averaging the objective values over a number of samples, a large population is used in implicit averaging. When population size is large, there are many similar solutions and the influence of noise is compensated as the algorithm revisits the same region repeatedly. In selection modification, the ranking and selection procedures are modified such that a solution is judged better than another solution only if it satisfies certain conditions.

The ideas behind these three approaches can be easily applied to multi-objective optimization to suppress the effects of noise. But no matter which approach is adopted, it is necessary to consider how noise affects the selection, elitism and diversity presevation processes in MOEA. There are two ways in which an inferior solution can be chosen over a better one in the selection process. Firstly, the selection mechanism can perceive an inferior solution to dominate a superior solution under the influence of noise. Secondly, as long as the inferior solution appears to be nondominated, it can be selected by virtue of a better perceived degree of diversity measure. Similarly, the archive can be deceived into storing inferior solutions. Not only can these archived solutions drive out superior solutions, they can also prevent good solutions from entering the archive. Recall that most state-of-the-art MOEAs are elitist in nature. The worst problem is that these archived inferior solutions can mislead the entire optimization process, resulting in suboptimal or unrealistic solution sets.

The straight forward approach is to remove elitism completely. However, there is no indication that non-elitist MOEAs will perform significantly better for noisy problems, and we can identify two difficulties in the absence of elitism. Firstly, good solutions will be lost along the evolution and we can expect slower convergence since elite solutions are not exploited. Furthermore, when noise intensity is sufficiently high, what will happen is that the MOEA will favor a particular set of solution in one generation while favoring another set of solutions in another generation. This will result in the algorithm oscillating between different regions in the search space without ever converging.

With noise affecting how the solutions are perceived, it is highly probable that the selection pressure is directed towards inferior solutions while the truly good solutions are lost quickly along the evolution. Babbar et al [4] incorporates the explicit averaging approach into the nondominated sorting genetic algorithm II (NSGAII) and modified the nondominated sorting procedure to allow seemingly dominated solutions into the first nondominated front. This modified NSGAII also incorporates a procedure to remove unreliable solutions from the final set of nondominated solutions. Although explicit averaging can reduce the uncertainty of the selection process, it is not feasible to use a large number of samples. Therefore it is highly likely that the first nondominated front will comprise of both dominated and nondominated solutions. To prevent the loss of potentially useful solutions, Babbar et al suggested a clustering mechanism to induce solutions from the inferior nondominated fronts into the first layer. This mechanism works by comparing the distance between solutions from the first nondominated front and the other fronts, and re-assigning a higher ranked solution to the first nondominated front if the perceived inferior solution is located in close proximity to a perceived nondominated solution. At the end of the evolutionary process, the clustering mechanism is applied once again. However, in this instance, it is applied to remove solutions that are significantly different from the other archived solutions.

Teich [68] proposed the estimate strength Pareto evolutionary algorithm (ESPEA), which is extended from the strength Pareto evolutionary algorithm (SPEA) [79]. The general algorithmic stucture of SPEA is retained in ESPEA and the main difference is that each noisy objective value is described by a property interval. A probablity of dominance is defined such that a solution $\boldsymbol{F}_a$ dominates solution $\boldsymbol{F}_b$ when the worst case $\boldsymbol{F}_a$ is better than the best case $\boldsymbol{F}_b$. On the other hand, $\boldsymbol{F}_a$ does not dominate $\boldsymbol{F}_b$ as long as the upper bound $f_{b,i}^U$ is smaller than the lower bound $f_{a,i}^L$ for any one objective. When there is an overlap between the property intervals of both solutions overlap, $\boldsymbol{F}_a \preceq \boldsymbol{F}_b$

with a certain degree of probability. The probability of dominance allows dominated solutions in the deterministic scheme a chance to enter the archive. The idea is to ensure that all solutions with a decent probability of being nondominated are copied into the archive while solutions with high probability of being dominated are removed from the archive.

Hughes [36,37] suggested a multi-objective probabilistic selection evolutionary algorithm (MOPSEA) which also employs probabilistic dominance to account for noise. The probability of dominance introduced by Hughes is different from that implemented by Teich in the sense that noise is assumed to be normally distributed. However, the standard deviation of noise, σ_n, is typically not known *a priori* and it can only be estimated during the optimization process. If computational resources and time permits, σ_n can be estimated for each and every solution by evaluating the solution over a few samples. But this is not the case for many real-world applications. If variance is assumed to be constant throughout the search space, Hughes suggested estimating σ_n from a random solution at the start of the optimization process and using it for all subsequent comparisons. Fieldsend and Everson [24] considered the efficient computation of probabilistic ranking and suggested an online Bayesian learning algorithm to estimate the noise variance σ_n. More significantly, it is demonstrated that the algorithm is capable of tracking noise variance under different conditions such as unknown noise properties, independent noise for each objectives and etc. MOPSEA maintains only one evolving population on which genetic operations are performed to generate new solutions. Elitism is implemented by retaining a significant portion of the best solutions. Nonetheless, Hughes suggested (but did not simulate) an unique approach of using a probabilistic tournament selection to update an archive in [38]. The basic algorithmic structure of MOPSEA is kept simple, with stochastic universal sampling, intermediate crossover and uniformly distributed mutation operators, probably to highlight the contributions of probabilistic dominance in noisy MO optimization.

To reduce the influence of inferior archived solutions on the optimization process, Buche *et al* [14] modified the archive updating mechanism of the SPEA [79]. Buche *et al* particularly concerned about the influence of outliers, which are the result of instrumentation failure and it can be of several magnitudes larger than measurement noise. Under the influence of an outlier, an inferior solution may be perceived to be unrealistically good, driving out the true nondominated solutions and misleading the optimization process. To this end, the NTSPEA incorporates the notion of a lifetime to all archived solutions so the impact of inferior solutions on the optimization process will be limited. The lifetime of each nondominated individual is dependent on the fraction of archived solutions it dominates. At every generation, nondominated solutions with expired lifetime are removed from the archive, added to the evolving population and re-evaluated. After the re-evaluation, it is unlikely that an outlier will remain nondominated and, hence, it will not be updated into the archive again. On the other hand, a good solution is likely to remain nondominated.

Limbourg [48] extended a probabilistic theory known as the Dempster-Shafer framework of evidence to model epistemic uncertainties, and suggested a set of guidelines for the design of selection and archive updating mechanisms in noisy multi-objective optimization. Associated with this probabilistic framework is the notion of belief and plausibility, which are used to characterize the lower and upper bounds of the k-nearest neighbour distances calculated during density estimation. A weak uncertain dominance relation is used in the selection process. Although probability of making the wrong decision increases with the degree of overlap between the objective intervals, this relation has a low degree of indifference. The rationale of making the accuracy tradeoff is that a high degree of indifference in the comparison of solutions will result in a random search. On the other hand, in the archiving process, all solutions that are not dominated in the strong certain sense are added to the archive. Note that such an updating scheme allows both perceived dominated and nondominated solutions exist in the archive simulataneously and we can expect the archive to be filled up quickly with solutions. When the size of the archive reaches the limit, truncation is performed to keep the archive size under control. As in the case of MOEAs for deterministic optimization, only solutions located in the least crowded regions are kept. Density assessment is estimated using the $k\text{-}th$ neighbour distance according to the values of belief and plausibility. After calculating the belief and plausibility pairs for all the solutions in the archive, binary tournament selection is conducted to remove the more crowded solutions. During the tournament selection, two solutions are chosen at random and their $k\text{-}th$ neighbour distance intervals are compared using the uncertain dominance criterion. The winner of the tournament remains in the archive while the loser is removed. In the event that the belief and plausibility pair of the two solutions are indifferent, the survivor is selected randomly.

Basseur and Zitler [5] considered how noise can be handled in indicator-based evolutionary algorithms (IBEA). In this particular implementation of IBEA, it is a steady-state algorithm and the ϵ_+-indicator is used to guide the evolutionary process and the expected indicator values are used as fitness values to improve robustness. In IBEA, each solution is associated with a probability distribution over the objective space and the expected indicator value is calculated based on the average of indicator values with respect to the set of objective vectors. One drawback is that the calculation of expected $I_\epsilon+$ for each solution is computational expensive. To minimmize the computation cost, the authors exploited that fact that the minimun $I_\epsilon+$ value determines the actual value, and hence, not all combinations of objective vectors have to be considered. They suggested using a bucket sort to reduce computational complexity of the sorting performed.

3.2   Multi-Objective Evolutionary Algorithm with Robust Features

We suggested three noise-handling features [27] and incorporated them into a simple MOEA, which we call MOEA-RF. The three features include two heuristics, namely the experiential learning directed perturbation (ELDP) which exploits better decision-making at the early stage of evolution and the gene

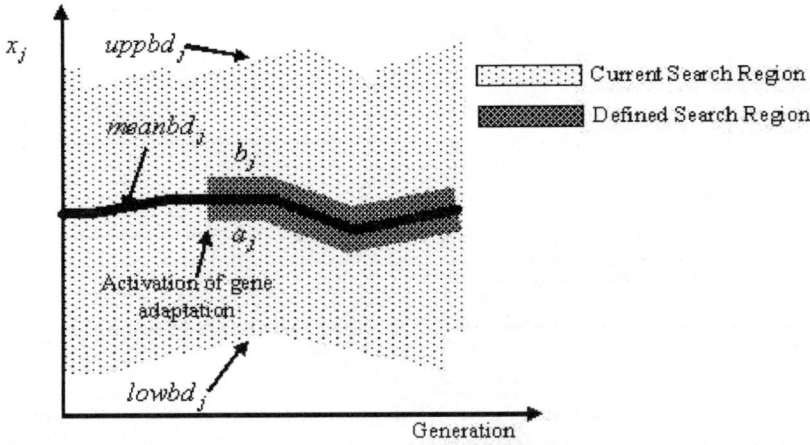

Fig. 2. Search range defined by divergence model

adaptation selection strategy (GASS) which exploits the observation that the mean location of archived individuals remains relatively constant under different noise conditions. The third feature is the possibilistic archiving methodology which helps to perform effective archiving of noisy objective vectors.

In ELDP, changes in chromosome are ordered and variation can be performed either in genotype or phenotype space. The actual adaptation in ELDP is based on posterior knowledge of favorable movements in the search space and it is inspired by the role of momentum term in back-propagation for neural networks; accelerating movement in the direction of improvement while restricting movement otherwise. The ELDP defines a two-mode operation to impose the necessary control for directed variation in the phenotype space and to perform bit-flip mutation in the genotype space for genetic diversity. ELDP operates as a bit-flip mutation, either when the magnitude of directed perturbation is too large or too small, ensuring that a new search direction is initiated through bit-flip mutation to reduce the impact of outliers or whenever the evolutionary search process has stalled.

The GASS attempts to manipulate population distribution so that the evolutionary algorithm exhibits certain desirable search characteristics. It first builds a posterior model of the desired population distribution and then adapts part of the selected individual's chromosome. Under the influence of noise, the evolutionary process may degenerate into a random search which is characterized by a non-convergent population distribution. The aim of the model is to reduce stochastic change in gene structure due to random selection of individuals by providing a stable search range as shown in Fig. 2. Note that the interval specifying the location of new individuals is only a rough deduction of the search region based on the available information.

The possibilistic archive includes two archiving models namely, the necessity-possible (NP-) model and necessity (N-) model, which are based on the concept of

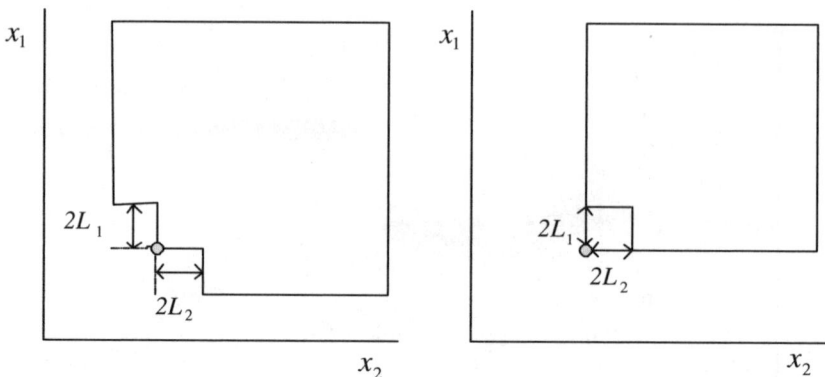

Fig. 3. Region of dominance based on (a) NP-dominance relation, and (b) N-dominance relation. The shaded region represents the area dominated by the individual marked by a circle.

possibilistic Pareto dominance relation. Fig. 3 illustrates the different dominance relation for a minimization problem. The NP-model behaves similarly to existing archiving models, but allows decision-maker to reject certain non-dominated individuals in the evolution if necessary. This archiving model compares and updates individuals according to the NP-dominance relation. As shown in Fig. 3(b), the width of the fuzzy membership function associated with the i-th objective is denoted by L_i, which represents the tolerance level of inferiority for each objective. As L_i tends to zero, the behavior of NP-dominance approaches that of Pareto-dominance relation. The pruning criterion is based upon some degree of crowding or niche count, which helps to maintain population diversity in the archive. A tagging system is implemented to allow both the archiving models to co-exist in the situation where the uncertainty level is low. Each individual is assigned either a NP-tag or N-tag that defines the behavior it will experience during the archiving process, e.g., an individual assigned with the NP-tag is regarded as if only the NP-model is implemented. The assignment of tags is based on a probability distribution. Intuitively, the probability of being assigned a N-tag increases with noise.

3.3   Simulation Results

In this section, a comparative study is conducted between MOEA-RF, NTSPEA, MOPSEA, SPEA2, NSGAII and PAES is presented. A basic MOEA incorporated with explicit averaging is also included in the study. Fig. 4 shows that the performance of the algorithms deteriorates with the increase of noise level; particularly there is a drastic performance change in PAES when the noise level is increased to 5%. Fig. 4(c) shows that the MOEA-RF, NTSPEA and MOPSEA are capable of evolving better solutions in a noisy environment as compared to algorithms without any noise compensation techniques. With the exception of

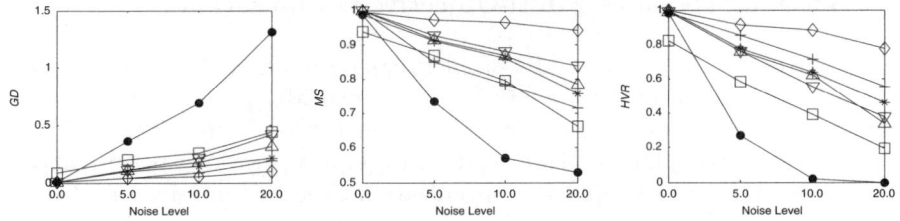

Fig. 4. Performance metric of (a) GD, (b) MS, and (c) HVR for ZDT1 attained by MOEA-RF ($\Diamond$), RMOEA ($\Box$), NTSPEA($|$), MOPSEA ($*$), SPEA2 (∇), NSGAII ($\triangle$) and PAES ($\bullet$) under the influence of different noise levels

RMOEA, most algorithms encountered no problem in converging and maintaining a diverse set of solutions for ZDT1 under noiseless environment. Among the conventional MOEAs, i.e., SPEA2, NSGAII and PAES, it is apparent that the PAES is worst affected by the noise. MOEA-RF, NTSPEA and MOPSEA produce competitive results since various features are included in these algorithms to deal with the noise. On the other hand, the performance of RMOEA is the worst among all algorithms except for PAES. As can be observed in Fig. 4, the MOEA-RF is capable of evolving a more diverse and uniformly distributed Pareto front for ZDT1 in the presence of noise as compared to other algorithms.

4   Dynamic Multi-Objective Optimization

Apart from noise, many real-world problems involve time-dependent components. Instances of such problems can be found in the areas of control, scheduling, vehicle routing, autonomous path planning, and economics, just to name a few. For such problems, it is unlikely that the optimal Pareto set and the Pareto front will remain invariant and the previous solution must be adapted to reflect the current requirements. Therefore, the optimization goal is not only to evolve a near optimal and diverse PF^A but also to track it as it changes with time.

The fitness topology of dynamic multi-objective problems may change but the objective value is deterministic at any one time. In this context, the term *static* is more appropriate than deterministic for denoting multi-objective problems without explicit consideration of its dynamism. For dynamic multi-objective problems, the PF^* and the PS^* is unlikely to remain invariant and the optimization algorithm must be capable of tracking the PS^* over time. The dynamic multi-objective problem can be described as

$$\min_{\boldsymbol{x} \in \boldsymbol{X}^{n_x}} \boldsymbol{F}(\boldsymbol{x}, t) = \{f_1(\boldsymbol{x}, t), f_2(\boldsymbol{x}, t), ..., f_M(\boldsymbol{x}, t)\} \tag{3}$$

where t is typically measured in terms of solution evaluations. The terms PF_t^* and PS_t^* refer to the desired Pareto front and solution set at time t while the set of tradeoffs and nondominated solutions evolved by the dynamic MOEA at time t will be termed as PF_t^A and PS_t^A respectively.

4.1   Handling Dynamic Multi-Objective Optimization

In a certain sense, the dynamic MO problem can be considered simply as the consecutive optimization of different time-constrained MO problems with varying complexities. On the other hand, one of the challenges of evolutionary dynamic optimization is to exploit past information to improve tracking performance. It is simply inefficient to restart the optimization process everytime a change in landscape is detected, especially when the new PS* is somewhat similar to the previous solutions. It is also imperative that the MOEA must be capable of high convergence speed in order to find the optimal solution set before it changes. However, when MOEA converges to the PS*, the problem is that there will be a lack of search space diversity necessary to explore the search space for the new PF* and PS* when landscape changes.

The application of MOEAs to dynamic multi-objective problems is explored only recently. Nonetheless, from the available literature, it is clear that EAs for dynamic optimization in any problem domain must be capable of detecting the change in fitness landscape and maintaining diversity within the evolving population. Many dynamic EAs also incoporate some form of memory to store past solutions in anticipation of eventual reuse to improve performance. Adaptation to the MOEA design must be made to account for outdated elitist solutions and diversity loss. One potential problem of MOEAs in dynamic environment is their exploitation of nondominated solutions. For the former issue, the current solution set may not be indicative of the optimal Pareto front and will misguide the optimization process when the landscape changes. For the latter issue, diversity preservation tehniques are adopted in MOEAs only to maintain diversity in the objective space to obtain a well-distributed and well-spread solution set. Unless the new optimal solution set is within the vicinity of the previous optimal solution set, it is unlikely that MOEA is able to track any landscape changes. Different techniques proposed to handle the issue of diversity are based on the following three classes, namely diversity introduction, diversity maintainance, and multiple population.

In diversity introduction, diversity is introduced using techniques such as random restart or reinitialization upon the detection of landscape change [15,32,73]. This approach can be easily extended to MOEAs. The main drawback is that information gained is lost after the introduction of diversity. The approach of diversity maintainance sought to maintain diversity throughout the run [25,31,52]. While the conventional MOEA tends to converge quickly to PS_t^*, the incorporation of a diversity maintainance scheme will ensure the presence of individuals in other regions of the search space. One of the techniques that can be easily incorporated in MOEAs is the random immigrant. Diversity preservation techniques such as niching and crowding can also be used to encourage the growth of unique solutions. Applying the multiple population approach allows the algorithm to conduct simultaneous exploration in different regions to track any change or emergence of new optimal solutions [13,72,74]. Typically, this approach involves a population which exploits the current optimal solution while the other populations are encouraged to explore the search space.

Farina *et al* [22] suggested a hybridization between evolutionary strategies (ES) and deterministic local search to improve the speed of convergence to track the dynamic PS^{*t} and PF^{*t}. This direction-based dynamic evolutionary multi-objective optimization algorithm (DB-DEMOA) involves a two-stage optimization procedure. The first stage searches for the payoff matrix, utopia and nadir points of the PF^{*t} while the second stage searches for a set of uniformly distributed solutions among the Utopia points.

Deb *et al* [20] extended the NSGAII for the online optimization of a dynamic hydro-thermal power scheduling problem. The dynamic NSGAII (DNSGAII) randomly selects and re-evaluates 10% of the parent population in each generation to detect possible changes in the fitness landscape. In the event of landscape variation, all parent solutions are re-evaluated before combining with the offspring population. The authors explored two means of introducing diversity into the evolving population, either by inserting new individuals or performing mutation on existing solutions. The two schemes have their own strength and weakness. Reinitializing new solutions will allow DNSGAII to adapt faster to drastic changes in the landscape but performs poorly in environments with small changes. On the other hand, mutating existing solutions will not introduce too much diversity into the population. But this will allow NSGAII to handle small variations in the problem quickly at the expense of poor performances when faced with great changes in the environment.

Zeng *et al* [75] suggested a dynamic orthogonal multi-objective evolutionary algorithm (DOMOEA) that incorporates orthogonal design methodology to improve convergence. Periodic checks are performed to detect landscape changes and the DOMOEA treats the dynamic multi-objective problem as a new problem instance after every landscape change. However, it exploits past information by using the PS_t^A prior to change as the new initial population. The selection process of DOMOEA is very similar to the crowding-sort procedure of NSGAII. However, the clustering algorithm of SPEA is adopted instead of crowding. The DOMOEA applies two crossover operators, namely, the linear crossover and orthogonal crossover operator. Diversity is maintained in the evolving population through the linear crossover operator, which generates an offspring different from its parents. On the other hand, the orthogonal crossover is used to perform local fine-tuning and can give very good results if the region bounded by the parents is linear or quadratic.

Hatzakis and Wallace [35] presented a dynamic queuing multi-objective optimizer (D-QMOO) which exploits past information to predict the future behavior of the dynamic multi-objective problem. An autoregressive model is employed to estimate the location of PS_{t+1}^* and the predicted individuals are used to seed the population when a change in the problem landscape is detected. Diversity is also maintained throughout the evolution. The D-QMOO is based on a fairly elaborate algorithm, the queuing multi-objective optimizer. Clustering is performed in the decision space to promote diversity and to handle the possible presence of multiple discontinuous fronts. Both clustering and ranking are only performed periodically to reduce computational cost. Genetic operators to be executed on each individual are also embedded as parameters within the chromosome as in the case of mutation strength in evolutionary strategies.

Zhang [76] suggested an immune algorithm approach for dynamic multi-objective optimization. The multi-objective immune algorithm (MOIA) employs different cell types, 1) antigens (Ag) models the dynamic problem 2) B-cells represents the potential solution to the problem, 3) T_h cells represents the non-dominated or archived solutions and 4) memory cells are the collection of all previous nondominated solutions. Upon a landscape change, the MOIA searches the environmental memory for nondominated solutions of past environments that satisfy two conditions, 1) the solutions have the same number of objectives with the current environment, and 2) the solutions exhibit similar behavior in it's previous and current environment. The MOIA first searches the memory for past solutions with the same dimensions with the current enviornment. If such solutions are found, the first set of solution that is also located within the neighborhood will be used to seed the new population. In the event that no suitable solutions are found, the evolving population will be re-initialized.

Most algorithms presented here maintains a mechanism to re-evaluate solutions periodically to check for discrepancies as indications of problem variation. Since sufficient diversity is the key to tracking the dynamic solution set, all dynamic MOEAs incorporate some form of genetic operator to perform exploration immediately after a problem change. In addition, the DB-DEMOA and the DO-MOEA implements local search operators that exploits similarities between the landscape prior and after change. The D-QMOO also applied an autoregressive model to predict the location of the next solution set to improve convergence.

4.2   Competitive-Cooperative Coevolution for Dynamic Multi-Objective Optimization

The coevolutionary paradigm, inspired by the reciprocal evolutionary change driven by the cooperative [56] or competitive interaction [60] between different species, has been demonstrated to provide high speeds of convergence and diversity for multi-objective optimization recently [17,39,43,49,51,65]. However successful implementation of coevolution requires appropriate problem decomposition. The best way of handling problem decomposition may not be known *a priori* and may change with time in dynamic multi-objective problem. To solve this issue, we suggested a dynamic competitive-cooperative coevolutionary algorithm (dCOEA) [64] that incorporates both competitive and cooperative mechanisms observed in nature to track the Pareto front in a dynamic environment. The main idea of competitive-cooperation coevolution is to allow the decomposition process of the optimization problem to adapt and emerge rather than being pre-defined at the start of the optimization process. Each species subpopulation will compete to represent a particular subcomponent of the multi-objective problem while the eventual winners will cooperate to evolve the better solutions. Through this iterative process of competition and cooperation, the various subcomponents are optimized by different species subpopulations based on the optimization requirements of that particular time instant, enabling the MOEA to handle the dynamic multi-objective landscape.

Competitive-Cooperative Coevolution. *Cooperative process:* The cooperative mechanism exhibits a fine-grained search capability desirable in many applications and maintains good diversity across the subpopulations. At the start of the optimization process, the i-th subpopulation is initialized to represent the i-th variable. Concatenation between individuals in S_i and representatives from the other subpopulations is necessary to form a valid candidate solution for evaluation. To illustrate the cooperative process, consider the example of an 3-decision variable problem where subpopulations, S_1, S_2 and S_3, represent the variables, x_1, x_2 and x_3, respectively. When assessing the fitness of an individual in S_1, it will combine the individual with the representatives of S_2 and S_3 to form a valid candidate solution. Archive updating is conducted after the evaluation of each individual. The Pareto ranking and niche count of each individual are then calculated with respect to the archive. This cooperative process is carried out for all subpopulations in an iterative manner. Before proceeding to the evaluation of next subpopulation, the representative of the assessed subpopulation is updated to improve the speed of convergence. This updating process is based on a partial order such that ranks will be considered first, and followed by niche count if there is a tie in the rank. The rationale of selecting a nondominated representative with the lowest niche count is to promote diversity of the solutions via the approach of cooperation among multiple subpopulations.

Competitive process: One simple approach of conducting the competitive process is to allow the different subpopulations to take up the role of each problem subcomponent in a round-robin fashion. The most competitive subpopulation is then determined and the subcomponent will be optimized by the winning species in the next cooperative process. Ideally, the competition is performed such that all individuals from a particular subpopulation compete with all other individuals from the other subpopulations in order to determine the extent of its suitability. However, such an exhaustive approach requires extensive computational effort that is often practically infeasible. A more practical approach is thus to conduct competition with only selected individuals among a certain number of competitor subpopulations to estimate the species fitness and suitability.

For the $i - th$ variable, the representative of the associated subpopulation is selected along with random competitors from the other subpopulations to form a competition pool. These competitors will then compete via the cooperative mechanism described above to determine the extent of the cooperation achieved with representative of the other subpopulations. The winning species is determined by checking the originating subpopulation of the representative after the representative update. At the end of the competitive process, the subpopulaion representing variable i will remain unchanged if its representative wins the competition. In the case that a winner emerges from other subpopulations, the subpopulation will be replaced by the individuals from the winning subpopulation. The rationale of replacing the losing subpopulation instead of associating the winning subpopulation with the decision variable directly is that different variables may have close but not identical properties. Therefore it is more appropriate to seed the losing subpopulation with the desirable information and to allow it to evolve independently.

Introducing Diversity via Stochastic Competitors. The competitive process provides a natural conduit in which the introduction of diversity into the subpopulations can be regulated. Instead of re-initialization or subjecting the entire subpopulation to hypermutation, a set of stochastic competitors are introduced together with the competitors from the other subpopulations. The idea is to compare the potential of new regions in the search space and the past information to decide whether the subpopulation should be initialized. In the case that stochastic competitor emerges as the winner, the particular subpopulation is reinitialized in the region that the winner is sampled from. Hence, diversity is introduced into the subpopulations only when it presents an advantage over the current information at hand.

Handling Outdated Archived Solutions. If there is any environmental change in the evolution, it is likely that the archived solutions will not remain nondominated, and these outdated archived solutions will keep out the nondominated solutions if they are being left unchecked. A simple approach is to simply discard all the archived solutions but useful information about past PF_t cannot be exploited in the case where PS_t^* is cyclic in nature. In order to store the potentially useful information in dCOEA, an additional external population denoted as the temporal memory is used in conjunction with the archive. To store the outdated solutions, a fixed number R_{size} of the archive is added to the temporal memory upon a landscape change. When the upper bound of the temporal memory is reached, the oldest set of R_{size} outdated solutions is removed for newer solutions. To select the R_{size} outdated solutions, the dCOEA stores the extreme solutions along each dimension in the objective space. In the case where R_{size} is greater than the number of extreme solutions, the rest of the solutions to be stored are randomly selected from the archive. On the other hand, if R_{size} is smaller than the number of extreme solutions, then R_{size} extreme solutions will be randomly selected into the temporal memory.

4.3   Simulation Results

Two different dynamic MOEAs built upon a basic MOEA and the cooperative-coevolutionary evolutionary algorithm (CCEA) [65], are applied and compared against the dCCEA on the test function of FDA1. In both the dynamic MOEA (dMOEA) and dynamic CCEA (dCCEA), a fixed number of archived solutions are re-evaluated in every generation. In the case where a change in the landscape is detected, the temporal memory described previously will be applied, and random restart is incorporated to generate diversity within the evolving population. FDA1 challenges the dynamic MOEAs' ability to track and converge towards the PF_t^* with every landscape change. One interesting characteristic of this problem is that the distribution and diversity of the solutions along the PF_t

Table 1. Performance of MOEA, dCCEA and dCOEA for FDA1 at different settings of τ_T and n_T. The best results are highlighted in bold only if it is statistically different based on the KS test.

(τ_t, n_T)		VD$_{offline}$			MS$_{offline}$		
		MOEA	dCCEA	dCOEA	MOEA	dCCEA	dCOEA
	1st quartile	0.666	0.243	**0.107**	0.789	0.829	**0.939**
(5,10)	Median	0.683	0.255	**0.110**	0.801	0.834	**0.944**
	3rd quartile	0.695	0.264	**0.113**	0.801	0.841	**0.953**
	1st quartile	0.489	0.154	**0.034**	0.870	0.863	**0.963**
(10,10)	Median	0.508	0.163	**0.038**	0.878	0.873	**0.970**
	3rd quartile	0.521	0.167	**0.039**	0.890	0.882	**0.977**
	1st quartile	0.485	0.080	**0.001**	0.876	0.926	**0.979**
(25,10)	Median	0.528	0.091	**0.002**	0.894	0.939	**0.985**
	3rd quartile	0.583	0.102	**0.003**	0.914	0.947	**0.989**
	1st quartile	1.008	0.135	**0.020**	0.535	0.857	**0.973**
(10,1)	Median	1.031	0.149	**0.022**	0.585	0.866	**0.981**
	3rd quartile	1.064	0.156	**0.025**	0.599	0.883	**0.984**
	1st quartile	0.542	0.152	**0.039**	0.847	0.858	**0.970**
(10,20)	Median	0.584	0.162	**0.042**	0.868	0.875	**0.975**
	3rd quartile	0.606	0.171	**0.044**	0.881	0.888	**0.979**

are not affected by the landscape change. The simulation results for the metric of $VD_{offline}$ and $MS_{offline}$ with different settings of τ_T and n_T are summarized in Table 1. In general, the coevolutionary paradigm is shown to be more appropriate than canonical MOEAs in handling dynamic landscapes. As can be seen, the dCOEA outperforms dCCEA in both the aspect of tracking and finding a diverse solution set. Table 1 also shows a better convergence and diversity performance of dMOEA, dCCEA and dCOEA for larger value of τ_T or less frequent landscape changes. Although the dMOEA gives a better convergence for larger value of n_T or less severe landscape changes, it is observed that better results of dCCEA and dCOEA can be obtained with a more severe landscape change.

5 Robust Multi-Objective Optimization

Real-world applications may also involve the simultaneous optimization of several competing objectives thhat are susceptible to decision or environmental parameter variation which results in large or unacceptable performance variation. Apart from the multi-objective optimization goals of evolving a set of near optimal, diverse, and uniformly distributed Pareto solution set, robust multi-objective optimization sought to find a set of tradeoff solutions that is robust or reliable [19]. A robust solution can be defined as a solution that provides satisfactory performance in the face of parametric variations, i.e it is insensitive to small variations in design and/or enviroment variables.

In order to reduce the consequences of uncertainties on optimality and practicality of the solution set, factors such as decision variable variation, environmental variation, and modeling uncertainty have to be considered explicitly. Therefore, the minimization of multi-objective problem is redefined as follows,

$$\min \boldsymbol{F}(\boldsymbol{x}, \boldsymbol{\delta_x}, \boldsymbol{\delta_e}) = \{f_1(\boldsymbol{x}, \boldsymbol{\delta_x}, \boldsymbol{\delta_e}), f_2(\boldsymbol{x}, \boldsymbol{\delta_x}, \boldsymbol{\delta_e}), \quad\quad (4)$$
$$..., f_M(\boldsymbol{x}, \boldsymbol{\delta_x}, \boldsymbol{\delta_e})\}$$

$$\text{s.t. } \boldsymbol{g}(\boldsymbol{x}, \boldsymbol{\delta_x}, \boldsymbol{\delta_e}) \geq 0, \boldsymbol{h}(\boldsymbol{x}, \boldsymbol{\delta_x}, \boldsymbol{\delta_e}) = 0$$

where δ_x and δ_e represent, respectively, the uncertainties associated with $\boldsymbol{x}$ and environmental conditions. Both forms of uncertainties may be treated equivalently. Different noise models such as normal, Cauchy, and uniform distributions have been considered in the literature.

The optimal robust Pareto front and solution set are dependent on the noise model and the robust measure. Therefore, the notation should reflect the noise model and the robust measure used. In this chapter, the optimal robust Pareto front and optimal solution set are denoted as $\text{PF}^*_{rm,\delta}$ and $\text{PS}^*_{rm,\delta}$, respectively. The terms rm and δ refers to the robust measure and noise model in consideration, respectively. Accordingly, $\text{PF}^A_{rm,\delta}$ refers to the final set of non-dominated solutions evolved by robust MOEA based on the robust measure, rm and noise model, δ.

Table 2. Empirical Results of NSGAII and SPEA2 for the different robust multi-objective test functions

		NSGAII			SPEA2		
		Ratio	Space	Maximum Spread	Ratio	Space	Maximum Spread
rMOP1	2-D	0.8867	0.5411	1.0	0.7280	0.6582	1.0
	5-D	0.0	0.5431	1.0	0.0	0.6783	1.0
	10-D	0.0	0.5626	0.9999	0.0	0.6773	0.9999
rMOP2	2-D	0.9947	0.5314	1.0	0.9890	0.6362	1.0
	5-D	0.9883	0.5369	1.0	0.9840	0.6278	1.0
	10-D	0.9850	0.5781	0.9999	0.9807	0.6849	0.9999
rMOP3	2-D	0.9853	0.5332	1.0	0.9833	0.6354	1.0
	5-D	0.7193	0.4965	1.0	0.6203	0.6484	1.0
	10-D	0.0	0.5039	0.9999	0.0	0.6268	1.0
rMOP4	2-D	0.5250	0.5066	0.9999	0.4243	0.6500	0.9999
	5-D	0.0920	0.5012	0.9997	0.0	0.6442	0.9999
	10-D	0.0	04900	0.9997	0.0	0.6174	0.9998

Table 3. Definitions of the GTCO test suite

Problem	Type	Definition
GTCO1	Class 1	$f_1(\boldsymbol{x}_{\mathrm{D1}}) = x_1$ $g(\boldsymbol{x}) = 1 + \sum_{i=2}^{|\boldsymbol{x}_{\mathrm{D2}}|}\{(x_{\mathrm{D2},i}-0.4)^2 + b(\boldsymbol{x}_{\mathrm{R}},x_{\mathrm{D2},i})\}$ $h(f_1,g) = 1 - \sqrt{\frac{f_1}{g}}$ $b(\boldsymbol{x}_{\mathrm{R}},x_i) = 1 - \frac{1}{|\boldsymbol{x}_{\mathrm{R}}|}\sum_{j\in\boldsymbol{x}_{\mathrm{R}}}\exp[(\frac{x_{\mathrm{R},j}^5 - E(\sigma,s)}{W(x_i)})^2]$ $W(x_i) = 0.1 + 0.1\cos(20(x_i-0.4)\pi)\cdot(1-|x_i-0.4|)^5$ $E(\sigma,s) = U(-\sigma,\sigma),\ \boldsymbol{x}_{\mathrm{D1}},\boldsymbol{x}_{\mathrm{D2}},\boldsymbol{x}_{\mathrm{R}}\in[0,1]$
GTCO2	Class 2	$f_1(\boldsymbol{x}_{\mathrm{D1}}) = x_1$ $g(\boldsymbol{x}) = 1 + b(\boldsymbol{x}_{\mathrm{R}})$ $h(f_1,g) = 1 - \sqrt{\frac{f_1}{g}}$ $b(\boldsymbol{x}_{\mathrm{R}}) = 1 - \frac{1}{|\boldsymbol{x}_{\mathrm{R}}|}\sum_{i\in\boldsymbol{x}_{\mathrm{R}}}\max\{0.8\exp[(\frac{x_{\mathrm{R},i}-0.25E(\sigma,s)}{0.1})^2]$ $,\exp[(\frac{x_{\mathrm{R},i}-0.75E(\sigma,s)}{0.1})^2]\}$ $E(\sigma,s) = 1 + U(-\sigma,\sigma),\ \boldsymbol{x}_{\mathrm{D1}},\boldsymbol{x}_{\mathrm{R}}\in[0,1]$
GTCO3	Class 3	$f_1(\boldsymbol{x}_{\mathrm{D1}}) = x_1$ $g(\boldsymbol{x}) = 1 + 10(\sum_{i=2}^{|\boldsymbol{x}_{\mathrm{D2}}|}\frac{x_{\mathrm{D2},i}}{|\boldsymbol{x}_{\mathrm{D2}}|-1})^{1.25-b_1(\boldsymbol{x}_{\mathrm{R1}})} + b_2(\boldsymbol{x}_{\mathrm{R2}})$ $h(f_1,g) = 1 - \sqrt{\frac{f_1}{g}}$ $b_1(\boldsymbol{x}_{\mathrm{R1}}) = 1 - \frac{1}{|\boldsymbol{x}_{\mathrm{R1}}|}\sum_{i\in\boldsymbol{x}_{\mathrm{R}}}\exp[(\frac{x_{\mathrm{R1},i}^5 - E_1(\sigma,s)}{0.05})^2]$ $b_2(\boldsymbol{x}_{\mathrm{R2}}) = 1 - \frac{1}{|\boldsymbol{x}_{\mathrm{R2}}|}\sum_{i\in\boldsymbol{x}_{\mathrm{R2}}}\max\{0.8\exp[(\frac{x_{\mathrm{R2},i}-0.25E_2(\sigma,s)}{0.1})^2]$ $,\exp[(\frac{x_{\mathrm{R2},i}-0.75E_2(\sigma,s)}{0.1})^2]\}$ $E_1(\sigma,s) = U(-\sigma,\sigma),$ $E_2(\sigma,s) = 1 + U(-\sigma,\sigma),\ \boldsymbol{x}_{\mathrm{D1}},\boldsymbol{x}_{\mathrm{D2}},\boldsymbol{x}_{\mathrm{R1}},\boldsymbol{x}_{\mathrm{R2}}\in[0,1]$
GTCO4	Class 6	$f_1(\boldsymbol{x}_{\mathrm{D1}}) = x_1$ $g(\boldsymbol{x}_{\mathrm{D2}}) = 1 + 10\sum_{i=2}^{|\boldsymbol{x}_{\mathrm{D2}}|}x_{\mathrm{D2},i}$ $h(f_1,g,\boldsymbol{x}_{\mathrm{R}}) = 1 - (\frac{f_1}{g})^\alpha$ $\alpha = 0.5 + b(\boldsymbol{x}_{\mathrm{R}})$ $b(\boldsymbol{x}_{\mathrm{R}}) = 1 - \frac{1}{|\boldsymbol{x}_{\mathrm{R}}|}\sum_{i\in\boldsymbol{x}_{\mathrm{R}}}\max\{0.8\exp[(\frac{x_{\mathrm{R},i}-0.25E(\sigma,s)}{0.05})^2]$ $,\exp[(\frac{x_{textR,i}-0.75E_{ij}(\sigma,s)}{0.05})^2]\}$ $E(\sigma,s) = 1 + U(-\sigma,\sigma),\ \boldsymbol{x}_{\mathrm{D1}},\boldsymbol{x}_{\mathrm{D2}},\boldsymbol{x}_{\mathrm{R}}\in[0,1]$
GTCO5	Class 7	$f_1(\boldsymbol{x}_{\mathrm{D1}}) = x_1$ $g(\boldsymbol{x}) = 1 + \sum_{i=2}^{|\boldsymbol{x}_{\mathrm{D2}}|}\{x_{\mathrm{D2},i} + 5b_1(\boldsymbol{x}_{\mathrm{R1}})\}$ $h(f_1,g) = 1 + 2b_2(\boldsymbol{x}_{\mathrm{R2}},f_1) - \sqrt{\frac{f_1}{g}}$ $b_1(\boldsymbol{x}_{\mathrm{R1}},x_i) = 1 - \frac{1}{|\boldsymbol{x}_{\mathrm{R1}}|}\sum_{i\in\boldsymbol{x}_{\mathrm{R1}}}\exp[(\frac{x_{\mathrm{R1},i}-E(\sigma,s)}{W_1(x_i)})^2]$ $b_2(\boldsymbol{x}_{\mathrm{R2}},f_1) = 1 - \frac{1}{|\boldsymbol{x}_{\mathrm{R2}}|}\sum_{i\in\boldsymbol{x}_{\mathrm{R2}}}\exp[(\frac{x_{\mathrm{R2},i}-E(\sigma,s)}{W_2(f_1)})^2]$ $W_1(x_i) = 0.2,\text{if }x_i < 0.05$ $W_1(x_i) = 0.1x_i + 0.05,\text{otherwise}$ $W_2(f_1) = 0.2\cdot(1.1 - \sqrt{f_1})$ $E(\sigma,s) = U(-\sigma,\sigma),\ \boldsymbol{x}_{\mathrm{D1}},\boldsymbol{x}_{\mathrm{D2}},\boldsymbol{x}_{\mathrm{R1}},\boldsymbol{x}_{\mathrm{R2}}\in[0,1]$

Table 4. Characteristics of the GTCO test suite

Problem	PF* change	PS* change	Feature change
GTCO1	-	-	Increasing multi-modality
GTCO2	-	Global to local	-
GTCO3	-	Global to local	Decreasing density of solutions near PF*
GTCO4	Increasing nonconvexity	Global to local	-
GTCO5	Decreasing coverage of PF*	Range of x_1 reduces	Linear to deceptive

5.1   Robust Multi-Objective Problem Test Suite

Several desirable properties of deterministic benchmarks and test suites have been suggested in the EA literature. In addition to these guidelines, the following issues should be considered in the development of robust benchmark problems in the context of multi-objective optimization:

– Robust multi-objective problems are essentially multi-objective problems and guidelines for the construction of multi-objective benchmark problems established in the existing literature should be taken into account;
– The robust multi-objective test problems should not have any bias towards $PS^*_{rm,\delta}$;
– Some test problems should contain noise-induced features that pose more difficulty to the optimization algorithm;
– The benchmark problem component which determines how the problem behaves in the presence of noise should be scalable;
– Some test problems should contain possible tradeoffs between robustness and the different objectives.

Empirical investigations conducted in [29] to analyze the behavior of four existing benchmark problems found in the literature showed that many test functions may suffer from strong bias towards $PS^*_{rm,\delta}$. Three of the problems studied are extended from single-objective benchmark problems in [11,9,55] using the ZDT framework [?], which allows the easy incorporation of problem characteristics that hinder MOEA progress to the Pareto front. The fourth is a robust multi-objective problem proposed in [18]. The simulation results of NS-GAII and SPEA2 with respect to the metrics of ratio of convergence, spacing (S) and maximum spread (MS) are shown in Table 2. All four problems are characterized by at least one local Pareto set that becomes more desirable than PS^*_{det} when noise is introduced. Nonetheless, the failure to find PS^*_{det} clearly indicates that the two algorithms can converge to more robust regions readily even without the incorporation of any robust handling mechanism. This is because PS^*_{det} is located at a much narrower region and hence harder to locate as compared to the broader dips corresponding to the more robust solutions. This implies that rMOP1, rMOP3, and rMOP4 are not suitable for the evaluation of robust

MOEA techniques since it is not possible to ascertain whether robust solutions are found due to the adopted robust measures or due to the algorithms' failure to locate PS^*_{det} in the first place.

In Goh *et al*, we present a framework for the construction of robust continuous MO test functions characterized by different noise-induced features. These noise-induced features can pose different difficulties to the optimization algorithms and do not suffer the limitations of bias and scalability. The fundamental component of the robust test problems is a Gaussian landscape generator that facilitates the specification of robust optimization-specific characteristics such as noise-induced features, fitness landscape, and decision space variation. One further desirable property of the construction guidelines is that it provides a means to extend existing multi-objective test problems to robust multi-objective test functions without changing the original problem characteristics. The rationale is to allow researchers to investigate the impact of robust optimization on test functions with different characteristics such as deception, multi-modality and discontinuities. The test functions designed using this landscape generator are different from most previous works in two aspects:

- Any solution space or objective space transformation is a consequent of *environmental* variation. Although environmental parameter variation is rarely considered in the literature, it is definitely more flexible compared to decision parameter variation when it comes to the design of different possible scenarios.
- The basin of attraction of the various troughs to be very similar. This ensures that there is no initialization bias towards any particular region of the search space.

The definitions and characteristics of the proposed test suite are summarized in Table 3 and Table 4, respectively.

5.2   Evolutionary Robust Optimization Techniques

While MOEAs have been demonstrated to be capable of discovering good trade-off solutions for various multi-objective problems, it is necessary to ensure that these solutions are implementable in practice. Conventional MOEAs, without the necessary mechanisms to identify robust solutions, are not capable of finding $PS^*_{rm,\delta}$ unless it coincides with PS^*_{det}. Therefore, the selection of an appropriate robust measure is of utmost importance in robust optimization.

It can be noted that studies on evolutionary robust optimization are mainly conducted in the domain of robust single-objective optimization. Nevertheless, the robust measures and uncertainty handling mechanisms adopted in these works are generally applicable for robust multi-objective optimization. Specific issues such as diversity preservation and fitness assignment must be considered in robust MOEA design.

Based on the state-of-the-art, EAs for robust optimization can be classified into single-objective and multi-objective approaches depending on how the various measures are incorporated into the EA.

1. The *single-objective* approach optimizes the selected robust measures in place of the original objectives.
2. The *multi-objective* approach considers the selected robust measures as additional objectives to be optimized.

As noted by Jin and Branke [40], the former is the more popular approach. This is perhaps because of its ease of implementation whereas there is a need to consider the implications brought about by the increase in problem dimensionality for the latter.

Single-objective approach. Since it is usually difficult to compute the various robust measures analytically, this approach is also characterized by the stochastic evaluation of the adopted robust measure to account for uncertainties, i.e. these measures are usually estimated over a number of randomly sampled perturbations. The optimization of the expected objective values estimated from the mean of the sampled points is also known as *explicit averaging* and has been applied successfully for robust multi-objective optimization [18]. In the same work, the effects of sample size and noise level on $PF^A_{eff,\delta}$ are investigated. On the other hand, Hughes [36] introduced a probabilistic approach for Pareto ranking scheme to account for the presence of uncertainties.

Although simple to implement, stochastic evaluation is computationally intensive since additional solution evaluations are required. It can be expected that this situation would be exacerbated by the presence of multiple objectives in multi-objective problems. Therefore, suitable methods for reducing the number of evaluations will be required to lower the computational cost. To this end, the Latin hypercube sampling is applied by [10,18] to get a more efficient fitness estimate. Other methods in the robust single-objective optimization literature that are appropriate for reducing computational cost of robust MOEAs include:

– Allocation of more computational resource for the evaluation of Pareto optimal solutions,
– Use of similar individuals evaluated in the past to estimate the expected fitness,
– Adaptation of computational resource allocation for evaluation through the evolutionary process, and
– Use of approximate models in place of the original objective functions.

A viable option for the efficient optimization of expected objective values is the method of *implicit averaging* [71,69] where each solution is perturbed once before evaluation. This approach is based on the concept that solutions are implicitly averaged over a set of perturbed samples as the MOEA tends to revisit promising regions of the search space. Tsutsui and Ghosh also showed, by means of the Schema theorem, that an EA with infinite population size working on perturbed evaluations has the same effects as one working on the effective fitness.

Multi-objective approach. The multi-objective approach involves both deterministic and stochastic evaluation of the various objectives and robust measures, respectively. Therefore, computational cost is also an important issue as in the case of the single-objective approach. When the multi-objective approach is applied to solve robust single-objective problems, it allows us to consider the tradeoff between robustness and solution quality. On the other hand, when it is applied to solve robust multi-objective problems, this approach poses an interesting decision-making exercise particularly if there are also tradeoffs between the robustness of the different objectives. Furthermore, the scaling capability of the MOEA becomes extremely important, for example, a five-objective problem can become a ten-objective problem if the effective performance of each objective is considered explicitly.

At present, there are two variants of the multi-objective approach for robust MOEA. The first approach optimizes the selected robust measures on top of the existing deterministic objective functions and sought to discover the inherent tradeoff between optimality and robustness. This is also known as the multi-objective approach [42] and various combinations of different measures such as expected fitness and variance-based measures have been applied in the evolutionary robust single-objective optimization literature. In [59], Ray utilized three objectives, the nominal objective value, the effective objective value, and the standard deviation, to evolve designs that remain feasible under decision variable variations.

More recently, Goh and Tan [28] extended this method to evolve the tradeoff between Pareto optimality and the robust measure of worst case performance for constrained multi-objective problems. In order to improve computational efficiency, the mechanisms of Tabu-restriction and μGA are implemented for the computation of the worst case performance and constraint violation. Apart from the nominal objective values, Li et al [46] applied the concept of the worst case sensitivity region (WCSR) and sought to maximize the radius of this region. The WCSR approximates the worst violation that a particular design can absorb before it violates some predetermined threshold on performance variation.

In [47], Lim et al also presented a single-objective/multi-objective inverse evolutionary optimization methodology for robust design. In contrast to conventional forward robust optimization, the inverse approach avoids making assumptions about the uncertainty when insufficient field data exists for estimating its structure. Apart from the objectives of nominal fitness and robustness, Lim et al considered the possible benefits as the uncertainty prevails by introducing an opportunity criterion in the inverse search scheme as the third objective.

The second variant is proposed by Deb and Gupta [18] as a more practical approach to the single-objective method and treats the selected robust measures as hard constraints. Adopting a similar approach in an independent work, Gunawan and Azarm [33] considers a designer specified WCSR radius as a constraint to be satisfied. Therefore, in both works, the goal is to evolve the best PF_{det}^A that satisfies the tolerable bounds on performance deviation.

5.3   Solving Vehicle Routing Problem with Stochastic Demand

The vehicle routing problem with stochastic demand (VRPSD) as an instance of real-world robust multi-objective problem. The VRPSD differs from its deterministic counterparts in that the actual cost of a particular solution to the VRPSD cannot be known with certainty before the actual implementation of the solution. Optimization of the VRPSD deterministically may yield very good route schedules which are very sensitive to variations in customer demand. In [66], a hybrid MOEA is applied to solve the VRPSD problem. The algorithm incorporates two heuristics which exploits knowledge of the VRPSD route structures to improve algorithmic performances. In addition, an intuitive route simulation method (RSM) which is an elaborate form of explicit averaging is proposed to address the issue of evaluating the expected costs of solutions.

Local Search Heuristic. The role of local search is vital in multiobjective evolutionary optimization in order to encourage better convergence and to discover any gap in the Pareto front. The local search approach can contribute to the intensification of the optimization results, which is usually regarded as a complement to the evolutionary operators that mainly focus on global exploration.

1) Shortest Path Search: Shortest Path Search is designed to exploit the fact that route failures are more likely to occur at the end of a route. The SPS attempts to rearrange the order of customers in a particular route. For example, given a route that contains five customers, a new route is built by choosing the customer that is furthest from the depot as the first customer in the route, while the customer that is nearest to the depot is chosen as the last customer of the route. Next, the customer that is nearest to the first customer is chosen as the second customer, while the customer that is nearest to the last customer is chosen as the second last customer of the new route. This step continues until all the customers in the original route are re-routed. The new route will be compared against the original one and the better route will be retained. By re-routing customers in such a manner, customers that are further from the depot will be at the beginning of the route whereas those that are nearer to the depot will be at the end of the route. The rationale is to reduce the additional transportation cost that will be incurred by the recourse policy.

2) Which Directional Search: Which Directional Search (WDS) is designed to exploit the fact that the expected transportation cost of a route is dependent on the traversed direction. In contrast, for the deterministic VRP, the transportation cost of a route is the same regardless of the direction in which the route is traversed. To be specific, given a route, the WDS builds a new route that runs in the opposite direction. Similarly, the new route will replace the original one if it is better.

Route Simulation Method.   One of the main difficulties of solving the VRPSD is in finding an objective function that is able to define properly the expected transportation cost of a solution. The route simulation method (RSM) is an elaborate form of explicit averaging to evaluate the expected costs of solutions. Fig. 5 will be used to illustrate the operation of the RSM. The figure

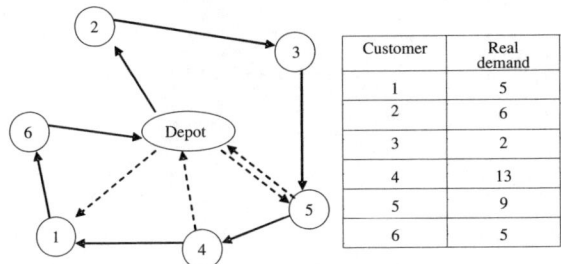

Customer	Real demand
1	5
2	6
3	2
4	13
5	9
6	5

Fig. 5. Example illustrating the operation of the RSM

shows a route sequence, Depot→2→3→5→4→1→6→Depot. The solid arrows indicate the route that the vehicle will take if this were a deterministic VRP. In the VRPSD, due to the recourse policies in the event of a route failure, the actual route taken by the vehicle cannot be known with certainty before the route is actually implemented. However, the implementation of the route can be simulated by generating a set of demands of all the customers based on their demand distributions and treating these demands as if they were the real demands revealed when a vehicle first arrives at the customer. The set of demands generated is tabulated in Fig. 5. For this particular example, it is assumed that the vehicle capacity is 15 and each arrow indicates a unit of distance.

The vehicle first leaves the depot and arrives at customer 2. It is able to satisfy its demand with a remaining capacity of 9. The vehicle then travels to customer 3 and satisfies its demand. The capacity of the vehicle is 7 when it reaches customer 5. The vehicle then finds that it is unable to satisfy the demand of customer 5, so it unloads all remaining goods and makes a return trip to the depot to restock. This recourse is indicated by the dashed arrows between the depot and customer 5. The vehicle then unloads two units of goods and leaves customer 5 for customer 4 with a capacity of 13. After serving customer 4, the vehicle is empty and returns to the depot to restock. Since the demand of customer 4 has been satisfied, the vehicle travels to customer 1 from the depot. The vehicle then satisfies the demands of customers 1 and 6 and returns to the depot. From this simulation, the total distance traveled by the vehicle (10 units for this example) and the remuneration for the driver for a particular realization of the set of customer demands can be obtained.

Due to the stochastic nature of the cost considered, there is a need to sample or repeat the above operation H times for every route of a particular solution, using a different set of demands randomly generated based on the demand distributions of the customers each time and then taking the average to obtain the expected transportation cost of the solution.

Simulation Results. Simulation studies are conducted using HMOEA to solve for three instances of the VRPSD. The PF_{det}^A for each of the test problems is obtained by treating the mean demand of each customer as its deterministic demand when evaluating the expected costs of solutions. This approach converts

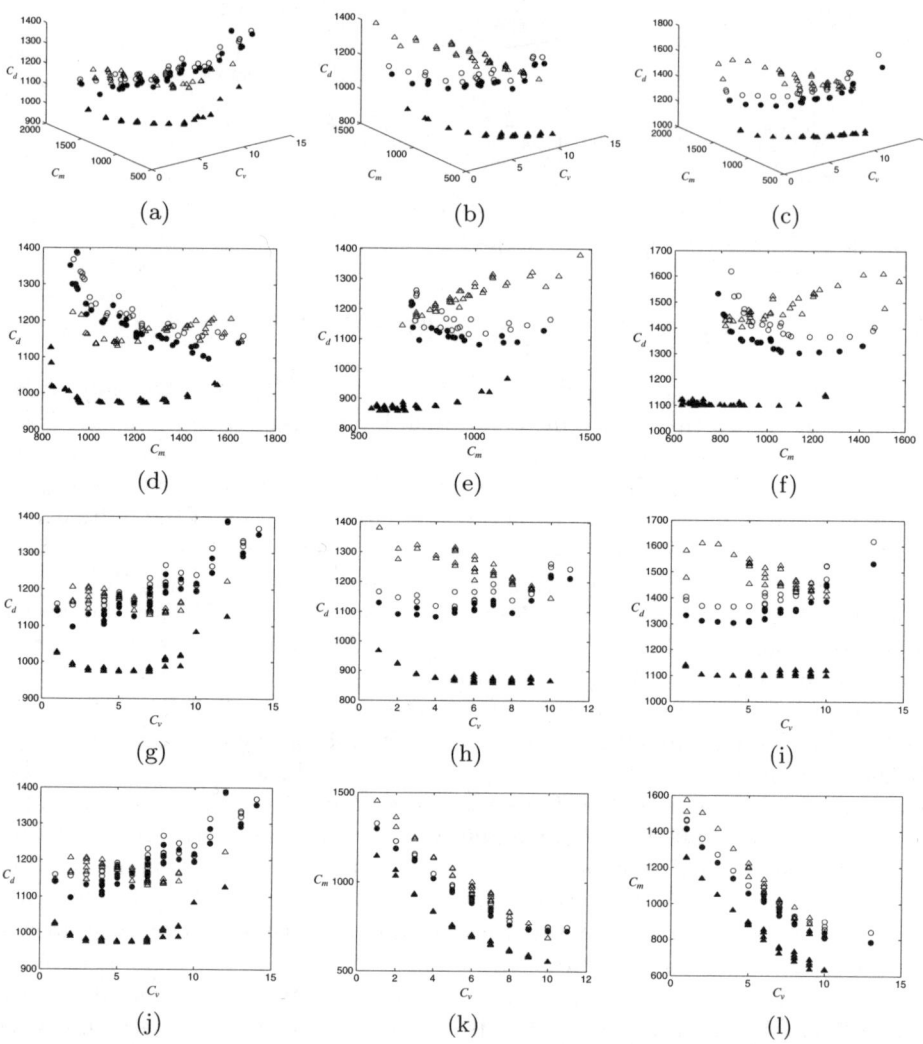

Fig. 6. Pareto fronts for VRPSD1 (first column), VRPSD2 (second column), VRPSD3 (third column) test problems. The first row shows the 3-dimensional Pareto fronts, the second row shows the same fronts along travel distance (C_d) and driver renumeration (C_m), the third row shows the same fronts along C_d and number of vehicles (C_v) and the fourth row shows the same front along C_m and C_v. ○ denote solutions evolved using averaging while △ denote solution evolved deterministically. ● and filled △ represent the implementation costs of the corresponding solutions.

the stochastic problem into a deterministic one. On the other hand, the $\mathrm{PF}^A_{\mathrm{eff},\sigma}$ is obtained by evaluating a VRPSD solution multiple times (H=10), each time using a different set of customer demands randomly generated based on their demand distributions. The costs are then averaged and taken as the expected cost of the solution. In Fig. 6, the filled points correspond to the respective fronts obtained immediately after the optimization process. These points are what the logistic manager sees when he is deciding which one of the solutions along the front to implement. The hollow points are obtained by implementing the corresponding filled points on 100 sets of randomly generated customer demands and then taking the average to get the implementation cost. This cost would be a good indication of the quality of the solution and the Euclidean distance between corresponding filled and hollow points would give an indication of the robustness of the solution. From Fig. 6(a)-(c), it is obvious that the solutions along $PF^A_{\text{eff},\sigma}$ are more robust than those along PF^A_{det} since their expected cost values do not differ much from their implementation costs and would provide the logistic manager with more accurate information about the quality of the solutions based on which decision will be made.

In order to further examine the behaviors of the PF^A_{det} and $PF^A_{\text{eff},\sigma}$ for each of the VRPSD test problems, separate two-dimensional graphs, each time considering only two of the objectives, are plotted in Fig. 6(d)-(l). It can be observed from the figures that the $PF^A_{\text{eff},\sigma}$ considering the expected and implementation costs have the same shape. The spacing between corresponding solutions is also uniform. This implies that all the solutions along $PF^A_{\text{eff},\sigma}$ have almost equal levels of robustness. This cannot be said of the solutions along PF^A_{det}. The front for the expected costs changes shape in most cases when the implementation costs of the solutions along the front are considered. In Fig. 6(f), the front for the expected costs is horizontal but curves upwards when the implementation costs are considered. This shows that solutions with higher driver renumeration (C_m) are less robust. Similarly, it can also be seen from Fig. 6(h), 6(i), 6(k), and 6(l) that solutions for the Type-CS and Type-RCS test problems become more robust as the number of vehicles (C_v) increases. It is also observed from Fig. 6(g) that the robustness of the solutions for the Type-RS test problem is the worst when there are four or five vehicles. All these observations point to the fact that the deterministic approach is not reliable in solving the VRPSD. This finding demonstrates that the stochastic nature of the problem is not trivial and that inadequate handling of the problem will give the logistic manager a false understanding of the situation at hand.

6   Conclusion

This chapter has provided a comprehensive treatment on multi-objective optimization in three different uncertain environments, namely, noise, dynamic and robustness. The three forms of uncertainties affect MOEA design in different ways, and the various MOEA design issues are discussed. Additionally, for noisy MO optimization, we have further described heuristical approaches to noise suppression and demonstrated it's effectiveness. In the case of dynamic MO optimization, we presented the effectiveness of the competitive-cooperative coevolutionary algorithm in tracking the dynamic solution set. On the topic of robust MO optimization, we highlighted some possible limitations of existing robust optimization test functions and presented a set of robust MO test problems with various noise-induced features. Finally, we described a hybrid multi-objective evolutionary algorithm which is capable of finding robust routes for the vehicle routing problem with stochastic demands.

References

1. Arnold, D.V., Beyer, H.G.: Local performance of the (1+1)-ES in a noisy environment. IEEE Transactions on Evolutionary Computation 6(1), 30–41 (2002)
2. Arnold, D.V., Beyer, H.G.: A General Noise Model and Its Effects on Evolution Strategy Performance. IEEE Transactions on Evolutionary Computation 10(4), 380–391 (2006)
3. Back, T., Hammel, U.: Evolution strategies applied to perturbed objective functions. In: Proceedings of the First IEEE Conference on Evolutionary Computation, vol. 1, pp. 40–45 (1994)
4. Babbar, M., Lakshmikantha, A., Goldberg, D.E.: Modified NSGA-II to solve Noisy Multi-objective Problems. In: Proceedings of the 2003 Genetic and Evolutionary Computation Conference, Late-Breaking Papers, pp. 21–27 (2003)
5. Basseur, M., Zitzler, E.: Handling Uncertainty in Indicator-Based Multiobjective Optimization. International Journal of Computational Intelligence Research 2(3), 255–272 (2006)
6. Bosman, P., Thierens, D.: The balance between proximity and diversity in multi-objective evolutionary algorithms. IEEE Transactions on Evolutionary Computation 7(2), 174–188 (2003)
7. Beielstein, T., Markon, S.: Threshold selection, hypothesis tests, and DOC methods. In: Proceedings of the 2002 Congress on Evolutionary Computation, vol. 1, pp. 777–782 (2002)
8. Beyer, H.G.: Evolutionary algorithms in noisy environments: Theoretical issues and guidelines for practice. Computer Methods in Applied Mechanics and Engineering 186, 239–267 (2000)
9. Branke, J.: Evolutionary Optimization in Dynamic Environments. Kluwer Academic Publishers, Dordrecht (2001)
10. Branke, J.: Reducing the sampling variance when searching for robust solution. In: Proceedings of the Genetic and Evolutionary Computation Conference, pp. 235–424 (2001)
11. Branke, J.: Creating robust solutions by means of evolutionary algorithms. In: Proceedings of the Fifth International Conference on Parallel Problem Solving from Nature, pp. 119–128 (1998)
12. Branke, J., Schmidt, C., Schmeck, H.: Efficient fitness estimation in noisy environments. In: Proceedings of Genetic and Evolutionary Computation, pp. 243–250. Morgan Kaufmann, San Francisco (2001)
13. Branke, J., Schmeck, H.: Designing evolutionary algorithms for dynamic optimization problems. In: Tsutsui, S., Ghosh, A. (eds.) Theory and Application of Evolutionary Computation: Recent Trends, pp. 239–262. Springer, Heidelberg (2002)
14. Buche, D., Stoll, P., Dornberger, R., Koumoutsakos, P.: Multiobjective Evolutionary Algorithm for the Optimization of Noisy Combustion Processes. IEEE Transactions on Systems, Man, and Cybernetics art C: Applications and Reviews 32(4) (2002)

15. Cobb, H.G.: An investigation into the use of hypermutation as an adaptive operator in genetic algorithms having continuous, time-dependent nonstationary environments. Technical Report AIC-90-001, Naval Research Laboratory, Washington, DC (1990)
16. Coello Coello, C.A., Aguirre, A.H.: Design of combinational logic circuits through an evolutionary multiobjective optimization approach. Artificial Intelligence for Engineering, Design, Analysis and Manufacture 16(1), 39–53 (2002)
17. Coello Coello, C.A., Sierra, M.R.: A Coevolutionary Multi-Objective Evolutionary Algorithm. In: Proceedings of the 2003 Congress on Evolutionary Computation, vol. 1, pp. 482–489 (2003)
18. Deb, K., Gupta, H.: Introducing robustness in multiobjective optimization. Kanpur Genetic Algorithms Lab. (KanGAL), Indian Institue of Technology, Kanpur, India, Technical Report 2004016 (2004)
19. Deb, K., Padmanabhan, D., Gupta, S., Kumar Mall, A.: Reliability-based multiobjective optimization using evolutionary algorithms. In: Proceedings of the Fourth International Conference on Evolutionary Multi-Criterion Optimization, pp. 66–80 (2007)
20. Deb, K., Udaya Bhaskara Rao, N., Karthik, S.: Dynamic Multi-Objective Optimization and Decision-Making Using Modified NSGA-II: A Case Study on Hydro-Thermal Power Scheduling. Kanpur Genetic Algorithms Lab. (KanGAL), Indian Institue of Technology, Kanpur, India, Technical Report 2006008 (2006)
21. Deb, K., Agrawal, S., Pratap, A., Meyarivan, T.: A fast and elitist multiobjective genetic algorithm: NSGA-II. IEEE Transactions on Evolutionary Computation 6(2), 182–197 (2002)
22. Farina, M., Deb, K., Amato, P.: Dynamic Multiobjective Optimization Problems: Test Cases, Aproximations, and Applications. IEEE Transactions on Evolutionary Computation 8(5), 425–442 (2004)
23. Farina, M., Amato, P.: A fuzzy definition of "optimality" for many-criteria optimization problems. IEEE Transactions on Systems, Man, and Cybernetics-Part A: Systems and Humans 34(3), 315–326 (2003)
24. Fieldsend, J.E., Everson, R.M.: Multi-objective Optimisation in the Presence of Uncertainty. In: Proceedings of the 2005 Congress on Evolutionary Computation, vol. 1, pp. 243–250 (2005)
25. Ghosh, A., Tstutsui, S., Tanaka, H.: Function optimization in non-stationary environment using steady state genetic algorithms with aging of individuals. In: Proceedings of 1998 IEEE Congress on Evolutionary Computation (CEC 1998), pp. 666–671 (1998)
26. Goldberg, D.E.: The Design of Innovation: Lessons from and for Competent Genetic Algorithms. Kluwer Academic Publishers, Dordrecht (2002)
27. Goh, C.K., Tan, K.C.: An investigation on noisy environments in evolutionary multiobjective optimization. IEEE Transactions on Evolutionary Computation 11(3), 354–381 (2007)
28. Goh, C.K., Tan, K.C.: Evolving the tradeoffs between Pareto-optimality and Robustness in Multi-Objective Evolutionary Algorithms. In: Yang, S., Ong, Y.S., Jin, Y. (eds.) Evolving the tradeoffs between Pareto-optimality and Robustness in Multi-Objective Evolutionary Algorithms, pp. 457–478. Springer, Heidelberg (2007)
29. Goh, C.K., Tan, K.C., Cheong, C.Y., Ong, Y.S.: Noise Induced Features in Robust Multiobjective Optimization Problems. In: Proceedings of the 2007 IEEE Congress on Evolutionary Computation, pp. 568–575 (2007)

30. Goh, C.K., Teoh, E.J., Tan, K.C.: Hybrid multiobjective evolutionary design for artificial neural networks. IEEE Transactions on Neural Networks (accepted)
31. Grefenstette, J.J.: Genetic algorithms for changing environments. In: Proceedings of the Second International Conference on Parallel Problem Solving from Nature, pp. 137–144 (1992)
32. Grefenstette, J.J., Ramsey, C.L.: An approach to anytime learning. In: Proceedings of the Ninth International Conference on Machine Learning, pp. 41–49 (1987)
33. Gunawan, S., Azarm, S.: Multi-objective robust optimization using a sensitivity region concept. Structural and Multidisciplinary Optimization 29, 50–60 (2005)
34. Gupta, H., Deb, K.: Handling constraints in robust multi-objective optimization. In: Proceedings of the 2005 IEEE Congress on Evolutionary Computation, pp. 25–32 (2005)
35. Hatzakis, I., Wallace, D.: Dynamic Multi-Objective Optimization with Evolutionary Algorithms: A Foward-Looking Approach. In: Proceedings of the 2006 Genetic and Evolutionary Computation Congress, pp. 1201–1208 (2006)
36. Hughes, E.J.: Evolutionary multi-objective ranking with uncertainty and noise. In: Proceedings of the First Conference on Evolutionary Multi-Criterion Optimization, pp. 329–343 (2001)
37. Hughes, E.J.: Constraint handling with uncertain and noisy multi-objective evolution. In: Proceedings of the 2001 Congress on Evolutionary Computation, vol. 2, pp. 963–970 (2001)
38. Hughes, E.J.: Multi-objective Probabilistic Selection Evolutionary Algorithm (MOPSEA). Technical Report No. DAPS/EJH/56/2000, Department of Aerospace, POwer & Sensors, Cranfield University (2000)
39. Iorio, A.W., Li, X.: A Cooperative Coevolutionary Multiobjective Algorithm Using Non-dominated Sorting. In: Proceedings of the 2004 Genetic and Evolutionary Computation Congress, pp. 537–548 (2004)
40. Jin, Y., Branke, J.: Evolutionary Optimization in Uncertain Environments–A Survey. IEEE Transactions on Evolutionary Computation 9(3), 303–317 (2005)
41. Jin, Y., Sendhoff, B.: Constructing Dynamic Optimization Test Problems Using the Multi-objective Optimization Concept. In: Proceedings of the 2004 EvoWorkshops, pp. 525–536 (2004)
42. Jin, Y., Sendhoff, B.: Tradeoff between performance and robustness: An evolutionary multiobjective approach. In: Proceedings of the Second Conference on Evolutionary Multi-Criterion Optimization, pp. 237–251 (2003)
43. Keerativuttiumrong, N., Chaiyaratana, N., Varavithya, V.: Multiobjective cooperative coevolutionary genetic algorithm. In: Proceedings of the Seventh International Conference on Parallel Problem Solving from Nature, pp. 288–297 (2002)
44. Khor, E.F., Tan, K.C., Lee, T.H., Goh, C.K.: A study on distribution preservation mechanism in evolutionary multi-objective optimization. Artificial Intelligence Review 23(1), 31–56 (2005)
45. Laumanns, M., Zitzler, E., Thiele, L.: A unified model for multi-objective evolutionary algorithms with elitism. In: Proceedings of the 2000 Congress on Evolutionary Computation, vol. 1, pp. 46–53 (2000)
46. Li, M., Azarm, S., Aute, V.: A multi-objective genetic algorithm for robust design optimization. In: Proceedings of the 2005 Genetic and Evolutionary Computation Conference, pp. 771–778 (2005)
47. Lim, D., Ong, Y.S., Lim, M.H., Jin, Y.: Single/Multi-objective Inverse Robust Evolutionary Design Methodology in the Presence of Uncertainty. In: Yang, S., Ong, Y.S., Jin, Y. (eds.) Evolutionary Computation in Dynamic and Uncertain Environments, Springer, Heidelberg (in press)

48. Limbourg, P.: Multi-objective optimization of Problems with Epistemic Uncertainty. In: Proceedings of the Third International Conference on Evolutionary Multi-Criterion Optimization, pp. 413–427 (2005)
49. Lohn, J.D., Kraus, W.F., Haith, G.L.: Comparing a coevolutionary genetic algorithm for multiobjective optimization. In: Proceedings of the 2002 Congress on Evolutionary Computation, pp. 1157–1162 (2002)
50. Lu, H., Yen, G.G.: Rank-based multiobjective genetic algorithm and benchmark test function study. IEEE Transactions on Evolutionary Computation 7(4), 325–343 (2003)
51. Maneeratana, K., Boonlong, K., Chaiyaratana, N.: Multi-objective Optimisation by Co-operative Co-evolution. In: Yao, X., Burke, E.K., Lozano, J.A., Smith, J., Merelo-Guervós, J.J., Bullinaria, J.A., Rowe, J.E., Tiňo, P., Kabán, A., Schwefel, H.-P. (eds.) PPSN 2004. LNCS, vol. 3242, pp. 772–781. Springer, Heidelberg (2004)
52. Mori, N., Kita, H., Nishikawa, Y.: Adaptation to a changing environment by means of the thermodynamical genetic algorithm. In: Ebeling, W., Rechenberg, I., Voigt, H.-M., Schwefel, H.-P. (eds.) PPSN 1996. LNCS, vol. 1141, pp. 513–522. Springer, Heidelberg (1996)
53. Nissen, V., Propach, J.: On the robustness of population based versus point-based optimization in the presence of noise. IEEE Transactions on Evolutionary Computation 2(3), 107–119 (1998)
54. Ong, Y.S., Nair, P.B., Lum, K.Y.: Min-Max Surrogate Assisted Evolutionary Algorithm for Robust Aerodynamic Design. IEEE Transactions on Evolutionary Computation 10(4), 392–404 (2006)
55. Paenke, I., Branke, J., Jin, Y.: Efficient Search for Robust Solutions by Means of Evolutionary Algorithms and Fitness Approximation. IEEE Transactions on Evolutionary Computation 10(4), 405–420 (2006)
56. Potter, M.A., Jong, K.A.: Cooperative coevolution: An architecture for evolving coadapted subcomponents. Evolutionary Computation 8(1), 1–29 (2000)
57. Rana, S., Whitney, D., Cogswell, R.: Searching in the presence of noise. In: Ebeling, W., Rechenberg, I., Voigt, H.-M., Schwefel, H.-P. (eds.) PPSN 1996. LNCS, vol. 1141, pp. 198–207. Springer, Heidelberg (1996)
58. Rattray, M., Shapiro, J.: Noisy fitness evaluations in genetic algorithms and the dynamics of learning. In: Belew, R.K., Vose, M.D. (eds.) Foundations of Genetic Algorithms, vol. 4, pp. 117–139. Morgan Kaufmann, San Francisco (1997)
59. Ray, T.: Constrained robust optimal design using a multiobjective evolutionary algorithm. In: Proceedings of the 2002 Congress on Evolutionary Computation, pp. 419–424 (2002)
60. Rosin, C.D., Belew, R.K.: New methods for competitive coevolution. Evolutionary Computation 5(1), 1–29 (1997)
61. Rudolph, G.: A partial order approach to noisy fitness functions. In: Proceedings of the 2001 Congress on Evolutionary Computation, vol. 1, pp. 318–325 (2001)
62. Sano, Y., Kita, H.: Optimization of noisy fitness functions by means of genetic algorithms using history of search with test of estimation. In: Proceedings of the 2002 Congress on Evolutionary Computation, vol. 1, pp. 360–365 (2002)
63. Singh, A.: Uncertainty based Multi-objective Optimization of Groundwater Remediation Design. Master's Thesis, University of Illinois at Urbana-Champaign (2003)
64. Tan, K.C., Goh, C.K.: A Competitive-Cooperation Coevolutionary Paradigm for Dynamic Multi-objective Optimization. IEEE Transactions on Evolutionary Computation (accepted)

65. Tan, K.C., Yang, Y.J., Goh, C.K.: A distributed cooperative coevolutionary algorithm for multiobjective optimization. IEEE Transactions on Evolutionary Computation 10(5), 527–549 (2006)
66. Tan, K.C., Cheong, C.Y., Goh, C.K.: Solving multiobjective vehicle routing problem with stochastic demand via evolutionary computation. European Journal of Operational Research 177, 813–839 (2007)
67. Tan, K.C., Lee, T.H., Khor, E.F., Ang, D.C.: Design and real-time implementation of a multivariable gyro-mirror line-of-sight stabilization platform. Fuzzy Sets and Systems 128(1), 81–93 (2002)
68. Teich, J.: Pareto-front exploration with uncertain objectives. In: Proceedings of the First Conference on Evolutionary Multi-Criterion Optimization, pp. 314–328. Springer, Heidelberg (2001)
69. Tsutsui, S., Ghosh, A.: A comparative study on the effects of adding perturbations to phenotypic parameters in genetic algorithms with a robust solution searching scheme. In: Proceedings of the 1999 IEEE International Conference on Systems, Man, and Cybernetics, pp. 585–591 (1999)
70. Tsutsui, S., Ghosh, A.: A comparative study on the effects of adding perturbations to phenotypic parameters in genetic algorithms with a robust solution searching scheme. In: Proceedings of the 1999 IEEE International Conference on Systems, Man, and Cybernetics, pp. 585–591 (1999)
71. Tsutsui, S., Ghosh, A.: Genetic algorithms with a robust solution searching scheme. IEEE Transactions on Evolutionary Computation 1(3), 201–208 (1997)
72. Ursem, R.K.: Multinational GA optimization techniques in dynamic environments. In: Proceedings of the 2000 Genetic and Evolutionary Computation Congress, pp. 19–26 (2000)
73. Vavak, F., Jukes, K., Fogarty, T.C.: Adaptive combustion balancing in multiple burner boiler using a genetic algorithm with variable range of local search. In: Proceedings of the Seventh International Conference on Genetic Algorithms, pp. 719–726 (1997)
74. Wineberg, M., Oppacher, F.: Enhancing the GA ability to cope with dynamic environments. In: Proceedings of the 2000 Genetic and Evolutionary Computation Congress, p. 310 (2000)
75. Zeng, S.Y., Chen, G., Zheng, L., Shi, H., Garis, H., Ding, L., Kang, L.: A Dynamic Multi-objective Evolutionary Algorithm Based on an Orthogonal Design. In: Proceedings of the 2006 IEEE Congress on Evolutionary Computation, pp. 573–580 (2006)
76. Zhang, Z.: Multiobjective optimization immune algorithm in dynamic environments and its application to greenhouse control. In: Proceedings of the 2005 IEEE Congress on Evolutionary Computation, pp. 714–719 (2005)
77. Zitzler, E., Laumanns, M., Thiele, L.: SPEA2: Improving the Strength Pareto Evolutionary Algorithm. Technical Report 103, Computer Engineering and Networks Laboratory (TIK), Swiss Federal Institute of Technology (ETH) Zurich, Switzerland (2001)
78. Zitzler, E., Deb, K., Thiele, L.: Comparison of multiobjective evolutionary algorithms: empirical results. Evolutionary Computation 8(2), 173–195 (2000)
79. Zitzler, E., Thiele, L.: Multiobjective evolutionary algorithms: a comparative case study and the strength Pareto approach. IEEE Transactions on Evolutionary Computation 3(4), 257–271 (1999)

VCV2 – Visual Cluster Validity

Jacalyn M. Huband[1] and James C. Bezdek[2]

[1] Computer Science Department
University of West Florida
Pensacola, FL 32514, USA
[2] Department of Computer Science and Software Engineering
University of Melbourne
Parkville, 3010, Melbourne
Victoria, Australia

Abstract. All clustering algorithms partition data into a specified or algorithmically determined number of clusters, whether or not that number of clusters actually exists in the data. Therefore, identifying a "best" solution amongst a set of candidate partitions is an important step in the clustering process. This paper presents a visual technique for comparing found partitions with a pre-clustering VAT (Visual Assessment of cluster Tendency) image of the unlabeled input data. The method is developed independent of any particular clustering algorithm, and then illustrated with numerical examples that use the fuzzy c-means clustering method. The experiments use samples from mixtures of bivariate normals, a bivariate uniform, and a small real data set to illustrate the efficacy of the method.

Keywords: cluster validity, cluster analysis, visual assessment of cluster tendency, visual cluster validty.

1 Introduction

Researchers have long recognized the challenge of determining how many clusters to look for in a set of unlabeled data. In 1953, R. L. Thorndike [1] asked, **"How then shall we decide upon the value of k – the number of families or clusters?** And once k has been determined, how shall we decide upon the boundaries and the centroids of the various clusters?" After considering these questions, Thorndike went on to say "Let us start with the second problem first [finding the clusters], because it looks somewhat more docile and amenable to attack [than deciding how many to look for]..." The technique that Thorndike employed (and which is still frequently used today) was to enumerate candidate solutions by applying a clustering algorithm to data for different values of k. Referring to this technique, Thorndike stated, "One would like some type of significance test of the change in the criterion as k increases from 2 to 3 to 4, and so on. But I am unable to produce such a test."

In effect, Thorndike raised two issues related to clustering: i) how can we *assess* the number of clusters in a data set *prior* to clustering? and ii) how can we *validate* solutions produced by a clustering algorithm *post* clustering). Assessment and validity

J.M. Zurada et al. (Eds.): WCCI 2008 Plenary/Invited Lectures, LNCS 5050, pp. 293–308, 2008.

have attracted much attention ever since, but continue to be problematical areas for clustering.

Consider a set of n objects $O = \{o_1,...,o_n\}$. The objects might be types of fish, malignant tumors, edible tubers, cigars, motorcycles - virtually anything. Numerical *object* data has the form $X = \{\mathbf{x}_1,...,\mathbf{x}_n\} \subset \Re^p$, where the coordinates of $\mathbf{x}_i$ provide feature values (e.g., weight, length, alcohol content, wrapper shape, exhaust pipes, etc.) describing object o_i. The second data structure commonly used to represent the objects in O is numerical *relational* data, which consist of n^2 similarities (or dissimilarities) between pair of objects in O, represented by an nxn relational matrix $R_n = [R_{ij} = (\text{dis-})\text{similarity}(o_i, o_j) \mid 1 \le i, j \le n]$. We can always convert X into *dissimilarity* data $D_n = D_n(X)$, where $d_{ij} = \left\| \mathbf{x}_i - \mathbf{x}_j \right\|$ in any vector norm on $\Re^p$, so most relational clustering algorithms are (implicitly) applicable to object data. However, there are both similarity and dissimilarity relational data sets that do not begin as object data, and for these, we have no choice but to use a relational algorithm. We will refer to these two types of data sets as X and D_n.

Clustering in unlabeled data X or D_n is the assignment of *labels* to the objects in O. To consider possible solutions for the clustering problem, let (c) be an integer, $1 \le c \le n$ (our c is Thorndike's k; amazingly, the use of an integer other than k for this dummy variable still causes controversy). We include $c = 1$ and $c = n$ so that algorithms such as the SAHN clustering methods, which begin or end with singleton clusters ($c = n$) or the universal cluster ($c = 1$) are included in the general discussion. The *c-partitions* of X are sets of (cn) values $\{u_{ik}\}$ that can be conveniently arrayed as a $c \times n$ matrix $U = [u_{ik}]$. There are three sets of partition matrices:

$$M_{pcn} = \{U \in \Re^{cn} \mid 0 \le u_{ik} \le 1 \forall i, k; \exists i \ni u_{ik} > 0 \forall k\} \tag{1}$$

$$M_{fcn} = \{U \in M_{pcn} \mid \sum_{i=1}^{c} u_{ik} = 1 \forall k\} \tag{2}$$

$$M_{hcn} = \{U \in M_{fcn} \mid u_{ik} \in \{0,1\} \forall i, k\} \tag{3}$$

Equations (1-3) define, respectively, the sets of non-degenerate possibilistic, constrained fuzzy *or* probabilistic, and crisp *c-partitions of X*. So, there are *four* kinds of label vectors (the k-th column of U is a label vector in $\Re^c$ for object o_k in O); and hence, four kinds of partitions, corresponding to the four types of models we use for clustering. But only three of the four kinds are mathematically distinct. (Fuzzy and probabilistic label vectors and c-partitions are mathematically indistinguishable, having entries between 0 and 1 that sum to 1 over each column. However, the assumptions we make and methods we use as a basis for these two kinds of clustering is very different.) If the label vectors are hard (crisp), we hope they identify c natural subgroups in X.

Each run of a clustering algorithm produces one or more U's in some M_{pcn} (this covers all cases, since $M_{hcn} \subset M_{fcn} \subset M_{pcn}$). For example, if D_n contains $n*(n-1)/2$ distinct dissimilarities, applying the single linkage algorithm to D_n produces exactly one U in M_{hjn}, j=n, n-1, …1. Or, fixing all control and model parameters except c, applying relational fuzzy c-means to D_n produces one U in M_{fjn} for each j=2, 3, …,n-1.

Other runs on Dn with variations of these or other clustering algorithms may produce other U's for consideration. Now we collect all the *candidate partitions* into a set $CP(D_n)$, and ask: which of these U's is the most satisfactory explanation of substructure in O? This is the cluster validity (more simply, "validation") problem. In what follows we refer to $CP(X)$ or $CP(D_n)$ more briefly as CP, the data set underlying any set of candidate partitions being understood.

The validation question is in a sense ill-posed, for we can easily demonstrate that more than one U in CP may be the "best answer". For example, how many clusters are there in a deck of cards, each object being one of 52 cards in a standard deck? Well, if the defining property of the clusters is card color, then c=2 (with 26 objects in each cluster); if the property is suit, c=4 (with 13 objects in each cluster); if rank, c=13 with 4 objects in each cluster. And so on. We will put this annoying complication aside for now, but will have more to say about it in the concluding section.

Visually validating the cluster structure suggested by some U in CP is possible only for U in M_{hcn} (crisp partitions), only for object data, and only for $p \leq 3$ (i.e., only when the data can be plotted on one-, two-, or three-dimensional graphs). So, direct visual validation is far too limited to be generally useful. Consequently, many researchers rely on mathematical techniques for validation.

One group of mathematical techniques uses validity functionals, $v : D_v \mapsto \Re$, where D_v, the domain of v, depends on the particular function at hand, but always depends on at least U in CP. The values of v are computed, and then used to rank the attractiveness of the candidates in CP. Usually, the minimum or maximum value of v points to the preferred choice.

When $D_v = M_{hcn}$, v is a *direct validity measure* because $v(U)$ assesses properties of crisp (real) clusters identified by U in the data set; otherwise, it is indirect. Bear in mind that these are mathematically defined clusters in a data set chosen to represent the $\{o_k\}$ - we hope the corresponding clusters in O are useful, but this still need not be the case. There have been countless studies of direct and indirect validity indices for crisp, probabilistic, fuzzy and possibilistic clustering methods [2-13]. Some establish theoretical properties of a particular index; others offer simulations that compare various candidates. There are seemingly countless methods because when X is not labeled, the true parameters of any model attempting to represent substructure in it are (and always will be) unknown. Consequently, validity functionals have little chance of being generally useful for identifying the "best" clustering solution, even within a restricted class of models and/or data processes. More typically, they are relied upon to eliminate egregiously *wrong* solutions; so, they are usually used as part of the processing done prior to final validation by humans and/or rule bases. Put another way, since none of us knows the right answer for the overwhelming majority of all unlabeled data, there is no set of parameters against which we can measure the "fit" of a set of clusters, nor is there structure we can assess visually. Nonetheless, many important applications (such as image segmentation, data mining and knowledge discovery) will tolerate a "best under these circumstances" solution, so we continue to study, and hopefully improve, methods based on validity indices. However, the technique we present here lies in a different direction.

Here is a preview of the new method. The VAT algorithm (Visual Assessment of cluster Tendency, [14]), reorders the row and columns of any $n \times n$ matrix R (where r_{ij} is a scaled dissimilarity value between objects o_i and o_j) with a modified version of

Prim's minimal spanning tree algorithm. If the image I(R*) of the reordered version of R has c dark blocks along its main diagonal, this suggests that R contains c (as yet unfound) clusters. The size of each block may even indicate the approximate size of the suggested cluster. VAT is well defined for X or D_n, has been generalized to n of arbitrary size (scalable sVAT, [15]), and can also be used for assessment of possible clusters in rectangular relational data (co-clusters coVAT, [16]).

After determining any set of candidate partitions of the objects by any method whatsoever (the U's can be crisp or soft, from data set X or D_n of any size), we re-order the columns of each U to $\widehat{U}$ (this amounts to permuting the objects in O) using the index array of R*. Next, we transform each $\widehat{U}$ to the square matrix $U^* = \mathbf{1}_n - (\widehat{U}^T \widehat{U} / \max\{(\widehat{U}^T \widehat{U})_{ij}\})$. Finally, we (visually) compare the fixed image I(R*) derived from the data *before* clustering to the images in the set of (modified) partitions, {I(U*): U∈CP}. Finally, we visually compare I(R*) to each I(U*). Our heuristic is that a good (visual) match indicates a "good" set of candidate clusters. In other words, if the clustering algorithm performs well, we expect I(U*) to show well-defined dark blocks down the main diagonal, similar those seen in the VAT image I(R*).

The remainder of this article is organized as follows. In Section 2 we review the VAT algorithm. Section 3 describes our approach for converting the partition matrix, U, to an nxn image matrix I(U*). Section 4 has several numerical examples. Section 5 contains our summary, and some ideas for future research.

2   The VAT Image

The VAT algorithm displays a grayscale image I(R*), where each element of I(R*) is a scaled dissimilarity value between objects o_i and o_j. Being dissimilarity measures, each element on the diagonal is zero. Off the diagonal, the scaled values range from 0 to 1. Values close to 0 represent small dissimilarities, and values close to 1 represent strong dissimilarities. If an object is a member of a cluster, then it also should be part of a submatrix of "similarly small" values, whose diagonal is superimposed on the diagonal of the image matrix. These submatrices are seen are dark blocks along the diagonal of the VAT image. We next give a concise statement of the VAT algorithm. In general, the functions arg max and arg min, in Steps 1 and 2, are set-valued, and when the sets contain more than one pair of optimal arguments, any optimal pair can be selected. Throughout, we will suppress n, the size of R or D, we will denote the result of applying VAT to R as R*=VAT(R), and we call the displayed output image I(R*).

Algorithm VAT: Visual Assessment of Cluster Tendency in Square Dissimilarity Data [14]

Input: An $n \times n$ matrix of dissimilarities R = $[r_{ij}]$ satisfying, for all $1 \le i, j \le n$: $r_{ij} = r_{ji}$, $r_{ij} \ge 0$, and $r_{ii} = 0$.

Step 1.   Set I = $\varnothing$; J = {1,2,...,n}; P = (0, 0, ..., 0)
            Select (i,j) $\in$ $\underset{p\in J, q\in J}{\arg\max}$ {r_{pq}} ; Set P(1) = j; Replace I $\leftarrow$ I $\cup$ {j} and J $\leftarrow$ J – {j}.

Step 2. For t = 2,..., n:

Select $(i,j) \in \underset{p \in I, q \in J}{\arg\min} \{r_{pq}\}$: Set P(t) = j; Replace $I \leftarrow I \cup \{j\}$ and $J \leftarrow J - \{j\}$.

Next t.

Step 3. Form the ordered dissimilarity matrix $R^* = [r_{ij}^*] = [r_{P(i)P(j)}]$ for $1 \leq i,j \leq n$.

Output: Image $I(R^*)$, scaled so that $\underset{i,j}{\max}\{r_{ij}^*\}$ corresponds to white and $\underset{i,j}{\min}\{r_{ij}^*\}$ to

black.   □

Figure (1a) is a scatterplot of n=1000 data points in $\Re^2$ drawn from a mixture of five normal distributions. The means, mixing proportions, and number of samples in each cluster (i.e., cardinality n_i, i = 1, 2, 3, 4, 5) are listed in Table 1. The covariance matrices are $\Sigma_1 = \Sigma_2 = \Sigma_3 = \Sigma_4 = \Sigma_5 = \sigma^2 I$, where I is the 2×2 identity matrix and $\sigma^2 = 1.2$.

These object data were converted to a 1000x1000 dissimilarity matrix R by computing $r_{ij} = \|\mathbf{x}_i - \mathbf{x}_j\|$ with the Euclidean norm. The c = 5 visually apparent clusters in Figure (1a) are suggested by the 5 distinct dark diagonal blocks in Figure (1c), which is the VAT image $I(R^*)$ of the data after reordering. Comparing this to view (1b), which is the image $I(R)$ of the dissimilarities in input order, it is clear that reordering is necessary to reveal the structure of the underlying data. Notice that nothing can be inferred about the cluster structure from the image $I(R)$ of the unordered matrix R.
>>>

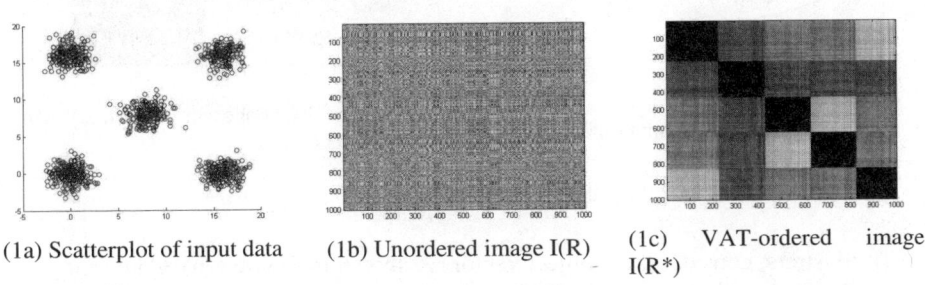

(1a) Scatterplot of input data   (1b) Unordered image I(R)   (1c) VAT-ordered image I(R*)

Fig. 1. An example of VAT that indicates c = 5 clusters reside in the data in view (1a)

Table 1. Data Set X shown in Scatterplot view, Figure (1a)

Mean	mixing proportions	n_i
$\mu_1 = (0,0)$	$\alpha_1 = 0.21$	225
$\mu_2 = (8,8)$	$\alpha_2 = 0.21$	203
$\mu_3 = (16,0)$	$\alpha_3 = 0.21$	197
$\mu_4 = (0,16)$	$\alpha_4 = 0.21$	200
$\mu_5 = (16,16)$	$\alpha_5 = 0.16$	175

3   Transforming the Partition Matrix

Let U be a cxn partition matrix generated by any clustering algorithm. Then $U^T U$ is an nxn matrix whose elements measure of the relationship between pairs of objects in each of the c clusters. To see this, recall that u_{st} is the membership of object o_t in cluster s, so $(u_{ki} u_{kj})$ = (membership of o_i in cluster k)(membership of o_j in cluster k). The ij-th element of $U^T U$ sums this product over the c clusters in U, $(U^T U)_{ij} = \sum_{k=1}^{c} u_{ki} u_{kj}$, and hence, is a measure of the binding between objects i and j over all c clusters. If U is crisp (in M_{hcn}), its elements are all 1's or 0's, and the c products comprising $(U^T U)_{ij}$ will all be zero unless objects i and j are in (say) the k-th cluster, in which case the ij-th entry will have the value 1. (See [17] for a more complete discussion of this idea for crisp and soft partitions.)

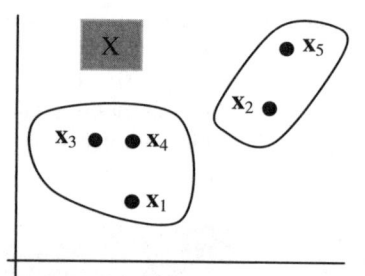

(2a) Scatterplot of input data

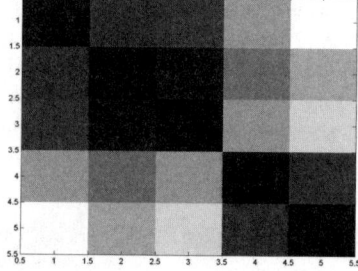

(2b) VAT image I(R*) showing 2 primary and 3 secondary clusters

Fig. 2. Five data points and the resulting VAT image

To illustrate, consider the 5 object vectors X shown in Figure (2a). Converting X to dissimilarity data R with the Euclidean norm enables us to construct the VAT image of R, shown in Figure (2b). This view suggests that the primary (or first order) interpretation of substructure in X is that c=2; and the secondary (or second order) interpretation shown by the imbedded 2×2 block in the upper 3×3 block is for c=3. Visually, this agrees with the input data; the primary partition is $\{x_1, x_3, x_4\} \cup \{x_2, x_5\}$, whereas the secondary partition is $\{x_1\} \cup \{x_3, x_4\} \cup \{x_2, x_5\}$.

A crisp clustering algorithm such as single linkage would generate the 2 - partition $U = \begin{bmatrix} 1 & 0 & 1 & 1 & 0 \\ 0 & 1 & 0 & 0 & 1 \end{bmatrix}$ on X. Taking $U^T U$, we have

$$U^T U = \begin{bmatrix} 1 & 0 \\ 0 & 1 \\ 1 & 0 \\ 1 & 0 \\ 0 & 1 \end{bmatrix} \begin{bmatrix} 1 & 0 & 1 & 1 & 0 \\ 0 & 1 & 0 & 0 & 1 \end{bmatrix} = \begin{bmatrix} 1 & 0 & 1 & 1 & 0 \\ 0 & 1 & 0 & 0 & 1 \\ 1 & 0 & 1 & 1 & 0 \\ 1 & 0 & 1 & 1 & 0 \\ 0 & 1 & 0 & 0 & 1 \end{bmatrix} \quad (4)$$

It is well known that *some* reordering of this matrix will yield a matrix with diagonal sub-blocks that are all 1's, corresponding to the (matrix of) the unique equivalence relation which is in 1-1 correspondence with this 2-partition of the 5 objects [17]. (If we denote the equivalence relation by ~, then $(U^T U)_{ij}=1 \Leftrightarrow x_i \sim x_j$, that is, if and only if objects o_i and o_j, are in the same cluster.)   For this simple example, the permutation array from the VAT reordering shown in Figure (2b) is exactly the reordering that reveals this block matrix. Applying the VAT permutation to (4), $(U^T U)_{perm} =$

$$\begin{bmatrix} 1 & 1 & 1 & 0 & 0 \\ 1 & 1 & 1 & 0 & 0 \\ 1 & 1 & 1 & 0 & 0 \\ 0 & 0 & 0 & 1 & 1 \\ 0 & 0 & 0 & 1 & 1 \end{bmatrix}$$. Finally, to produce an image which is visually comparable (black

is 0, white is 1) to the VAT image, we subtract $U^T U$ from the nxn matrix $\mathbf{1}_n$ (all ones):

$$U^* = \mathbf{1}_{5x5} - (U^T U)_{perm} = \begin{bmatrix} 0 & 0 & 0 & 1 & 1 \\ 0 & 0 & 0 & 1 & 1 \\ 0 & 0 & 0 & 1 & 1 \\ 1 & 1 & 0 & 0 & 0 \\ 1 & 1 & 0 & 0 & 0 \end{bmatrix}.$$

Comparing the ordered image of the transformed partition matrix $I(U^*)$ to the VAT image $I(R^*)$, we see very similar dark blocks along the two diagonals. This means that objects which were relationally close in the VAT analysis are placed in the same clusters during the partitioning process. Thus, we have a visual validation that the partition found is consistent with our pre-clustering estimate. This example captures the essentials of our new method, but we make two modifications of the procedure just described that make it more efficient and useful for soft partitions of data.

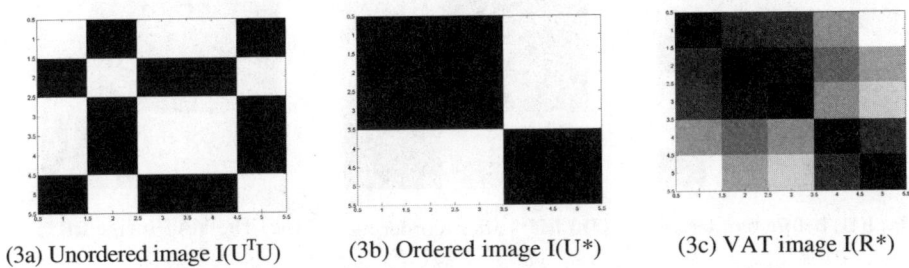

(3a) Unordered image $I(U^T U)$ (3b) Ordered image $I(U^*)$ (3c) VAT image $I(R^*)$

Fig. 3. Transforming a crisp partition to a VAT-like image

First, we point out that the end result of constructing U^* by reordering $U^T U$ with the VAT permutation will be identical to that found by permuting U *before* obtaining $U^T U$. This second method is more efficient than the first because we permute only the n columns of U, instead of the n rows and columns of $U^T U$. And second, when the partition U is not crisp, we normalize the values in $U^T U$ to obtain values in U^* that will provide visually comparable images. The normalization we choose is to divide $U^T U$ by its maximum element. We summarize the overall transformations:

$$(X \text{ or } R) \xrightarrow{\text{VAT permutation } \sigma_v} R^* \rightarrow \xrightarrow{(Pre-Clustering) \text{ Assessment Image}} I(R^*). \qquad (5a)$$

$$(X \text{ or } R) \xrightarrow{\text{Clustering}} U \xrightarrow{\sigma_v} \hat{U} \rightarrow \left(1_n - \left(\frac{\hat{U}^T \hat{U}}{\max_{i,j} \{(\hat{U}^T \hat{U})_{ij}\}} \right) \right) = U^* \xrightarrow{\text{Partition Image}} I(U^*)$$

$$(5b)$$

We illustrate the application of equations (5) for the same set of five objects shown in Figure (2a). A fuzzy c-means partition of X ($m = c = 2$, Euclidean norm) is

$$U = \begin{bmatrix} 0.84 & 0.13 & 0.91 & 0.91 & 0.09 \\ 0.16 & 0.87 & 0.09 & 0.09 & 0.91 \end{bmatrix}. \text{ Whence}$$

$$U^* = \left(1_n - \left(\frac{\hat{U}^T \hat{U}}{\max_{i,j} \{(\hat{U}^T \hat{U})_{ij}\}} \right) \right) = \begin{bmatrix} 0.13 & 0.70 & 0.08 & 0.07 & 0.74 \\ 0.70 & 0.08 & 0.76 & 0.76 & 0.04 \\ 0.08 & 0.76 & 0.01 & 0.01 & 0.80 \\ 0.07 & 0.76 & 0.01 & 0.01 & 0.80 \\ 0.74 & 0.04 & 0.80 & 0.80 & 0.00 \end{bmatrix}. \qquad (6)$$

The image of this transformed partition matrix will have dark gray intensities for numbers significantly close to zero and lighter gray intensities for numbers approaching one. The results are shown in Figure 4. Visual comparison of I(R*) to I(U*) in Figures (4b) and (4c) again suggests that the pre- and post-clustering results are in close agreement, allowing us to conclude that U is a fairly accurate representation of cluster structure suggested by VAT in X. A formal statement of the new *visual cluster validity* (VCV2) algorithm is nothing more than the use of equations (5) on a found partition coupled with the pre-clustering VAT image of the input data.

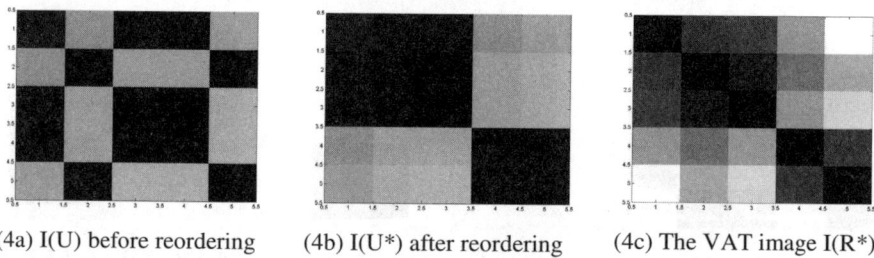

(4a) I(U) before reordering   (4b) I(U*) after reordering   (4c) The VAT image I(R*)

Fig. 4. Visual validation of a fuzzy partitioning of X in Figure (2a)

Algorithm VCV2: Visual cluster validity
Inputs: An unlabeled object data set
$$X = \{x_1, \ldots, x_n\} \subset \Re^P \rightarrow R = [r_{ij}] = \|x_i - x_j\|, 1 \le i, j \le n \qquad \text{(or)}$$

An unlabeled dissimilarity data set $R = [r_{ij}] \subset \Re^{nn}$ satisfying, for $1 \le i, j \le n$:
$r_{ij} = r_{ji}$, $r_{ij} \ge 0$, and $r_{ii} = 0$

The VAT permutation array $P = \sigma_v$ of indices of R and VAT image I(R*)

Compute:

$$(X \text{ or } R) \xrightarrow{\underset{\text{Algorithm}}{\text{Clustering}}} U \xrightarrow{\sigma_v} \hat{U} \rightarrow \left(\mathbf{1}_n - \left(\frac{\hat{U}^T \hat{U}}{\underset{i,j}{\max}\{(\hat{U}^T\hat{U})_{ij}\}} \right) \right) = U^* \xrightarrow{\text{Partition Image}} I(U^*)$$

Compare : $I(U^*)$ to $I(R^*)$.   ❏

If the dark blocks along the diagonal of $I(U^*)$ are not consistent with the dark blocks in the VAT image $I(R^*)$, we know that a problem exists either (i) prior to clustering, in the VAT assessment cluster tendency; or (ii) in the partition U of the data found by the chosen method of clustering. Different algorithms are used to generate $I(R^*)$ and $I(U^*)$, so visually similar images buoys our confidence in both processes. There is no doubt, however, than VAT and/or VCV2 can fail, and if the comparison reveals very different images, we may be sure that something is amiss, cueing us to investigate more fully.

4 Numerical Examples

To test the VCV2 algorithm, we consider four data sets, including one that has no visual cluster structure. We will generate a set CP of candidate partitions in each example by partitioning the object data with "correct" and incorrect numbers of clusters. All candidate partitions are generated by the fuzzy c-means (FCM) algorithm with m = 2, the Euclidean norm for the objective function, and various values of c. For each data set, the initial partition was chosen (somewhat arbitrarily) by assigning the jth object to the ith cluster using the formula $i = \text{mod}(j,c)+1$. The stopping tolerances were $\varepsilon = 0.00001$. The relational data inputted to VAT for each of these four test sets were computed with a scaled Euclidean norm, $R = [r_{ij}] = \dfrac{\left\| x_i - x_j \right\|}{\underset{i,j}{\max} \left\| x_i - x_j \right\|}$, $1 \le i, j \le n$,

where $\left\| x_i - x_j \right\| = \sqrt{x_i^T x_j}$.

Example 1: Five distinct clusters
The data for this example are the same as those shown in Figure (1a) and discussed in Section 2, Table 1. The statistics from the FCM runs are provided in Table 2. Figures (5a) and (5b) are Figures (1a) and (1c), repeated here for ease of visual comparison. The VAT image of the data, Figure (5b), clearly shows five dark blocks along the diagonal, suggesting again that the data possess c=5 clusters.

For the VCV2 images at c = 2 and c = 4, the dark blocks differ visibly from those in the VAT image, so we reject these two candidates. When we compare the VCV2 images in Figures (5e) and (5f) to the VAT image $I(R^*)$, we see (almost) the same dark blocks along the diagonal at c=5 and c=6. The principal difference between these two images is that the center block in the image for c=6 is somewhat lighter in color. This discrepancy between the number of blocks (6) and the number of clusters (5)

Table 2. FCM Statistics for the Five Distinct Clusters

c	Number of iterations	Step size at termination
2	114	9.8426e-006
4	34	5.841e-006
5	16	8.6705e-006
6	71	9.0335e-006

may indicate that a very small subset of points (possibly a singleton) was chosen as the sixth cluster. (Recall that a clustering algorithm will force a partition to exhibit the prescribed number of clusters.) This interesting result suggests that overestimates of c are less pervasive in obscuring substructure than underestimates are; the VCV2 image continues to suggest that c=5 is the preferred solution. Because the number (and even the relative size) of the dark blocks is the same in the VAT and VCV2 images for c=5, we are most confident that among the four choices in this set of CPs, the FCM partition found at c=5 is the best choice.

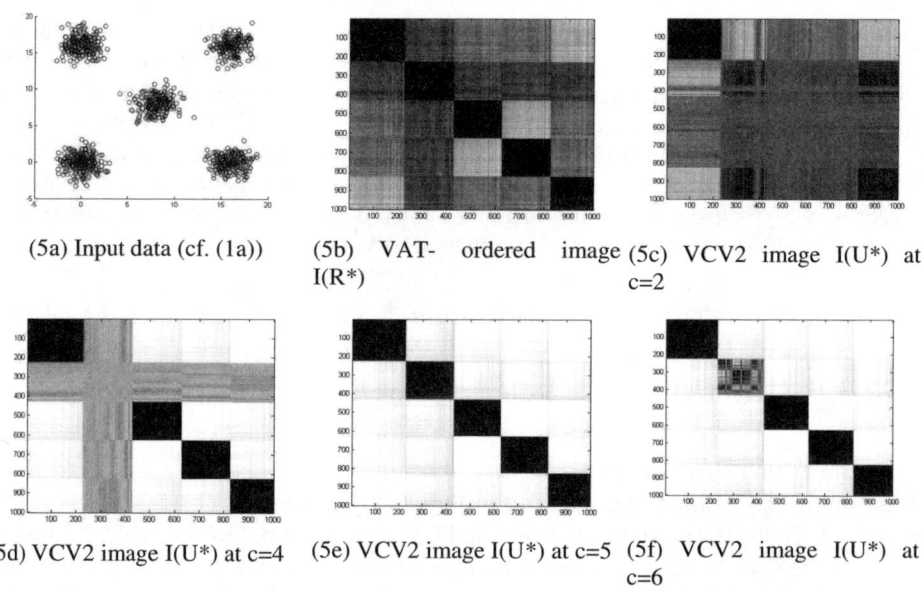

(5a) Input data (cf. (1a))   (5b) VAT- ordered image (5c) VCV2 image I(U*) at
I(R*)   c=2

(5d) VCV2 image I(U*) at c=4   (5e) VCV2 image I(U*) at c=5   (5f) VCV2 image I(U*) at
c=6

Fig. 5. VAT image I(R*) and VCV2 images I(U*) at c=2, 4, 5 and 6 for the Five Clusters dataset

Example 2: Three blended clusters

Next, we consider data where the clusters are not as distinct as those in Example 1. Figure (6a) shows a data set containing 5000 points in $\Re^2$ that form three blended clusters, drawn from a mixture of c=3 bivariate normal distributions with the following component parameters: mixing proportions $\alpha_1 = 0.15$, $\alpha_2 = 0.35$ $\alpha_3 = 0.50$; means $\mu_1 = (0,0)$, $\mu_2 = (3,4)$, $\mu_3 = (6,0)$; covariance matrices $\Sigma_1 = \Sigma_2 = \Sigma_3 = \sigma^2$ I, where I is the 2×2 identity matrix and $\sigma^2 = 1.5$. The statistics from the FCM runs are provided in Table 3. The VAT image I(R*) of the data, Figure (6b), shows three (somewhat) dark blocks along the diagonal, with medium gray off the diagonal. Because the VAT image lacks strong black and white intensities, we know that the three clusters are not as compact and well separated as those in Example 1. Instead, they are relatively close to each other, and most likely have a high degree of overlap. Nonetheless, the estimate of c suggested by I(R*) is c=3. Figures (6c)-(6f) show I(U*) for each of four candidate partitions found by FCM at c=2, 3, 4 and 5.

Table 3. FCM Statistics for the Three Blended Clusters

c	Number of iterations	Step size at termination
2	18	7.1991e-006
3	23	3.9098e-006
4	158	9.8369e-006
5	265	9.9233e-006

When we compare the VCV2 images to the VAT image, the most attractive visual match is Figure (6d) at c = 3. We see the same dark blocks along the diagonal, but more noticeable on the VCV2 image are the L-shaped gray bands edging the middle

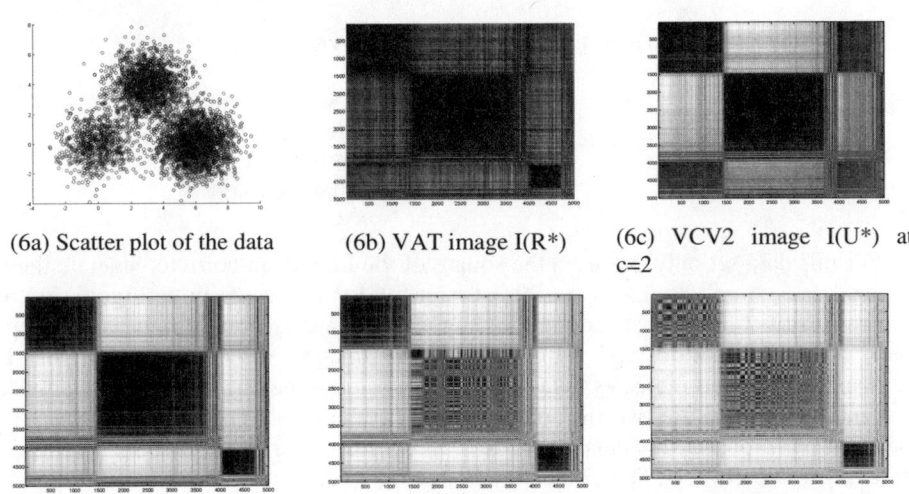

(6a) Scatter plot of the data   (6b) VAT image I(R*)   (6c) VCV2 image I(U*) at c=2

(6d) VCV2 image I(U*) at c=3   (6e) VCV2 image I(U*) at c=4   (6f) VCV2 image I(U*) at c=5

Fig. 6. VAT image I(R*) and VCV2 images I(U*) at c = 2, 3, 4 and 5 for the three clusters dataset

and bottom-right dark blocks. These bands are formed when data points are equally-likely (or near equally-likely) to belong to each of the clusters. The VCV2 images associated with the FCM partitions at c = 2, 4 and 5 all show some affinity to indicate that c = 3. Again, visual display of candidate partitions with overestimates (c = 4 and c=5) do not deteriorate into images with rough approximations to 4 and 5 diagonal blocks.

Example 3: Real-World data with an unknown number of clusters
Our next example involves a data set from the UCI Machine Learning Repository which is available at http://www.ics.uci.edu/~mlearn/MLSummary.html. The data are generated from the Congressional Voting Records Database, and consist of the 1984 records of the 435 United States House of Representatives on 16 key votes. We will refer to this data as the VOTE data set. The data in the original database consist of "y" for yea, "n" for nay, and "?" for unknown disposition. Also, the two identified classes are Republican (54.8%) and Democrat (45.2%), but this does not guarantee that the numerical data contain two geometrically well-defined clusters.

To represent the data numerically, we chose the values 0.5 for yea, -0.5 for nay, and 0 for unknown disposition. Thus, the voting record of each Congressman is represented using an object data vector in $\Re^{16}$. Because the data representing each object (a representative) are column vectors in $\Re^{16}$, we cannot plot the raw data to look for clusters. But, we can use the VAT image at Figure (7a) to estimate the number of clusters. The relational data needed by the VAT algorithm to construct I(R*) are generated from the object data vectors as pair-wise squared Euclidean distances. The statistics from the FCM runs are provided in Table 4.

Table 4. FCM Statistics for the VOTE data

c	Number of iterations	Step size at termination
2	20	5.6706e-006
3	79	9.4474e-006

For this data set only, we used the square of the Euclidean norm to generate the relational data, simply to improve contrast intensity for visual acuity. In Figure (7a), we see two distinct dark blocks along the diagonal and a gray area forming an L-shaped band along the bottom and right edge of the image. Again, assuming that the L-shaped gray band indicates data which do not clearly belong to any known cluster, we guess that there are two first-order clusters in the data. The VCV2 images I(U*) of FCM partitions of this data at c=2 and c=3 confirm our belief that the VOTE data consist of two clusters. Our conjecture is that the two apparent clusters correspond to Democrats and Republicans, and the badly defined regions in the bottom right corner of all three images correspond to voters crossing party lines.

Example 4: Data with no clusters
To understand what the VCV2 algorithm might suggest when candidate partitions for clusters that are not present in the data, we generated a set of 1000 points uniformly distributed in $[0,1] \times [0,1]$. Suppose that for these data, scatter plotted in Figure (8a),

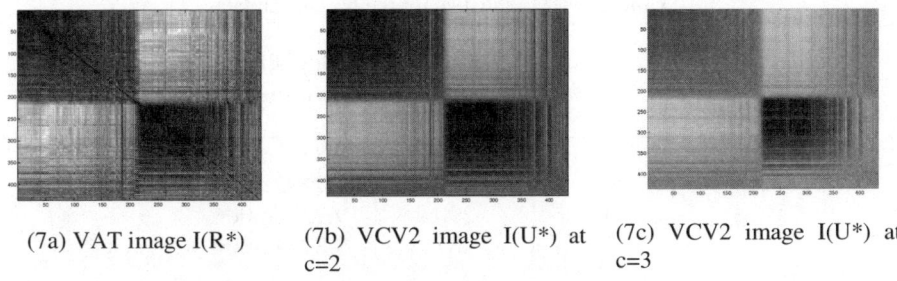

(7a) VAT image I(R*)   (7b) VCV2 image I(U*) at   (7c) VCV2 image I(U*) at
                             c=2   c=3

Fig. 7. VAT image I(R*) and VCV2 images I(U*) at c = 2 and 3 for the VOTE data

we have only the VAT image in Figure (8b) upon which to base an estimate of the number of clusters that may exist in the data. It looks as if there are several dark blocks in the lower right corner, so you might speculate that there is some type of cluster substructure in the data, albeit weak and perhaps, not distinguishable by the reordering procedure used by VAT.

The data were submitted to the FCM algorithm with c = 3, 5, 10, and 20. The statistics from the FCM runs are provided in Table 5. The VCV2 image at c = 3 in Figure (8c) is not a good visual match to the VAT image. What happens for larger values of c? Figures (8d), (8e) and (8f) show VCV2 images at c= 5, 10 and 20. Each of these images hints at substructure for c=4, but they are also increasingly composed of a nondescript shade of gray, indicating that most data points are neither distinct

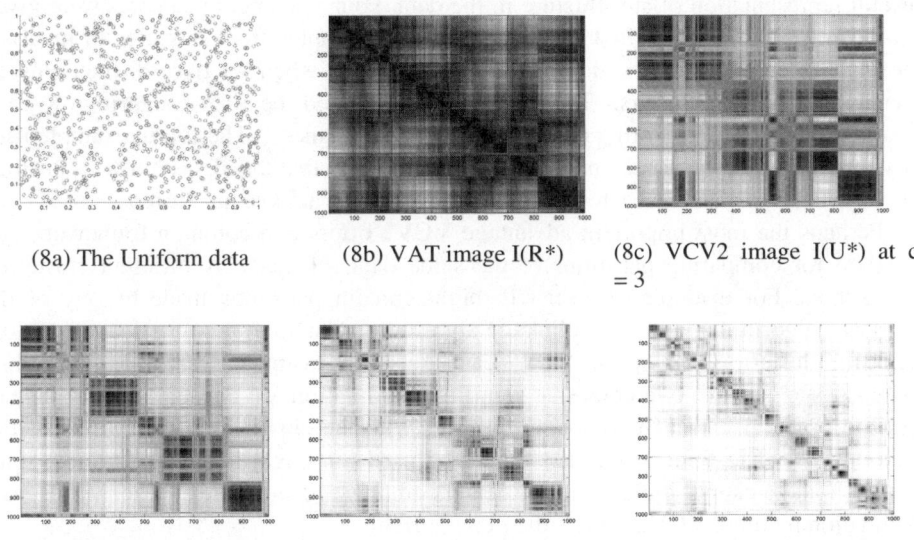

(8a) The Uniform data   (8b) VAT image I(R*)   (8c) VCV2 image I(U*) at c
                                                           = 3

(8d) VCV2 image I(U*) at c=5 (8e) VCV2 image I(U*) at c=10 (8f) VCV2 image I(U*) at
                                                           c=20

Fig. 8. VAT image I(R*) and VCV2 images I(U*) at c = 3, 5, 1 and 20 for the Uniform data

Table 5. FCM Statistics for the no-cluster data

c	Number of iterations	Step size at termination
3	200	9.8357e-006
5	109	9.7402e-006
10	290	9.7516e-006
20	487	9.8234e-006

members, nor nonmembers, of the given clusters. And the main point is that they also are increasingly less attractive matches to I(R*). Our conclusion is, and should be, that the data possesses no distinct clusters.

5 Discussion and Conclusions

We have developed and demonstrated VCV2, a visual cluster validity technique for validating a partition of any type generated by any clustering algorithm. The method is basically one of image matching. The base image is the pre-clustering VAT image I(R*). We use the VAT permutation to reorder each partition matrix in any set of candidate partitions of the data, and then transform each matrix into a VAT-like image I(U*). The images of candidate partitions are visually compared with the VAT image; the stronger the visual match, the more confident we are that the candidate partition is a useful representation of substructure in the data. Four numerical examples were given to illustrate VCV2. The first two examples used samples from mixtures of bivariate normals to demonstrate the differences in the images when we have distinct clusters versus overlapping clusters. The third example applied the VCV2 algorithm to real-world data (the VOTE voting records of 435 Congressmen). The fourth example used a sample from the bivariate uniform distribution verify that the algorithm will illustrate inconsistencies when we try to forcibly partition data that have no clusters.

Perhaps the most important advantage VCV2 offers is a common framework and method for comparing partitions of the same data set made by different clustering algorithms. For example, the set CP might contain partitions made by any of the c-means algorithms, any of the linkage algorithms, Gaussian mixture decomposition, spectral clustering, and so on. VCV2 can be used to evaluate every $U \in CP \subset M_{pcn}$ against I(R*). This overcomes a significant limitation of cluster validity indices $v : D_v \mapsto \Re$, viz., that the domain D_v varies with the clustering algorithm used. For example, the Xie-Beni index [12] cannot be used for single linkage partitions, since single linkage cannot include cluster centers amongst its outputs. In other words, as a validation method, VCV2 "levels the playing field".

A first question concerns our conjecture that VCV2 provides evidence for the correct partition when the presented partition contains more clusters than seem correct. This agrees with a general "rule of thumb" in the clustering community - that overestimation of c is less disastrous than underestimation. Could VCV2 be used to add computational gravitas to this tribal knowledge?

A second interesting and important question is how to extend VCV2 so that it still works on very large (VL or unloadable) data sets. Specifically, what happens if we cannot generate I(R*) and I(U*) for the complete data set? Can subsets of the data be used for the images? How do we know that a subset is representative of the entire data? These are all questions that require more investigation and experimentation.

Finally, we add this caveat to our comments. It is scientifically impossible to assert that a generally applicable validation procedure *cannot* be found. However, examples such as the deck of cards are pretty discouraging. How discouraging? Well, we shouldn't abandon the search for increasingly reliable and robust validation methods. But we should harbor realistic expectations about what we can accomplish in this domain. VCV2 is a clear improvement over current methodologies if only because it is equally applicable to all candidates. But we know it will fail, because we know clustering can fail. Moreover, we are certain that VAT can provide false estimates of c prior to clustering, but we have not been able to characterize situations when it might be expected to fail. And when VAT fails, VCV2 cannot succeed. So, can we devise a way to know a priori that VAT and/or VCV2 will lead us astray? If we could, then the validity problem would be "solved".

References

1. Thorndike, R.L.: Who belongs in the family? Psychometrika 18(4), 267–276 (1953)
2. Bezdek, J.C.: Cluster validity with fuzzy sets. Jo. Cyber. 3(3), 58–72 (1974)
3. Dubes, R., Jain, A.: Clustering techniques: the user's dilemma. Patt. Recognition 8, 247–260 (1977)
4. Bock, H.H.: On some significance tests in cluster analysis. Jo. Classification 2, 77–108 (1985)
5. Milligan, G., Cooper, M.C.: An examination of procedures for determining the number of clusters in a data set. Psychometrika 50, 159–179 (1985)
6. Soromenho, G.: Comparing approaches for testing the number of components in a finite mix-ture model. Comp. Statistics, 65–78 (1994)
7. Cutler, A., Windham, M.P.: Information-based validity functionals for mixture analysis. In: Bozdogan, H. (ed.) Proc. 1st US/Japan Conf. on the Frontiers of Statistical Modeling, pp. 149–170. Kluwer, Amsterdam (1994)
8. Celeux, G., Soromenho, G.: An entropy criterion for assessing the number of clusters in a mixture model. Jo. Classification 13(2), 195–212 (1996)
9. Bezdek, J.C., Windham, M., Ehrlich, R.: Statistical Parameters of Fuzzy Cluster Validity Functionals. Int. Jo. Comp. and Inf. Sci. 9(4), 232–336 (1980)
10. Windham, M.P.: Cluster validity for the fuzzy c-means clustering algorithm. IEEE Trans. PAMI 4(4), 357–363 (1982)
11. Pal, N.R., Bezdek, J.C.: On cluster validity for the fuzzy c-means model. IEEE Trans. Fuzzy Systems 3(3), 370–376 (1995)
12. Bezdek, J.C., Li, W.Q., Attikiouzel, Y.A., Windham, M.P.: A geometric approach to cluster validity for normal mixtures. Soft Computing 1, 166–179 (1997)

13. Bezdek, J.C., Pal, N.R.: Some new indices of cluster validity. IEEE Trans. SMC 28(3), 301–315 (1998)
14. Bezdek, J.C., Hathaway, R.J.: VAT: A tool for visual assessment of (cluster) tendency. In: Proc. IJCNN 2002, pp. 2225–2230. IEEE Press, Piscataway (2002)
15. Hathaway, R.J., Bezdek, J.C., Huband, J.M.: Scalable visual assessment of cluster tendency for large data sets. Pattern Recognition 39, 1315–1324 (2006)
16. Bezdek, J.C., Hathaway, R.J., Huband, J.M.: Visual Assessment of Clustering Tendency for Rectangular Dissimilarity Matrices. IEEE Trans. Fuzzy Systems (in press, 2007)
17. Bezdek, J.C., Harris, J.D.: Fuzzy partitions and relations: an axiomatic basis for clustering. Fuzzy Sets and Systems 1, 111–127 (1978)

Data Management by Self-Organizing Maps

Teuvo Kohonen

Helsinki University of Technology, Centre of Adaptive Informatics, P.O. Box 5400,
02015 HUT, Finland
teuvo.kohonen@tkk.fi

Abstract. The self-organizing map (SOM) is an automatic data-analysis method. It is widely applied to clustering problems and data exploration in industry, finance, natural sciences, and linguistics. The most extensive applications, exemplified in this paper, can be found in the management of massive textual data bases. The SOM is related to the classical vector quantization (VQ), which is used extensively in digital signal processing and transmission. Like in VQ, the SOM represents a distribution of input data items using a finite set of models. In the SOM, however, these models are automatically associated with the nodes of a regular (usually two-dimensional) grid in an ordered fashion such that more similar models become automatically associated with nodes that are adjacent in the grid, whereas less similar models are situated farther away from each other in the grid. This organization, a kind of similarity diagram of the models, makes it possible to obtain an insight into the topographic relationships of data, especially of high-dimensional data items. If the data items belong to certain predetermined classes, the models (and the nodes) can be calibrated according to these classes. An unknown input item is then classified according to that node, the model of which is most similar with it in some metric used in the construction of the SOM. A new finding introduced in this paper is that an input item can even more accurately be represented by a linear mixture of a few best-matching models. This becomes possible by a least-squares fitting procedure where the coefficients in the linear mixture of models are constrained to nonnegative values.

1 Introduction

There is no end in view in the production and proliferation of masses of information in electronic media. For their automatic organization, analysis, and retrieval, new methods will be necessary.

Notwithstanding, the mass information not only consists of texts but also of statistical data, measurements, images, and acoustical signals as well. Apparently, no traditional indexing is possible for these. Still some sort of *directory* would be necessary for their identification, retrieval, and browsing.

One traditional approach to that end is to *cluster* the (natural or experimental) data items with respect to some *similarity* or *distance measure*. If then an unknown item is given, the best-matching stored item is identified first, e.g., by

J.M. Zurada et al. (Eds.): WCCI 2008 Plenary/Invited Lectures, LNCS 5050, pp. 309–332, 2008.
© Springer-Verlag Berlin Heidelberg 2008

an exhaustive comparison of the given item with all of the stored ones, after
which the wanted number of more or less similar items can be traced from the
local cluster structure.

The classical clustering algorithms [2], [16], [18], [61], however, are usually
rather heavy computationally, since in most approaches, every data item must
be compared with all of the other ones, maybe reiteratively. For masses of data
this is obviously no longer possible. A remedy is to represent the set of all
data items by a much smaller set of *models*, each of which stands for a subset of
similar or almost similar data items. In the classification of the input data items,
their comparison with the models can be at least an order of magnitude lighter
operation. This idea was already applied in the classical *vector quantization
(VQ)*, as described in [12], [13], [40], [42], and [69]. However, the VQ method
as such is not yet suitable for the management of large data bases. This paper
discusses innovations to the latter problem.

2   The Classical Vector Quantization (VQ)

In the classical *vector quantization (VQ)*, the space of vector-valued input data
is partitioned into a finite number of contiguous regions, and each region is represented by a single *model vector*, called the *codebook vector* in the VQ. (The
latter term ensues from digital signal transmission, where the VQ is used for
the encoding and decoding of the transmitted information.) In an optimal partitioning, called (in two- or three-dimensional spaces) the *Dirichlet tessellation*
[10] and for arbitrary dimensionalities the *Voronoi tessellation* [66], the codebook vectors are constructed such that the mean distance of each input data
item from its respective closest codebook vector is minimized, i.e., the *average
quantization error* is minimized.

The general principle of the *VQ learning* may be expressed, without a detailed
mathematical analysis, as follows:

*Principle 1 (The VQ learning principle): Every input data item shall select and
modify only the codebook vector that matches best with it, in such a direction
that the degree of matching is increased.*

Steps of the above kind shall be reiterated for all of the input data items.
With time, when all of the inputs have been applied reiteratively, eventually a
great number of times, the codebook vectors will have converged to reasonably
good approximations of the optimal values.

If the set of the input data items is finite, a *batch computation method* is also
feasible. It is called the *Linde-Buzo-Gray (LBG)* algorithm [40]. First a copy of
each input data item is added to a list associated with the codebook vector that
matches best with it. After that, the *mean* of each list obtained in this way is
taken for the new value of the respective codebook vector. These steps shall be
reiterated until the values of the codebook vectors become steady. Usually this
takes only a finite and relatively small number of batch iterations.

As mentioned above, the vector-quantization method has been applied extensively and with great success in digital signal transmission. Its main disadvantage

from the point of view of data organization and retrieval is that the set of code-book vectors obtained in this way is not yet *ordered*, so the codebook vectors are not able to reflect any *structures* of the data. It also has a principal characteristic weakness: the final state may depend, even significantly, on the *initialization* of the codebook vectors in the above updating process. That is, the optimum thereby reached is generally not yet *global* but only *local*. It may be mentioned that the SOM method discussed in this paper can be used, among many other applications, for the definition of the initial values of the vector quantization in a robust way.

It may also be noted that certain much-credited competitive-learning neural-network models implicitly apply VQ principles that were invented in digital signal processing almost a decade earlier.

3   The Self-Organizing Map (SOM)

3.1   General

Around 1981-82 this author introduced a new method, called the *self-organizing map (SOM)*, which otherwise resembles the VQ, but in which the models (corresponding to the codebook vectors) become *spatially ordered* [22], [23], [25], [27]. They are associated with the *nodes* of a regular, usually two-dimensional *grid* (Fig.1) such that more similar models will be associated with nodes that are closer in the grid, whereas less similar models will be situated farther away in the grid.

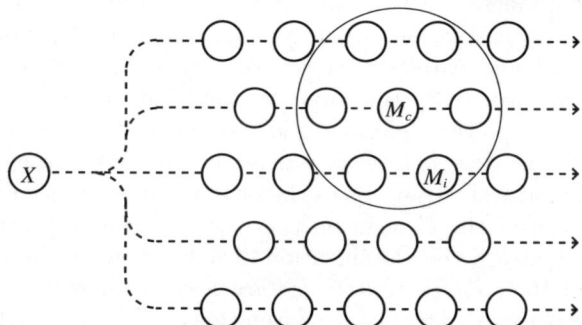

Fig. 1. Illustration of a self-organizing map. An input data item X is broadcast to a set of models M_i, of which M_c matches best with X. All models that lie in the neighborhood (larger circle) of M_c in the grid match better with X than the rest.

3.2   Calibration of the SOM

If the input items fall in a finite number of predetermined classes, the different models can be made to stand for these classes and be provided with corresponding *symbolic class labels*. This kind of *calibration* of the models can be made in either of the following two ways: 1. If the number of input items is sufficiently large, one can first study the distributions of matches that each input data item makes with the various models. A particular model is labeled according to that class that occurs in the majority of matches with this model. In the case of a tie, one may carry out, e.g., a majority voting over a larger neighborhood of the model. 2. If, in relation to the number of nodes, there is only a small number of input data items available so that the above majority voting makes no sense (e.g., there are only few hits of the input data items with the models, or no hits at all with some of the models), one can apply the so-called k *nearest neighbors (kNN)* method. For each model, those k input data items that are most similar with it (in some metric) are searched. Usually the integer k is selected from the range of half a dozen to a hundred, depending on the number of input data items. At any rate, k must be much larger than unity but much smaller than the number of input items. The majority voting and labeling are then carried out among these k nearest neighbors. In the case of a tie, the value of k is increased in this case until the tie is resolved.

When a new, unknown input item is compared with all of the models, it will be identified with the best-matching model. The classification of the input item is then regarded as that of the best-matching model.

3.3   Comparison and Classification of Input Items on the Basis of Features

Before proceeding further, it will be necessary to emphasize a basic fact. An input data item, often given as a set of structural elements such as pixels or other pattern components, will usually not be applicable as such as an input vector. Only in rare cases, e.g., when an input item is described by a set of manually evaluated and carefully chosen statistical indicators, can the set of the latter be used directly as the input vector. Contrary to that, it has been realized a long time ago that, e.g., an image of, or a set of signals emitted by a natural object as such does not yet constitute a good basis for the identification of the latter. The natural variations in the observations are usually so wide that a direct comparison of the objects on the basis of their appearances does not make any sense. Instead, the classification of natural items should be based on the extraction of their characteristic *features*, which must be as *invariant* as possible with respect to natural transformations of the primary observations. The biological sensory systems operate in this way. If one can describe the input objects by a finite and rather small set of such features, the dimensionality of the input representations, and the computing load are reduced drastically.

The selection of a set of characteristic features, and even the automatic extraction of such features from the primary observations must often be based on heuristic rules. In biology, various *feature detectors* have been developed in the very long course of evolution. Sometimes, however, certain mathematical functions or transforms (e.g. spectra, principal components, or other orthonormal basis vectors) of the input items may serve as reasonably good features [14], [68].

So, also when constructing a SOM for natural items, the first task is to try to extract a good set of features for each item and to use the vector formed of them as the input vector to the SOM.

One intriguing question is whether one should *normalize* the feature scales in any way. Experience has shown that a rather good first strategy is to scale every feature dimension in such a way that the *variance* of every feature variable becomes the same. Alternatively, one may make the *range* (from minimum to maximum) of every feature variable the same.

When an ordered SOM has then been constructed, it can be directly used for the statistical classification of the input items, or alternatively, it is applicable as a *directory* or gateway to the files, in the searching or browsing of the data items.

3.4   Main Application Areas of the SOM

Before looking into the details, one may be interested in knowing the justification of the SOM method. Briefly, by the end of the year 2005 we had listed over 7700 scientific publications that analyze, develop, or apply the SOM [21], [49], [51]. The following short list exemplifies the main application areas:

1. Statistical methods
    (a) exploratory analysis of multivariate statistical data at large
    (b) statistical analysis and organization of texts, e.g. [32], [37]
2. Industrial analyses and control, cf. [30]
3. Telecommunications, cf. [30], [27]
4. Finance analyses at large [8]
5. Biomedical analyses and applications at large

In addition to these, one may mention some specific important applications, e.g., retrieval of photographs from very large image data bases [35]. Another example is the categorization of galaxies [46], whereby a classification scheme, still unknown at the time when this project was launched, of thousands of galaxies observed by the Hubble telescope at moderate red shifts (distances) emerged. Further one may mention criminological applications, e.g. the system named the CATCH for the computer-aided tracking of homicides and sexual assaults, as developed by the Battelle Pacific Northwest Division in cooperation with the Attorney General of the state of Washington.

It is not possible to give a full account of the theory, different versions, and applications of the SOM in this article. We can only refer to the extensive lists of publications [21], [49], [51] and to more than a dozen textbooks, monographs, or edited books written on the SOM, e.g.[24], [25], [27], [1], [44], [47], [48], [53], [57], [59], [60], [63]. In addition to these, numerous doctoral theses, many of them published later as monographs, have concentrated on the SOM.

4   Learning Principles of the SOM

4.1   General

It may be easier to understand the rather involved learning principles and mathematics of the SOM, if its central idea is first expressed as the following general principle.

Principle 2 (The SOM learning principle): Every input data item shall select the model that matches best with it, and this model, as well as a subset of its spatial neighbors in the grid, shall be modified for better matching.

Like in the VQ, the modification of the models is thus concentrated on a location in the grid where it is probably most effective. On the other hand, since a whole *spatial neighborhood in the grid* is modified at a time, the degree of *local ordering* of the models in this neighborhood, due to a smoothing action, is increased. Due to successive corrections caused by different inputs in different locations, the smoothing and ordering actions will take place in different subsets of models, overlap, and are gradually propagated over the grid. This may be intuitively clear, but the underlying real mathematical processes are somewhat more complicated than that.

The actual computations to produce the ordered set of the SOM models, like in the VQ, can be implemented by either of the following main types of algorithms: 1. The models in the original SOM algorithm can be computed in a *stepwise, recursive* approximation process in which the input data items are applied to the algorithm one at a time, in a periodic or random sequence, for as many steps as it will be necessary to reach a reasonably stable state. 2. In the *batch-type* process, all of the available input data items can be applied to the algorithm as *one batch*, and all of the models are updated in a single concurrent operation. This batch process usually needs to be reiterated a few times, after which it will usually be stabilized exactly. This requires normally an order of magnitude less computations than the stepwise algorithm.

More detailed descriptions of the SOM algorithms will be given in the following subchapters.

Many versions of these algorithms have been suggested over the years. They are too numerous to be reviewed here; cf. the extensive bibliographies mentioned as references [21], [49], [51]. On the other hand, while the basic SOM algorithms were defined for real-valued vectors only, we shall see that there exist versions for the ordering of *symbolic data items* as well. A special problem is connected with how to make a self-organizing map for *sequences* (e.g., time series) of input data items.

The two-dimensional organization of the models is usually already effective for the representation of similarity relations of high-dimensional data items, like in some earlier methods called the *multidimensional scaling (MDS)* [34], [56]. One should notice, however, that in the MDS methods, *every data item* must be displayed geometrically, whereas the SOM uses only a relatively small set of *models* that it displays on a regular grid. In this way, plenty of computations will be saved.

For special purposes, grids with dimensionality higher than two may be used, too. Naturally, a simple rank ordering can be made to take place along a one-dimensional grid. However, such a simple order will not be able to display any more general topological relationships between the data items.

4.2   The Original, Stepwise Recursive SOM Algorithm

The original formulation of the SOM algorithm resembles a stochastic gradient-step procedure [54]. It must be emphasized, however, that the stepwise recursive version of the SOM algorithm was introduced only heuristically, when trying to materialize the Principle 2 above. It has not yet been shown to be deriv-able from any *objective function*. Its convergence to the globally ordered state has been proven mathematically in some simple low-dimensional cases [6], [7], while there also exists a proof for that a *local order* can be reached for more general dimensionalities of the vectors, if the distribution of the input vectors is *discrete-valued* [53]. Certain modified SOM algorithms [41], [17], [26] can be shown to be derivable from *energy functions*, whereby their convergence is better defined. Below, only the original form of the SOM is explained and used, because it is computationally simplest and lightest, and in practice it will produce use-ful results for input vectors of arbitrary dimensionality. One should also stress the fact that there exist numerous methods for the analysis of the topological correctness, optimality, and quality of the produced maps [27].

Consider again Fig. 1. Let now the input data items X in it stand for a sequence $\{\mathbf{x}(t)\}$ of real n-dimensional Euclidean vectors $\mathbf{x}$, where t, an integer, defines a step in the sequence. Let $\{\mathbf{m}_i(t)\}$ be another sequence of n-dimensional real vectors that now represent the successively computed approximations of the model denoted by M_i in Fig. 1. Here i is the spatial index of the grid node with which $\mathbf{m}_i(t)$ is associated. It is assumed that the following original SOM algorithm converges and produces the wanted ordered values for the models:

$$\mathbf{m}_i(t+1) = \mathbf{m}_i(t) + h_{ci}(t)[\mathbf{x}(t) - \mathbf{m}_i(t)], \tag{1}$$

where $h_{ci}(t)$ is called the *neighborhood function*. This function resembles the kernel that is applied in usual smoothing processes. The subscript c is the index of a particular node in the grid, namely, the one with the model $\mathbf{m}_c(t)$ ("winner") that has the smallest Euclidean distance from $\mathbf{x}(t)$ in the input space:

$$c = \arg\min_i\{||\mathbf{x}(t) - \mathbf{m}_i(t)||\}. \tag{2}$$

Equations (1) and (2) define a recursive step where first the input data item $\mathbf{x}(t)$, according to (2), defines or selects the *best-matching model* ("winner") in the grid. Then, according to (1), the models at this "winner" node as well as at its *spatial neighbors* in the grid are modified. The modifications always take place in such a direction and with such a magnitude that the modified models will match better with the input. The rates of the modifications at different nodes depend on how the mathematical form of the function $h_{ci}(t)$ is defined, and that can be determined best experimentally. For an increasing step index

t, this function shall always *decrease* monotonically. For $i = c$ it shall attain the *maximum value* in the grid, and the value shall *decrease* with the *increasing spatial distance of the nodes* i and c in the grid. Moreover, in order to reach a global ordering of the models as promptly as possible, the *spatial width of the neighborhood function* ought to be rather large in the beginning, say, a little less than half of the diameter of the grid. At the end of the process, when the neighborhood function already attains small values and the corrections are small, the width should be only a small fraction of the diameter of the grid in order to define a "map" with a high resolution. The detailed forms for the learning rates and neighborhood functions can best be defined by experience.

A much applied choice for the neighborhood function $h_{ci}(t)$ is

$$h_{ci}(t) = \alpha(t)\exp[-\text{sqdist}(c,i)/2\sigma^2(t)], \tag{3}$$

where $\alpha(t)$ is a monotonically (e.g., hyperbolically, exponentially, or piecewise linearly) decreasing scalar function of t, sqdist(c, i) is the square of the geometric distance between the nodes c and i in the grid, and $\sigma(t)$ is another monotonically decreasing function of t, respectively. Note that σ is proportional to the *mean radius of the neighborhood function*. The true mathematical form of $\sigma(t)$ is not crucial, as long as the value of σ is fairly large (e.g. on the order of half of the width of the grid) in the beginning, and decreases to a value that is a fraction of the initial value, when a rough order of the models has been achieved (which usually takes about 1000 steps or more).

On the other hand, after this initial phase of *rough ordering*, the *final convergence* to nearly optimal values of the models takes, say, an order of magnitude more steps. For a sufficient statistical accuracy, every model must be updated sufficiently often.

For reasonably small grids, say, with a few hundred nodes, the radius of the final neighborhood function can be one grid spacing, and then the number of final convergence steps may be roughly 1000 times the number of nodes, totalling, maybe, a hundred thousand recursive steps. For very large grids, such as those used in Sec. 5.2, the final radius could be larger. Then, however, with very large grids it is no longer reasonable to use the simple stepwise recursive algorithm, but the *batch computation* explained in the next subsection.

What is said above may sound somewhat sophisticated, but the SOM method and its details have gradually grown up from much simpler versions. These final forms have been tested over many years in practice with numerous types and dimensions of real-world data.

A special question concerns the selection of the *initial values* for the $\mathbf{m}_i$. It has been demonstrated [27] that they can be selected even as *random vectors*, whereas a much faster ordering and convergence follows if the initial values are selected as a *regular, two-dimensional sequence of vectors taken along a hyperplane spanned by the two largest principal components of* $\mathbf{x}$ [27]. With this choice, the expression of h_{ci} can be chosen as simpler.

There are plenty of detailed aspects, not only associated with the parameter values and the convergence, but also with the selection of the *dimensionalities of the vectors and the SOM grid* to be taken into account. All of them cannot be surveyed here. The reader is strongly advised to use some of the ready-made complete software packages such as the SOM_PAK [31] or SOM Toolbox [64], [65], where default values for the parameters are given to guarantee a good learning process of the SOM. Both of these packages are freely downloadable from *www.cis.hut.fi/research/*, and especially the SOM Toolbox, which uses Matlab functions, is provided with good graphics tools. On the other hand, the SOM_PAK, which is written in the C language, may be better suitable for very large SOMs, such as those of extremely large document collections.

Also the following procedure for the batch computation of the SOM, which is included in the SOM Toolbox, is a very viable alternative in practice, especially since it contains fewer parameters than the stepwise recursive algorithm, and converges faster, especially with a good initialization of the models.

4.3   The Batch Computation of the SOM

In the continuation we shall prefer the *batch computation* of the SOM. Then, however, it will be very advantageous to initialize the $\mathbf{m}_i$ as a regular two-dimensional sequence along with the two largest principal components, as described in the previous subsection.

Consider Fig. 2, where a two-dimensional hexagonal array of nodes, depicted by the small circles, is shown. With each node i, a model $\mathbf{m}_i$ is associated. Also a list, containing copies of certain input vectors, is associated with each node.

Then consider the set of input data vectors $\{\mathbf{x}(t)\}$, where t is the integer-valued sample index of a vector. Compare each $\mathbf{x}(t)$ with all of the models and make a copy of $\mathbf{x}(t)$ into the sublist associated with that node, the model vector of which matches best with $\mathbf{x}(t)$ in the Euclidean metric.

When all of the $\mathbf{x}(t)$ have been assigned to the sublists in this way, consider the neighborhood set N_i around node i. Here N_i consists of all of the nodes up to a certain radius from the node i in the grid. Next, compute the *mean* of the $\mathbf{x}(t)$ in N_i, that is, of all the $\mathbf{x}(t)$ that have been copied into the *union* of all of the sublists in N_i. A similar mean is computed *for every node* i, i.e. over the neighborhoods around all of the nodes. *Updating* of the $\mathbf{m}_i$ is then carried out in one concurrent computing operation over all of the nodes of the grid, whereupon all of the old values of the $\mathbf{m}_i$ are replaced by the respective means. This concludes one updating cycle.

This updating cycle is reiterated, always distributing copies of the original input vectors under those nodes, the (updated) models of which match best with the due input vectors. The updated model values, sooner or later, become steady and are no longer changed in continued iterations, whereupon the training is stopped.

A process that complies even better with the stepwise recursive learning is obtained if the means are formed as *weighted averages*, where the weights are related to each other like the h_{ci} in (1). Here c is the index of the node, the model of which is updated, and i stands for the indices of the nodes in its neighborhood.

The batch-computing method can be generalized for *nonvectorial* data, too, if a *generalized median* over the samples $\mathbf{x}(t)$ can be defined. This kind of median

Inputs

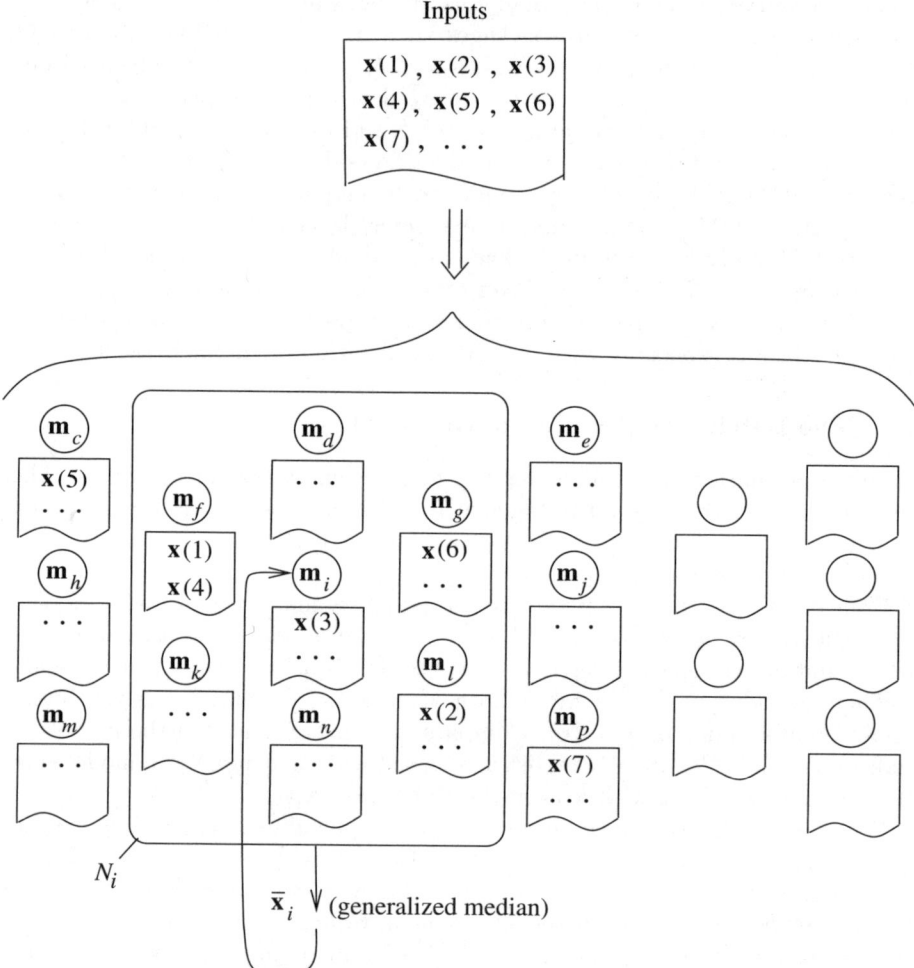

Fig. 2. Illustration of one cycle in the batch process. The input data items $\mathbf{x}(t)$ are first distributed into sublists associated with their best-matching models, and then the new values of the models are determined as means, or more generally, as "generalized medians" over the sets of sublists, each one concentrated on the neighborhood N_i of node i.

may be defined as an item *that has the smallest sum of any defined distances from the items in N_i.* Where we earlier used the mean over the union of the sublists in N_i, we shall now use the generalized median in the same role [33]. Such SOMs have been constructed, e.g., for protein sequences [33] and virus-type DNA [50].

A discussion of the convergence and ordering of the above method has been presented by Cheng [5].

5   Applications of the SOM

5.1   Ordering of Countries on the Basis of Sets of Socioeconomic Indicators

To start with, we take an example [19] in which a set of statistical indicators directly defines an input vector to the SOM. Consider the following socioeconomic data, obtained from the World Development Record of the year 1992, as published by the World Bank [67]. There were 39 indicators, most of them given relative to the population, that describe the *welfare* of the 126 countries listed. Some of the indicators were economic, such as the gross national product per capita, while others describe the social services such as the number of doctors per capita. Some of the indicators describe the standard of education etc. All of the input vectors formed of these features were normalized to unit length.

For 78 countries, 12 or more indicators were given, and these data were used in the training of the SOM. In the calibration of the resulting SOM, the nodes corresponding to these countries were labeled in Fig. 3 by capital symbols. The symbols of the rest of the countries were written in lower case, and they were not included in training, they were only mapped to the best-matching nodes. Nonetheless, notice that the dimensionality of the models was always 39, corresponding to all of the 39 indicators. However, if some of the indicator values were *missing* from an input data item, the comparison of this item with all of

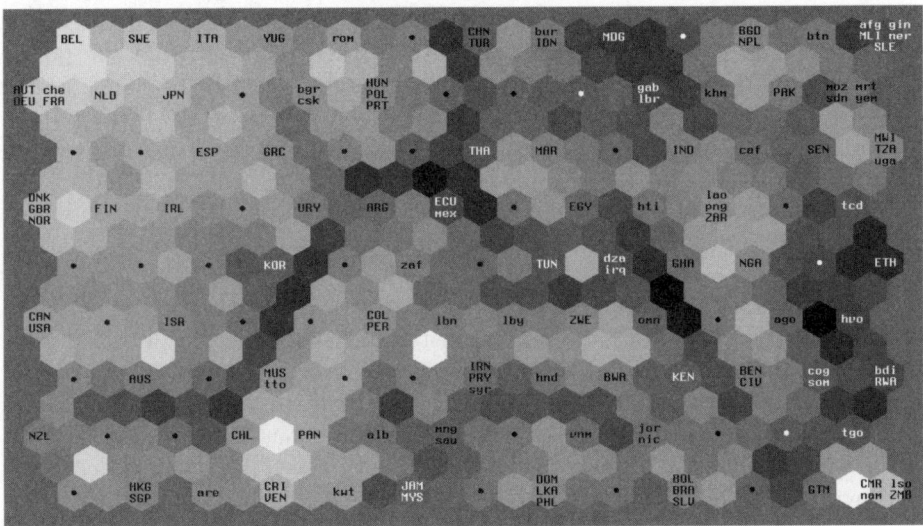

Fig. 3. The welfare map of 126 countries, based on socioeconomic indicators published by the World Bank in 1992. The symbols of the countries may be self explanatory; cf. also [27]. One should emphasize that the distances between the countries on the map are not proportional to any measure of dissimilarity. The ordering relations are only reflected in the *relative topographic positions* of the items.

the models was made *on the basis of the given indicators* only, i.e., with lower-dimensional vectors.

For extra clarity, the nodes of the map (actually, the small areas around the nodes) can be colored according to the so-called *U matrix* [62]. The shade of gray indicates the *average distance between neighboring vectors*; so if the neighboring models are similar, the shade is light, whereas greater jumps in welfare, when moving from one node to another, are indicated by a darker color. Fig. 3 shows the resulting welfare map. It may be understandable that a two-dimensional display gives a more qualitative insight into the relative status of the countries than any socio-econometric measure can do, and the differences between countries that are adjacent in the map are usually determined by different subsets of indicators in different areas.

5.2   SOMs of Very Large Document Collections

In this paper we shall emphasize the benefits of SOM methods in massive problems such as the *organization of documents* and *searching* of them from large document corpora. For the identification of documents, the *relative frequencies of words* and other *terms* (word combinations) in them have been found to constitute a possible set of characteristic features. It may be necessary to forestall eventual criticisms by stating that although *semantic analyses* of the texts would be even more accurate for the classification of the true textual contents, we could not afford such methods. Our objective was to deal with *really large collections of textual documents*, whereupon the standard contemporary mainframe computers, albeit rather large and advanced ones, already had to operate at the limits of their capacities. Our choice for the features then demonstrated the power of the (weighted) *word histograms* as features, combined with the SOM, in the statistical exploration of very large text collections.

The main problem with the use of word histograms as input feature vectors is that the vocabularies are usually vast, often on the order of hundreds of thousands of words. However, it is helpful to observe that the various words have a very different power in distinguishing the texts [55]. If a word occurs in many documents, eventually in all of them, it has a rather *low distinctive value*. On the other hand, a word that occurs only in very few documents has a large distinctive value. The words in a word histogram shall thus be provided with different *weights*. Then, words with low weights, such as stopwords, can simply be neglected, whereupon the dimensionality of the histogram of the remaining words can already be acceptable for mass computations. The simplest choice for the weight, which is often rather effective, is the so-called *inverse document frequency (IDF)*. The document frequency here means in how many documents a particular word occurs.

A bit more developed weighting scheme, introduced by Manning and Schütze [43], is the following. In it, the word w belonging to document d has the weight $W_{w,d}$:

$$W_{w,d} = (1 + \ln f_d(w))\ln(N/f_w(d)), \qquad (4)$$

where $f_d(w)$ is the frequency of word w in document d, $f_w(d)$ ("document frequency") tells in how many documents word w appears, and N is the total number of documents.

It is also possible to resort to the information-theoretic *Shannon entropy* ("negentropy") in weighting. Let the corpus of the documents be divided into groups g, $g = 1, ..., N$, which are of sufficient size so that the following probabilities can be evaluated. Denote by $P_g(w)$ the probability for word w belonging to group g. Then the weight W_w of word w is (cf. [27])

$$W_w = \ln N + \sum_g P_g(w) \ln P_g(w). \qquad (5)$$

In all of the above three approaches, the dimensionality of the input feature vector was reduced by reducing the number of words in the word histograms. A quite different approach is to *transform* the set of histograms into another *basis* that has a smaller dimensionality, without essential loss of the distinguishing information. A traditional method is based on the so-called *latent semantic indexing (LSI)* [9], in which an ordered set of *factors* is computed for the matrix formed of the histograms. Only the most significant factors are used as features.

S. Kaski from our group found out [20] that a certain dimensionality-reducing *projection* [52] of the word histograms is essentially as effective as the LSI but it is computationally very much lighter, especially after finding shortcuts to the computing operations [32]. Let $\mathbf{R}$ be a matrix, the elements of which are random numbers such that each row has a constant norm. Let the number of rows in $\mathbf{R}$ be much smaller than the number of columns. If we denote the original input vector formed of the histograms, regarded as a column vector, by $\mathbf{x}'$, then the matrix transformation $\mathbf{R}\mathbf{x}'$ produces the vector $\mathbf{x}$ with a much lower dimensionality, but with only little loss in information content. Here $\mathbf{R}$ shall be *the same for all histograms*.

In numerous applications, all of the above weighting strategies have produced results, the accuracies of which do not seem very different in practice. It may be emphasized that our main efforts were concentrated on proving that the SOM principles can be used in exploratory document analysis.

The SOM of the Encyclopaedia Britannica. From the electronic version of the *Encyclopaedia Britannica*, 68,000 articles were picked up. In addition, 43,000 textual entities such as summaries, updates, and other miscellaneous material were collected. Very long articles were split into several sections. A total of 115,000 "documents" were thus defined. An average "document" contained 490 words.

These documents were first preprocessed to remove the HTML markups, links, and images. Inflected word forms were converted to their base forms using a standard morphological analyzer; otherwise, every inflected form would have been regarded as a different word. The baseform words were then weighted by the inverse document frequency. By ignoring words that had a very low IDF value (such as stopwords), the size of the finally accepted vocabulary was 39,058 words. After that, a random projection of each of the 39,058-dimensional feature vectors onto a 1000-dimensional space was formed. This was the dimensionality of the SOM models, too.

The size of the SOM grid in this application was 12,096 nodes. Several speedups in the SOM computation were applied [32]: e.g., by first computing an SOM with a much smaller grid, then introducing interstitial nodes to form a larger grid, interpolating values for the approximations of their model vectors, and finally carrying out a relatively short fine tuning of all of the models of the larger grid.

When an SOM has been constructed [37] and the models have been fixed, each document will be mapped into one and only one node, namely, the one whose model vector matches best with the feature vector of the given document. One can store the original documents in a *separate database* in the computer memory system, and associate with each SOM node only an *address pointer* to the respective document. *Usually several documents will be mapped into each SOM node, and all of them will be activated simultaneously when clicking this node.* There exist now several modes of use of this system.

The document map is presented as a series of HTML pages and clickable parts of the images, which enable the exploration of the nodes of the grid. A mouse click on a node brings to the view the links to all of the documents associated with that grid node. However, a large map can first be *zoomed* electronically to view parts of it with higher resolution. For the largest maps we have used several zooming levels.

One simple mode of use of the SOM is *browsing*. When a particular node in the zoomed display is clicked, all of the address pointers associated with that node are activated, and the titles of the corresponding documents and eventually some additional information such as descriptive words of the documents at this node are written out. After that, the wanted document can be selected and read out in full.

Fig. 4 shows a close-up view of a part of the large SOM. The *keywords* written into the vicinity of some locations on the map have been determined by an automatic procedure [36].

The map, also in the above example, is usually provided with a *form field* into which the user can type a query. We shall demonstrate the use of the form field in connection with the next example.

The SOM of nearly seven million patent abstracts. Next we explain how a real *content-addressable searching* operation is implemented [32]. As mentioned earlier, to that end the interface of the map must be provided with a form field into which the user can type a query, e.g., in the form of a short "document." The weighted vocabulary of this query is supposed to match, approximately at least, with the vocabulary of some of the documents. Actually it will suffice that the match, even a coarse one, is better for some documents than the others, and this can certainly be guaranteed. In this sense, *the query may even consist of a few keywords only* that are supposed to be characteristic of a document.

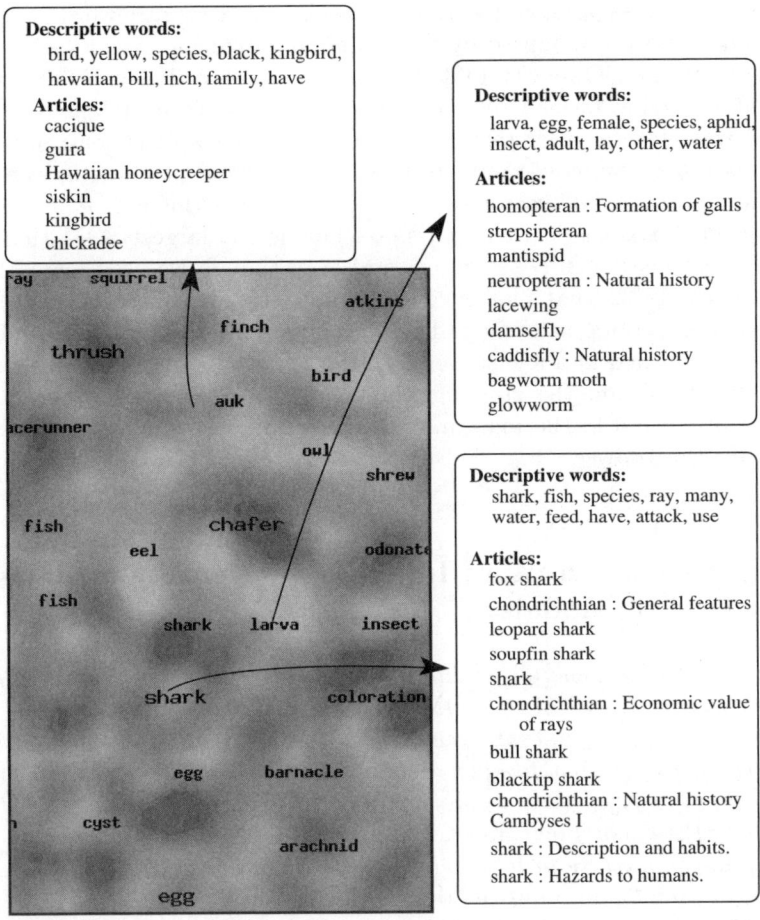

Descriptive words:
 bird, yellow, species, black, kingbird,
 hawaiian, bill, inch, family, have
Articles:
 cacique
 guira
 Hawaiian honeycreeper
 siskin
 kingbird
 chickadee

Descriptive words:
 larva, egg, female, species, aphid,
 insect, adult, lay, other, water
Articles:
 homopteran : Formation of galls
 strepsipteran
 mantispid
 neuropteran : Natural history
 lacewing
 damselfly
 caddisfly : Natural history
 bagworm moth
 glowworm

Descriptive words:
 shark, fish, species, ray, many,
 water, feed, have, attack, use
Articles:
 fox shark
 chondrichthian : General features
 leopard shark
 soupfin shark
 shark
 chondrichthian : Economic value
  of rays
 bull shark
 blacktip shark
 chondrichthian : Natural history
 Cambyses I
 shark : Description and habits.
 shark : Hazards to humans.

Fig. 4. A close-up view of the map of Encyclopaedia Britannica articles. When the user clicks a node in the map region with the label "shark," he or she obtains the listing of articles on sharks. The node "larva" contains articles on insects and larvae, and a node in the middle of the area of birds contains articles on birds. A bit more remote areas (not shown here) contain articles on whales and dolphins, and areas beyond them describe whaling, harpoons, and eskimos, which proves that the topics change smoothly and continuously on the map. By clicking the searched titles, the complete articles can be read out.

The text corpus in this example was about 16 times that of Encyclopaedia Britannica. It consisted of 6,840,568 patent abstracts written in English. They were available on some 200 CD ROMs and were obtained from U.S., European, and Japan patent offices as two databases: the "first page" database (1970-1997), and the "Patent Abstracts of Japan" (1976-1997). The average length of the abstracts was 132 words, thus the number of documents was about 60 times that in Encyclopaedia Britannica. The size of the vocabulary which was finally accepted after omission of stopwords, numerals, and very rare words was 43,222 words, and they were weighted by the Shannon entropy.

The size of the SOM was 1,002,240 nodes and the dimensionality of each model (after having formed the random projections of the weighted word histograms) was 500. In order to fit a map of this size into the central computer of our laboratory, we could only reserve one byte for each vector component of each model. With 500 vector components, however, this coarse accuracy was still sufficient for statistical comparisons. This is the largest SOM that to our knowledge has ever been constructed, and even though we used many different speedup methods [32] that shortened the computation time by several decades, it took six weeks to obtain the map. The searching operations, on the other hand, can be accomplished in a few seconds, so the map can be used in real time.

After having found the starting points by the content-addressable search, further items stored in the map and the collection as a whole can be explored using a WWW browser. Fig. 5 exemplifies a case of the content-addressable search.

6   Approximation of an Input Data Item by a Linear Mixture of Models

An analysis hitherto generally unknown is introduced in this chapter. The purpose is to extend the use of the SOM by showing that instead of a single winner model, one can approximate the input data item more accurately by means of a set of *several models* that *together* define the input data item more accurately. It shall be emphasized that we do not mean k winners that are rank-ordered according to their matching. Instead, the input data item is approximated by an *optimized linear mixture of the models, using a nonlinear constraint*, which will be shown to provide an improved description of it.

Consider the n-dimensional SOM models $\mathbf{m}_i, i = 1, 2, \ldots, p$, where p is the number of nodes in the SOM. Their general linear mixture is written as

$$k_1\mathbf{m}_1 + k_2\mathbf{m}_2 + \ldots + k_p\mathbf{m}_p = \mathbf{M\,k}, \tag{6}$$

where the k_i are scalar-valued weighting coefficients, $\mathbf{k}$ is the p-dimensional column vector formed of them, and $\mathbf{M}$ is the matrix with the $\mathbf{m}_i$ as its columns. Now $\mathbf{M\,k}$ shall be the *estimate* of some input vector $\mathbf{x}$. The vectorial fitting error is then

$$\mathbf{e} = \mathbf{M\,k} - \mathbf{x}. \tag{7}$$

Our aim is to minimize the norm of $\mathbf{e}$ in the sense of least squares. However, the special constraint must then be taken into account.

6.1   Fitting with the Nonnegativity Constraint

Much attention has recently been paid to least-squares problems where the fitting coefficients are constrained to *nonnegative values*. Such a constraint is natural,

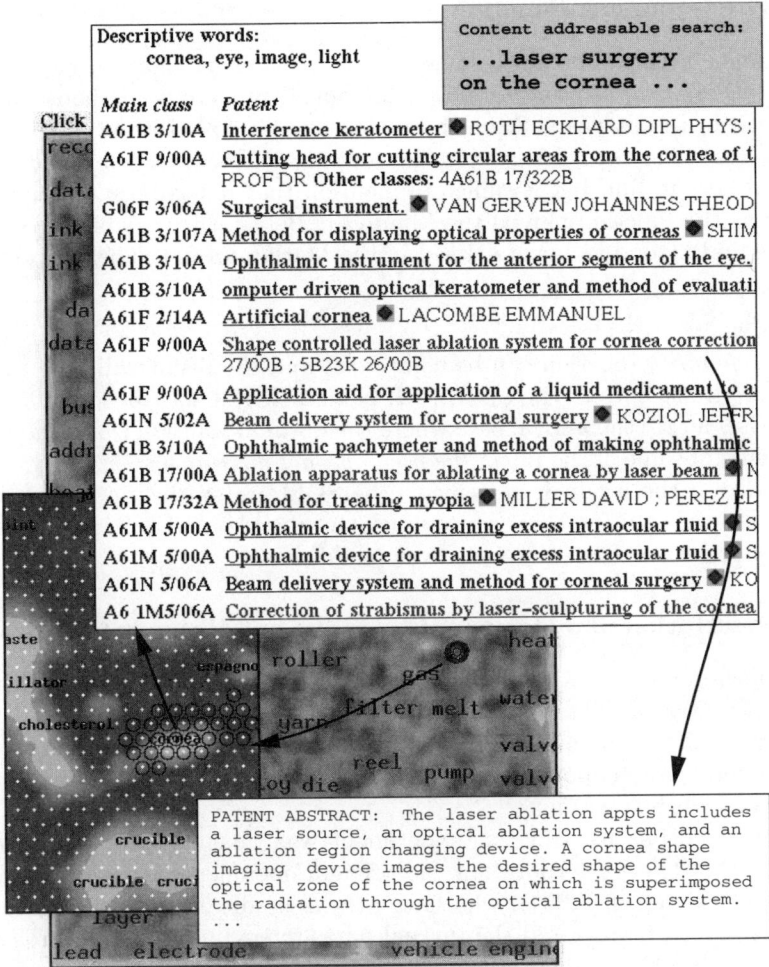

Fig. 5. The document map of nearly 7 million patent abstracts is depicted in the background, and the form field for the query is shown in the upper right-hand corner. The query was "laser surgery on the cornea," and twenty best-matching nodes have been marked with small circles. In the main map, these circles are fused together. An enlarged portion of the map, in which the twenty nodes are visible separately, can be obtained by digital zooming, as shown in the lower left-hand corner by the small circles. By clicking a particular node, the list of the titles associated with this node is obtained, and by clicking a selected title, the corresponding abstract can be read out.

when the *negatives* of the items have no meaning, for instance, when the input item consists of statistical indicators that can have only nonnegative values, or is a weighted word histogram of a document. In these cases at least, the constraint contains additional information that is expected to make the fits more meaningful.

6.2   The lsqnonneg Function

The present fitting problem belongs to the broader category of *quadratic programming* or *quadratic optimization*, for which numerous methods have been developed in recent years. A much-applied one-pass algorithm is based on the *Kuhn-Tucker theorem* (Lawson & Hanson, 1974), but it is too involved to be reviewed here in full. Let it suffice to mention that it has been implemented in Matlab as the function named the *lsqnonneg*. Below, the variables $\mathbf{k}$, $\mathbf{M}$, and $\mathbf{x}$ must be understood as being defined in the Matlab format. Then we obtain the weight vector $\mathbf{k}$ as

$$\mathbf{k} = \text{lsqnonneg}(\mathbf{M}, \mathbf{x}). \tag{8}$$

The *lsqnonneg* function can be computed, and the result will be meaningful, for an *arbitrary rank* of the matrix $\mathbf{M}$. Nonetheless it has to be admitted that there exists a rare theoretical case where the optimal solution is not *unique*. This case occurs, if some of the $\mathbf{m}_i$ *in the final optimal mixture* are *linearly dependent*. In practice, if the input data items to the SOM are stochastic, the probability for the optimal solution being not unique is negligible. At any rate, the locations of the nonzero weights are unique even in this case!

6.3   Description of a Document by a Linear Mixture of SOM Models

The following analysis applies to most of the SOM applications. Here it is exemplified by textual data bases.

In text analysis, one possible task is to find out whether a text comes from different sources, whereupon its word histogram is expected to be a linear mixture of other known histograms.

The text corpus used in this experiment was taken from a collection published by the Reuters corporation. No original documents were made available; however, Lewis et al. (2004), who have prepared this corpus for benchmarking purposes, have preprocessed the textual data, removing the stop words and reducing the words into their stems. Our work commenced with the ready word histograms. J. Salojärvi from our laboratory selected a 4000-document subset from this preprocessed corpus, restricting only to such articles that were assigned to one of the following classes:

1. Corporate-Industrial.
2. Economics and Economic Indicators.
3. Government and Social.
4. Securities and Commodities Trading and Markets.

There were 1000 documents in each class. Salojärvi then picked up those 1960 words that appeared at least 200 times in the selected texts. In order to carry out *statistically independent experiments*, a few documents were set aside for testing. The 1960-dimensional word histograms were weighted by factors used by Manning and Schütze [43]. Using the weighted word histograms of the rest of the 4000 documents as input, a 2000-node SOM was constructed.

Fig. 6 shows the four distributions of the hits on the SOM, when the input items from each of the four classes were applied separately to the SOM. It is clearly discernible that the map is *ordered*, i.e., the four classes of documents are segregated to a reasonable accuracy, and the mappings of classes 1, 3, and 4 are even singly connected, in spite of their closely related topics.

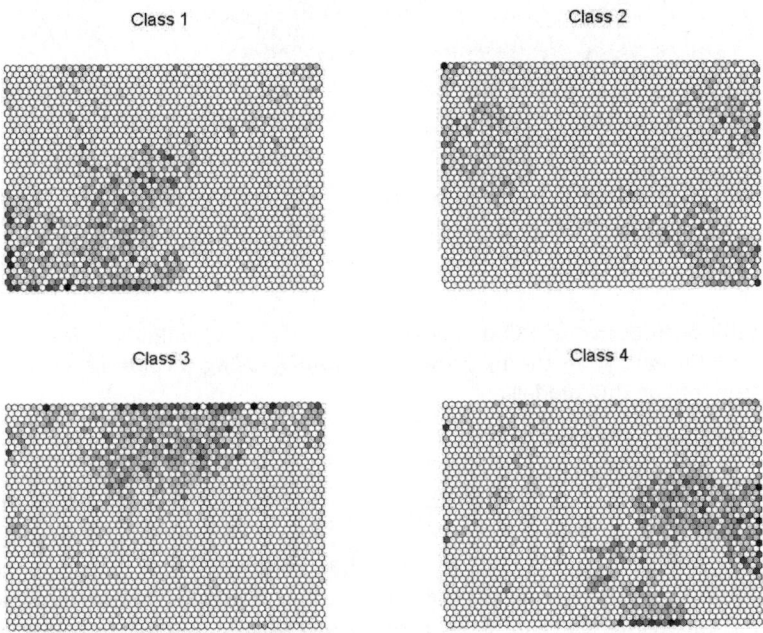

Fig. 6. Mapping of the four Reuters document classes onto the SOM. The densities of the "hits" are shown by shades of gray.

Fig. 7 shows a typical example, where a linear mixture of SOM models was fitted to a new, unknown document. The values of the weighting coefficients k_i in the mixture are shown by dots with relative shades of gray in the due positions of the SOM models. It is to be emphasized that this fitting procedure also defines the optimal *number* of the nonzero coefficients. In the experiments with large document collections, this number was usually very small, less than a per cent of the number of models.

When the models fall in classes that are known a priori, the weight of a model in the linear mixture also indicates the *weight of the class label associated with that model*. Accordingly, by summing up the weights of the various types of class labels one then obtains the *class-affiliation* of the input with the various classes.

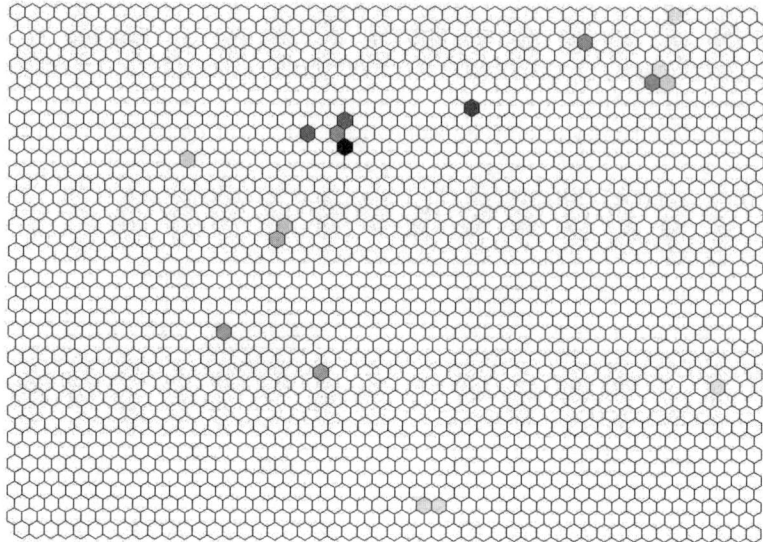

Fig. 7. A linear mixture of SOM models fitted to a new, unknown document. The weighting coefficients k_i in the mixture are shown by using a coloring with a relative shade of gray of the due models.

7   Discussion

The self-organizing map (SOM) principle has been used extensively as an analytical and visualization tool in exploratory data analysis. It has had plenty of practical applications ranging from industrial process control and finance analyses to the management of very large document collections. New, promising applications exist in bioinformatics.The largest applications so far have been in the management and retrieval of textual documents, of which this paper contains two examples.

Several commercial software packages as well as plenty of freeware on the SOM are available. This author strongly encourages the use of two public-domain software packages developed by us, namely, the *SOM_PAK* and *SOM Toolbox*, both downloadable from our homepages at *www.cis.hut.fi/research/*. Both packages contain auxiliary analytical procedures, and especially the SOM Toolbox, which makes use of the Matlab functions, is provided with good and versatile graphics as well as thoroughly proven statistical analysis programs of the results.

This paper has applied the basic version of the SOM, on which the majority of applications is based. Nonetheless there may exist at least theoretical interest in different versions of the SOM, where some of the following modifications have been introduced.

Above the SOM grid was always taken as two dimensional and regular, preferably as hexagonal. This form of the array is advantageous if the purpose is to *visualize* the structure of the data base and to span the data distribution smoothly by the SOM grid in the data space. One of the different versions of the grid is *cyclic*, where the network formed of the nodes is either *toroidal* or *spherical* [58]. These "topologies" may have some meaning if the data themselves have a cyclic distribution, or if the purpose is to avoid border effects of the rectangular array of nodes. This may be the case if the SOM is used for process control, for the description of process states.

Another, often suggested version of the SOM is to replace the regular grid by a *structured graph* of nodes, where the structure and the number of nodes are determined dynamically, in an attempt to optimize the representation accuracy [11]. There are cases in which this leads to interesting examples, especially if the purpose is to describe the details of the data structure accurately. However, it is then no longer possible to use a two-dimensional display for visualization, as before.

An important task is to represent *dynamic phenomena* by the SOM. This becomes possible, if the models are made to represent *dynamic states*. A very important discussion of dynamic SOMs has been presented by Hammer et al. [15].

Then, of course, there arises a question whether one could define a SOM-like system based on quite different mathematical principles. One of the interesting suggestions is the *generative topographic mapping (GTM)* [3], [4]. It is based on direct computation of the topological relations of the nodes in the grid. A different, theoretically deep approach has been made by Van Hulle [63], using information-theoretic measures in the construction of the SOM topology.

Notwithstanding, we must also state that our original motivation of the SOM research was actually an attempt to explain the many somatotopic and abstract feature maps found in the *biological central nervous systems*. This aspect has been totally ignored in the paper in presentation, not because it were unimportant, but mainly because the research on the detailed brain models is still in progress. The primary goal in the recent SOM research has been to develop algorithms and computational procedures for engineering and other practical applications. An example of modified approaches in the biologically relevant direction is the attempt to explain the adaptive formation of visual feature detectors for the models of *biological vision*, as described in [45].

References

1. Allinson, N., Yin, H., Allinson, L., Slack, J. (eds.): Advances in Self-Organizing Maps. Springer, London (2001)
2. Anderberg, M.: Cluster Analysis for Applications. Academic, New York (1973)
3. Bishop, C.M., Svensen, M., Williams, C.K.I.: Developments of the generative topographic mapping. Neurocomputing 21, 203–224 (1998)
4. Bishop, C.M., Svensen, M., Williams, C.K.I.: GTM: The generative topographic mapping. Neural Computation 10, 215–234 (1998)
5. Cheng, Y.: Convergence and ordering of Kohonen's Batch map. Neural Computation 9, 1667–1676 (1997)
6. Cottrell, M., Fort, J.C.: Étude d'un processus d'auto-organization. Ann. Inst. Henri Poincaré 23, 1–20 (1987)

7. Cottrell, M., Fort, J.C., Pagés, G.: Theoretical aspects of the SOM algorithm. In: Proc. WSOM 1997, Workshop on Self-Organizing Maps, Helsinki University of Technology, Neural Networks Research Centre, Espoo, Finland, pp. 246–267 (1997)
8. Deboeck, G., Kohonen, T. (eds.): Visual Explorations in Finance with Self-Organizing Maps. Springer, London (1998)
9. Deerwester, S., Dumais, S., Furnas, G., Landauer, K.: Indexing by latent semantic analysis. J. Am. Soc. Inform. Sci. 41, 391–407 (1990)
10. Dirichlet, G.L.: Über die Reduktion der positiven quadratischen Formen mit drei unbestimmten ganzen Zahlen. J. Reine und Angew. Math. 40, 209–227 (1850)
11. Fritzke, B.: Growing cell structures - a self-organizing network for unsupervised and supervised learning. Neural Networks 7, 1441–1460 (1994)
12. Gersho, A.: On the structure of vector quantizers. IEEE Trans. Inform. Theory IT 25, 373–380 (1979)
13. Gray, R.M.: Vector quantization. IEEE ASSP Mag. 1, 4–29 (1984)
14. Grenander, U.: Abstract Inference. Wiley, New York (1981)
15. Hammer, B., Micheli, A., Sperduti, A., Strickert, M.: Recursive self-organizing network models. Neural Networks 17, 1061–1085 (2004)
16. Hartigan, J.: Clustering Algorithms. Wiley, New York (1975)
17. Heskes, T.M., Kappen, B.: Error potential for self-organization. In: Proc. ICNN 1993, Int. Conf. on Neural Networks, vol. III, pp. 1219–1223. IEEE Service Center, Piscataway (1993)
18. Jain, A.K., Dubes, R.C.: Algorithms for Clustering of Data. Prentice-Hall, Englewood Cliffs (1988)
19. Kaski, S., Kohonen, T.: Exploratory data analysis by the self-organizing map: Structures of welfare and poverty in the world. In: Refenes, A.-P., Abu-Mostafa, Y., Moody, J., Weigand, A. (eds.) Neural Networks in Financial Engineering. Proc. Third Int. Conf. on Neural Networks in the Capital Markets, London, England, October 11-13, 1995, pp. 498–507. World Scientific, Singapore (1996)
20. Kaski, S.: Dimensionality reduction by random mapping. In: Proc. IJCNN 1998, Intl. Joint Conf. on Neural Networks, pp. 413–418. IEEE Press, Los Alamitos (1998)
21. Kaski, S., Kangas, J., Kohonen, T.: Bibliography of self-organizing map (SOM) papers: 1981-1997. Neural Computing Surveys 1, 1–176 (1998), (Available in electronic form, pp. 102–350), http://www.cse.ucsc.edu/NCS/vol1.html
22. Kohonen, T.: Self-organized formation of topologically correct feature maps. Biol. Cyb. 43, 59–69 (1982)
23. Kohonen, T.: Clustering, taxonomy, and topological maps of patterns. In: Proc. Sixth Int. Conf. on Pattern Recognition, Munich, Germany, pp. 114–128 (1982)
24. Kohonen, T.: Self-Organization and Associative Memory, 3rd edn. Springer, Heidelberg (1989)
25. Kohonen, T.: The self-organizing map. Proc. IEEE 78, 1464–1480 (1990)
26. Kohonen, T.: Self-organizing maps: optimization approaches. In: Kohonen, T., Mäkisara, K., Simula, O., Kangas, J. (eds.) Artificial Neural Networks, vol. II, pp. 981–990. North-Holland, Amsterdam (1991)
27. Kohonen, T.: Self-Organizing Maps, 3rd edn. Springer, Heidelberg (2001)
28. Kohonen, T.: Description of Input Patterns by Linear Mixtures of SOM Models, Report E8. Espoo, Finland: Helsinki University of Technology, Laboratory of Computer and Information Science (2007)
29. Kohonen, T.: Description of input patterns by linear mixtures of SOM models. In: WSOM 2007 CD-ROM Proceedings, Bielefeld University, Bielefeld, Germany (2007), http://biecoll.ub-bielefeld.de

30. Kohonen, T., Oja, E., Simula, O., Visa, A., Kangas, J.: Engineering applications of the self-organizing map. Proc. IEEE 84, 1358–1384 (1996)
31. Kohonen, T., Hynninen, J., Kangas, J., Laaksonen, J.: The Self-Organizing Map Program Package, Report A31. Espoo, Finland: Helsinki University of Technology, Laboratory of Computer and Information Science (1996)
32. Kohonen, T., Kaski, S., Lagus, K., Salojärvi, J., Honkela, J., Paatero, V., Saarela, A.: Self organization of a massive document collection. IEEE Trans. on Neural Networks 11, 574–585 (2000)
33. Kohonen, T., Somervuo, P.: How to make large self-organizing maps for nonvectorial data. Neural Networks 15, 945–952 (2002)
34. Kruskal, J.B., Wish, M.: Multidimensional Scaling. Sage University Paper Series on Quantitative Applications in the Social Sciences No. 07-011. Sage Publications, Newbury Park (1978)
35. Laaksonen, J., Koskela, M., Oja, E.: PicSOM - self-organizing image retrieval with MPEG-7 content descriptors. IEEE Trans. Neural Networks 13, 841–853 (2002)
36. Lagus, K., Kaski, S.: Keyword selection method for characterizing text document maps. In: Proc. ICANN 1999, Ninth Int. Conf. on Artificial Neural Networks, vol. 1, pp. 371–376. IEE, London (1999)
37. Lagus, K., Kaski, S., Kohonen, T.: Mining massive document collections by the WEBSOM method. Inf. Sciences 163, 135–156 (2004)
38. Lawson, C.L., Hanson, R.J.: Solving Least-Squares Problems. Prentice-Hall, Englewood Cliffs (1974)
39. Lewis, D.D., Yang, Y., Rose, T.G., Li, T.: RCV1: A new benchmark collection for text categorization research. J. Mach. Learn. Res. 5, 361–397 (2004)
40. Linde, Y., Buzo, A., Gray, R.M.: An algorithm for vector quantization. IEEE Trans. Communications COM 28(1080), 84–95
41. Luttrell, S.P.: Technical Report 4669. Malvern, UK: DRA (1992)
42. Makhoul, J., Roucos, S., Gish, H.: Vector quantization in speech coding. Proc. IEEE PROC-73, 1551–1588 (1985)
43. Manning, C.D., Schütze, H.: Foundations of Statistical Natural Language Processing. MIT Press, Cambridge (1999)
44. Miikkulainen, R.: Subsymbolic Natural Language Processing: An Integrated Model of Scripts, Lexicon, and Memory. MIT Press, Cambridge (1993)
45. Miikkulainen, R., Bednar, J.A., Choe, Y., Sirosh, J.: Computational Maps in the Visual Cortex. Springer, New York (2005)
46. Naim, A., Ratnatunga, K.U., Griffiths, R.E.: Galaxy morphology without classification: Self-organizing maps. Astrophys. J. Suppl. Series 111, 357–367 (1997)
47. Obermayer, K., Sejnowski, T.: Self-Organizing Map Formation: Foundations of Neural Computation. MIT Press, Cambridge (2001)
48. Oja, E., Kaski, S. (eds.): Kohonen Maps. Elsevier, Amsterdam (1999)
49. Oja, M., Kaski, S., Kohonen, T.: Bibliography of self-organizing map (SOM) papers; 1998-2001 addendum. Neural Computing Surveys 3, 1–156 (2003), Available in electronic form at http://www.cse.ucsc.edu/NCS/vol3.html
50. Oja, M., Somervuo, P., Kaski, S., Kohonen, T.: Clustering of human endogeneous retrovirus sequences with median self-organizing map. In: Proc. WSOM 2003, Workshop on Self-Organizing Maps, Hibikino, Japan (2003)
51. Pöllä, M., Honkela, T., Kohonen, T.: Bibliography of SOM papers, http://www.cis.hut.fi/research/sombibliography
52. Ritter, H., Kohonen, T.: Self-organizing semantic maps. Biol. Cyb. 61, 241–254 (1989)

53. Ritter, H., Martinetz, T., Schulten, K.: Neural Computation and Self-Organizing Maps: An Introduction. Addison-Wesley, Reading (1992)
54. Robbins, H., Monro, S.: A stochastic approximation method. Ann. Math. Statist. 22, 400–407 (1951)
55. Salton, G., McGill, M.J.: Introduction to Modern Information Retrieval. McGraw-Hill, New York (1983)
56. Sammon, J.W.: A nonlinear mapping for data structure analysis. IEEE Trans. Computers C-18, 401–409 (1969)
57. Seiffert, U., Jain, L.C. (eds.): Self-Organizing Neural Networks: Recent Advances and Applications. Physica-Verlag, Heidelberg (2002)
58. SOM Japan Co., Ltd.: The "BLOSSOM" software package, http://www.somj.com
59. Tokutaka, H., Kishida, S., Fujimura, K.: Application of Self-Organizing Maps - Two-Dimensional Visualization of Multi-Dimensional Information (in Japanese). Kaibundo, Tokyo (1999)
60. Tokutaka, H., Ookita, M., Fujimura, K.: SOM and the Applications (in Japanese). Springer, Japan (2007)
61. Tryon, R., Bailey, D.: Cluster Analysis. McGraw-Hill, New York (1973)
62. Ultsch, A.: Self-organizing neural networks for visualization and classification. In: Opitz, O., Lausen, B., Klar, R. (eds.) Information and Classification, pp. 307–313. Springer, Berlin (1993)
63. Van Hulle, M.: Faithful Representations and Topographic Maps: From Distoryion-to Information-Based Self-Organization. Wiley, New York (2000)
64. Vesanto, J., Himberg, J., Alhoniemi, E., Parhankangas, J.: Self-organizing map in Matlab: the SOM Toolbox. In: Proc. Matlab DSP Conference 1999, Espoo, Finland, November 16-17, pp. 35–40 (1999)
65. Vesanto, J., Alhoniemi, E., Himberg, J., Kiviluoto, K., Parviainen, J.: Self-organizing map for data mining in Matlab: the SOM Toolbox. Simulation News Europe, 25(54) March (1999)
66. Voronoi, G.: Nouvelles applications des paramétres continus á la théorie des formes quadratiques. J. Reine und Angew. Math. 133, 97–178 (1907)
67. World Bank: World Development Report 1992. Oxford Univ. Press, New York (1992)
68. Young, T.Y., Fu, K.S. (eds.): Handbook of Pattern Recognition and Image Processing. Academic, Orlando (1986)
69. Zador, P.L.: Asymptotic quantization error of continuous signals and the quantization dimension. IEEE Trans. Inform. Theory IT-28, 139–149 (1982)

Cocktail Party Processing

DeLiang Wang[1] and Guoning Hu[2]

[1] Department of Computer Science and Engineering &
Center for Cognitive Science
The Ohio State University
Columbus, OH 43210
dwang@cse.ohio-state.edu
[2] Biophysics Program
The Ohio State University
Columbus, OH 43210
hu.117@osu.edu

Abstract. Speech segregation, or the cocktail party problem, has proven to be an extremely challenging problem. This chapter describes a computational auditory scene analysis (CASA) approach to the cocktail party problem. This monaural approach performs auditory segmentation and grouping in a two-dimensional time-frequency representation that encodes proximity in frequency and time, periodicity, amplitude modulation, and onset/offset. In segmentation, our model decomposes the input mixture into contiguous time-frequency segments. Grouping is first performed for voiced speech where detected pitch contours are used to group voiced segments into a target stream and the background. In grouping voiced speech, resolved and unresolved harmonics are dealt with differently. Grouping of unvoiced segments is based on the Bayesian classification of acoustic-phonetic features. This CASA approach has led to major advances towards solving the cocktail party problem.

1 Introduction

The acoustic environment we live in consists of sound energy from multiple sources. Take, for example, the first author's "quiet" study where he is currently writing this chapter: There are air blowing from the heater, a car passing by, and voices in the house, not to mention the noise of the computer he is writing on. The problem of hearing in such an environment is epitomized as the *cocktail party problem* - the term coined by Cherry in 1953 [10].

> One of our most important faculties is our ability to listen to, and follow, one speaker in the presence of others. This is such a common experience that we may take it for granted; we may call it "the cocktail party problem." ([9], p. 280)

Cherry further observed that "no machine has yet been constructed to do just that." ([9], p. 280). His assessment on machine performance is, unfortunately, as accurate today in 2008 as it was back in 1957 (see also [16] [42]).

J.M. Zurada et al. (Eds.): WCCI 2008 Plenary/Invited Lectures, LNCS 5050, pp. 333–348, 2008.
© Springer-Verlag Berlin Heidelberg 2008

Auditory perception is determined by the waveform that reaches our ears, or more precisely the air vibration impinging on the eardrum inside the ear canal. From this input of eardrum vibrations, how does the auditory system solve the cocktail party problem? Decades of psychophysical research provides much insight into this fundamental question. In particular, Bregman's auditory scene analysis theory gives the most coherent and comprehensive account to date [4]. Auditory scene analysis (ASA) views the acoustic environment as a time-frequency scene, and organizes the input into mental representations of sound sources, called *streams*. In this context, the cocktail party problem is considered the ASA problem for speech. Conceptually speaking, ASA takes place in two main stages: *Segmentation* and *grouping*. In segmentation, the acoustic input is decomposed into sensory elements or segments, each of which should primarily originate from a single source. In grouping, the segments that are likely to arise from the same source are grouped together. Segmentation and grouping are guided by ASA cues that characterize intrinsic sound properties, including harmonicity, onset and offset, and location, as well as prior knowledge of specific sounds. ASA is divided into *primitive organization* and *schema-based organization*. Primitive organization is regarded as an innate and bottom-up process, relying on auditory features, whereas schema-based organization is regarded as a top-down process relying on prior knowledge or trained models.

A solution to the cocktail party problem, or segregation of target speech from background interference, is critically important for many applications, such as automatic speech and speaker recognition, hearing aid design, and audio information retrieval. Extensive research effort has been made to develop computational systems for speech segregation. When multiple microphones are available, one may use beamforming [26] or independent component analysis [24] to perform spatial filtering that removes or attenuates interference from nontarget directions. Spatial filtering techniques are obviously not applicable when target and interference originate from the same direction or close directions, or when only monaural recordings are available. In monaural situations, one must consider intrinsic properties of target or interference in order to distinguish and separate them. Two main approaches for monaural speech segregation are speech enhancement and computational auditory scene analysis (CASA). The speech enhancement approach usually assumes certain properties (or models) of interference and then enhances speech or attenuates interference based on these assumptions [3]. This approach has limited capacity in dealing with the variability of interference. The CASA approach aims to perform speech segregation based on ASA principles.

This chapter describes a monaural CASA approach to the cocktail party problem. Section 2 gives a high-level introduction to CASA. Section 3 describes peripheral analysis and feature extraction. Section 4 discusses auditory segmentation. Sections 5 and 6 describe the methods for voiced speech and unvoiced speech segregation, respectively. Further discussions are given in Section 7.

2   Computational Auditory Scene Analysis

CASA is the computational study of auditory scene analysis [42]. In the context of this chapter, CASA may be viewed as the study of constructing a machine system that achieves human performance in cocktail party processing. Compared to other approaches to sound separation, a major advantage of CASA is that it does not make strong assumptions about interference. A typical CASA system is shown in Fig. 1, which has four stages: Peripheral analysis, feature extraction, segmentation, and grouping. Peripheral analysis processes the input signal using an auditory peripheral model, resulting in a *cochleagram* which is a two-dimensional time-frequency (T-F) representation. A cochleagram is composed of T-F units, each of which corresponds to the response of a specific auditory filter within a time frame. The second stage extracts auditory features, producing a number of feature representations. In the segmentation stage, the system generates a collection of segments or contiguous regions in a cochleagram. On the basis of extracted features and segments, the grouping stage produces streams corresponding to individual sound sources. The grouping stage includes *simultaneous grouping* which organizes segments overlapping in time into *simultaneous streams*, and *sequential grouping* which organizes segments or simultaneous streams across time into complete streams.

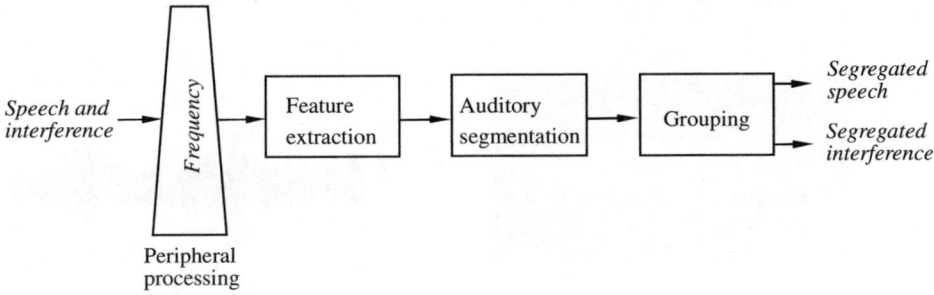

Fig. 1. Schematic diagram of a typical CASA system. A mixture of speech and interference is processed in four stages until target speech is segregated.

An important notion in CASA concerns its computational goal. With the peripheral representation of a cochleagram, we have suggested that a main goal of CASA system should be to retain the T-F units within which the target source is more intense than interference and remove the other T-F units [18] [19] (see [39] for an extensive discussion). In other words, the goal is to identify a binary T-F mask, referred to as the *ideal binary mask* (IBM), where 1 indicates that target is stronger than interference within the corresponding T-F unit and 0 otherwise. The target source can then be resynthesized from the ideal binary mask by retaining the acoustic energy from T-F regions corresponding to 1's and rejecting the acoustic energy corresponding to 0's. Fig. 2 illustrates the IBM notion. Fig. 2(a) and 2(b) show the cochleagram and the waveform of a speech utterance from the TIMIT database [15]. The cochleagram is generated using an auditory filterbank with 128 gammatone filters and 20-ms rectangular time windows with 10-ms window shift (see Section 3 for details). Fig. 2(a) shows the energy distribution within T-F units, where a brighter pixel indicates stronger energy. Fig. 2(c) and 2(d) show the cochleagram and the waveform of this utterance mixed with a crowd noise, at the overall signal-to-noise ratio (SNR) of 0 dB. Fig. 2(e) shows the ideal binary mask for the mixture in Fig. 2(c), where 1 is indicated by black and 0 by white. The speech resynthesized from the ideal binary mask is shown in Fig. 2(f). By comparing Fig. 2(f) and Fig. 2(d) with Fig. 2(b), it is clear that the speech resynthesized from the IBM is much closer to clean speech than the mixture.

The concept of IBM is directly motivated by the auditory masking phenomenon: Within a critical band, a weaker signal tends to be masked by a

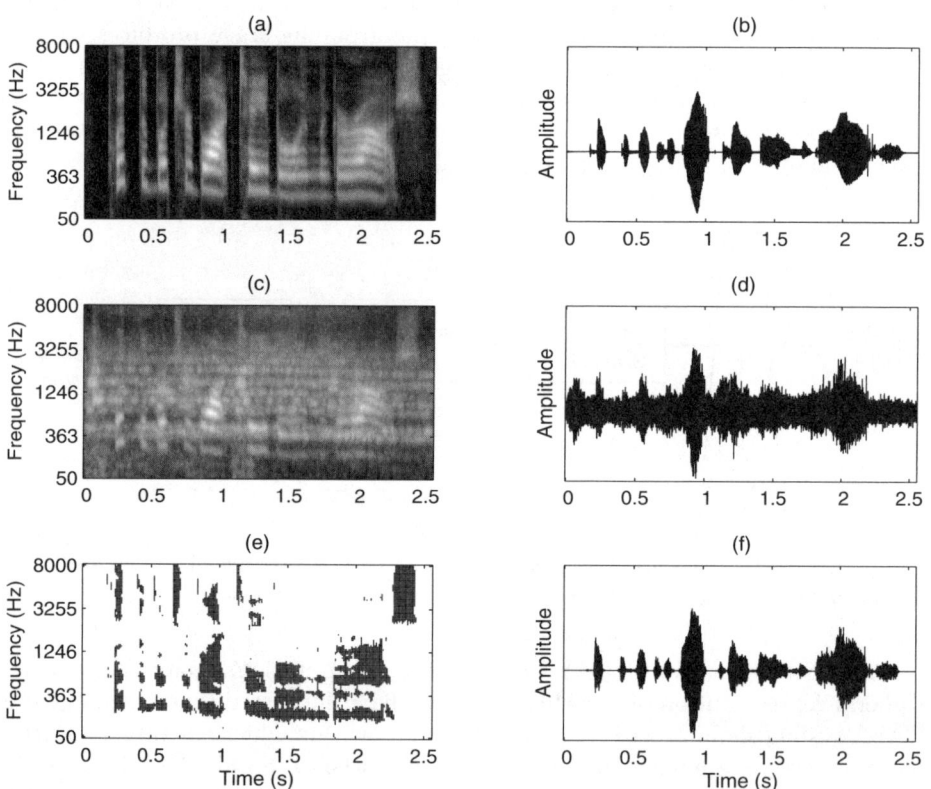

Fig. 2. Ideal binary mask. (a) Cochleagram of a female utterance, "Put the butcher block table in the garage." (b) Waveform of the utterance. (c) Cochleagram of the utterance mixed with a crowd noise at 0-dB SNR. (d) Waveform of the mixture. (e) Ideal binary mask where black regions indicate T-F units with the mask value of 1 and white regions the mask value of 0. (f) Waveform resynthesized from the ideal binary mask.

stronger one [29]. Ideal binary masking leads to dramatic improvements in human speech intelligibility [32] [8] [1] and automatic speech recognition [12] [32]. Under certain conditions the ideal binary mask has the optimal SNR gain among all the binary masks [39] [27].

Wang and Brown recently edited a book on CASA that gives a comprehensive review on various CASA approaches, systems, and applications [42]. Instead of another review, this chapter mainly describes our systematic effort on monaural speech segregation.

3   Peripheral Analysis and Feature Extraction

We describe below early auditory processing that first decomposes the input in the T-F domain, and then extracts auditory features corresponding to ASA cues.

A gammatone filterbank is used to perform peripheral analysis that decomposes the input in the frequency domain [30]. The impulse response of a gammatone filter centered at frequency f is:

$$g(f, t) = \begin{cases} b^a t^{a-1} e^{-2\pi bt} \cos(2\pi ft), & t \geq 0, \\ 0, & \text{else,} \end{cases} \tag{1}$$

where $a = 4$ is the order of the filter, and b is the equivalent rectangular bandwidth which increases as the center frequency f increases. For a filter channel c with center frequency f_c, its response to input signal $x(t)$ is

$$x(c, t) = x(t) * g(f_c, t) \tag{2}$$

where "$*$" denotes convolution. The response of a filter channel can be further processed by a model of auditory nerve transduction (see e.g. [28]), the output of which represents the firing rate of an auditory nerve fiber, denoted by $h(c, t)$. The output of each filter channel is divided into 20-ms time frames with 10-ms frame shift, producing a cochleagram as shown in Fig. 2(a) and 2(c).

A correlogram is a commonly used periodicity representation, composed of autocorrelations of filter responses across all the filter channels. Let u_{cm} denote a T-F unit for frequency channel c and time frame m. The corresponding autocorrelation of the filter response is given by

$$A_H(c, m, \tau) = \sum_n h(c, mT - n) h(c, mT - n - \tau) \tag{3}$$

where τ denotes the time lag, n denotes discrete time, and T the frame shift. The above summation is over a time frame.

Cross-channel correlation measures the similarity between the responses of two adjacent filter channels, which indicates whether the filters respond to the same sound component. For u_{cm}, its cross-channel correlation with $u_{c+1,m}$ is given by

$$C_H(c, m) = \sum_\tau \tilde{A}_H(c, m, \tau) \tilde{A}_H(c + 1, m, \tau) \tag{4}$$

where $\tilde{A}_H(c, m, \tau)$ denotes $A_H(c, m, \tau)$ normalized to zero mean and unity variance, and L the maximum delay for A_H.

The amplitude modulation (AM) information can be captured by analyzing a response envelope. A general way to obtain a response envelope is to perform half-wave rectification and lowpass filtering. Since we are interested in the envelope fluctuations corresponding to target pitch, we perform bandpass filtering instead, where the passband corresponds to the plausible F0 range of target speech. Let $h_E(c, t)$ denote the resulting envelope. Given such a response envelope, we can compute envelope autocorrelation, $A_E(c, m, \tau)$, and cross-channel correlation of response envelopes, $C_E(c, m)$, following (3) and (4), respectively.

Onsets and offsets correspond to sudden intensity increases and decreases. A standard way to identify such intensity changes is to take the first-order derivative of the intensity with respect to time and then find the peaks and valleys of the derivative. Because of intrinsic intensity fluctuations, many peaks and valleys of the derivative do not correspond to actual onsets and offsets. To reduce such fluctuations, we smooth the intensity over time, which can be performed through Gaussian smoothing [40] [21]. Onsets then correspond to the peaks of the derivative above a certain threshold, and offsets the valleys below a certain threshold.

4   Auditory Segmentation

Segmentation is an important stage in ASA. A segment as a contiguous region of T-F units contains more global information of the source that is missing from individual T-F units, and this information could be key for distinguishing sounds from different sources. We describe two approaches below for auditory segmentation.

Cross-Channel Correlation Based Approach

A speech signal lasts for a period of time, within which it has good temporal continuity. Therefore, neighboring T-F units in time tend to originate from the same source. In addition, because the passbands of adjacent channels have significant overlap, a harmonic usually activates a number of adjacent channels, which leads to high cross-channel correlation. Therefore, one can perform segmentation by merging T-F units based on temporal continuity and cross-channel correlation [41] [19]. More specifically, only units with sufficiently high cross-channel correlation (see Section 3) are aggregated, and neighboring aggregated units are iteratively merged into segments. To account for AM effects of unresolved harmonics, we separately aggregate and merge high-frequency units on the basis of cross-channel correlation of response envelopes. The cross-channel correlation based approach is suitable mainly for segmenting voiced speech.

Onset/Offset Based Segmentation

Unvoiced speech lacks harmonic structure, and as a result is more difficult to segment. A different approach for segmentation that is applicable to both unvoiced and voiced speech is based on analyzing event onsets and offsets [21]. This method has three steps: Smoothing, onset/offset detection, and multiscale integration. In smoothing, the intensity is first smoothed over time in order to reduce insignificant fluctuations and then over frequency to enhance synchronized onsets and offsets across frequency. The degree of smoothing is referred to as the *scale* [33]: A larger scale leads to smoother intensity. In the step of onset/offset detection and matching, we detect onsets and offsets in each filter channel and merge them into onset and offset *fronts* if they are sufficiently close in time. A front corresponds to a boundary along the frequency (vertical) axis on a 2-D cochleagram. Individual onset and offset fronts are matched, and a matching pair encloses a segment. In the multiscale analysis step, we detect and localize events at different scales. This step starts at a large scale and then gradually moves to the finest scale. At each scale, the system generates new segments from within the current background and locates more accurate onset and offset fronts for existing segments.

Fig. 3 illustrates the bounding contours of obtained segments for the mixture in Fig. 2(c). The background is represented by gray. Compared with the ideal binary mask in Fig. 2(e), the obtained segments capture a majority of target speech. Some segments for the interference are also formed. Note that the segmentation stage does not distinguish between segments corresponding to target and those corresponding to interference, which is the task of grouping described in the next two sections.

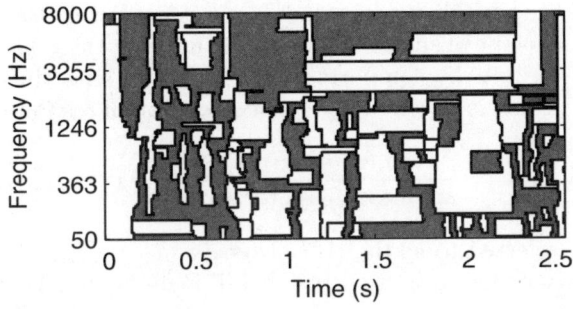

Fig. 3. Auditory segmentation. Obtained segments correspond to white regions with black bounding contours, and the background is indicated by gray. The input is the mixture shown in Fig. 2(c).

5   Voiced Speech Segregation

To group voiced speech, we use the segments obtained by the cross-channel correlation based approach described in Sect. 4. An important task of voiced speech segregation is to track pitch contours of the target speech. To perform pitch tracking, we apply the Wang and Brown algorithm [41] in initial grouping. The grouping in their algorithm is based on the dominant pitch of each time frame, and it eliminates many T-F units that unlikely belong to the target. With this initial grouping, we track a *target pitch contour* by pooling autocorrelations from the remaining T-F units. Although this initial grouping is not accurate in the high-frequency range, we use it only for the purpose of pitch tracking, which requires only a subset of harmonics. To further enhance the reliability of target pitch tracking, we first determine the consistency of an estimated pitch based on its coherence with the periodicity patterns of the retained T-F units in initial grouping, and then use pitch continuity to interpolate unreliable pitch points on the basis of reliable ones (for details see [19]). This pitch tracking algorithm can be applied iteratively to handle a general situation when the target utterance contains multiple pitch contours separated by unvoiced speech or silence [20].

Given pitch contours from the pitch tracking method described above, we label each T-F unit as target dominant or interference dominant according to detected target pitch. For a T-F unit, we first compare the periodicity of its response with the estimated pitch. Specifically, a T-F unit u_{cm} is labeled as target if the correlogram response at the estimated pitch period $\tau_S(m)$ is close to the maximum of the autocorrelation within the plausible pitch range, Γ:

$$\frac{A_H(c, m, \tau_S(m))}{\arg\max_{\tau \in \Gamma} A_H(c, m, \tau)} > \theta_T \tag{5}$$

The above criterion, referred to as the *periodicity criterion*, works well for re-solved harmonics.

For T-F units responding to multiple harmonics, their responses are amplitude-modulated, and the periodicity criterion does not work well. Based on the observation that the envelope of such a response fluctuates at the F0 rate of the source, we label these T-F units by comparing their AM rates with the estimated pitch. A straightforward way is to check the autocorrelation of response envelopes:

$$\frac{A_E(c, m, \tau_S(m))}{\arg\max_{\tau \in \Gamma} A_E(c, m, \tau)} > \theta_A \tag{6}$$

This criterion is referred to as the *AM criterion*.

The periodicity criterion is used to label T-F units that belong to segments formed using the cross-channel correlation approach. Such units primarily correspond to resolved harmonics. The remaining units are labeled by the AM criterion.

With unit labels, we group a segment into the target stream if the acoustic energy corresponding to its T-F units labeled as target exceeds half of the total energy of the segment. Furthermore, significant T-F regions not labeled as target are removed from further consideration. Finally, to group more target energy we expand each target segment by iteratively grouping its neighboring units that do not belong to any segment. When this expansion ends, the system yields a target stream and its background that consists of the remaining T-F units.

Fig. 4(a) and 4(b) show the segregated target stream and the corresponding resynthesized speech for the mixture in Fig. 2(d). Comparing to the IBM in Fig. 2(e), this stream contains a majority of the T-F units where voiced target

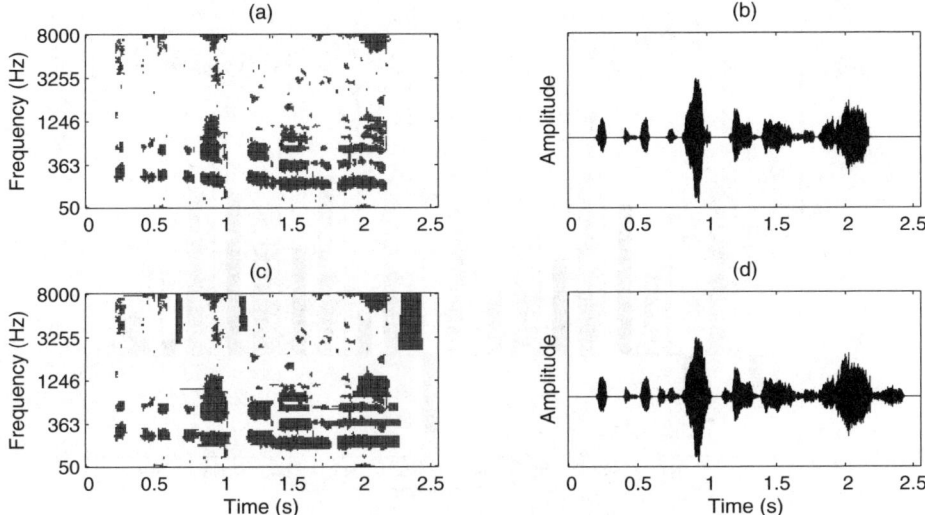

Fig. 4. Segregation results for the mixture in Fig. 2(d). (a) Binary mask corresponding to segregated voiced target. (b) Waveform resynthesized from the mask in (a). (c) Binary mask corresponding to final segregated speech. (d) Waveform resynthesized from the mask in (c).

speech dominates. In addition, a relatively small number of units where intrusion dominates are incorrectly included. The segregated speech waveform in Fig. 4(b) within voiced speech sections is similar to that of the clean speech in Fig. 2(b).

The performance of our voiced speech segregation system has been evaluated using a corpus of 100 mixtures composed of 10 voiced utterances mixed with 10 intrusions collected by Cooke [11]. This corpus has been used to test previous CASA systems [11] [6] [41] [13]. The intrusions have a considerable variety; specifically they are: N0 - 1kHz pure tone, N1 - white noise, N2 - noise bursts, N3 - "cocktail party" noise, N4 - rock music, N5 - siren, N6 - trill telephone, N7 - female speech, N8 - male speech, and N9 - female speech. As discussed in Sect. 2, our computational goal is to estimate the ideal binary mask. Therefore, our evaluation compares the segregated speech, $\hat{s}(n)$, against the speech waveform resynthesized from the ideal binary mask, $s(n)$. We can measure SNR of segregated speech in decibels:

$$\text{SNR} = 10 \log_{10}\{\sum_n s^2(n)/\sum_n [s(n) - \hat{s}(n)]^2\} \tag{7}$$

The SNR for each intrusion averaged across 10 target utterances is shown in Fig. 5, together with the SNR of the original mixtures and the results from the Wang-Brown system [41], whose performance is representative of previous CASA systems, and spectral subtraction [23], a standard method for speech enhancement. Our system shows significant improvements. In particular, it yields

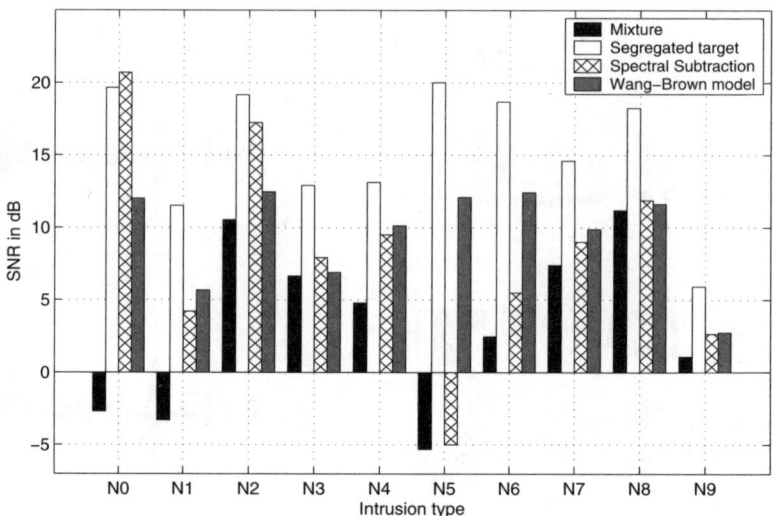

Fig. 5. Results for segregated speech and original mixtures (from [20]). Black bars show the SNRs of original mixtures, and white bars the SNR results from our system. Cross bars and gray bars show the results from spectral subtraction and the Wang-Brown system, respectively.

on average a 5.8 dB improvement over the Wang-Brown model and a 7.0 dB gain over the spectral subtraction method.

6 Unvoiced Speech Segregation

In English, unvoiced speech is composed of a subset of stops, fricatives, and affricates. With the exception of the fricative /h/, stops, fricatives, and affricates are called obstruents in phonetics. To simplify terminology, we refer to all of them as *expanded obstruents*. Unvoiced speech segregation is a more difficult problem than voiced speech segregation because of two reasons. First, unvoiced speech lacks the harmonic cue and is often noise-like acoustically. Second, sound energy of unvoiced speech is usually much weaker than that of voiced speech; as a result, unvoiced speech is more susceptible to interference. Our approach to unvoiced speech segregation first segments an input mixture using the onset/offset based method (see Sect. 4), which is applicable to both unvoiced and voiced speech, and then groups segments dominated by unvoiced speech. Due to the lack of an effective technique for sequential grouping, we focus on segregating unvoiced speech from non-speech interference in this section.

A segment may be dominated by voiced target, unvoiced target, or interference. Our goal is to group segments dominated by unvoiced target. As voiced speech is expected to be easier to segregate, we first employ voiced speech segregation described in the previous section and use its results to identify the segments dominated by voiced speech. We consider a segment to be dominated by voiced target if more than half of its total energy is included in the voiced time frames of the segment, and more than half of its energy in the voiced frames is included in segregated voiced speech. All the segments dominated by voiced target are grouped into a voiced stream. Note that the voiced stream may include some unvoiced speech because an unvoiced consonant is often strongly coarticulated with a neighboring voiced phoneme, hence included in a segment dominated by voiced target.

Once segments dominated by voiced speech are grouped, the remaining segments will be dominated by either unvoiced speech or interference. Consequently, we formulate unvoiced speech segregation as a classification problem [43] [22]. Let s denote a remaining segment, which lasts from frame m_1 to m_2, and $\mathbf{X}_s = [X_s(m_1), X_s(m_1 + 1), , X_s(m_2)]$ its corresponding T-F region on the cochleagram. $H_0(m_1, m_2)$ denotes the hypothesis that s is dominated by speech and $H_1(m_1, m_2)$ the hypothesis that it is dominated by interference. Furthermore, let $H_{0,a}(m_1, m_2)$ be the hypothesis that this region is dominated by an expanded obstruent and $H_{0,b}(m_1, m_2)$ by any other speech sound. We classify s as dominated by unvoiced speech if:

$$P(H_{0,a}(m_1, m_2)|\mathbf{X}_s) > P(H_1(m_1, m_2)|\mathbf{X}_s) \tag{8}$$

Because the durations of segments are varied, direct evaluation of the probabilities in the above inequality is unfeasible computationally. Therefore, we assume that each time frame is statistically independent. With this frame independence assumption, (8) becomes [22],

$$\prod_{m=m_1}^{m_2} P(H_{0,a}(m)|X_s(m)) > \prod_{m=m_1}^{m_2} P(H_1(m)|X_s(m)) \tag{9}$$

By applying the Bayes rule and a further assumption that the prior and the posterior probabilities of a frame do not depend on the frame index within a given segment, we have,

$$\left[\frac{P(H_{0,a})}{P(H_1)}\right]^{m_2-m_1+1} \prod_{m=m_1}^{m_2} \frac{p(X_s(m)|H_{0,a})}{p(X_s(m)|H_1)} > 1 \tag{10}$$

The prior probability ratio of $P(H_{0,a})$ and $P(H_1)$ obviously depends on the SNR of the acoustic mixture, and this relationship can be approximated by a linear function [17]. Moreover, one can estimate mixture SNR from segregated voiced speech [17]. Specifically, the result of voiced speech segregation allows us to estimate the total amount of interference by assuming that per-frame interference in unvoiced frames is the same as that in voiced frames. The total amount of speech can be approximated by estimating the average energy ratio of unvoiced speech and voiced speech at the sentence level, which is 0.09 for the training part of the TIMIT corpus [15].

In (10), the likelihood ratio between $p(X_s(m)|H_{0,a})$ and $p(X_s(m)|H_1)$ is estimated by training a multilayer perceptron (MLP) whose desired output is 1 if the corresponding frame is dominated by an expanded obstruent and 0 otherwise. Note that the trained MLP gives a good estimate of the probability [5]. With the MLP estimate of the likelihood ratio and the SNR-based estimate of the prior probability ratio, (10) is used to label a segment as either expanded obstruent or interference. All the segments labeled as unvoiced speech are grouped to the voiced stream to produce the final segregated speech stream.

Fig. 4(c) and 4(d) show the final segregated target and the corresponding synthesized waveform for the mixture in Fig. 2(d). Compared with the IBM in Fig. 2(e) and the corresponding synthesized waveform in Fig. 2(f), our system segregates most of the target energy and rejects most of the interfering energy. The target utterance, "Put the butcher block table in the garage," includes 8 stops (/p/ and /t/ in "put", /b/ in "butcher", /b/ and /k/ in "block", /t/ and /b/ in "table", and /g/ in "garage"), 2 fricatives (/θ/ in two occurrences

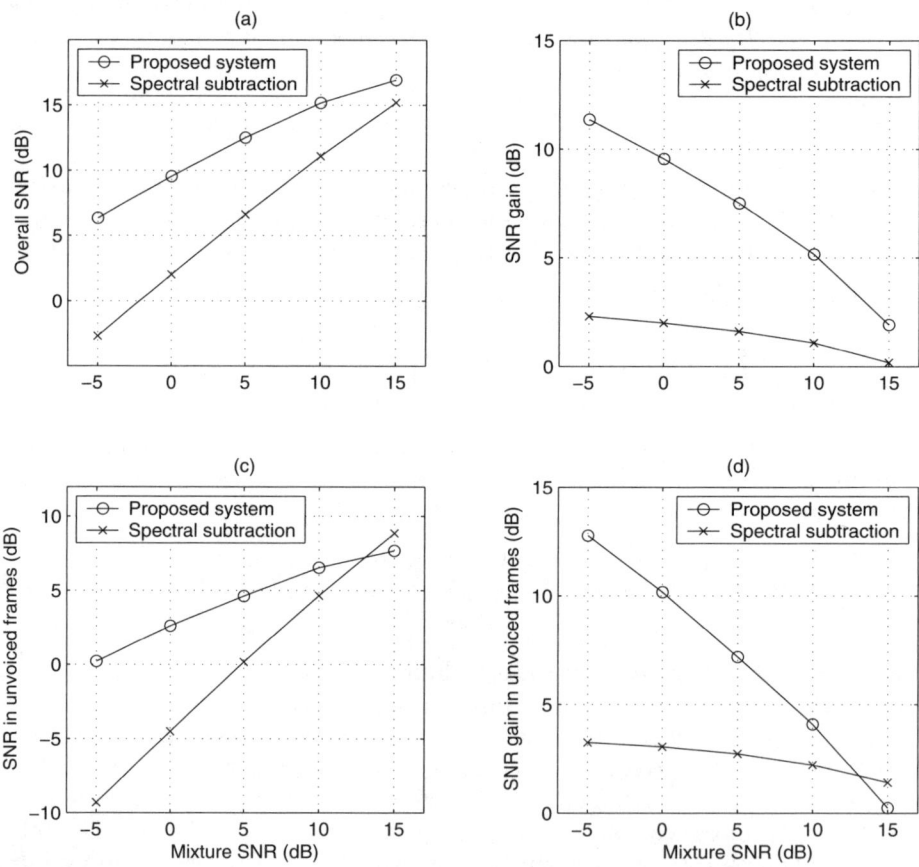

Fig. 6. SNR results of our proposed system and spectral subtraction (from [22]). (a) SNRs of segregated speech at different mixture SNR levels. (b) SNR gains of segregated targets. (c) SNRs of segregated targets at unvoiced frames. (d) SNR gains of segregated targets at unvoiced frames.

of "the"), and 2 affricates (/tʃ/ in "butcher" and /ʤ/ in "garage"). Unvoiced parts of some consonants that are coarticulated with voiced speech, such as /b/ in "block" and "table", are segregated due to their inclusion in segments dominated by voiced speech. The unvoiced consonants of /tʃ/ in "butcher" and /t/ in "table" and the affricate /ʤ/ in "garage" are partly segregated during segment classification. Some expanded obstruents, such as /p/ and /t/ in "put", are severely corrupted by the intrusion and therefore not segregated.

We have systematically evaluated segregation performance in terms of the SNR metric of (7). The evaluation uses a test corpus containing 20 target utterances randomly selected from the test part of the TIMIT database mixed with 15 nonspeech intrusions [22]. The intrusions have not been used during training, and represent a broad range of nonspeech sounds encountered in typical acoustic environments. Each speech utterance is mixed with every intrusion at the SNR levels of −5 dB, 0 dB, 5 dB, 10 dB, and 15 dB. Hence the test corpus contains 300 mixtures at each SNR level and 1500 mixtures in total. Fig. 6 shows the systematic results at different SNRs. Fig. 6(a) and 6(b) display the average SNR of segregated target and the corresponding SNR gain. Fig. 6(c) and 6(d) display the results at unvoiced frames separately. The figure clearly shows that our system produces significant SNR improvements. To put our performance in perspective, Fig. 6 also shows the SNR results of a spectral subtraction method. It is clear from the figure that our system performs substantially better for both voiced and unvoiced speech than spectral subtraction, with the only exception that occurs for unvoiced speech segregation at the input SNR of 15 dB. The amount of improvement increases with decreasing mixture SNR.

7 Discussion

Our speech segregation system deals with voiced speech first, which is easier to segregate than unvoiced speech, and uses its results to assist in unvoiced speech segregation. The results of voiced speech segregation are used in several parts of unvoiced segregation. First, segregated voiced speech marks the time intervals in which unvoiced speech may occur. Second, the unvoiced speech that is coarticulated with voiced speech tends to be segmented together, and grouped with the voiced stream. Third, segregated voiced speech provides the basis on which mixture SNR is estimated, and estimated SNR plays a significant role in classifying unvoiced segments. Our study demonstrates that this two-stage processing is an effective strategy for cocktail party processing.

We should point out that our approach to cocktail party processing is primarily feature-based [40]. The features used by the system, such as periodicity, AM, and onset, are general properties of sound. Our system does not employ trained models of speech or interference, except in unvoiced speech grouping where training is used as part of segment classification. Prior knowledge helps ASA in the form of schema-based grouping [4]. Model-based organization is emphasized by Ellis [14], and is a subject of several recent studies [34] [2] [38] [36].

These model-based approaches are expected to further enhance the performance of a feature-based system.

A natural speech utterance contains silent gaps and other sections masked by interference. In practice, one needs to group the utterance across such time intervals. This is the problem of sequential grouping, and by assuming non-speech interference we have sidestepped sequential organization in this chapter. The assumption of nonspeech interference is obviously not applicable to mixtures of multiple speakers. Recently, Shao and Wang proposed to perform sequential grouping using trained speaker models [36], and reported promising results (see also [35]). Room reverberation is another important issue that must be addressed before speech segregation systems can be deployed in real world environments, and recent studies have begun to address this challenging issue [31] [7] [25].

The speech segregation algorithm described in this chapter has been used by Srinivasan et al. [37] in a speech separation and recognition challenge hosted in the 2006 INTERSPEECH conference. The competition contains two tasks, one on two-talker mixtures and another on mixtures of one talker and speech-shaped noise (SSN) which is a stationary noise whose long-term spectrum matches that of speech. Their system uses our segregation algorithm before performing automatic speech recognition (ASR). For two-talker mixtures, they obtain uniform ASR improvements for all the SNR levels specified in the challenge; for example, at 0-dB SNR, Srinivasan et al.'s system increases ASR accuracy by 20% in absolute terms. For the SSN task, the system produces even greater improvements and gives the best performance in the competition. These evaluation results underscore the progress made in CASA since Weintraub developed the first CASA model in 1985, which did not show convincing ASR improvement [44].

In summary, we have described a monaural CASA approach to cocktail party processing. Our approach first segregates voiced speech on the basis of periodicity and amplitude modulation as well as temporal continuity. We then segregate unvoiced speech by performing onset/offset based segmentation and segment classification. Evaluation results show that our approach performs well for both voiced and unvoiced speech segregation. In particular, our study on unvoiced speech segregation represents the first systematic effort on addressing this challenge.

Acknowledgement

This research described here was supported in part by an AFOSR grant (FA9550-04-01-0117), an AFRL grant (FA8750-04-1-0093), and an NSF grant (IIS-0534707). We thank Yipeng Li for his assistance in formatting.

References

1. Anzalone, M.C., Calandruccio, L., Doherty, K.A., Carney, L.H.: Determination of the potential benefit of time-frequency gain manipulation. Ear & Hearing 27, 480–492 (2006)
2. Barker, J., Cooke, M., Ellis, D.: Decoding speech in the presence of other sources. Speech Communication 45, 5–25 (2005)

3. Benesty, J., Makino, S., Chen, J. (eds.): Speech enhancement. Springer, New York (2005)
4. Bregman, A.S.: Auditory scene analysis. MIT Press, Cambridge (1990)
5. Bridle, J.: Probabilistic interpretation of feedforward classification network outputs, with relationships to statistical pattern recognition. In: Fogelman-Soulie, F., Herault, J. (eds.) Neurocomputing: Algorithms, architectures, and applications, pp. 227–236. Springer, New York (1989)
6. Brown, G.J., Cooke, M.P.: Computational auditory scene analysis. Computer Speech and Language 8, 297–336 (1994)
7. Brown, G.J., Palomäki, K.J.: Reverberation. In: Wang, D.L., Brown, G.J. (eds.) Computational auditory scene analysis: Principles, algorithms, and Applications, pp. 209–250. Wiley & IEEE Press, Hoboken NJ (2006)
8. Brungart, D., Chang, P.S., Simpson, B.D., Wang, D.L.: Isolating the energetic component of speech-on-speech masking with ideal time-frequency segregation. Journal of the Acoustical Society of America 120, 4007–4018 (2006)
9. Cherry, E.C.: On human communication. MIT Press, Cambridge (1957)
10. Cherry, E.C.: Some experiments on the recognition of speech, with one and with two ears. Journal of the Acoustical Society of America 25, 975–979 (1953)
11. Cooke, M.: Modelling auditory processing and organization. Cambridge University Press, Cambridge (1993)
12. Cooke, M., Green, P., Josifovski, L., Vizinho, A.: Robust automatic speech recognition with missing and unreliable acoustic data. Speech Communication 34, 267–285 (2001)
13. Drake, L.A.: Sound source separation via computational auditory scene analysis (CASA)-enhanced beamforming. Ph.D. Dissertation, Northwestern University Department of Electrical Engineering (2001)
14. Ellis, D.P.W.: Prediction-driven computational auditory scene analysis. Ph.D. Dissertation, MIT Department of Electrical Engineering and Computer Science (1996)
15. Garofolo, J., Lamel, L., Fisher, W., Fiscus, J., Pallett, D., Dahlgren, N.: DARPA TIMIT acoustic-phonetic continuous speech corpus. Technical Report NISTIR 4930, National Institute of Standards and Technology (1993)
16. Haykin, S., Chen, Z.: The cocktail party problem. Neural Computation 17, 1875–1902 (2005)
17. Hu, G.: Monaural speech organization and segregation. Ph.D. Dissertation, The Ohio State University Biophysics Program (2006)
18. Hu, G., Wang, D.L.: Speech segregation based on pitch tracking and amplitude modulation. In: Proceedings of IEEE Workshop on Applications of Signal Processing to Audio and Acoustics, pp. 79–82 (2001)
19. Hu, G., Wang, D.L.: Monaural speech segregation based on pitch tracking and amplitude modulation. IEEE Transactions on Neural Networks 15, 1135–1150 (2004)
20. Hu, G., Wang, D.L.: An auditory scene analysis approach to monaural speech segregation. In: Hansler, E., Schmidt, G. (eds.) Topics in acoustic echo and noise control, pp. 485–515. Springer, Heidelberg (2006)
21. Hu, G., Wang, D.L.: Auditory segmentation based on onset and offset analysis. IEEE Transactions on Audio, Speech, and Language Processing 15, 396–405 (2007)
22. Hu, G., Wang, D.L.: Segregation of unvoiced speech from nonspeech interference. Journal of the Acoustical Society of America (conditionally accepted, 2008)
23. Huang, X., Acero, A., Hon, H.W.: Spoken language processing: A guide to theory, algorithms, and system development. Prentice Hall PTR, Upper Saddle River (2001)

24. Hyvarinen, A., Karhunen, J., Oja, E.: Independent component analysis. Wiley, New York (2001)
25. Jin, Z., Wang, D.L.: A supervised learning approach to monaural segregation of reverberant speech. Proceedings of IEEE ICASSP IV, 921–924 (2007)
26. Krim, H., Viberg, M.: Two decades of array signal processing research: The parametric approach. IEEE Signal Processing Magazine 13, 67–94 (1996)
27. Li, Y., Wang, D.L.: On the optimality of ideal binary time-frequency masks. In: Proceedings of IEEE ICASSP (in press, 2008)
28. Meddis, R.: Simulation of auditory-neural transduction: Further studies. Journal of the Acoustical Society of America 83, 1056–1063 (1988)
29. Moore, B.C.J.: An introduction to the psychology of hearing, 5th edn. Academic Press, London (2003)
30. Patterson, R.D., Holdsworth, J., Nimmo-Smith, I., Rice, P.: SVOS final report, part B: Implementing a gammatone filterbank. Technical Report Rep. 2341, MRC Applied Psychology Unit (1988)
31. Roman, N., Wang, D.L.: Pitch-based monaural segregation of reverberant speech. Journal of the Acoustical Society of America 120, 458–469 (2006)
32. Roman, N., Wang, D.L., Brown, G.J.: Speech segregation based on sound localization. Journal of the Acoustical Society of America 114, 2236–2252 (2003)
33. Romeny, B.H., Florack, L., Koenderink, J., Viergever, M. (eds.): Scale-space theory in computer vision. Springer, New York (1997)
34. Roweis, S.T.: One microphone source separation. In: Advances in Neural Information Processing Systems (NIPS 2000), vol. 13. MIT Press, Cambridge (2001)
35. Shao, Y.: Sequential organization in computational auditory scene analysis. Ph.D. Dissertation, Ohio State University Department of Computer Science & Engineering (2007)
36. Shao, Y., Wang, D.L.: Model-based sequential organization in cochannel speech. IEEE Transactions on Audio, Speech, and Language Processing 14, 289–298 (2006)
37. Srinivasan, S., Shao, Y., Jin, Z., Wang, D.L.: A computational auditory scene analysis system for robust speech recognition. In: Proceedings of INTERSPEECH, pp. 73–76 (2006)
38. Srinivasan, S., Wang, D.L.: A schema-based model for phonemic restoration. Speech Communication 45, 63–87 (2005)
39. Wang, D.L.: On ideal binary mask as the computational goal of auditory scene analysis. In: Divenyi, P. (ed.) Speech separation by humans and machines, pp. 181–197. Kluwer Academic, Norwell (2005)
40. Wang, D.L.: Feature-based speech segregation. In: Wang, D.L., Brown, G.J. (eds.) Computational auditory scene analysis: Principles, algorithms, and Applications, pp. 81–114. Wiley & IEEE Press, Hoboken, NJ (2006)
41. Wang, D.L., Brown, G.J.: Separation of speech from interfering sounds based on oscillatory correlation. IEEE Transactions on Neural Networks 10, 684–697 (1999)
42. Wang, D.L., Brown, G.J. (eds.): Computational auditory scene analysis: Principles, algorithms, and applications. Wiley & IEEE Press, Hoboken, NJ (2006)
43. Wang, D.L., Hu, G.: Unvoiced speech segregation. In: Proceedings of IEEE ICASSP, pp. 953–956 (2006)
44. Weintraub, M.: A theory and computational model of auditory monaural sound separation. Ph.d. dissertation, Stanford University Department of Electrical Engineering (1985)

Similarities in Fuzzy Data Mining:
From a Cognitive View to Real-World Applications

Bernadette Bouchon-Meunier, Maria Rifqi, and Marie-Jeanne Lesot

UPMC Univ Paris 06, CNRS,UMR 7606, LIP6, F-75005, Paris, France
{bernadette.bouchon-meunier, maria.rifqi,
marie-jeanne.lesot}@lip6.fr

Abstract. Similarity is a key concept for all attempts to construct human-like automated systems or assistants to human task solving since they are very natural in the human process of categorization, underlying many natural capabilities such as language understanding, pattern recognition or decision-making. In this paper, we study the use of similarities in data mining, basing our discourse on cognitive approaches of similarity stemming for instance from Tversky's and Rosch's seminal works, among others. We point out a general framework for measures of comparison compatible with these cognitive foundations, and we show that measures of similarity can be involved in all steps of the data mining process. We then focus on fuzzy logic that provides interesting tools for data mining mainly because of its ability to represent imperfect information, which is of crucial importance when databases are complex, large, and contain heterogeneous, imprecise, vague, uncertain or incomplete data. We eventually illustrate our discourse by examples of similarities used in real-world data mining problems.

Keywords: similarity, data mining, categorization, prototype, fuzzy sets.

1 Introduction

Since similarities are very natural in the human process of categorization underlying many natural capabilities such as language understanding, pattern recognition or decision-making, they are naturally fundamental for all attempts to construct human-like automated systems or assistants to human task solving, and particularly in data mining and information retrieval.

Those domains are difficult to cope with for various reasons. First, most of the databases are complex, large, and contain heterogeneous, imprecise, vague, uncertain or incomplete data. Furthermore, the queries may be imprecise or subjective in the case of information retrieval; the mining results must be easily understandable by a user.

Fuzzy logic provides interesting tools for such tasks, mainly because of its capability to manage imprecise categories, to represent imperfect information, for instance by means of fuzzy sets, graduality, measures of resemblance or aggregation methods.

J.M. Zurada et al. (Eds.): WCCI 2008 Plenary/Invited Lectures, LNCS 5050, pp. 349–367, 2008.

We propose to explore the capabilities of similarities to interact with fuzzy methods in several steps of the data mining process, information retrieval or knowledge discovery, such as clustering, construction of prototypes, utilization of expert or association rules or fuzzy querying, for instance.

With this object, we present a view of the concept of similarity rooted in cognitive psychology, and we discuss its utilization in the framework of fuzzy data mining. The paper is organized as follows: in Section 2, we consider the cognitive point of view on similarity and the related categorization notion, pointing out the elements that could be of interest in data mining. In Section 3, we examine the use of similarity in the data mining framework, underlining its central role in all steps of this process. In Section 4, we focus on the case of fuzzy logic: after recalling existing measures, we describe a general framework for comparison measures that is compatible with the cognitive foundations, and state properties that can be desired from similarity measures. These properties provide guides for selecting a measure appropriate for a given problem. In Section 5, we eventually present some real-world applications where these paradigms have been exploited among others to manage various types of data, such as image retrieval or risk analysis.

2   Similarity and Categorization in Cognitive Science

Similarities, in a general sense, have been widely studied in cognitive psychology, from different points of view, and in particular for the categorization task. The latter aims at reducing the amount of information in order to decrease our cognitive effort and at reflecting the structure of the real world [1][2]. Data mining issues are connected to these objectives: likewise, it aims at extracting, from large data bases, relevant information that still reflects the structure of the whole base.

If bridges have been made between data mining and cognitive science regarding similarity, many aspects studied by one of the communities remain unknown by the other one. In this section, we would like to point out some of the elements raised on the subject of similarities that could be of interest in data mining and related topics. Our purpose in this brief cursory glance at similarities in cognitive psychology is to show the various possible angles we can choose to treat them and to leave doors open to new visions of similarities in the data mining domain.

The concepts of categorization, similarities and prototypes are intrinsically connected, even though their relationships are not uniformly accepted and various approaches of these concepts intertwine. For the sake of simplicity, we consider them successively, pointing some of their interrelations.

2.1   Categorization

In psychology, several approaches of categorization [3] can be distinguished, according to the underlying structure of the categories they assume: one approach assumes that there exist rules used for the recognition of categories; another one is based on the knowledge of properties shared by members of a category. A third one considers that categories are based on the recognition of similarities.

One vision of categorization supposes an all-or-none relationship between categories and objects: an object belongs to a category or it does not. This way, categories are defined in terms of necessary conditions, and not of similarities. For instance [4] points out the existence of categories that are based on explicit definitions, such as the category of triangles defined by a list of geometric properties, or on an ad hoc process gathering objects for a given purpose, such as Valentine's day gifts.

Under the assumption that similarities are the basis of natural categorization, there exists a variety of points of view. The notion of family resemblance introduced by Wittgenstein [5] in 1953 (for instance the family of games) does not use an explicit definition of the similarity involved in the categorization but a subjective judgment, which may be dependent on the context.

Exemplar models [4] can be regarded along the same lines and consider that categories are represented in terms of individual instances. A new object or element is then classified in a category if it is more similar to elements already stored in this category than to elements of other categories. This theory does not take into account any representative of a category.

On the opposite, another stream dealing with family resemblance considers a category by means of a central representative of the category called a prototype, and a graded structure around it [2], formed by objects similar to the prototype. Furthermore, a category can be associated to several prototypes in case of a diversity of subcategories.

More precisely, Rosch [2] considers that an object can better represent its category than another one. The typicality of an object for a given category depends on its resemblance to the other members of the category and its differences to the members of other categories. In other words they are spread on a scale, or a gradient, of typicality: the more typical an object, the more attributes it shares with other members of its category and the less attributes it shares with members of the other categories.

Kleiber [6] extends Rosch's approach, underlining the notion of fuzzy frontiers and the fact that the belonging to a category is based on the degree of similarity with the prototype of this category. Furthermore, the notions of typicality degree and membership degree are clearly distinct whereas they were not in Rosch's approach. It means that, even if an object is less typical of a category than another one, it does not necessarily belong to this category to a smaller extent: although an ostrich is not a typical bird, it still totally belongs to the bird category. This example moreover highlights the fact a category can be binary defined and still characterized in terms of typicality.

In the studies regarding similarity and typicality, there have been several attempts to prove that a differentiation can be made between them [3]. One reason is that frequency of instantiation can be regarded as involved in the identification of typicality in addition to similarity. The extreme option considers that frequency is the most important factor in typicality [7]. Familiarity with exemplars may appear to be involved in the construction of a graded structure of a category [8], and familiarity is both related to frequency and context. Another reason to differentiate similarity and typicality deals with causality, and takes into account the variability and the existence of changes in the identification of categories, for instance related to the history of changes, with a consideration of time in the identification of categories [9].

It can be considered that human beings naturally form concepts through this prototype mechanism. For instance Posner [10] considers that prototypes play a part in the formation of abstract ideas, taking into account the variability of instances expressed in terms of distances between patterns.

A different view of categories can be based on intrinsic coherence, regarded as the existence of links between properties that constitute a kind of conceptual core [4]. This view is not incompatible with the notion of typicality but it describes atypical elements by means of the non-existence or co-existence of some properties. This leads to the identification of hybrid categories and does not accept any graduality. There is no explicit reference to similarities in such a theory. Category variability [11] is pointed out as a motivation for this different vision of categorization, since properties satisfied by a category may depend on the context.

2.2 Similarity

As it is accepted that most natural categories are structured in terms of family resemblance or centered around prototypes, similarity plays a central role in category structure.

The seminal work by Tversky [12] rejects the classic assumption of the need of a metric space to define similarities. He assumes in particular that symmetry is not a necessity for a similarity judgment, since an observed object can be compared to a reference object in an asymmetric way due to the status of the two objects. He also rejects the necessity of the transitivity property. He introduces the so-called contrast model, defining a measure of similarity of two objects as a function increasing with respect to the features common to these objects and decreasing with respect to their distinctive features. He suggests more properties to require from measures of similarity, and he mentions the importance of context, reducing it to a choice of features. He observes that similarity has two faces: the first one is causal in that it serves as a basis for the classification of objects, and the second one is derivative as it is influenced by the existing classification.

He proposes the so-called ratio model as a particular case of the contrast model, defining similarity measures by the following form, for two given non-negative parameters α and β :

$$s_{\alpha,\beta}(A, A') = \frac{f(A \cap A')}{f(A \cap A') + \alpha f(A \ominus A') + \beta f(A' \ominus A)} \qquad (1)$$

It is to be noted that Tversky considers features as basic granules of the description of objects: for instance, for the description of a human face, a feature can be the presence or absence of a smiling mouth, a frowning mouth or a straight eyebrow. A particular case corresponds to features considered as attributes with values in universes of discourse, for instance "mouth" with values {smiling, frowning, neutral}.

A more shaded approach [13] suggests that one can observe a difference between similarity judgments and categorization tasks when deeper features than perceptual elements are used for the categorization and this difference could be rubbed out if several levels of similarity were taken into account, from perceptual similarities to conceptual similarities based on domain theories. This approach does not seem to have been much explored.

Recently, attempts to take into account changes and variations of categories in time have given rise to a dynamic similarity processing formalization [14] as opposed to the classical static similarities we described above. Several views of dynamic similarities are possible [15], either concerned by the history of perceptual patterns, or approaching previsions of categories expected in the future, for instance through an adaptation process.

2.3 Related Concepts

There exist various interpretations of the general concept of similarity; words like similarity, analogy, proximity or closeness are often used in an undifferentiated way, even though they refer to different definitions.

In a cognitive sense, analogy is formalized in a simple representation by "as A is to B, so C is to D", and is based on an identity of relations between situations or objects which can be of a completely different nature, involving the idea of structure or function. On the contrary, similarity, simply expressed as "A is similar to B", identifies a resemblance between two objects. Similarity and analogy are still connected in several aspects. The most obvious connection lies on the fact that the recognition of analogy is often based on similarities. Furthermore, the so-called alignment model of similarity [16] is based on the assumption that mental representations are structured and evaluating the similarity between elements or objects takes into account relations in the structure, for instance relations between perceptual units or classic semantic relations such as meronymy or holonymy relations. This model stems from representations of analogy in addition to semantic descriptions.

The concepts of proximity and closeness are related to distance measures. In many cases, a similarity measure can be defined as the dual of a dissimilarity measure or a distance, on the basis of a rule of the form "the less distant, the more similar". Nevertheless, similarity and dissimilarity are concepts which can be considered as antinomic or complementary, depending on the angle: dissimilarities, called differences, are recognized as different from the opposite of similarities by Tversky [12], whereas in the case of prototypes, similarity and dissimilarity are two complementary components of a global approach of classification.

Inclusion is another related concept that has been mainly attached to the idea of implication or inference in cognitive psychology. It is involved in the identification of similarities, for instance in a property-based categorization [4].

3 Similarities in Data Mining

Similarities (or dissimilarities) have been widely used in artificial intelligence. Rissland [17] points out their importance and underlines their central role, explaining it by the difficulty of representing real-world concepts. She considers that real-world concepts are "messy" in the sense that they have grey areas of interpretation, which leads to an open-textured property, they change and are submitted to a non-stationary property and they have exceptions. She suggests that representing messy concepts in artificial intelligence presents a challenge that can be braved thanks to the notion of similarity. We complete this assumption in claiming that these properties of real-world concepts lead to fuzzy–set based representation.

3.1 Standard Data Mining

The well-known description of the data mining process given by Fayyad [18] presents a succession of four steps:

(i) from databases or data warehouses, a selection process extracts relevant data,

(ii) these relevant data are submitted to cleaning or transformation operations in order to construct a training set,

(iii) on this training set, a machine learning method is used to elicit a model of information,

(iv) this model is submitted to an interpretation in order to obtain knowledge understandable from the user or expert.

Now all these four steps can benefit from the use of some types of similarities.

In step (i), the selection can be achieved thanks to a matching between query and data, on the basis of similarities.

In step (ii), data cleaning and data reduction strategies are various and similarities, among others, bring solutions to these processes. There exist for instance various approaches to the management of missing data [19], and some of them exploit the notion of similarity or distance, especially those based on the use of clusters of similar observations to assign a value replacing a missing one [20] [21]. Methods to simplify data by means of attribute selection and dimensionality reduction can also be based on the use of distances.

In step (iii), many machine learning methods can be related to the concept of similarity. Clustering is for instance based on the principle of grouping elements as close as possible to each other with regard to attribute values and also (for some methods) as far as possible from elements of other groups. Statistical techniques such as Support Vector Machines lie on kernel functions that are nothing but similarities.

Similarities are explicitly used in non-classical reasoning approaches, such as case-based reasoning, analogical reasoning, similarity-based reasoning, where they constitute the core of the methods themselves.

Inductive learning is an exemplar-based construction of rules describing categories and the similarity of instances belonging to a category is action-oriented, an action being either the identification of a class or a decision to make. In the case of decision tree construction, it can be considered that the conditional entropy used to elicit the rules corresponds to a probabilistic version of similarity.

In step (iv), the passage from abstract models to knowledge needs an interpretation phase. In this part, again, similarities can be used for several purposes, for instance for rule base simplification [22].

3.2   Similarities in Fuzzy Data Mining

If we focus on data mining in a fuzzy set theory setting, specific needs of similarity management arise.

In database management, similarities are useful to compare an approximate value involved in a query to all possible solutions stemming from the database.

In the construction of a model, fuzzy association rules can be managed thanks to similarities [23]. In fuzzy inductive learning, the discretization phase splitting the attribute values in two or more fuzzy classes is based on similarities underlying this process of categorization [24]. The choice of the best attribute in the construction of a fuzzy decision tree relies on the optimization of a measure of the discriminating power of an attribute with regard to a class: the measure of classification ambiguity [25] proposed by Yan and Shaw is for instance based on fuzzy similarities.

When fuzzy decision trees are used to classify an example, similarity is used to compare its attribute values to those associated with edges of the tree, whereas a simple binary matching step is applied if a standard decision tree is used [26].

At the final level of the interpretability, expressivity of rules can be improved by means of linguistic modifiers closely related to similarities [27][28] or for a linguistic expression of categories [29]. In the case of a model taking the form of if-then rules for instance, similarities can be used to merge several rules and to simplify the model [30].

We have presented examples of situations where measures of similarities are useful. This is the reason why we present how they are represented in a fuzzy setting.

4   Similarities in a Fuzzy Setting

From the rapid presentations of similarity and categorization issues in psychology described in Section 2, the links with fuzzy sets appear clearly. The recurrent occurrence of variability in categories and their graded structure incline us to take advantage of the graduality and flexibility inherent in fuzzy sets to define measures of similarity and to model categories.

This has obviously been achieved from the early beginning of fuzzy set theory since L.A. Zadeh [31] has introduced the concept of similarity relation as an extension of equivalence relation, that presents the advantage of providing a crisp partitioning of data on the basis of a fuzzy knowledge of their relations. It should be remarked that the introduced softness is limited, properties inherited from classic relations such as transitivity and symmetry being preserved. Attempts to go further in the flexibility have led to indistinguishability relations [32] accepting a version of transitivity less constrained than similarity relations.

Such a fuzzy relation is defined on a given universe and provides the degree of similarity of any pair of elements in this universe. For instance, if a discrete universe of colors is considered, orange is more similar to red than to blue. In the case where one wants to compare fuzzy sets of the universe rather than precise elements, we can use extensions to sets of fuzzy sets of these similarity or indistinguishability relations. For instance, compared colors are regarded as fuzzy sets of a continuous universe. In this framework, measures of similarity or resemblance have been proposed.

4.1 Measures of Similarity

Measures of similarity (or dissimilarity) have obviously been introduced out of the scope of fuzzy set theory in a wider framework. The measures proposed by Jaccard [33], Dice [34] or Ochiai [35] have been extensively used in many domains, and they belong to a set-based view of similarities, taking into account the numbers of elements common to the compared objects or distinct between them.

Many set-theoretical measures actually correspond to particular cases of Tversky ratio model [12], as defined in Equation (1), Jaccard coefficient being for instance associated with parameters $\alpha = \beta = 1$, and Dice with parameters $\alpha = \beta = 1/2$.

Besides, many distances have also been introduced in a geometric vision of the descriptions of objects to compare (e.g. Minkowski, Chebychev, Hausdorff, Mahalonobis) and they have been used in various areas.

Such measures have given rise to a variety of measures of comparison of fuzzy sets, evaluating either their resemblance or their difference. Many attempts have been made to compare them [36] and to get them into some kind of order, in such a way that they could efficiently be used in practical domains, such as image processing, pattern recognition or data mining [37][38][39][40]. Most of the proposed classifications are based on a distinction between set-theoretic and geometric procedures, and some of them add the third class of logic-based procedure [41].

A thorough analysis of the existing so-called measures of similarity points out very different forms of measures of comparison of fuzzy sets, going farther than the simple notion of similarity. The most common are distances, measures of dissimilarity and inclusion. Generic terms such as compatibility measures [41], comparison indices [42], are proposed to take into account all measures of "matching" between fuzzy sets.

Tversky's contrast model is sometimes mentioned to study set-theoretic similarity measures in a fuzzy framework [41], and fuzzy versions of Tversky's contrast model have been proposed [43][44][45] in specific areas. In the following section, we describe a different connection between Tversky's approach and similarity of fuzzy sets, in which a general classification framework for comparison measures is introduced.

4.2 General Framework for Measures of Comparison

Working with various real-world applications requiring diverse measures to evaluate how fuzzy descriptions of objects differ or are similar, we have had a double concern: first, to help to classify such measures, embracing as many kinds as possible in a unique framework, in order to propose to the user the most convenient solutions to his problem; secondly, to remain compatible with cognitive psychology views of measures of similarity. Tversky's model has been chosen because of its degree of generality and the wide spectrum of potential instantiation. So-called general measures of comparison [46][47] have been introduced, encompassing the main measures of similarity, dissimilarity, satisfiability, resemblance, inclusion.

We briefly recall the principles of this framework before showing how it has been used in practical applications.

Let Ω be a given universe and $F(\Omega)$ the set of its fuzzy sets, equipped with the classical inclusion $\subseteq$, a fuzzy set measure $M : F(\Omega) \to \Re^+$, such that such that

$M(\varnothing) = 0$ and M is monotonous with regard to $\subseteq$, and two operations on $F(\Omega)$, namely an intersection $\cap$ and a difference Θ such that $A \subseteq B$ implies $A\Theta B = \varnothing$. We define a measure of comparison on Ω as a mapping:

$$S : F(\Omega) \times F(\Omega) \rightarrow [0,1] . \qquad (2)$$

of the form:

$$S(A, B) = F_S\left(M(A \cap B), M(B\Theta A), M(A\Theta B)\right), \qquad (3)$$

for a mapping $F_S : \mathfrak{R}^3 \rightarrow [0,1]$.

It must be remarked that they only follow the basic Tversky's requirement of matching. Interesting properties may be required from measures of comparison in order to capture all necessary behaviors involved in the comparison of elements in practical applications. Reflexivity and symmetry are simple extensions of classical notions. Exclusiveness is satisfied if $S(A, B) = 0$ for any A and B such that $A \cap B = \varnothing$.

Four main types of measures of comparison are identified to help the user in various kinds of processes. Regarding similarity assessments, three processes stem from the applications, in agreement with psychological studies: either an object is compared to a reference and measures of satisfiability or inclusion are introduced, or two objects with the same status are compared and measures of resemblance are presented. We also distinguish dissimilarity from the negation of similarity, introducing so-called measures of dissimilarity. The following classes are thus defined:

- Measures of *resemblance* are reflexive and symmetrical, increasing in $M(A \cap B)$, decreasing in $M(A\Theta B)$ and $M(B\Theta A)$.
- Measures of *satisfiability* are reflexive, exclusive, and independent of $M(A\Theta B)$, not necessarily symmetrical, increasing in $M(A \cap B)$, decreasing in $M(B\Theta A)$.
- Measures of *inclusion* are reflexive, exclusive and independent of $M(B\Theta A)$, not necessarily symmetrical, increasing in $M(A \cap B)$, decreasing in $M(A\Theta B)$.
- Measures of *dissimilarity* are independent of $M(A \cap B)$, non-decreasing in $M(B\Theta A)$ and $M(A\Theta B)$, such that $S(A, B) = 0$ for any A and B such that $A\Theta B = \varnothing$ and $B\Theta A = \varnothing$. They indicate the degree of difference between features.

Measures of resemblance represent the "similarity" between elements of the same kind or level and can be used for instance in clustering or data mining, while satisfiability and inclusion measures evaluate the "similarity" of a new element with a reference. A satisfiability measure evaluates to which extent B is compatible with A and it can be used in decision trees or case-based reasoning, for instance. An inclusion measure evaluates to which extent B can be considered as a particular case of A and it is useful when working on databases, semantic networks or relations between properties for instance. A dissimilarity measure is important for the construction of prototypes or in some clustering methods.

Measures of resemblance, satisfiability and inclusion are proved to be in agreement with Tversky's requirements of monotonicity, independence, solvability, invariance [46] for measures of similarity. Although there exist measures of similarity in Tversky's contrast model which are not in any of the above categories, the latter correspond to most of the needs in practical applications. For more details about this framework, see [46] [47][48].

Looking at Tversky's similarity measures $r_{\alpha,\beta}$ as defined in Equation (1), it can be noted that they have properties of measures of resemblance when $\alpha = \beta$, measures of satisfiability when $\alpha = 0$, and measures of inclusion when $\beta = 0$.

To point out the interest of differentiating those different measures, let us consider three particular measures, denoting f_A the membership function of a fuzzy set A:

- Measure of resemblance

$$S(A, B) = \frac{\int_\Omega f_{A \cap B}(u)du}{\int_\Omega f_{A \cup B}(u)du} \tag{4}$$

corresponding to Jaccard coefficient in a classical framework.

- Measure of satisfiability

$$S(A, B) = \frac{\int_\Omega f_{A \cap B}(u)du}{\int_\Omega f_B(u)du} \tag{5}$$

- Measure of inclusion

$$S(A, B) = \frac{\int_\Omega f_{A \cap B}(u)du}{\int_\Omega f_A(u)du} \tag{6}$$

The difference between them is very subtle for the user, and the two last ones can seem equivalent at a first glance. It must be noted that they have a very different nature and using one or the other is not neutral.

4.3 Properties of Measures of Comparison

Among all properties of measures of comparison that could be presented to help the user in his choice of one of them, we stress on two main studies providing guidelines to make a choice among the jungle of measures, concerning the discrimination power of measures on the one hand, ranking properties on the other hand.

The sensitivity of measures of comparison with respect to small variations of the compared elements is an important component in the choice of one of them. Choosing a representation of similarity measures avoiding the scaling problem [49][50], it is possible to show differences of behavior between measures very discriminating for small feature values variations, or for large feature values. Based on this study of the discriminating power of similarity measures [49], a new measure of similarity has been defined, the so-called Fermi-Dirac measure defined as follows:

$$S(A, B) = \frac{F_{FD}(\varphi) - F_{FD}(\Pi / 2)}{F_{FD}(0) - F_{FD}(\Pi / 2)} \qquad (7)$$

with

$$F_{FD}(\varphi) = \frac{1}{1 + \exp\left(\dfrac{\varphi - \varphi_0}{\Gamma}\right)}, \quad \varphi = \arctan\left(\frac{M(B\Theta A) + M(A\Theta B)}{M(A \cap B)}\right) \qquad (8)$$

$\Gamma \in \Re^+$ and $\varphi_0 \in [0, \Pi / 2]$ are parameters balancing the selectivity of Fermi-Dirac measure. It presents the particular property of being discriminating around the specific value φ_0 chosen by the user, with an intensity defined by the Γ parameter.

The second property of measures of comparison we put forward is more specific of problems where values of a similarity measure are not important as such, and where only the induced order matter [43]. This for instance occurs in the case of document retrieval systems, where the user is interested in the list of documents more similar to his request, ignoring the similarity score of each document [51]. Likewise, in case-based reasoning, the n first candidates are of importance, irrespective on their similarity values.

Classes of measures of resemblance providing the same ranking have been pointed out [52]. Three definitions of the equivalence of resemblance measures have been proposed, which appear to be themselves equivalent.

Given a reference object A, two resemblance measures S and S' are order equivalent if and only if any element B provides a value $S(A, B)$ greater than the value $S(A, C')$ corresponding to another object C' whenever $S'(A, B)$ is greater than $S'(A, C')$. Formally, this can be written as:

$$\forall A, B, C, \qquad S(A, B) \le S(A, C) \Leftrightarrow S'(A, B) \le S'(A, C). \qquad (9)$$

Such a condition is equivalent to the existence of a strictly increasing function from the image of S to the image of S':

$$f : \mathrm{Im}(S) \to \mathrm{Im}(S'). \qquad (10)$$

such that:

$$S' = f \circ S \qquad (11)$$

Another equivalent definition of the equivalence of measures of resemblance can be considered, based on a common structure in level sets for S and S', in such a way that, for any α in the image of S, there exists a unique β in the image of S' such that the α-level set of S is identical with the β-level set of S'.

It is to be remarked that, if we use a threshold α to select all objects B best resembling A at a level at least equal to α, the collection of objects we obtain is different if we use S or S'. If S and S' are equivalent, we obtain the same collection of objects if we consider a threshold α when using S and a threshold $\beta = f(\alpha)$ when using S'. If we fix the cardinality of the collection of objects we look for, we obtain the same collection when using equivalent resemblance measures.

If resemblance measures S and S' are not equivalent, for a given value of S', there may exist several values of S, which means that for one object resembling A at the level β for S', it is possible to find several objects resembling A when using S.

Considering the basic measures we have recalled, Jaccard (see Equation (4)) and Fermi-Dirac (see Equation (7)) measures are for instance equivalent, while Jaccard and Ochiai are not equivalent. Moreover, Tversky's measures $s_{\alpha,\beta}$ and $s_{\alpha',\beta'}$ (1) are equivalent whenever $\alpha.\beta' = \alpha'.\beta$ [51].

4.4 Similarity-Based Prototypes

Beyond similarity, the fuzzy setting makes it possible to implement other notions related to similarity and categorization in the cognitive framework. In particular, prototypes, viewed as significant representatives of the categories, can be built in agreement with the cognitive principles mentioned in Section 2.1, on the basis of similarity judgments. Since the graded character of prototypes has been emphasized, it is indeed natural to choose a fuzzy knowledge-based representation of prototypes, avoiding the choice of crisp representatives of a category such as the mean value for an attribute.

Several approaches have been proposed since the seminal one introduced by Zadeh [51], for instance based on fuzzy expected values [54] or fuzzy summaries [55]. Fuzzy prototype construction has often been considered as identical with fuzzy clustering, which is yet somewhat different. Fuzzy clustering forms categories of objects similar with regard to attribute values, while fuzzy prototypes propose abstract representatives of categories, generally not belonging to the categories of objects, associated with the most representative fuzzy value of each attribute. An extensive study of fuzzy prototypes has been presented in [56]. The principles of their construction are summarized [46][57], coherent with Rosch's vision of prototypes [1].

The basic concept is the degree of typicality of an object with respect to a category. It can be regarded as the aggregation of a degree of internal resemblance measure of an attribute value with regard to all other values of the same attribute for objects of the same category on the one hand, and a degree of external dissimilarity measure of this value with regard to all other values of the attribute for objects of other categories on the other hand. The most typical value of the attribute for a category corresponds to the maximum degree of typicality. A prototype is then an abstract object characterized by the most typical value of each attribute.

The variety of aggregation operators used to combine internal resemblance and external dissimilarity provides a flexible definition of a prototype balancing the relative importance of the common points of the category members and their distinctive properties as opposed to other categories.

It is also possible to consider objects globally and not attribute by attribute. The same principle leads to the computation of the internal resemblance of an object with respect to other members of the same category, and the external dissimilarity of this object with respect to members of other categories [58].

The particular case of numerical data has been considered in [58][59]. Exceptions have been less studied in cognitive science. They can be regarded as elements with a small typicality degree in all categories. Taking them into account in the construction of categories is a difficult problem and they are often considered as a nuisance in clustering methods. Some real-world problems require them to be considered as very informative elements and clustering methods have been settled to take them into account [60].

5 Examples of Utilization in Real Word Applications

Similarities and prototypes have been extensively used in fuzzy data mining. Similarity measures are obviously present in most of the systems based on fuzzy learning, fuzzy rule-based systems or fuzzy database management, for the evaluation of the degree of matching between a reference (attribute value in a decision tree, a rule, a query…) and all possible instances, examples or answers. We focus on systems in which the management of similarity is more complex.

To illustrate the above-mentioned use of similarities and dissimilarities, we will restrict this section to real-world applications that have been tackled in our research team [61]. We will distinguish mining in large amounts in data from information retrieval.

5.1 Image Interpretation

An example of environment where prototypes have been used to represent imprecise knowledge usually managed by medical doctors is the identification of microcalcifications [46][62]. Prototypes have been used to establish a link between linguistic criteria used by specialists to describe spots in mammographical images, for instance "round" or "small", and technical attributes, such as surface, convexity or elongation of the objects. Prototypes of "round" spots can for instance be described by fuzzy values of attributes [62], with simplified interpretations of the form "circularity is approximately 5 or 6, no more than 10" and "circumference is either approximately 6 or approximately 12" and…"

The resemblance measure that has proven to be the most successful with regard to the tests of classification is the following:

$$S(A, B) = \frac{2}{\pi} \arctan(2 * M(A \cap B) * \sup(A \cap B) / M(A \cup B)) \tag{12}$$

where M is the surface under the membership function.

5.2 Defect Forecasting

Fuzzy association rules can be chosen to extract knowledge from large databases. We have used this approach to forecast defects in gas pipelines. In-line inspections by means of smart pigs are used by gas operators, but they are not satisfying with regard to the exact dimensioning of the defects and real defects do not exactly correspond to those forecasted by the smart pig. We have used association rules to compare forecasted defects and real ones [23]. Fuzzy descriptions have been introduced to represent linguistic expertise. A measure of satisfiability has been chosen to compare observed data and fuzzy descriptions, as follows:

$$S(A, B) = \frac{2}{\pi} \arctan\left(\frac{M(A \cap B)}{M(B \Theta A)} \right) \tag{13}$$

5.3 Risk Rating

Risk prediction and analysis is a complex topic, subject to imprecision and uncertainty in data and to linguistic expert knowledge. We have faced the problem of country risk ratings and a methodology to assess internal conflict risk has been proposed, with various components [63]. In the case of dynamic early warning, scenarios have been elaborated, taking into account temporal constraints, on the basis of human expertise [64]. For a given piece of information regarding a country, a satisfiability measure is used to compare its description with a scenario and the obtained results provide hypotheses that will be confirmed or refuted in the future. In order to obtain automatically elements of the scenarios, prototypes have been constructed and the following resemblance and dissimilarity measures have been chosen for the quality of the obtained results:

$$S(A, B) = \frac{M(A \cap B)}{M(A \cup B)}, \tag{14}$$

$$D(A, B) = 1 - \exp\left(-\frac{M(A \Theta B) + M(B \Theta A)}{\Gamma} \right). \tag{15}$$

The interpretation of a prototype of the category "ethnic conflict" is for instance of the form "number of extended military aid approximately between 3 and 4, and number of ultimatum approximately between 4 and 5, and ..." The obtained fuzzy descriptions provide a prototypical vision of the category.

5.4 Web Usage Mining

Web usage mining requires recording and management of large amounts of web log files. A method has been conceived to select informative data and to construct prototypes of the activity of users on a website in order to provide a meaningful visualization of categories of users with similar navigation behavior [65]. Such a tool can help to improve the quality of a website through a better understanding of the expectations of prototypical users, or to provide an adaptive pedagogical support to learners according to the category they belong to, if the website is used in e-learning for instance.

In this case, the considered data are not fuzzy, thus the applied similarity measure does not belong to the framework described in the previous section. More precisely, the similarity must compare user web sessions, described in terms of the accessed web pages: two sessions are then considered as similar if they access similar pages. The similarity between web pages is based on their url addresses, and not on their content, so as to avoid an extraction and indexation step with high computational cost.

This approach relies on the assumption that the structure of the web site directory reflects its content. The similarity between two urls $S_{url}(u_1,u_2)$ is computed as the weighted sum of the elements identical in both paths. The normalized similarity between two sessions, s_1 and s_2, is then defined as

$$S(s_1,s_2) = \frac{\tilde{S}(s_1,s_2)}{\sqrt{\tilde{S}(s_1,s_1)\tilde{S}(s_2,s_2)}} \text{ with } \tilde{S}(s_1,s_2) = \sum_{u_1 \in s_1} \sum_{u_2 \in s_2} S_{url}(u_1,u_2) \qquad (16)$$

It is to be noticed that although the considered data are not fuzzy, the fuzzy similarity-based typicality framework can be used to identify the most typical user of each cluster, to characterize the identified clusters, i.e. the identified navigation behaviors.

5.5 Content-Based Image Retrieval

Image retrieval on large databases can be based on annotated documents or on a comparison of images on the basis of their content. An experimental platform has been proposed [66] for a retrieval of images similar to a given one considered as a reference. This image is automatically segmented into regions; attributes of regions such as colour or position in the image are taken into account. Colour histograms can be regarded as fuzzy sets of a universe of colors. The user chooses the regions of the reference image he wants to retrieve, to indicate the attributes he wants to consider and their importance in his query. Various measures of satisfiability were proposed and the result of the query was a list of images satisfying the query in a decreasing order of the satisfiability degree. This result was obviously dependent on the equivalence class of satisfiability measures [46], allowing the choice of a measure by the user to be restricted to those providing distinct orders.

6 Conclusion

Studies about similarity are countless and artificial intelligence takes advantage of studies in cognitive science in the construction of automated systems. In particular, the main streams of research on similarity and categorization in psychology have given rise to interesting foundations for developments in data mining. Because of the graded structure of natural categories and their variability, fuzzy set theory has been a privileged component of formalized versions of similarities and categories.

Our purpose has been to point out the richness of the concepts of similarity and category and some of their utilizations in fuzzy data mining. More directions remain to be explored.

Some of them have already been approached in artificial intelligence. Wittgenstein's concepts [5] in linguistics have been used by M. Sugeno and his colleagues as a semiotic base of an everyday language computing system, for instance in [67].

Let us just mention two other directions worth to develop in fuzzy data mining, dealing with dynamic similarities presented in Section 2.

In document retrieval, the choice of a similarity providing interesting answers to a user's queries is not simple. This choice is generally left to the expert. Determining the best measure of similarity for a user can also be done in an adaptive manner, on the basis of his interactions with the system. An example of such a method is based on a representation of retrieved images by means of Self Organizing Maps and a manual assignment of images to classes by the user [63]. The system learns such assignments to adapt its behaviour to the user's preferences, in an adaptive similarity management.

Another utilization of dynamic similarities concerns evolving categories. Methods of novelty detection, incremental classification methods or adaptive classification algorithms are based on the reorganization of classes or clusters according to the incoming data, in [69][70][71]for instance.

References

1. Rosch, E.: Principles of categorization. In: Rosch, E., Lloyd, B. (eds.) Cognition and Categorization. Lawrence Erlbaum, Mahwah (1978)
2. Rosch, E., Mervis, C.: Family resemblance: studies of the internal structure of categories. Cognitive psychology 7, 573–605 (1975)
3. Tijus, C.: Introduction à la psychologie cognitive. Nathan Université (2001)
4. Hampton, J.A.: The role of similarity in natural categorization. In: Hahn, U., Ramscar, M. (eds.) Similarity and Categorization, pp. 13–28. Oxford University Press, Oxford (2001)
5. Wittgenstein, L.: Philosophical Investigations. Blackwell Publishing, Malden (1953/2001)
6. Kleiber, G.: Prototype et prototypes. In: Sémantique et cognition. CNRS, Paris (1991)
7. Barsalou, L.W.: Ideals, Central Tendency, and Frequency of Instantiation as Determinants of Graded Structure. Journal of Experimental Psychology: Learning, Memory and Cognition 11, 629–654 (1985)
8. Nosofsky, R.M.: Similarity, frequency, and category representations. Journal of Experimental Psychology: Learning, Memory, and Cognition 14(1), 54–65 (1988)
9. Poitrenaud, S., Richard, J.-F., Tijus, C.: Properties, categories and categorization. Thinking and reasoning 11(2), 151–208 (2005)
10. Posner, M.I., Keele, S.W.: On the genesis of abstract ideas. Journal of Experimental Psychology 77, 353–363 (1968)
11. Barsalou, L.W.: The instability of graded structure: Implications for the nature of concepts. In: Neisser, U. (ed.) Concepts and conceptual development: Ecological and intellectual factors in categorization, pp. 101–140. Cambridge University Press, Cambridge (1987)
12. Tversky, A.: Features of similarity. Psychological Rev. 84(4), 327–352 (1977)
13. Hahn, U., Ramscar, M.: Introduction: similarity and categorization. In: Hahn, U., Ramscar, M. (eds.) Similarity and categorization, pp. 1–11. Oxford University Press, Oxford (2001)
14. Medin, D.L., Goldstone, R.L., Gentner, D.: Respects for similarity. Psychological Review 100(2), 254–278 (1993)
15. Keane, M.T., Smyth, B., O'Sullivan, F.: Dynamic similarity: A processing perspective on similarity (2001)
16. Markman, A.B., Gentner, D.: Structural alignment during similarity comparisons. Cognitive Psychology 25, 431–467 (1993)
17. Rissland, E.: AI and similarity, IEEE Int. Systems 21, 39–49 (2006)
18. Fayyad, U., Piatetsky-Shapiro, G., Smyth, P.: From data mining to knowledge discovery in databases. AI magazine 17(3), 37–54 (1996)

19. Delavallade, T., Dang, T.H.: Using Entropy to Impute Missing Data in a Classification Task. In: IEEE International Conference on Fuzzy Systems, London, pp. 1–6 (2007)
20. Timm, H., Döring, C., Kruse, R.: Differentiated treatment of missing values in fuzzy clustering. In: De Baets, B., Kaynak, O., Bilgiç, T. (eds.) IFSA 2003. LNCS, vol. 2715, pp. 354–361. Springer, Heidelberg (2003)
21. Song, Q., Shepperd, M.: A new imputation method for small software project data sets. Journal of Systems and Software 80(1), 51–62 (2007)
22. Setnes, M., Babuska, R., Kaymak, U., van Nauta Lemke, H.R.: Similarity measures in fuzzy rule base simplification. IEEE Transactions on Systems, Man, and Cybernetics, Part B 28(3), 376–386 (1998)
23. Pichlova, M., Bouchon-Meunier, B. : Using fuzzy association rules for defect forecasting in pipelines. Rencontres Francophones sur la Logique Floue et ses Applications, Nantes, Cépaduès-Editions, 305–312 (2004)
24. Marsala, C.: Fuzzy partitioning methods. In: Pedrycz, W. (ed.) Granular Computing: An Emerging Paradigm, pp. 163–186. Physica-Verlag GmbH, Heidelberg (2001)
25. Yuan, Y., Shaw, M.J.: Induction of Fuzzy Decision Trees. Fuzzy Sets and systems 69, 125–139 (1995)
26. Marsala, C., Bouchon-Meunier, B.: An adaptable system to construct fuzzy decision trees. In: Proceedings of the 18th International Conference of the North American Society, pp. 223–227 (1999)
27. Bouchon-Meunier, B., Marsala, C.: Linguistic modifiers and measures of similarity or resemblance. In: 9th IFSA World Congress, Vancouver, pp. 2195–2199 (2001)
28. Laurent, A., Marsala, C., Bouchon-Meunier, B.: Improvement of the Interpretability of Fuzzy Rule Based Systems: Quantifiers, Similarities and Aggregators. In: Davenport, J.H. (ed.) On the Integration of Algebraic Functions. LNCS (LNAI), pp. 102–123. Springer, Heidelberg (1981)
29. Hüllermeier, E.: Fuzzy-Methods in Machine Learning and Data Mining: Status and Prospects. Fuzzy Sets and Systems 156(3), 387–407 (2005)
30. Guillaume, S.: Designing Fuzzy Inference Systems from Data: An Interpretability-Oriented Review. IEEE Transactions on Fuzzy Systems 9(3), 426–444 (2001)
31. Zadeh, L.A.: Similarity relations and fuzzy ordering. Information Science, 177–200 (1971)
32. Valverde, L.: On the structure of F-indistinguishability operators. Fuzzy Sets and Systems 17, 313–328 (1985)
33. Jaccard, P.: Nouvelles recherches sur la distribution florale. Bulletin de la Société Vaudoise des Sciences Naturelles 44, 223–270 (1908)
34. Dice, L.R.: Measures of the amount of ecological association between species. Ecology 26, 297–302 (1945)
35. Ochiai, A.: Zoogeographic studies on the soleoid fishes found in Japan and its neighbouring regions. Bulletin of the Japanese Society for Science and Fisheries 22, 526–530 (1957)
36. Zwick, R., Carlstein, E., Budescu, D.V.: Measures of similarity among fuzzy concepts: A comparative analysis. International Journal of Approximate Reasoning 1, 221–242 (1987)
37. Chen, S., Yeh, M., Hsiao, P.: A comparison of similarity measures of fuzzy values. Fuzzy Sets Systems 72(1), 79–89 (1995)
38. Xuzhu, W., De Baets, B., Kerre, E.: A comparative study of similarity measures. Fuzzy Sets and Systems 73(2), 28, 259–268 (1995)
39. Jain, R., Murthy, S.N.J., Chen, P.L.-J., Chatterjee, S.: Similarity measures for image databases. In: IEEE International Conference on Fuzzy Systems, pp. 1247–1254 (1995)

40. Li, Y., Liu, J.-M., Li, J., Deng, W., Ye, C.-X., Wu, Z.-F.: The fuzzy similarity measures for content-based image retrieval. In: Proceedings of the International Conference on Machine Learning and Cybernetics, vol. 5, pp. 3224–3228 (2003)
41. Cross, V.V., Sudkamp, T.A.: Similarity and Compatibility in Fuzzy Set Theory: Assessment and Applications. Physica-Verlag (2002)
42. Dubois, D., Prade, H.: A unifying view of comparison indices in a fuzzy set-theoretic framework. In: Yager, R.R. (ed.) Fuzzy and possibility theory, pp. 3–13. Pergamon Press, Oxford (1982)
43. Shiina, K.: A fuzzy-set-theoretic feature model and its application to asymmetric data analysis. Japanese psychological research 30(3), 95–104 (1988)
44. Santini, S., Jain, R.: Similarity measures. IEEE Transactions on Pattern Analysis and Machine Intelligence 21(9), 871–883 (1999)
45. Tolias, Y.A., Panas, S.M., Tsoukalas, L.H.: Generalized fuzzy indices for similarity matching. Fuzzy Sets Systems 120(2), 255–270 (2001)
46. Rifqi, M. : Mesures de comparaison, typicalité et classification d'objets flous: théorie et pratique. PhD thesis, Université Paris VI (1996)
47. Bouchon-Meunier, B., Rifqi, M., Bothorel, S.: Towards general measures of comparison of objects. Fuzzy Sets and Systems 84(2), 143–153 (1996)
48. Bouchon-Meunier, B., Rifqi, M.: OWA operators and an extension of the contrast model. In: Yager, R.R., Kacprzyk, J. (eds.) The Ordered Weighted Averaging Operators: Theory, Methodology, and Applications, pp. 29–35. Kluwer Academic Publishers, Dordrecht (1997)
49. Rifqi, M., Berger, V., Bouchon-Meunier, B.: Discrimination power of measures of comparison. Fuzzy Sets and Systems 110(2), 189–196 (2000)
50. Rifqi, M., Detyniecki, M., Bouchon-Meunier, B.: Discrimination power of measures of resemblance. In: De Baets, B., Kaynak, O., Bilgiç, T. (eds.) IFSA 2003. LNCS, vol. 2715, Springer, Heidelberg (2003)
51. Omhover, J.-F., Detyniecki, M., Rifqi, M., Bouchon-Meunier, B.: Image Retrieval using Fuzzy Similarity: measure equivalence based on invariance in ranking. In: Proceedings of the IEEE International Conference on Fuzzy Systems, Budapest, Hungary, pp. 1367–1372 (2004)
52. Omhover, J.-F., Rifqi, M., Detyniecki, M.: Ranking Invariance based on Similarity Measures in Document Retrieval. In: Detyniecki, M., Jose, J.M., Nürnberger, A., van Rijsbergen, C.J. (eds.) AMR 2005. LNCS, vol. 3877, pp. 55–64. Springer, Heidelberg (2006)
53. Zadeh, L.A.: A note on prototype theory and fuzzy sets. Cognition 12, 291–297 (1982)
54. Friedman, M., Ming, M., Kandel, A.: On the theory of typicality. Int. Journ. of Uncertainty, Fuzziness and Knowledge-based Systems 3(2), 127–142 (1995)
55. Kacprzyk, J., Yager, R.: Linguistic summaries of data using fuzzy logic. Int. Journ. of General Systems 30, 133–154 (2001)
56. Lesot, M.-J., Rifqi, M., Bouchon-Meunier, B.: Fuzzy prototypes: from a cognitive view to a machine learning principle. In: Bustince, H., Herrera, F., Montero, J. (eds.) Fuzzy Sets and Their Extensions: Representation, Aggregation and Models Studies. Springer, Heidelberg (2007)
57. Rifqi, M.: Constructing prototypes from large databases. In: Proc. International Conference IPMU 1996, Granada, pp. 301–306 (1996)
58. Lesot, M.-J., Mouillet, L., Bouchon-Meunier, B.: Fuzzy prototypes based on typicality degrees. In: Proc. of Fuzzy Days 04, Springer, Advances on Soft Computing, pp. 125–138. Dortmund, Allemagne (2006)

59. Lesot, M.-J.: Similarity, typicality and fuzzy prototypes for numerical data. In: Res-Systemica, 5 (Special issue on the 6th European Congress on Systems Science, Paris 2005) (2005)
60. Lesot, M.-J.: Typicality-based clustering. Int. Journal of Information Technology and Intelligent Computing 1(2), 279–292 (2006)
61. Bouchon-Meunier, B., Detyniecki, M., Lesot, M.-J., Marsala, C., Rifqi, M.: Real world fuzzy logic applications in data mining and information retrieval. In: Wang, P.P., Ruan, D., Kerre, E.E. (eds.) Fuzzy Logic - A Spectrum of Theoretical and Practical Issues, Studies in Fuzziness, pp. 219–247. Springer, Heidelberg (2007)
62. Rifqi, M., Bothorel, S., Bouchon-Meunier, B., Muller, S.: Similarity and prototype-based approach for classification of microcalcifications. Int. J. General Systems 29(4), 623–636 (2000)
63. Delavallade, T., Mouillet, L., Bouchon-Meunier, B., Collain, E.: Monitoring event flows and modelling scenarios for crisis prediction, application to ethnic conflicts forecasting. Int. J. of Uncertainty, Fuzziness and knowledge-based systems, 15, 83–110 (2007)
64. Mouillet, L., Bouchon-Meunier, B., Collain, E.: Automated identification of political conflicts with a scenario recognition technique. In: 10th International Conference IPMU, Perugia, Italy, vol. 3, pp. 1609–1616 (2004)
65. Labroche, N., Lesot, M.-J., Yaffi, L.: A new web usage mining and visualization tool. In: IEEE International Conference on Tools with Artificial Intelligence (ICTAI), Patras, Greece, pp. 321–328 (2007)
66. Omhover, J.-F., Detyniecki, M.: STRICT: an Image Retrieval Platform for Queries Based on Regional Content. In: Enser, P.G.B., Kompatsiaris, Y., O'Connor, N.E., Smeaton, A.F., Smeulders, A.W.M. (eds.) CIVR 2004. LNCS, vol. 3115, pp. 473–482. Springer, Heidelberg (2004)
67. Kobayashi, I., Sugeno, M.: An approach to a dynamic system simulation based on human information processing. Int. J. Uncertain. Fuzziness Knowl.-Based Syst. 10(6), 611–633 (2002)
68. Detyniecki, M., Nürnberger, A.: Adaptive multimedia retrieval: from data to user interaction. In: Gabrys, B., Leiviska, K., Strackeljan, J. (eds.) Do Smart adaptive systems exist – Best practice for selection and combination of intelligent methods. Series on Studies on Fuzziness and Soft Computing, pp. 341–370. Springer, Heidelberg (2004)
69. Utgoff, P.E.: Incremental Induction of Decision Trees. In: Machine Learning, vol. 4, pp. 161–185 (1989)
70. Wang, T., Li, Z., Yan, Y., Chen, H.: An Incremental Fuzzy Decision Tree Classification Method for Mining Data Streams. In: Perner, P. (ed.) MLDM 2007. LNCS (LNAI), vol. 4571, pp. 91–103. Springer, Heidelberg (2007)
71. Prehn, H., Sommer, G.: An Adaptive Classification Algorithm Using Robust Incremental Clustering. In: 18th International Conference on Pattern Recognition, vol. 1, pp. 896–899 (2006)

Attaining Fault Tolerance through Self-adaption: The Strengths and Weaknesses of Evolvable Hardware Approaches

Garrison W. Greenwood

Department of Electrical & Computer Engineering, Portland State University,
Portland OR 97207, USA
greenwd@ece.pdx.edu

Abstract. Self-adaptive systems autonomously change their behavior to compensate for faults or to improve their performance. Evolvable hardware, which combines evolutionary algorithms with reconfigurable hardware, is often proposed as the cornerstone for systems that use self-adaption for fault recovery. Although evolvable hardware was first introduced over 15 years ago, there are few, if any, fault tolerant self-adaptive systems in operation today. One primary reason why these unfortunate circumstances have arisen is many designers—and not limited to just designers from the computational intelligence community—do not really understand how to build a basic fault tolerant system, let alone a self-adaptive fault tolerant system. This chapter describes how fault tolerant systems are built. A model for designing fault tolerant systems that rely on evolvable hardware for fault recovery is presented.

1 Introduction

Embedded systems are frequently used in remote and harsh operational environments. Examples include deep-space probes, unmanned aerial vehicles, deep-sea exploration and remote sensor systems on the battlefield. Embedded systems are often used in safety critical applications. Proper operation, over extended time periods, require these systems to be fault tolerant, which means they can autonomously recover from faults and continue to provide some level of operation. These requirements pose particular challenges for engineers.

Unfortunately, many people do not fully understand what is involved in designing a fault tolerant system. Simply providing spare resources or reconfigurable hardware is not enough to claim a system is fault tolerant. Yet, this belief persists throughout the computational intelligence community and is even occasionally found among engineers in industry.

The purpose of this chapter is to dispel the myths associated with designing fault tolerant systems. The reader will learn how fault tolerant systems are designed and what fault recovery methods are appropriate in specific circumstances. The benefits and limitations of evolvable hardware (EH) as a fault recovery method are particularly scrutinized.

J.M. Zurada et al. (Eds.): WCCI 2008 Plenary/Invited Lectures, LNCS 5050, pp. 368–387, 2008.

2   Background

EH is the primary fault recovery method addressed in this chapter. This method can only be properly implemented if the real-time aspects of fault tolerance systems are considered. A brief overview of both topics is given in this section. More detailed information on EH and real-time systems can be found in [1] and [2] respectively.

2.1   Basics of Evolvable Hardware

Essentially we can think of EH as

$$EH \quad = \quad \text{reconfigurable hardware} + \text{reconfiguration method}$$

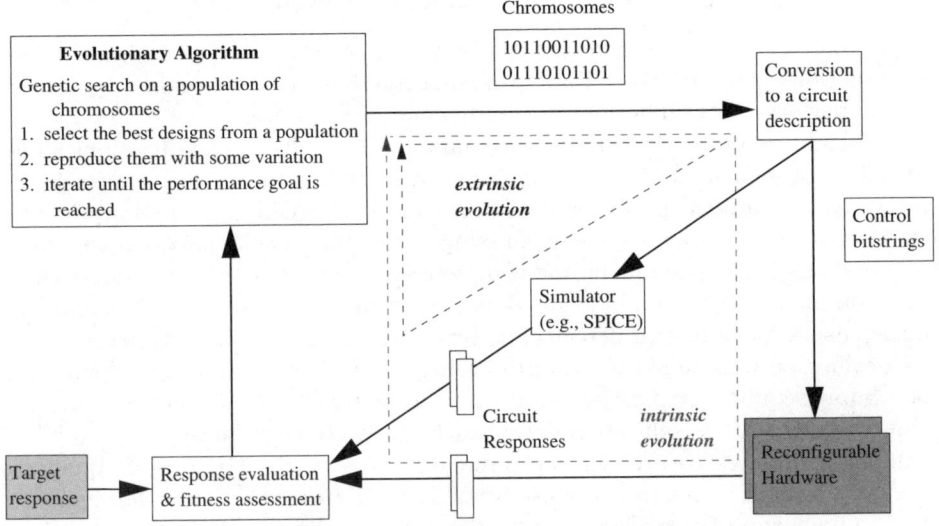

Fig. 1. Intrinsic versus extrinsic evolution of a circuit

Reconfigurable hardware can be just an electronic device, such as a field-programmable gate array, or it can be an entire system. But it doesn't even have to be an electronic device at all. For example, Linden [3] used EH techniques to design antennas for wireless communications systems. The reconfiguration method must intelligently search for a good configuration. Almost exclusively this method is some type of evolutionary algorithm (EA).

EH can be used for original design applications (synthesis) or for adapting existing designs to improve performance or correct faults. Adapting hardware—particularly in FT applications–can be tricky because that adaption is done on faulty hardware or in a hostile and largely undefined operational environment.

Under such conditions it is not possible to know *a priori* what type of performance evolution can achieve. Nevertheless, evolving a new circuit configuration may be the only viable means of restoring some operation.

The evolution can be done intrinsically (in hardware) or extrinsically (in software). Most circuit synthesis work done today relies on simulators. Unfortunately simulators do not always scale well, which makes extrinsic design poorly suited for large designs. Fault recovery has to be done intrinsically in deployed embedded systems. (This aspect is discussed in depth later in the chapter.) The difference between intrinsic and extrinsic reconfiguration is shown in Figure 1.

2.2   Real-Time Systems

Definition: (*real-time system*)

Any system that is both logically and temporally correct [2].

Logical correctness means the system correctly performs all of its assigned tasks and functions without error. Temporal correctness means the system is guaranteed to perform those functions within explicit timeframes.

Real-time systems are classified as hard or soft. These classifications indicate the consequences if temporal correctness requirements are not met. In soft systems degraded performance is the only consequence of missing a deadline. Conversely, in hard systems missing a deadline could have catastrophic consequences—up to and including complete system destruction. The exact classification for fault tolerant systems depends on the nature of the faults and the consequences for failing to detect and correct them in a timely manner.

Consider an unmanned underwater vehicle (UUV) exploring the ocean bottom. Suppose any over-pressure condition can be tolerated for no more than ten minutes or the UUV self-destructs. Clearly fault recovery must be completed within ten minutes to prevent the impending loss of the UUV. This qualifies as a hard system. However, suppose after five minutes a particular fault causes only a minor loss of some sensor data, but this lost data could be recovered using extrapolation techniques. Fault recovery in this case could take considerably longer than five minutes without any dire consequences. This qualifies as a soft system.

3   What Is Fault Tolerance?

All systems eventually fail. Some failures cause no perceptible affect on the system's behavior whereas in other cases failures can produce very obvious changes ranging from mild annoyance to total system destruction. The best way to improve system availability is to make the system *fault tolerant* (FT).

Definition: (*fault tolerant*)

A system is considered fault tolerant if it can continue to operate in the presence of failures (albeit, with perhaps degraded performance).

Making a system FT means it can identify when errors occur and it can do something about them. It may not always be possible to completely restore all of the behavior that existed prior to the failure, but *some* level of restoration is always better than none.

The terms "fault" and "failure" are often used interchangeably although these terms mean different things. Undesirable events, called *hazards*, can cause defects somewhere in a system. Component defects are commonly called *faults*. Left untreated, faults eventually lead to component *failures*, where the component can no longer perform as designed. In the worst case failures can propagate, leading to even more failures until a total system breakdown occurs. Catastrophic system breakdowns—i.e., where severe damage or injury happens—are called *mishaps*. The various terms just described are related by

$$\textbf{fault} \rightarrow \textbf{failure} \rightarrow \textbf{hazard} \rightarrow \textbf{mishap}$$

where "$\rightarrow$" indicates a causal relationship.

It is often not possible to make a system mishap-free. In other words, there is always some level of mishap risk and there may be little a designer can do about it. (There is no way to guarantee a car will never be in an accident. However, the risk is negligible with a good driver.) Some examples of hazards include complete loss of flight control in a spacecraft or the release of a hazardous material such as a toxic gas or liquid. Failures do not always cause hazards. For example, diagnostic circuitry failures may no effect on a system's performance when it is in normal operating mode. In most cases, however, failures result in undesirable system behavior where something must be done to correct the problem.

Fault tolerant systems must perform two independent functions: *fault detection and isolation* (FDI) and *fault recovery* (FR). FDI operations determine that a fault has occurred and attempt to isolate the fault to a particular subsystem or (hopefully) component within a subsystem. FR methods attempt to correct, mitigate or, in the worst case, at least contain the failure so no further problems can happen. One important characteristic of all fault tolerant systems is FDI and FR operations execute autonomously—i.e., without any human intervention.

Since FDI and FR inherently have deadlines, FT systems naturally qualify as real-time systems. For example, a fault must be detected and isolated within a certain period of time after it occurs, and the fault must be corrected within a certain period of time after it is detected, otherwise the failures might propagate. Fault recovery may also have an expected start time.

3.1   Some Common Myths

Myth #1:

A fault tolerant system can survive any failure (otherwise, you couldn't call it "fault tolerant").

Fault tolerant systems can withstand some pretty adverse conditions, but claiming they can survive *any* failure is overly optimistic. Complex systems have many potential failure modes. Some failures produce nearly immediate effects. Since fault recovery cannot happen in zero time, failures have an opportunity to propagate, possibly causing even more failures. It is not hard to imagine a situation where a succession of failures quickly overwhelms any fault recovery method.

Don't forget there are also situations where the event leading to a fault is completely external to the system and is therefore unaffected by any FR method local to that system. For example, consider an amplifier circuit, equipped with a very sophisticated FDI scheme and a comprehensive FR method capable of dealing with any single component failure. But suppose a component fails because of an over-temperature condition caused by a cooling system failure. The FR method is designed only to handle on-board component failures and was never intended to deal with cooling system failures. Components will continue to fail so long as the over-temperature condition persists, rendering the amplifier's local FR method ineffective.

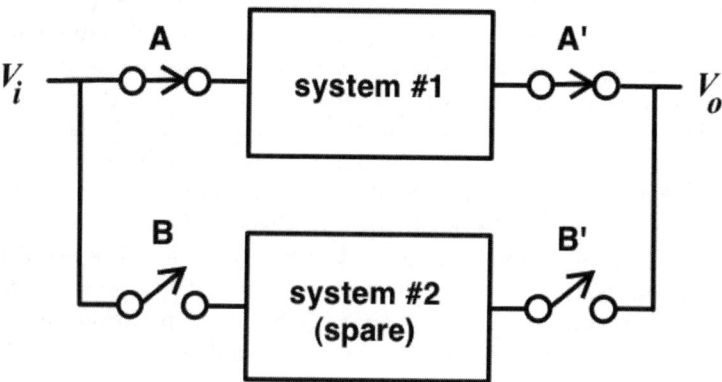

Fig. 2. Basic concept of redundancy as a FR method. Under normal circumstances switch A and A' are closed and system #1 is online. Should that system fail, switches A and A' are opened and switches B and B' are closed, which makes the spare system #2 the online system. Systems #1 and #2 are functionally identical.

Myth #2:

Redundancy is always the most effective fault recovery method

Redundancy is probably the best known FR method. The concept is simple: the faulty system is taken offline and a spare, duplicate system is then brought online (See Figure 2). Redundant hardware has several benefits when used as a FR recovery method. First, it performs FR very quickly, taking only as much time as necessary to switch the bad hardware offline and switch the new operational hardware online. Second, since it is an exact duplicate of the failed system, it can recovery from any internal failure mode. But like most systems there are design tradeoffs to consider. Spare hardware occupies space, which is not always available. Indeed, severe space and weight restrictions in many embedded systems preclude even considering redundancy as a viable FR method. Moreover, redundancy is completely ineffective when the hazard conditions are external to the system. The over-temperature condition described above is a good example.

Myth #3:

Fast $\Longleftrightarrow$ Real-Time

The whole concept of real-time is habitually misunderstood. Many researchers (and more or less everyone in a marketing department) believe real-time means really, really fast. This interpretation is completely wrong. Real-time does not necessarily mean fast and fast does not necessarily mean real-time. The real-time system definition does not say something has to be done really, really fast. Temporal correctness only requires the event take place by a specified deadline. Nothing more. A simple example should help clear up any confusion.

Suppose a document must be sent from Chicago to London and two delivery systems are available: surface mail with a guaranteed 3-day delivery time or e-mail with a guaranteed 5-minute delivery time. Note that both delivery systems guarantee delivery so logical correctness holds. Now we need to verify that temporal correctness holds. The e-mail delivery is orders of magnitude faster than surface mail, but that does not necessarily mean it qualifies as a real-time delivery system and the surface mail delivery system does not. It is the required delivery deadline that determines if temporal correctness is satisfied. Both delivery systems qualify as real-time systems if the deadline is six days because both systems are temporally correct. However, neither one is a real-time system if the deadline is three minutes because neither one is temporally correct.

Myth #4:

Reconfigurable $\Longrightarrow$ Fault Tolerance

The ability to reconfigure hardware does *not*, by definition, make a system FT. This misconception is all too common in the computational intelligence community. Reconfiguration is nothing more than one of many possible FR methods. Remember FR is just one piece of a FT system. Fault tolerance is a much broader concept that also includes FDI and real-time considerations. Reconfiguration should be considered as a viable FR method if—and only if—it can meet any temporal requirements. This aspect will be discussed in detail shortly.

4   Putting Fault Tolerance into Practice

To set the stage for an in-depth discussion of FDI and FR it helps to consider a typical embedded system. Figure 4 shows a typical embedded system. The computer, usually a microprocessor or microcontroller, issues commands to the plant under control through actuators and measures the plant's response through the

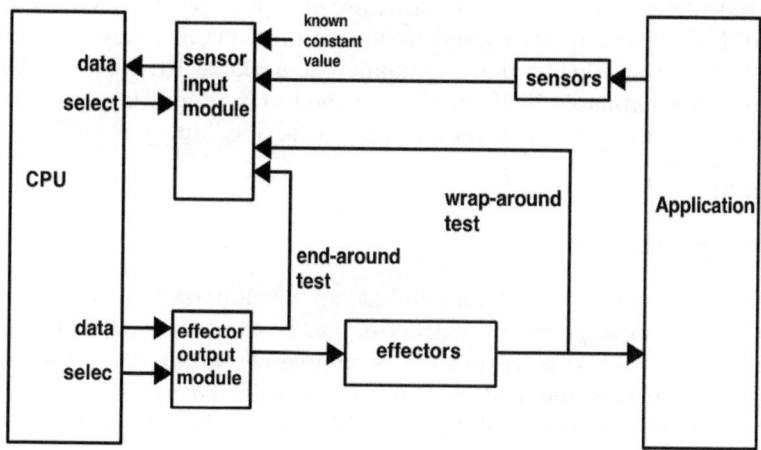

Fig. 3. Sensor-Input/Effector-output Module failure detection

sensors. Both sensors and actuators are analog devices so some type of interface circuitry is required. For example, the sensors (or transducers) convert physical quantities such as position or velocity into electrical voltages. These voltages must be converted into binary strings to be processed by the computer. Similarly binary commands issued by the computer must be converted into high voltages or currents to drive the actuators.

4.1   Fault Detection Methods

The onset of a failure normally causes some undesirable behavior change. However, FT systems do not have to wait for failures to physically occur before taking corrective action. Many systems have on-board diagnostics that are run periodically to help identify potential problems. They can also be run when a failure is suspected to exist. There are a variety of diagnostic programs that can be run to detect computer system failures. For example, checksums are often used to detect memory system failures. Sensor and effector failures are detected using *wraparound* and *end-around* tests (see Figure 3). The basic idea is simple. For sensor failures a known input signal is input on a spare channel. This signal could be a constant DC voltage from a zener diode. Since the signal is known, its expected data value processed by a correctly operating sensor input module is also known. Any deviation would indicate a sensor input module problem. To check effector problems, the diagnostic software generates a known command to a spare channel on the effector output module. This output signal is routed to a spare input on the sensor input module where it can be compared with a stored value. The actual effector output signal that is sent to the application can be similarly routed to a spare sensor input module channel for comparison. As before, any deviations from an expected value would indicate an effector or effector output module problem.

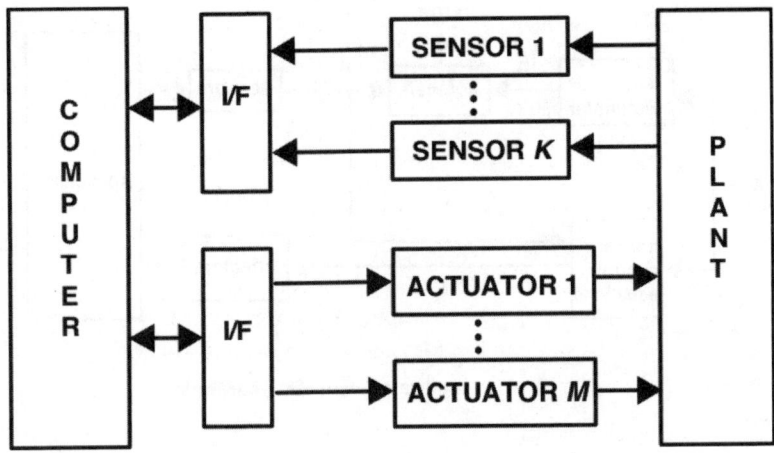

Fig. 4. A typical embedded system

It is also possible to use state estimators to check for sensor failures (see Figure 5). The commanded state (X_C is the desired state of the application— e.g., the position of a robot arm. This commanded state is furnished to the control equations, which generates the effector commands using the effector output module circuitry. The state estimator, implemented in software, uses the same commanded state X_C input and estimates what the output should be from the sensor input module. This estimate is compared against the actual output. Any excessive difference is flagged as an error. Of course the efficacy of this fault detection approach depends on the accuracy of the state simulator computer model.

4.2 Fault Recovery Methods

Fault recovery methods try to fix the failures. Perhaps the most widely used method for hardware fault recovery is *redundancy* [4], where a faulty component is replaced with an identical, fully operational spare component. One of the common myths is redundancy is always an effective FR method. (That myth was dispelled earlier).

Another fault recovery method—and this is where EH particularly excels—is *reconfiguration*. Reconfiguration methodically changes the failed system's design, both in component values and component interconnections, until a desired functionality is restored. Although there is no guarantee full functionality can be restored in all cases, reconfiguration is a viable and powerful fault recovery technique. In particular, reconfiguration can adapt existing hardware to work in unanticipated and potentially hostile environments—something redundant hardware can't do.

Some recent work has shown EH can be quite effective for reconfiguring existing hardware to overcome faults [6,7,8]. But the ability of EAs to find good

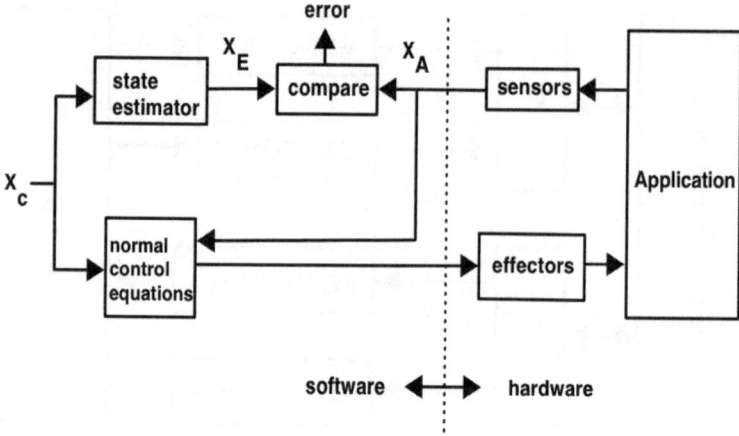

Fig. 5. Sensor fault detection using a state estimator. The state command signal (X_C) is applied to both the normal control equations and the state estimator. The estimated sensor value (X_E) is compared against the actual sensor value (X_A). The difference is the error. Notice the partition between software and hardware implementations.

reconfigurations is not at issue here. Instead, the issue is time. Most EH-based studies rely on device simulators rather than physical hardware. Simulators, which are used for extrinsic reconfigurations, hide the actual time it takes to conduct fitness evaluations, which can be considerable. For example, a numerical simulation can compute a circuit's settling time quite rapidly[1]. Intrinsic reconfigurations can take a significant amount of time. For instance, settling times of two minutes are not unheard of [11].

In this paper I restrict the discussion to evolving new hardware configurations implemented by reprogrammable devices. These include field-programmable gate arrays (FPGAs) and field-programmable analog arrays (FPAAs). An evolutionary algorithm, perhaps running on a microcontroller dedicated to other tasks in the embedded system, searches for a device configuration that constitutes a FR. Either extrinsic or intrinsic evolution could be used. However, in many instances intrinsic evolution is necessary because the only real way to evaluate a configuration is to implement it and have it actually operate in the physical environment.

Online systems—particularly safety-critical online systems—cannot operate indefinitely with faults. It is imperative that any faults be detected promptly and fault recovery operations commence before other things can go wrong. Evolutionary algorithm running time now becomes a factor in choosing the fault recovery method. *This suggests EH, despite its power and flexibility, may not always be the best choice for a fault recovery method.* It all depends on whether or not the EH approach can meet the recovery deadline.

---

[1] Settling time is the time it takes for any oscillation to die out. Normally the system is considered to have settled if the output is within a few percent of its final value.

It is not difficult to estimate the intrinsic reconfiguration time. The EA manipulates a genome that encodes all of the information needed to create a hardware configuration. During each generation λ offspring are created and each must undergo a fitness evaluation. There are three primary factors that contribute to the running time of a generation: (1) the time to program the reconfigurable device (t_{pgm}), (2) the time to conduct the fitness evaluation (t_{fit}), and (3) the overhead time of the EA (t_{oh}). By far t_{fit} is the predominant factor, particularly for analog systems. Hence, the discussion that follows will be restricted to just t_{fit}. (The interested reader should see [10] for a discussion on the contribution of t_{pgm} and t_{oh}).

In an intrinsic evolution t_{fit} is the test time, which is application dependent. The total intrinsic reconfiguration time for an EA run for k generations while procreating λ offspring per generation is

$$T_r(k, \lambda) = (k)(\lambda)(t_{fit}) \tag{1}$$

Of course T_r says nothing about the quality of the circuits produced after k generations.

t_{fit} is responsible for the rather long search times often encountered in analog applications. The fitness evaluation lasts as long as it takes for the analog circuit to process an input signal and reach some stable state where measurements can be taken. Even if this lasts only a few milliseconds per reconfiguration, it won't hake a very large population size nor a large number of generations to make a reconfiguration last for hours or even days.

Example:

An FPAA is used to compensate for aging effects in a control system responsible for positioning a satellite's communications antenna. A generational GA run for 500 generations with a population size of 100 does the reconfiguration search. The system's step response is measured to determine if the compensation is correct. This step response test takes $t_{fit} = 625$ milliseconds to conduct. Hence, $\lambda = 100$, $k = 500$. Substituting into the above Eq. (1) yields an intrinsic reconfiguration time of approximately 8.7 hours.

Reconfiguration times are meaningless until they are put into context. For instance, take the example above. Suppose brief communication sessions with the satellite are scheduled at 10-hour intervals. A session may be skipped, but skipping two sessions in a row is not permitted. If a fault is detected just prior to a scheduled session, and if the error results in missing the session, then the fault recovery deadline is 10 hours. This is the worst-case scenario[2]. An almost 9 hour reconfiguration time may seem quite long, but in this case it is perfectly acceptable because it is less than the mandatory 10 hour requirement. On the

---

[2] Missing one session is permitted. If the fault were detected just after a scheduled session, the fault recovery deadline would be 20 hours.

other hand, it would not be acceptable if communication sessions were scheduled at 6-hour intervals.

The only way to determine if there is a problem is to compare the reconfiguration time against the fault recovery deadline. This latter quantity is system dependent. No problem exists so long as the reconfiguration time is less than the recovery deadline. *Reconfiguration becomes a real-time process whenever it is used as a fault recovery method.* It does no good for researchers to talk about how an EA is able to restore a circuit's functionality. Such statements may show logical correctness, but without comparing the reconfiguration time against a deadline they say nothing about temporal correctness. Just reporting an algorithm's running time doesn't say anything about temporal correctness either. The key point is expressed by the following first principle:

First Principle of Fault Recovery

No fault recovery method can legitimately proclaim efficacy until it is proven to be both logically and temporally correct

---

The validity of this principle is easy to see. If the recovery method isn't logically correct, then the problem can't be fixed. If it isn't temporally correct, then the problem can't be fixed soon enough to prevent other things from going wrong. Without proving logical <u>and</u> temporal correctness, there is no basis for claiming any fault recovery method is effective.

It is easy to prove if a fault recovery method is logically correct—try it and see if it fixes the problem. Proving temporal correctness, however, is more complicated. Fortunately one outcome of any failure analysis is the effects those failures produce in the system, and how long before those effects take place, become known. That knowledge naturally defines the recovery deadlines. Temporal correctness is proven if a logically correct recovery is guaranteed to finish prior to the recovery deadline. Unfortunately, no intrinsic EH-based fault recovery method is temporally correct.

> **Comment:** The reader is cautioned not to draw any premature conclusions from this last sentence. Indeed, I will shortly show how EH-based techniques can satisfy just about any FR deadline associated with a specific mishap.

4.3 Fault Masking

Fault masking is not a traditional FT method in the sense it does not use FDI nor fault recovery. Nevertheless, it is discussed here for completeness.

The main concept is shown in Figure 6. Typically three hardware systems—functionally identical but not necessarily configured identically—all receive the

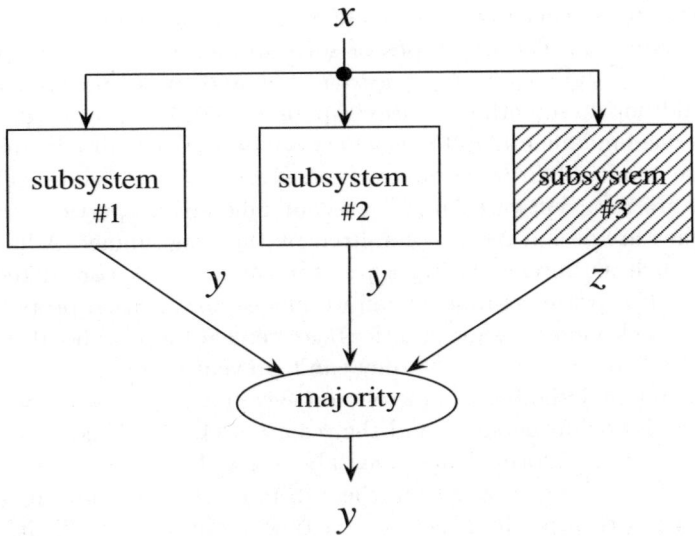

Fig. 6. Fault masking uses at least three functionally equivalent hardware configurations. Each subsystem performs the same function on the same input (x). Subsystem #3 has a failure and therefore generates an output different from the other correctly operating subsystems. Only the majority output is delivered.

same input. The individual outputs are presented to a "voter circuit" that outputs the majority value. The system continues to operate as designed so long as a majority of the systems are fault-free. Notice failures are allowed to persist. Indeed, no fault recovery takes place. This can have severe consequences: left untreated, some failures can propagate thereby inducing even more failures.

Fault masking is rarely (if ever) used in embedded systems. It is simply too expensive to provide triple-redundancy hardware. Besides, space is often at a premium, which means room for spare hardware is seldom available.

5   How to Design a Fault Tolerant System

The first step in designing a FTS actually has nothing to do with fault detection or fault recovery methods. This statement may seems somewhat surprising given the topics discussed so far, but any talk about detection methods or tradeoffs among FR methods is way too premature at this point.

Definition: (*failure mode*)

Any way in which a component or system can fail to perform its designed functions.

Hardware failure modes are inherent to the component and system. They are properties of the devices used and arise from their physical makeup. For example, diodes short-circuit and relay contacts weld together when they see excessive current flows. Several active devices undergo parametric changes when subjected to nuclear radiation. The two types of software failure modes are commission failures (where it did something it *wasn't* supposed to do) or omission failures (where it did not do something it *was* supposed to do).

There is one fundamental difference between hardware failure modes and software failure modes: hardware failure modes are inherent while software failure modes are not. Software "bugs" are not inherent properties of a software module—they are design errors committed by the programmer. A hardware design error, such as miscalculating a resistor value, is not considered a failure mode either. Hardware component failure modes are physical properties of the component itself. Good design practices can reduce the likelihood of hardware failure modes from occurring, but they can't prevent them.

The first step in designing a fault tolerant system is to find out how the system can fail—i.e., its failure modes—and the resultant effects of those failures. Only then can one make informed decisions about the best way to detect failures and an effective way to recover from them. Simply put, you have to know what can go wrong with something before you talk about ways to fix it. There are several ways of identifying and evaluating failure modes and the interested reader can find more comprehensive information elsewhere (e.g., Dunn [14] contains an excellent introduction). Here I will only give a brief overview.

Systems are normally evaluated with one of the following methods:

- *failure modes and effects analysis* (FMEA) A bottom-up method that looks at each component's failure modes and how those failures affect system performance.
- *fault tree analysis* (FTA)
  A top-down method that assumes a specific failure has happened and then analyzes each subsystem to look for the cause.

Figure 7 shows a typical FMEA spreadsheet. The analysis begins at the lowest level, the component level. Every possible failure mode for a component is listed in the spreadsheet. Engineers then analyze the circuit to see how a given component failure mode affects the circuit behavior. All possible effects are listed in the spreadsheet. If there is no effect, that too is indicated. Each effect is scrutinized to see what possible faults might be induced in other components. Often a risk is associated with each failure mode to indicate how crucial that particular failure is with respect to a total system failure. The analysis proceeds to higher levels in the design until either the search terminates (no more faults are induced) or a mishap is identified. It should be obvious conducting a thorough FMEA is time-consuming. Nevertheless, it is sometimes necessary particularly when dealing with safety-critical systems.

FMEAs are normally conducted for each system operating mode and in the context of a specific operational environment. This author, while working for the Eldec Corporation in Seattle, Washington, conducted a complete FMEA on a power switching amplifier used to control the tail-fins on a Tomahawk cruise missile. The purpose of this FMEA was to see how the amplifier behaved in a particular (classified) nuclear radiation environment. It is well known that passive components,

model number: ABC100
subsystem: power switching amplifier

Component	Failure Mode	Failure Effect	Risk Category
clamping diode (D2)	shorted	input grounded	5
	open	no input over-voltage protection; amplifier operation unaffected	5
switching transistor (Q4)	shorted	relay K5 cannot be turned off	3
	open	relay K5 cannot be turned on	4
Resistor (R10)	none		0

prepared by: Suzy Smith
date: 2/22/2008

Fig. 7. A typical FMEA spreadsheet. All failure modes for each component are listed along with their effects on the system behavior. Null effects must still be identified. The risk category indicates how destructive the effect is on overall system behavior. Failure modes with high risk are targeted for fault recovery.

such as carbon resistors, are immune to radiation whereas active component parameters are heavily effected. For instance, the input offset voltage and current of operational amplifiers increase in high radiation environments. The gate-to-source voltage switching levels for MOSFETs are also affected [12].

The FTA is a top down approach where you start with a defined mishap and then work down through the various subsystems to identify all possible faults that could have caused that specific mishap. Figure 8 shows a portion of a fault tree. Notice the root node has a specific mishap identified and lower levels in the tree show only those components (both hardware and software) that are relevant to that specific mishap. A separate fault tree is needed for each individual mishap.

Some people claim FMEAs can analyze software modules [16,17,18,19], but a survey of the technical literature shows a dearth of any concrete examples. The problem is the FMEA is a bottom-up approach starting at the component level, which is the lowest level in the design hierarchy. All hardware components have inherent failure modes that can be independently studied to determine their effect on the overall system. Any causal relationships can be quickly identified. Software components, on the other hand, have failure modes that lack any significance unless considered in context with the hardware system status—something hard to describe while working at the component level. (That old expression "you can't see the forest for the trees" is apropos here.) It therefore should come as

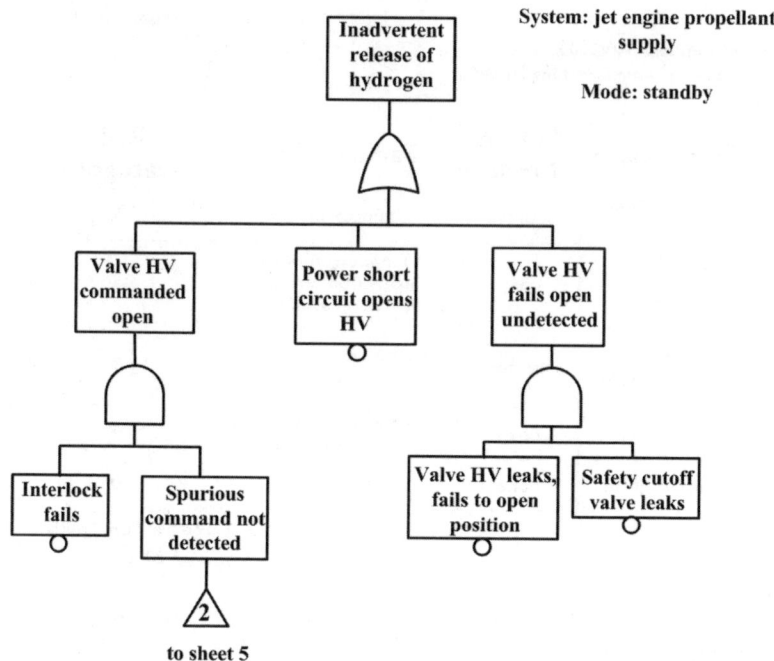

Fig. 8. A typical fault tree showing a specific mishap at the root node and possible causes at lower levels. The OR-gate and AND-gate symbols have their normal disjunction and conjunction interpretations. For example, looking at the left side of the tree, it takes both an interlock failure and a spurious command not detected to cause a valve HV commanded open situation. Blocks with a bubble at the bottom are terminal and have no lower tree levels. Triangular blocks point to continuations on other drawings. Notice that the mode of the system is always identified.

no surprise that, in practice, FMEAs are only conducted on hardware or on processes (e.g., manufacturing) [20]. Conversely, the FTA works well with systems that have both hardware and software components because it is a top-down approach. A top-down approach naturally defines the hardware environment as it works down the fault tree. Hence, any software components encountered will be analyzed in the context of a specific operational and/or hardware environment.

For hardware only systems the FMEA and FTA should, in principle, produce the same results. Done properly, either one should identify all potential faults and their effects on system performance [15]. One advantage of the FMEA is it can identify mishaps never envisioned by the design engineer.

Be that as it may, the FMEA is often not worth the time and effort! Not all failure modes ultimately lead to mishaps. It is therefore not necessary—nor cost-effective—to provide FR for every possible failure mode. More often than not the design engineer has a specific mishap (or a small set of mishaps) in mind that must be avoided at all costs if the system is going to be operational when

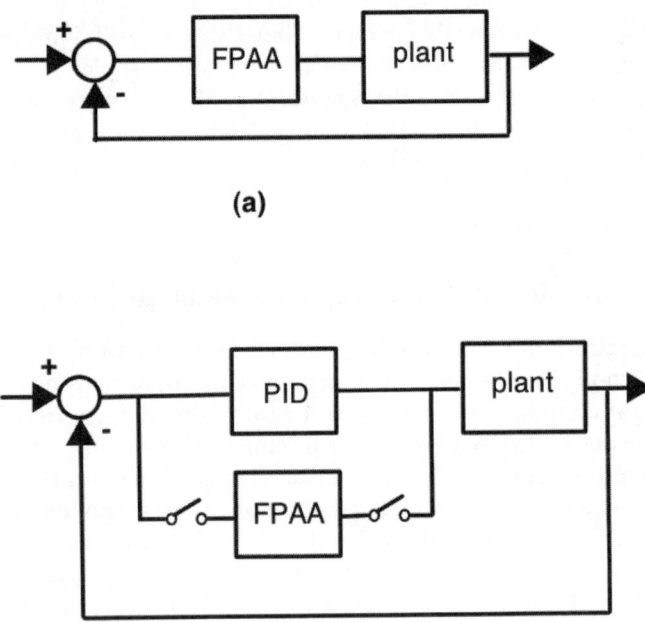

(a)

(b)

Fig. 9. Two FR methods where the fault in the plant is not directly corrected. In (*a*) the reconfigurable hardware is directly inserted into the command path while in (*b*) it is switched in to augment an existing controller (shown as a PID controller). In both cases the reconfigurable hardware only alters the plant input to compensate for the fault. The reconfigurable hardware, shown here as an FPAA, is intrinsically evolved, while online, by a microprocessor which was omitted for clarity.

needed. It is far more efficient and cost-effect to develop FR methods only for those critical failures that lead to those critical, predefined mishaps. Besides, the FTA is the only analysis method useful for systems with both hardware and software components. *It is for these reasons I believe the FTA is much better than the FMEA for designing FT systems.*

Once the FTA or FMEA is completed, critical mishaps that must be dealt with are known and their root causes identified. Now is the time—and not before—to tradeoff various FR techniques. Whichever FR method is chosen, it must be both logically and temporally correct. Both the FTA and FMEA entail conducting an effects analysis for each fault. It is during this effects analysis that temporal deadlines are determined. For example, one failure mode for a MOSFET is a short circuit between the drain and source. Suppose the effects analysis shows this failure overloads the power supply and five minutes afterwards the power supply itself fails. Clearly any FR method dealing with the MOSFET failure has a five minute deadline.

One fundamental question remains: how do you prove a particular FR method is both logically and temporally correct? That's where the *failure modes and effects testing* (FMET) comes into play. The concept is simple. The test engineer physically injects a failure into the system and then observes what the effect is. This testing is normally conducted in a laboratory where the operational environment can be regulated and any undesirable effects can be strictly monitored and controlled. Once the failure is injected, the FR method is executed to ascertain its logical and temporal correctness.

5.1   How Evolvable Hardware Supports Fault Recovery

Circuit reconfiguration is one way of handling failures and it can be a very effective FR method. For instance, Keymeulen et al. [6] successfully reconfigured a field-programmable transistor array to restore functionality that had deteriorated when the device was placed in a high temperature environment. Reconfiguration should also be effective for countering gradual behavioral changes such as those caused by aging effects or a slowly changing operational environment. But in general the thought of reconfiguring already failed circuitry suffers from one potentially fatal flaw: it assumes a new configuration exists that restores some level of acceptable behavior, if we can just find it. With unanticipated failures— i.e, failures never envisioned by the original designers—that assumption may not be valid. There is no way of knowing *a priori* if the search space contains any circuit configurations that could effectively deal with the existing failure. If not, an evolutionary algorithm could expend a great deal of computational effort and, in the end, accomplish nothing.

One way of dealing with unanticipated failures is to "work around" the failure instead of trying to physically remove it. The basic idea, shown in Figure 9, is analogous to compensation in linear control systems. The plant, where the failure exists, is left intact and EH tries to adjust the input to neutralize the failure effects. Of course this method won't handle hard plant failures, but it can deal with aging effects and other minor behavioral changes in a general way [9].

Intrinsic reconfiguration has one big advantage and one big disadvantage. It allows a system to deal with not just known mishaps, but also unanticipated or poorly defined fault conditions (the advantage). But, the running time can be large, making it hard to meet a recovery deadline (the disadvantage). There is, however, a way to get the advantage without the disadvantage.

As stated previously, engineers already know what mishap conditions to avoid. For each mishap condition an FTA can be conducted to determine the root cause(s). Now the engineer can tradeoff various FR methods including both extrinsic and intrinsic EH. Clearly extrinsic reconfiguration poses no real problem with running time because only computer models are involved, although the issue remains as to whether or not the models can accurately capture the plant failure. Intrinsic reconfiguration is therefore preferred because there are no computer models to deal with. An FMET setup is used to inject the fault and start the intrinsic reconfiguration process. The evolutionary algorithm running time is no longer a factor because the process is conducted in a regulated environment where failure effects can be controlled and any physical damage can be repaired. Once a logically correct configuration is found, it can be stored for later use. (Temporal correctness will be covered shortly.) This process is repeated for each mishap, thereby creating a database of circuit configurations developed specifically for fault recovery.

Each configuration in the database is stored as a programming stream in non-volatile memory. Of course, configurations don't have to be stored for just failures that lead to mishaps. Configurations can be stored for any failure no matter what its effect is. The only limitation is the size of available memory.

This database of circuit configurations forms the heart of a fault tolerant embedded system and is deployed along with the embedded system into a remote operating environment. FDI operations are conducted as normal—i.e., executed whenever faults are suspected. Fault recovery is now very straight forward. Once the fault is isolated, the system goes to the database and downloads the appropriate circuit configuration for that particular failure. Programming time is extremely short (typically a few milliseconds at best), which should easily comply any finite recovery deadline. Hence, temporal correctness is as good as guaranteed with this approach.

Of course creating a database of configurations probably isn't going to work if the embedded system has to contend with unanticipated failures. On-board, full intrinsic reconfiguration may be the only viable FR method. But that introduces some new challenges not previously discussed.

A survey of recent publications describing EH research shows the majority of work is done in a laboratory environment. The EA typically runs on a computer with a clock speed exceeding 1 GHz and with virtually limitless memory. Unfortunately, few deployed embedded systems have computing platforms even approaching these capabilities, particularly if that deployed system operates in a hostile environment. For example, an upcoming (2009) NASA mission to Mars uses an instrument with a rad-hard 8051 microcontroller [21]—a processor with a maximum clock speed of 20 MHz and which can only address 64 KB of external data memory. This type of processor poses two problems: (1) the slow clock speeds makes the reconfiguration time long, which in turn makes it difficult to meet a short recovery deadline, and (2) the limited memory means large EA populations cannot exist, which affects the search ability. This latter issue also increases the reconfiguration running time. The solution here may entail using reduced population EAs such as the mini-pop GA[22] or even a compact GA[23]. Some preliminary work has been done these reduced population EAs and it appears like the compact GA may show some promise[24].

6   Conclusions

In this chapter I have discussed what makes a system fault tolerant and how these type of systems are designed. FDI cannot be ignored because it is an integral part of any FT system. Several FDI methods were discussed.

Of particular importance to the computational community is the requirement that FT systems be treated as real-time systems. Both FDI and FR have deadlines. This imposes strict logical and temporal requirements on the fault recovery method, which possibly limits intrinsically evolved approaches as a FR method. However, a very promising approach, at least for recovery from anticipated faults, is to create a database of intrinsically evolved configurations. An FMET can be used to verify logical and temporal correctness. Fortunately, very few systems have to contend with unanticipated fault recovery, so the described approach should have broad appeal.

References

1. Greenwood, G., Tyrrell, A.: Introduction to Evolvable Hardware: A Practical Guide for Designing Self-Adaptive Systems. Wiley-IEEE Press (2006)
2. Burns, A., Wellings, A.: Real-Time Systems and Programming Languages, 3rd edn. Addison-Wesley-Longmain, Reading (2001)
3. Linden, D.: Optimizing signal strength in-situ using an evolvable antenna system. In: Proceedings of the 2002 NASA/DOD Conf. On Evolvable Hardware, pp. 147–151 (2002)
4. Avizienis, A.: Towards systematic design of fault tolerant systems. IEEE Computer 30(4), 51–58 (2004)
5. Belk, C., Robinson, J., Alexander, M., Cooke, W., Pavelitz, S.: Meteoroids and orbital debris: effects on spacecraft. In: NASA Reference Bulletin, p. 1408 (1997)
6. Keymeulen, D., Zebulum, R., Jin, Y., Stoica, A.: Fault tolerant evolvable hardware using field-programmable transistor arrays. IEEE Transactions on Reliability 49(3), 305–316 (2000)
7. Mange, D., Sipper, M., Stauffer, A., Tempesti, G.: Embryonics: a new methodology for designing field programmable gate arrays with self-repair and self-replicating properties. Proceedings of the IEEE 88(4), 416–541 (2000)
8. Sekanina, L., Drabek, V.: Relation between fault tolerance and reconfiguration in cellular systems. In: Proceedings of the 6th IEEE online testing workshop, pp. 25–30 (2000)
9. Greenwood, G., Hunter, D., Ramsden, E.: Fault recovery in linear systems via intrinsic evolution. In: Proc. 2004 NASA/DOD Conf. on Evol. Hdwe, pp. 115–122 (2004)
10. Greenwood, G.: On the practicality of using intrinsic reconfiguration for fault recovery. IEEE Transactions on Evolutionary Computation 9(4), 398–405 (2005)
11. NGST yardstick mission, NGST Monograph No. 1, Next Generation Space Telescope Project Study Office, Goddard Space Flight Center (1999)
12. Hughes, H., Benedetto, J.: Radiation effects and hardening of MOS technology: devices and circuits. IEEE Transactions on Nuclear Science 50(3), 500–521 (2003)
13. Wismer, M.: Steady-state operation of a high-voltage multiresonant converter in a high-temperature environment. IEEE Transactions on Power Electronics 18(3), 740–774 (2003)
14. Dunn, W.: Practical Design of Safety-Critical Computer Systems. Reliability Press (2002)
15. NASA-STD-8719.7, Facility system safety guidebook (January 1998)
16. Goddard, P.: Software FMEA techniques. In: Proc. 2000 Reliab. & Maintain. Symp., pp. 118–123 (2000)

17. Bowles, J., Wan, C.: Software failure modes and effects analysis for a small embedded control system. In: Proc. 2001 Reliab. & Maintain. Symp. pp. 1–6 (2001)
18. Craig, J.: A software reliability methodology using software sneak analysis: SW FMEA and the integrated system analysis approach. In: Proc. 2003 Reliab. & Maintain. Symp. pp. 12–18(2003)
19. Fadlovich, E.: Performing failure modes and effect analysis. Embedded Technology Magazine (January 2008)
20. Stamatis, D.: Failure Mode and Effect Analysis: FMEA from Theory to Execution. American Society for Quality (2003)
21. Didier Keymeulen, NASA JPL (private communication) (2007)
22. Vigraham, S., Gallagher, J.: A case for using minipop as the evolutionary engine in a CTRNN-EEH control device: an analysis of area requirements and search efficacy. In: Proc. 2005 NASA/DoD Conf. of Evol. Hdwe pp. 221–228 (2005)
23. Harik, G., Lobo, F., Goldberg, D.: The compact genetic algorithm. IEEE Trans. on Evol. Comput. 3(4), 287–297 (1999)
24. Jorgensen, C., Greenwood, G., Arefi, P.: Practical considerations for implementing intrinsic fault recovery in embedded systems. In: Proc. Congress on Evol. Comput. (to appear, 2004)

Author Index

Printing: Mercedes-Druck, Berlin
Binding: Stein+Lehmann, Berlin

Lecture Notes in Computer Science

Sublibrary 1: Theoretical Computer Science and General Issues

For information about Vols. 1– 4706
please contact your bookseller or Springer

Vol. 4917: P. Stenström, M. Dubois, M. Katevenis, R. Gupta, T. Ungerer (Eds.), High Performance Embedded Architectures and Compilers. XIII, 400 pages. 2008.

Vol. 4915: A. King (Ed.), Logic-Based Program Synthesis and Transformation. X, 219 pages. 2008.

Vol. 4912: G. Barthe, C. Fournet (Eds.), Trustworthy Global Computing. XI, 401 pages. 2008.

Vol. 4910: V. Geffert, J. Karhumäki, A. Bertoni, B. Preneel, P. Návrat, M. Bieliková (Eds.), SOFSEM 2008: Theory and Practice of Computer Science. XV, 792 pages. 2008.

Vol. 4905: F. Logozzo, D.A. Peled, L.D. Zuck (Eds.), Verification, Model Checking, and Abstract Interpretation. X, 325 pages. 2008.

Vol. 4904: S. Rao, M. Chatterjee, P. Jayanti, C.S.R. Murthy, S.K. Saha (Eds.), Distributed Computing and Networking. XVIII, 588 pages. 2007.

Vol. 4878: E. Tovar, P. Tsigas, H. Fouchal (Eds.), Principles of Distributed Systems. XIII, 457 pages. 2007.

Vol. 4875: S.-H. Hong, T. Nishizeki, W. Quan (Eds.), Graph Drawing. XIII, 402 pages. 2008.

Vol. 4873: S. Aluru, M. Parashar, R. Badrinath, V.K. Prasanna (Eds.), High Performance Computing – HiPC 2007. XXIV, 663 pages. 2007.

Vol. 4863: A. Bonato, F.R.K. Chung (Eds.), Algorithms and Models for the Web-Graph. X, 217 pages. 2007.

Vol. 4860: G. Eleftherakis, P. Kefalas, G. Păun, G. Rozenberg, A. Salomaa (Eds.), Membrane Computing. IX, 453 pages. 2007.

Vol. 4855: V. Arvind, S. Prasad (Eds.), FSTTCS 2007: Foundations of Software Technology and Theoretical Computer Science. XIV, 558 pages. 2007.

Vol. 4854: L. Bougé, M. Forsell, J.L. Träff, A. Streit, W. Ziegler, M. Alexander, S. Childs (Eds.), Euro-Par 2007 Workshops: Parallel Processing. XVII, 236 pages. 2008.

Vol. 4851: S. Boztaş, H.-F.(F.) Lu (Eds.), Applied Algebra, Algebraic Algorithms and Error-Correcting Codes. XII, 368 pages. 2007.

Vol. 4848: M.H. Garzon, H. Yan (Eds.), DNA Computing. XI, 292 pages. 2008.

Vol. 4847: M. Xu, Y. Zhan, J. Cao, Y. Liu (Eds.), Advanced Parallel Processing Technologies. XIX, 767 pages. 2007.

Vol. 4846: I. Cervesato (Ed.), Advances in Computer Science – ASIAN 2007. XI, 313 pages. 2007.

Vol. 4838: T. Masuzawa, S. Tixeuil (Eds.), Stabilization, Safety, and Security of Distributed Systems. XIII, 409 pages. 2007.

Vol. 4835: T. Tokuyama (Ed.), Algorithms and Computation. XVII, 929 pages. 2007.

Vol. 4818: I. Lirkov, S. Margenov, J. Waśniewski (Eds.), Large-Scale Scientific Computing. XIV, 755 pages. 2008.

Vol. 4800: A. Avron, N. Dershowitz, A. Rabinovich (Eds.), Pillars of Computer Science. XXI, 683 pages. 2008.

Vol. 4783: J. Holub, J. Žďárek (Eds.), Implementation and Application of Automata. XIII, 324 pages. 2007.

Vol. 4782: R. Perrott, B.M. Chapman, J. Subhlok, R.F. de Mello, L.T. Yang (Eds.), High Performance Computing and Communications. XIX, 823 pages. 2007.

Vol. 4771: T. Bartz-Beielstein, M.J. Blesa Aguilera, C. Blum, B. Naujoks, A. Roli, G. Rudolph, M. Sampels (Eds.), Hybrid Metaheuristics. X, 202 pages. 2007.

Vol. 4770: V.G. Ganzha, E.W. Mayr, E.V. Vorozhtsov (Eds.), Computer Algebra in Scientific Computing. XIII, 460 pages. 2007.

Vol. 4769: A. Brandstädt, D. Kratsch, H. Müller (Eds.), Graph-Theoretic Concepts in Computer Science. XIII, 341 pages. 2007.

Vol. 4763: J.-F. Raskin, P.S. Thiagarajan (Eds.), Formal Modeling and Analysis of Timed Systems. X, 369 pages. 2007.

Vol. 4759: J. Labarta, K. Joe, T. Sato (Eds.), High-Performance Computing. XV, 524 pages. 2008.

Vol. 4750: M.L. Gavrilova, C.J.K. Tan (Eds.), Transactions on Computational Science I. XI, 181 pages. 2008.

Vol. 4746: A. Bondavalli, F. Brasileiro, S. Rajsbaum (Eds.), Dependable Computing. XV, 239 pages. 2007.

Vol. 4743: P. Thulasiraman, X. He, T.L. Xu, M.K. Denko, R.K. Thulasiram, L.T. Yang (Eds.), Frontiers of High Performance Computing and Networking ISPA 2007 Workshops. XXIX, 536 pages. 2007.

Vol. 4742: I. Stojmenovic, R.K. Thulasiram, L.T. Yang, W. Jia, M. Guo, R.F. de Mello (Eds.), Parallel and Distributed Processing and Applications. XX, 995 pages. 2007.

Vol. 4739: R. Moreno Díaz, F. Pichler, A. Quesada Arencibia (Eds.), Computer Aided Systems Theory – EUROCAST 2007. XIX, 1233 pages. 2007.

Vol. 4736: S. Winter, M. Duckham, L. Kulik, B. Kuipers (Eds.), Spatial Information Theory. XV, 455 pages. 2007.

Vol. 4732: K. Schneider, J. Brandt (Eds.), Theorem Proving in Higher Order Logics. IX, 401 pages. 2007.

Vol. 4731: A. Pelc (Ed.), Distributed Computing. XVI, 510 pages. 2007.

Vol. 4728: S. Bozapalidis, G. Rahonis (Eds.), Algebraic Informatics. VIII, 291 pages. 2007.

Vol. 4726: N. Ziviani, R. Baeza-Yates (Eds.), String Processing and Information Retrieval. XII, 311 pages. 2007.

Vol. 4719: R. Backhouse, J. Gibbons, R. Hinze, J. Jeuring (Eds.), Datatype-Generic Programming. XI, 369 pages. 2007.

Vol. 4711: C.B. Jones, Z. Liu, J. Woodcock (Eds.), Theoretical Aspects of Computing – ICTAC 2007. XI, 483 pages. 2007.

Vol. 4710: C.W. George, Z. Liu, J. Woodcock (Eds.), Domain Modeling and the Duration Calculus. XI, 237 pages. 2007.

Vol. 4708: L. Kučera, A. Kučera (Eds.), Mathematical Foundations of Computer Science 2007. XVIII, 764 pages. 2007.

Vol. 4707: O. Gervasi, M.L. Gavrilova (Eds.), Computational Science and Its Applications – ICCSA 2007, Part III. XXIV, 1205 pages. 2007.